INFORMATICS IN CONTROL, AUTOMATION AND ROBOTICS I

Informatics in Control, Automation and Robotics I

Edited by

JOSÉ BRAZ
Escola Superior de Tecnologia de Setúbal, Portugal

HELDER ARAÚJO
University of Coimbra, Portugal

ALVES VIEIRA
Escola Superior de Tecnologia de Setúbal, Portugal

and

BRUNO ENCARNAÇÃO
*INSTICC - Institute for Systems and Technologies of Information,
Control and Communication, Setúbal, Portugal*

A C.I.P. Catalogue record for this book is available from the Library of Congress.

ISBN-10 1-4020-4136-5 (HB)
ISBN-13 978-1-4020-4136-5 (HB)
ISBN-10 1-4020-4543-3 (e-books)
ISBN-13 978-1-4020-4543-1 (e-books)

Published by Springer,
P.O. Box 17, 3300 AA Dordrecht, The Netherlands.

www.springer.com

Printed on acid-free paper

Printed in the Netherlands.

TABLE OF CONTENTS

PART 3 – SIGNAL PROCESSING, SYSTEMS MODELING AND CONTROL

PREFACE

The present book includes a set of selected papers from the first "International Conference on Informatics in Control Automation and Robotics" (ICINCO 2004), held in Setúbal, Portugal, from 25 to 28 August 2004.

The conference was organized in three simultaneous tracks: *"Intelligent Control Systems and Optimization"*, *"Robotics and Automation"* and *"Systems Modeling, Signal Processing and Control"*. The book is based on the same structure.

Although ICINCO 2004 received 311 paper submissions, from 51 different countries in all continents, only 115 where accepted as full papers. From those, only 29 were selected for inclusion in this book, based on the classifications provided by the Program Committee. The selected papers also reflect the interdisciplinary nature of the conference. The diversity of topics is an important feature of this conference, enabling an overall perception of several important scientific and technological trends. These high quality standards will be maintained and reinforced at ICINCO 2005, to be held in Barcelona, Spain, and in future editions of this conference.

Furthermore, ICINCO 2004 included 6 plenary keynote lectures and 2 tutorials, given by internationally recognized researchers. Their presentations represented an important contribution to increasing the overall quality of the conference, and are partially included in the first section of the book. We would like to express our appreciation to all the invited keynote speakers, namely, in alphabetical order: Wolfgang Arndt (Steinbeis Foundation for Industrial Cooperation/Germany), Albert Cheng (University of Houston/USA), Kurosh Madani (Senart Institute of Technology/France), Nuno Martins (MIT/USA), Rosalind Picard (MIT/USA) and Kevin Warwick (University of Reading, UK).

On behalf of the conference organizing committee, we would like to thank all participants. First of all to the authors, whose quality work is the essence of the conference and to the members of the program committee, who helped us with their expertise and time.

As we all know, producing a conference requires the effort of many individuals. We wish to thank all the members of our organizing committee, whose work and commitment were invaluable. Special thanks to Joaquim Filipe, Paula Miranda, Marina Carvalho and Vitor Pedrosa.

José Braz
Helder Araújo
Alves Vieira
Bruno Encarnação

CONFERENCE COMMITTEE

Conference Chair

Joaquim Filipe, Escola Superior de Tecnologia de Setúbal, Portugal

Program Co-Chairs

Helder Araújo, I.S.R. Coimbra, Portugal

Alves Vieira, Escola Superior de Tecnologia de Setúbal, Portugal

Program Committee Chair

José Braz, Escola Superior de Tecnologia de Setúbal, Portugal

Secretariat

Marina Carvalho, INSTICC, Portugal

Bruno Encarnação, INSTICC, Portugal

Programme Committee

Aguirre, L. (BRAZIL)

Allgöwer, F. (GERMANY)

Arató, P.(HUNGARY)

Arsénio, A. (U.S.A.)

Asama, H. (JAPAN)

Babuska, R. (THE NETHERLANDS)

Balas, M. (U.S.A.)

Balestrino, A. (ITALY)

Bandyopadhyay, B. (INDIA)

Bars, R. (HUNGARY)

Bemporad, A. (ITALY)

Birk, A. (GERMANY)

Bonyuet, D.(U.S.A.)

Boucher, P.(FRANCE)

Bulsari, A. (FINLAND)

Burke, E. (U.K.)

Burn, K. (U.K.)

Burrows, C. (U.K.)

Buss, M. (GERMANY)

Camarinha-Matos, L. (PORTUGAL)

Campi, M. (ITALY)

Cañete, J. (SPAIN)

Carvalho, J. (PORTUGAL)

Cassandras, C. (U.S.A.)

Chatila, R. (FRANCE)

Chen, T. (CANADA)

Cheng, A. (U.S.A.)

Choras, R. (POLAND)

Christensen, H. (SWEDEN)

Cichocki, A. (JAPAN)

Coello, C. (MEXICO)

Cordeiro, J. (PORTUGAL)

Correia, L. (PORTUGAL)

Costeira, J. (PORTUGAL)

Couto, C. (PORTUGAL)

Crispin, Y. (U.S.A.)

Custódio, L. (PORTUGAL)

Dillmann, R. (GERMANY)

Dochain, D. (BELGIUM)

Dourado, A. (PORTUGAL)

Duch, W. (POLAND)

Erbe, H. (GERMANY)

Espinosa-Perez, G. (MEXICO)

Feliachi, A. (U.S.A.)

Feng, D. (HONG KONG)

Ferrier, J. (FRANCE)

Ferrier, N. (U.S.A.)

Figueroa, G. (MEXICO)

Filip, F. (ROMANIA)

Filipe, J. (PORTUGAL)

Fyfe, C. (U.K.)

Gamberger, D. (CROATIA)

Garção, A. (PORTUGAL)

Gheorghe, L. (ROMANIA)

Ghorbel, F. (U.S.A.)
Gini, M. (U.S.A.)
Goldenberg, A. (CANADA)
Gomes, L.(PORTUGAL)
Gonçalves, J. (U.S.A.)
Gray, J. (U.K.)
Gustafsson, T.(SWEDEN)
Halang, W. (GERMANY)
Hallam, J.(U.K.)
Hammoud, R. (U.S.A.)
Hanebeck, U. (GERMANY)
Henrich, D. (GERMANY)
Hespanha, J. (U.S.A.)
Ho, W. (SINGAPORE)
Imiya, A. (JAPAN)
Jämsä-Jounela, S. (FINLAND)
Jarvis, R. (AUSTRALIA)
Jezernik, K. (SLOVENIA)
Jonker, B. (THE NETHERLANDS)
Juhas, G. (GERMANY)
Karcanias, N. (U.K.)
Karray, F. (CANADA)
Katayama, T. (JAPAN)
Katic, D. (YUGOSLAVIA)
Kavraki, L. (U.S.A.)
Kawano, H. (JAPAN)
Kaynak, O. (TURKEY)
Kiencke, U. (GERMANY)
Kihl, M. (SWEDEN)
King, R. (GERMANY)
Kinnaert, M. (BELGIUM)
Khessal, N. (SINGAPORE)
Koivisto, H. (FINLAND)
Korbicz, J. (POLAND)
Kosko, B. (U.S.A.)
Kosuge, K. (JAPAN)
Kovacic, Z. (CROATIA)
Kunt, M. (SWITZERLAND)
Latombe, J. (U.S.A.)
Leite, F. (PORTUGAL)
Leitner, J. (U.S.A.)
Leiviska, K. (U.S.A.)
Lightbody, G. (IRELAND)
Ligus, J. (SLOVAKIA)
Lin, Z. (U.S.A.)

Ljungn, L. (SWEDEN)
Lückel, J. (GERMANY)
Maione, B. (ITALY)
Maire, F. (AUSTRALIA)
Malik, O. (CANADA)
Mańdziuk, J. (POLAND)
Meirelles, M. (BRAZIL)
Meng, M. (CANADA)
Mertzios, B. (GREECE)
Molina, A. (SPAIN)
Monostori, L. (HUNGARY)
Morari, M. (SWITZERLAND)
Mostýn, V. (CZECH REPUBLIC)
Murray-Smith, D. (U.K.)
Muske, K. (U.S.A.)
Nedevschi, S. (ROMANIA)
Nijmeijer, H. (THE NETHERLANDS)
Ouelhadj, D. (U.K.)
Papageorgiou, M. (GREECE)
Parisini, T. (ITALY)
Pasi, G. (ITALY)
Pereira, C. (BRAZIL)
Pérez, M. (MEXICO)
Pires, J. (PORTUGAL)
Polycarpou, M. (CYPRUS)
Pons, M. (FRANCE)
Rana, O. (NEW ZEALAND)
Reed, J. (U.K.)
Ribeiro, M. (PORTUGAL)
Richardson, R. (U.K.)
Ringwood, J. (IRELAND)
Rist, T. (GERMANY)
Roffel, B. (THE NETHERLANDS)
Rosa, A. (PORTUGAL)
Rossi, D. (ITALY)
Ruano, A. (PORTUGAL)
Sala, A. (SPAIN)
Sanz, R. (SPAIN)
Sarkar, N. (U.S.A.)
Sasiadek, J. (CANADA)
Scherer, C. (THE NETHERLANDS)
Schilling, K. (GERMANY)
Sentieiro, J. (PORTUGAL)
Sequeira, J. (PORTUGAL)
Sessa, R. (ITALY)

Siciliano, B. (ITALY)
Silva, C. (CANADA)
Spong, M. (U.S.A.)
Stahre, J. (SWEDEN)
van Straten, G. (THE NETHERLANDS)
Sznaier, M. (U.S.A.)
Tarasiewicz, S. (CANADA)
Tettamanzi, A. (ITALY)
Thalmann, D. (SWITZERLAND)
Valavani, L. (U.S.A.)
Valverde, N. (MEXICO)
van Hulle, M. (BELGIUM)

Varkonyi-Koczy, A. (HUNGARY)
Veloso, M. (U.S.A.)
Vlacic, L. (GERMANY)
Wang, J. (CHINA)
Wang, L. (SINGAPORE)
Yakovlev, A. (U.K.)
Yen, G. (U.S.A.)
Yoshizawa, S. (JAPAN)
Zhang, Y. (U.S.A.)
Zomaya, A. (AUSTRALIA)
Zuehlke, D. (GERMANY)

Invited Speakers

Kevin Warwick, University of Reading, UK

Kurosh Madani, PARIS XII University, France

F. Wolfgang Arndt, Fachhochschule Konstanz, Germany

Albert Cheng, University of Houston, USA

Rosalind Picard, Massachusetts Institute of Technology, USA

Nuno Martins, Massachusetts Institute of Technology, USA

INVITED SPEAKERS

ROBOT-HUMAN INTERACTION
Practical experiments with a cyborg

Kevin Warwick

Department of Cybernetics, University of Reading,
Whiteknights, Reading, RG6 6AY, UK
Email: k.warwick@reading.ac.uk

Abstract: This paper presents results to indicate the potential applications of a direct connection between the human nervous system and a computer network. Actual experimental results obtained from a human subject study are given, with emphasis placed on the direct interaction between the human nervous system and possible extra-sensory input. An brief overview of the general state of neural implants is given, as well as a range of application areas considered. An overall view is also taken as to what may be possible with implant technology as a general purpose human-computer interface for the future.

1 INTRODUCTION

There are a number of ways in which biological signals can be recorded and subsequently acted upon to bring about the control or manipulation of an item of technology, (Penny et al., 2000, Roberts et al., 1999). Conversely it may be desired simply to monitor the signals occurring for either medical or scientific purposes. In most cases, these signals are collected externally to the body and, whilst this is positive from the viewpoint of non-intrusion into the body with its potential medical side-effects such as infection, it does present enormous problems in deciphering and understanding the signals observed (Wolpaw et al., 1991, Kubler et al., 1999). Noise can be a particular problem in this domain and indeed it can override all other signals, especially when compound/collective signals are all that can be recorded, as is invariably the case with external recordings which include neural signals.

A critical issue becomes that of selecting exactly which signals contain useful information and which are noise, and this is something which may not be reliably achieved. Additionally, when specific, targeted stimulation of the nervous system is required, this is not possible in a meaningful way for control purposes merely with external connections. The main reason for this is the strength of signal required, which makes stimulation of unique or even small subpopulations of sensory receptor or motor unit channels unachievable by such a method.

A number of research groups have concentrated on animal (non-human) studies, and these have certainly provided results that contribute generally to the knowledge base in the field. Unfortunately actual human studies involving implants are relatively limited in number, although it could be said that research into wearable computers has provided some evidence of what can be done technically with bio-signals. We have to be honest and say that projects which involve augmenting shoes and glasses with microcomputers (Thorp, 1997) are perhaps not directly useful for our studies, however monitoring indications of stress or alertness by this means can be helpful in that it can give us an idea of what might be subsequently achievable by means of an implant. Of relevance here are though studies in which a miniature computer screen was fitted onto a standard pair of glasses. In this research the wearer was given a form of augmented/remote vision (Mann, 1997), where information about a remote scene could be relayed back to the wearer, thereby affecting their overall capabilities. However, in general, wearable computers require some form of signal conversion to take place in order to interface external technology with specific human sensory receptors. What are clearly of far more interest to our own studies are investigations in which a direct electrical link is formed between the nervous system and technology.

Quite a number of relevant animal studies have been carried out, see (Warwick, 2004) for a review. As an example, in one study the extracted brain of a lamprey was used to control the movement of a small-wheeled robot to which it was attached (Reger et al., 2000). The innate response of a lamprey is to position itself in water by detecting and reacting to external light on the surface of the water. The lam-

3

J. Braz et al. (eds.), Informatics in Control, Automation and Robotics I, 1–10.
© 2006 *Springer. Printed in the Netherlands.*

prey robot was surrounded by a ring of lights and the innate behaviour was employed to cause the robot to move swiftly towards the active light source, when different lights were switched on and off.

Rats have been the subjects of several studies. In one (Chapin et al., 1999), rats were trained to pull a lever in order that they received a liquid reward for their efforts. Electrodes were chronically implanted into the motor cortex of the rats' brains to directly detect neural signals generated when each rat (it is claimed) thought about pulling the lever, but, importantly, before any physical movement occurred. The signals measured immediately prior to the physical action necessary for lever pulling were used to directly release the reward before a rat actually carried out the physical action of pulling the lever itself. Over the time of the study, which lasted for a few days, four of the six implanted rats learned that they need not actually initiate any action in order to obtain the reward; merely thinking about the action was sufficient. One problem area that needs to be highlighted with this is that although the research is certainly of value, because rats were employed in the trial we cannot be sure what they were actually thinking about in order to receive the reward, or indeed whether the nature of their thoughts changed during the trial.

In another study carried out by the same group (Talwar et al., 2002), the brains of a number of rats were stimulated via electrodes in order to teach them to be able to carry out a maze solving problem. Reinforcement learning was used in the sense that, as it is claimed, pleasurable stimuli were evoked when a rat moved in the correct direction. Although the project proved to be successful, we cannot be sure of the actual feelings perceived by the rats, whether they were at all pleasurable when successful or unpleasant when a negative route was taken.

1.1 Human Integration

Studies which focus, in some sense, on integrating technology with the Human Central Nervous System range from those considered to be diagnostic (Deneslic et al., 1994), to those which are clearly aimed solely at the amelioration of symptoms (Poboronuic et al., 2002, Popovic et al., 1998, Yu et al., 2001) to those which are directed towards the augmentation of senses (Cohen et al., 1999, Butz et al., 1999). By far the most widely reported research with human subjects however, is that involving the development of an artificial retina (Kanda et al., 1999). In this case small arrays have been attached directly onto a functioning optic nerve, but where the person concerned has no operational vision. By means of

stimulation of the nerve with appropriate signal sequences the user has been able to perceive shapes and letters indicated by bright light patterns. Although relatively successful thus far, the research does appear to still have a long way to go, in that considerable software modification and tailoring is required in order to make the system operative for one individual.

Electronic neural stimulation has proved to be extremely successful in other areas which can be loosely termed as being restorative. In this class, applications range from cochlea implants to the treatment of Parkinson's disease symptoms. The most relevant to our study here however is the use of a single electrode brain implant, enabling a brainstem stroke victim to control the movement of a cursor on a computer screen (Kennedy et al., 2000). In the first instance extensive functional magnetic resonance imaging (fMRI) of the subject's brain was carried out. The subject was asked to think about moving his hand and the fMRI scanner was used to determine where neural activity was most pronounced. A hollow glass electrode cone containing two gold wires was subsequently positioned into the motor cortex, centrally located in the area of maximum-recorded activity. When the patient thought about moving his hand, the output from the electrode was amplified and transmitted by a radio link to a computer where the signals were translated into control signals to bring about movement of the cursor. The subject learnt to move the cursor around by thinking about different hand movements. No signs of rejection of the implant were observed whilst it was in position (Kennedy et al., 2000).

In the human studies described thus far, the main aim is to use technology to achieve some restorative functions where a physical problem of some kind exists, even if this results in an alternative ability being generated. Although such an end result is certainly of interest, one of the main directions of the study reported in this paper is to investigate the possibility of giving a human extra capabilities, over and above those initially in place.

In the section which follows a MicroElectrode Array (MEA) of the spiked electrode type is described. An array of this type was implanted into a human nervous system to act as an electrical silicon/biological interface between the human nervous system and a computer. As an example, a pilot study is described in which the output signals from the array are used to drive a range of technological entities, such as mobile robots and a wheelchair. These are introduced merely as an indication of what is possible. A report is then also given of a continuation of the study involving the feeding of signals obtained from ultrasonic sensors down onto the nervous system, to bring about sensory enhancement, i.e. giving a human an ultrasonic sense. It is worth emphasising here that what is described in this article is an actual application study rather than a computer simulation or mere speculation.

2 INVASIVE NEURAL INTERFACE

There are, in general, two approaches for peripheral nerve interfaces when a direct technological connection is required: Extraneural and Intraneural. In practical terms, the cuff electrode is by far the most commonly encountered extraneural device. A cuff electrode is fitted tightly around the nerve trunk, such that it is possible to record the sum of the single fibre action potentials apparent, this being known as the compound action potential (CAP). In other words, a cuff electrode is suitable only if an overall compound signal from the nerve fibres is required. It is not suitable for obtaining individual or specific signals. It can though also be used for crudely selective neural stimulation of a large region of the nerve trunk. In some cases the cuff can contain a second or more electrodes, thereby allowing for an approximate measurement of signal speed travelling along the nerve fibres.

For applications which require a much finer granularity for both selective monitoring and stimulation however, an intraneural interface such as single electrodes either inserted individually or in groups can be employed. To open up even more possibilities a MicroElectrode Array (MEA) is well suited. MEAs can take on a number of forms, for example they can be etched arrays that lie flat against a neural surface (Nam et al., 2004) or spiked arrays with electrode tips. The MEA employed in this study is of this latter type and contains a total of 100 electrodes which, when implanted, become distributed within the nerve fascicle. In this way, it is possible to gain direct access to nerve fibres from muscle spindles, motor neural signals to particular motor units or sensory receptors. Essentially, such a device allows a bi-directional link between the human nervous system and a computer (Gasson et al., 2002, Branner et al., 2001, Warwick et al., 2003).

2.1 Surgical Procedure

On 14 March 2002, during a 2 hour procedure at the Radcliffe Infirmary, Oxford, a MEA was surgically implanted into the median nerve fibres of my left arm. The array measured 4mm x 4mm with each of the electrodes being 1.5mm in length. Each electrode was individually wired via a 20cm wire bundle to an electrical connector pad. A distal skin incision marked at the distal wrist crease medial to the *palmaris longus* tendon was extended approximately 4 cm into the forearm. Dissection was performed to identify the median nerve. In order that the risk of infection in close proximity to the nerve was reduced, the wire bundle was run subcutaneously for 16 cm before exiting percutaneously. For this exit a second proximal skin incision was made distal to the elbow 4 cm into the forearm. A modified plastic

Figure 1: A 100 electrode, 4X4mm MicroElectrode Array, shown on a UK 1 pence piece for scale.

shunt passer was inserted subcutaneously between the two incisions by means of a tunnelling procedure. The MEA was introduced to the more proximal incision and pushed distally along the passer to the distal skin incision such that the wire bundle connected to the MEA ran within it. By removing the passer, the MEA remained adjacent to the exposed median nerve at the point of the first incision with the wire bundle running subcutaneously, exiting at the second incision. At the exit point, the wire bundle linked to the electrical connector pad which remained external to the arm.

The perineurium of the median nerve (its outer protective sheath) was dissected under microscope to facilitate the insertion of electrodes and to ensure that the electrodes penetrated the nerve fibres to an adequate depth. Following dissection of the perineurium, a pneumatic high velocity impact inserter was positioned such that the MEA was under a light pressure to help align insertion direction. The MEA was pneumatically inserted into the radial side of the median nerve allowing the MEA to sit adjacent to the nerve fibres with the electrodes penetrating into a fascicle. The median nerve fascicle was estimated to be approximately 4 mm in diameter. Penetration was confirmed under microscope. Two Pt/Ir reference wires were also positioned in the fluids surrounding the nerve.

The arrangements described remained permanently in place for 96 days, until 18th June 2002, at which time the implant was removed.

2.2 Neural Stimulation and Neural Recordings

Once it was in position, the array acted as a bi-directional neural interface. Signals could be transmitted directly from a computer, by means of either a hard wire connection or through a radio transmitter/receiver unit, to the array and thence to directly bring about a stimulation of the nervous system. In addition, signals from neural activity could be detected by the electrodes and sent to the computer and thence onto the internet. During experimentation, it was found that typical activity on the median nerve fibres occurs around a centroid frequency of approximately 1 KHz with signals of apparent interest occurring well below 3.5 KHz. However noise is a distinct problem due to inductive pickup on the wires, so had to be severely reduced. To this end a fifth order band limited Butterworth filter was used with corner frequencies of $f_{low}= 250$ Hz and $f_{high} = 7.5$ KHz.

To allow freedom of movement, a small wearable signal processing unit with Radio Frequency communications was developed to be worn on a gauntlet around the wrist. This custom hardware consisted of a 20 way multiplexer, two independent filters, two 10bit A/D converters, a microcontroller and an FM radio transceiver module. Either 1 or 2 electrodes from the array could be quasi-statically selected, digitised and sent over the radio link to a corresponding receiver connected to a PC. At this point they could either be recorded or transmitted further in order to operate networked technology, as described in the following section. Onward transmission of the signal was via an encrypted TCP/IP tunnel, over the local area network, or wider internet. Remote configuration of various parameters on the wearable device was also possible via the radio link from the local PC or the remote PC via the encrypted tunnel.

Stimulation of the nervous system by means of the array was especially problematic due to the limited nature of existing results prior to the study reported here, using this type of interface. Published work is restricted largely to a respectably thorough but short term study into the stimulation of the sciatic nerve in cats (Branner et al., 2001). Much experimental time was therefore required, on a trial and error basis, to ascertain what voltage/current relationships would produce a reasonable (i.e. perceivable but not painful) level of nerve stimulation.

Further factors which may well emerge to be relevant, but were not possible to predict in this experimental session were firstly the plastic, adaptable nature of the human nervous system, especially the brain – even over relatively short periods, and secondly the effects of movement of the array in relation to the nerve fibres, hence the connection and associated input impedance of the nervous system was not completely stable.

After experimentation lasting for approximately 6 weeks, it was found that injecting currents below 80μA onto the median nerve fibres had little perceivable effect. Between 80μA and 100μA all the functional electrodes were able to produce a recognisable stimulation, with an applied voltage of around 20 volts peak to peak, dependant on the series electrode impedance. Increasing the current above 100μA had little additional effect; the stimulation switching mechanisms in the median nerve fascicle exhibited a non-linear thresholding characteristic.

In all successful trials, the current was applied as a bi-phasic signal with pulse duration of 200μsec and an inter-phase delay of 100μsec. A typical stimulation waveform of constant current being applied to one of the MEAs implanted electrodes is shown in Fig. 2.

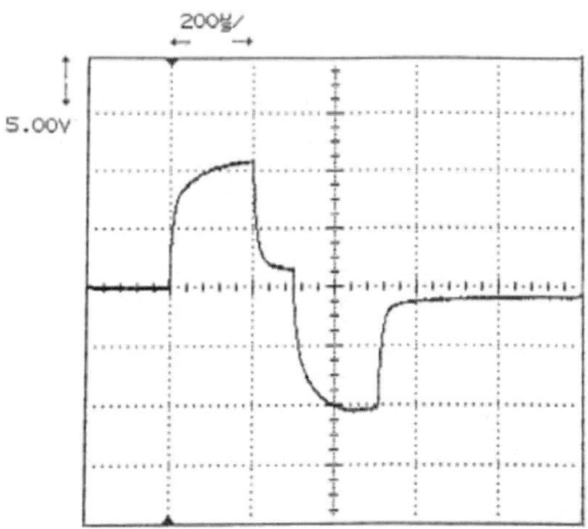

Figure 2: Voltage profile during one bi-phasic stimulation pulse cycle with a constant current of 80µA.

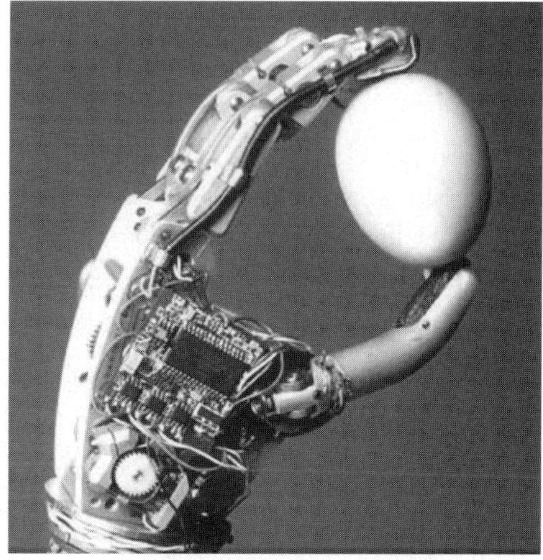

Figure 3: Intelligent anthropomorphic hand prosthesis.

It was therefore possible to create alternative sensations via this new input route to the nervous system, thereby by-passing the normal sensory inputs. The reasons for the 6 weeks necessary for successful nerve stimulation, in the sense of stimulation signals being correctly recognised, can be due to a number of factors such as (1) suitable pulse characteristics, (i.e. amplitude, frequency etc) required to bring about a perceivable stimulation were determined experimentally during this time, (2) my brain had to adapt to recognise the new signals it was receiving, and (3) the bond between my nervous system and the implant was physically changing.

3 NEURAL INTERACTION WITH TECHNOLOGY

It is apparent that the neural signals obtained through the implant can be used for a wide variety of purposes. One of the key aims of the research was, in fact, to assess the feasibility of the implant for use with individuals who have limited functions due to a spinal injury. Hence in experimental tests, neural signals were employed to control the functioning of a robotic hand and to drive an electric wheelchair around successfully (Gasson et al., 2002, Warwick et al., 2003). The robotic hand was also controlled, via the internet, at a remote location (Warwick et al., 2004).

In these applications, data collected via the neural implant were directly employed for control pur-

poses, removing the need for any external control devices or for switches or buttons to be used. Essentially signals taken directly from my nervous system were used to operate the technology. To control the electric wheelchair, a sequential-state machine was incorporated. Neural signals were used as a real-time command to halt the cycle at the intended wheelchair action, e.g. drive forwards. In this way overall control of the chair was extremely simple to ensure, thereby proving the general potential use of such an interface.

Initially selective processing of the neural signals obtained via the implant was carried out in order to produce discrete direction control signals. With only a small learning period I was able to control not only the direction but also the velocity of a fully autonomous, remote mobile platform. On board sensors allowed the robot to override my commands in order to safely navigate local objects in the environment.

Once stimulation of the nervous system had been achieved, as described in section 2, the bi-directional nature of the implant could be more fully experimented with. Stimulation of the nervous system was activated by taking signals from fingertips sensors on the robotic hand. So as the robotic hand gripped an object, in response to outgoing neural signals via the implant, signals from the fingertips of the robotic hand brought about stimulation. As the robotic hand applied more pressure the frequency of stimulation increased. The robotic hand was, in this experiment, acting as a remote, extra hand.

By passing the neural signals not simply from computer to the robot hand, and vice versa, but also via the internet, so the hand could actually be lo-

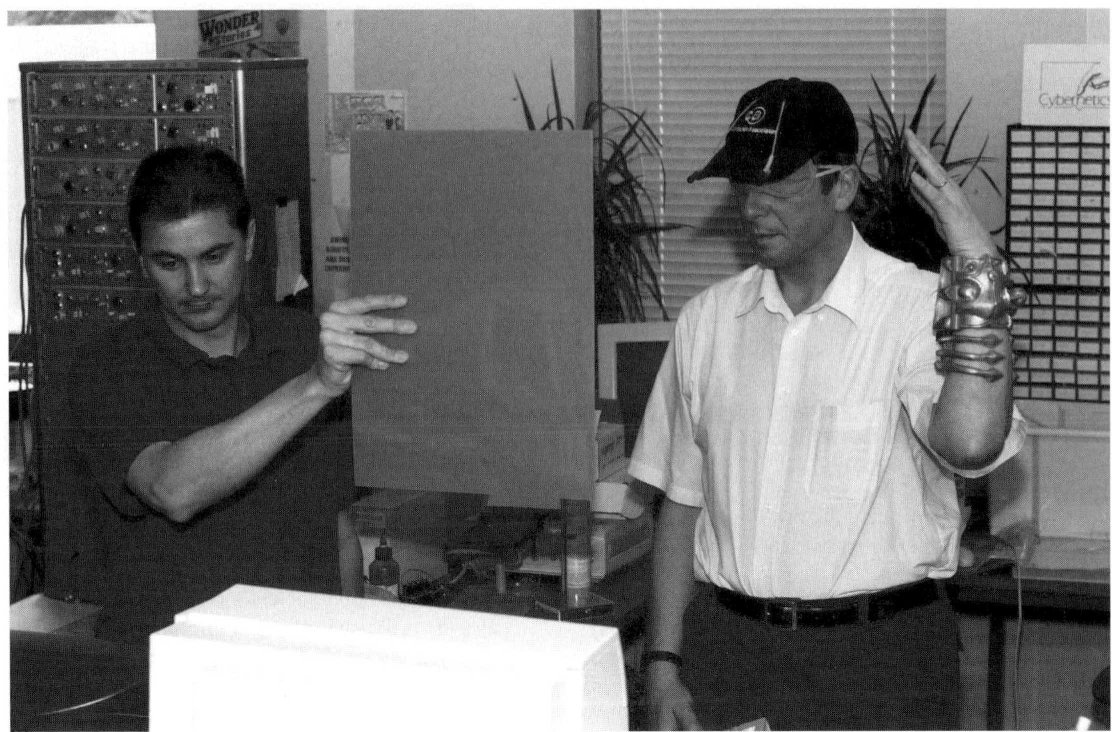

Figure 4: Experimentation and testing of the ultrasonic baseball cap.

cated remotely. In a test (Warwick et al., 2004) signals were transmitted between Columbia University in New York City and Reading University in the UK, with myself being in New York and the robot hand in the UK. Effectively this can be regarded as extending the human nervous system via the internet. To all intents and purposes my nervous system did not stop at the end of my body, as is the usual case, but rather went as far as the internet would take it, in this case across the Atlantic Ocean.

In another experiment, signals were obtained from ultrasonic sensors fitted to a baseball cap. The output from these sensors directly affected the rate of neural stimulation. With a blindfold on, I was able to walk around in a cluttered environment whilst detecting objects in the vicinity through the (extra) ultrasonic sense. With no objects nearby, no neural stimulation occurred. As an object moved relatively closer, so the stimulation increased proportionally (Gasson et al., 2005).

It is clear that just about any technology, which can be networked in some way, can be switched on and off and ultimately controlled directly by means of neural signals through an interface such as the implant used in this experimentation. Not only that, but because a bi-directional link has been formed, feedback directly to the brain can increase the range

of sensory capabilities. Potential application areas are therefore considerable.

4 CONCLUSIONS

Partly this study was carried out in order to assess the implant interface technology in terms of its usefulness in helping those with a spinal injury. As a positive result in this sense it can be reported that during the course of the study there was no sign of infection or rejection. In fact, rather than reject the implant, my body appeared to accept the device implicitly to the extent that its acceptance may well have been improving over time.

Clearly the implant would appear to allow for the restoration of some movement and the return of body functions in the case of a spinally injured patient. It would also appear to allow for the patient to control technology around them merely by neural signals alone. Further human experimentation is though clearly necessary to provide further evidence in this area.

Such implanted interface technology would however appear to open up many more opportunities. In the case of the experiments described, an articulated robot hand was controlled directly by neural signals.

For someone who has had their original hand amputated this opens up the possibility of them ultimately controlling an articulated hand, as though it were their own, by the power of their own thought.

Much more than this though, the study opens up the distinct possibility of humans being technically enhanced and upgraded, rather than merely repaired. One example of this was the extra sensory (ultra sonic) experiment that was far more successful than had been expected. Although this does open up a number of distinct ethical questions, as to what upgrades are acceptable and for whom, it also opens up an exciting period of experimentation to see how far the human brain can be expanded in a technical sense.

The author accepts the fact that this is a one off study based on only one implant recipient. It may be that other recipients react in other ways and the experiments carried out would not be so successful with an alternative recipient. In that sense the author wishes this study to be seen as evidence that the concept can work well, although it is acknowledged that further human trials will be necessary to investigate the extent of usefulness.

As far as an implant interface is concerned, what has been achieved is a very rudimentary and primitive first step. It may well prove to be the case that implants of the type used here are not ultimately those selected for a good link between a computer and the human brain. Nevertheless the results obtained are felt to be extremely encouraging.

ACKNOWLEDGEMENTS

Ethical approval for this research to proceed was obtained from the Ethics and Research Committee at the University of Reading and, with regard to the neurosurgery, by the Oxfordshire National Health Trust Board overseeing the Radcliffe Infirmary, Oxford, UK.

My thanks go to Mr. Peter Teddy and Mr. Amjad Shad who performed the neurosurgery at the Radcliffe Infirmary and ensured the medical success of the project. My gratitude is also extended to NSIC, Stoke Mandeville and to the David Tolkien Trust for their support.

REFERENCES

Penny, W., Roberts, S., Curran, E., and Stokes, M., 2000, "EEG-based communication: A pattern recognition approach", IEEE Transactions on Rehabilitation Engineering., Vol. 8, Issue.2, pp. 214-215.

Roberts, S., Penny, W., and Rezek, I., 1999, "Temporal and spatial complexity measures for electroencephalogram based brain-computer interfacing", Medical and Biological Engineering and Computing, Vol. 37, Issue.1, pp. 93-98.

Wolpaw, J., McFarland, D., Neat, G. and Forneris, C., 1991, "An EEG based brain-computer interface for cursor control", Electroencephalography and Clinical Neurophysiology, Vol. 78, Issue.3, pp. 252-259.

Kubler, A., Kotchoubey, B., Hinterberger, T., Ghanayim, N., Perelmouter, J., Schauer, M., Fritsch, C., Taub, E. and Birbaumer, N., 1999, "The Thought Translation device: a neurophysiological approach to communication in total motor paralysis", Experimental Brain Research, Vol. 124, Issue.2, pp. 223-232.

Thorp, E., "The invention of the first wearable computer", 1997, In Proceedings of the Second IEEE International Symposium on Wearable Computers, pp. 4–8, Pittsburgh.

Mann, S., 1997, "Wearable Computing: A first step towards personal imaging", Computer, Vol. 30, Issue.2, pp. 25-32.

Warwick, K., 2004, "I, Cyborg", University of Illinois Press.

Reger, B., Fleming, K., Sanguineti, V., Simon Alford, S., Mussa-Ivaldi, F., 2000, "Connecting Brains to Robots: The Development of a Hybrid System for the Study of Learning in Neural Tissues", Artificial Life VII, Portland, Oregon.

Chapin, J., Markowitz, R., Moxon, K., and Nicolelis, M., 1999, "Real-time control of a robot arm using simultaneously recorded neurons in the motor cortex". Nature Neuroscience, Vol. 2, Issue.7, pp. 664-670.

Talwar, S., Xu, S., Hawley, E., Weiss, S., Moxon, K., Chapin, J., 2002, "Rat navigation guided by remote control". Nature, Vol. 417, pp. 37-38.

Denislic, M., Meh, D., 1994, "Neurophysiological assessment of peripheral neuropathy in primary Sjögren's syndrome", Journal of Clinical Investigation, Vol. 72, 822-829.

Poboroniuc, M.S., Fuhr, T., Riener, R., Donaldson, N., 2002, "Closed-Loop Control for FES-Supported Standing Up and Sitting Down", Proc. 7th Conf. of the IFESS, Ljubljana, Slovenia, pp. 307-309.

Popovic, M. R., Keller, T., Moran, M., Dietz, V., 1998, "Neural prosthesis for spinal cord injured subjects", Journal Bioworld, Vol. 1, pp. 6-9.

Yu, N., Chen, J., Ju, M.; 2001, "Closed-Loop Control of Quadriceps/Hamstring activation for FES-Induced Standing-Up Movement of Paraplegics", Journal of Musculoskeletal Research, Vol. 5, No.3.

Cohen, M., Herder, J. and Martens, W.; 1999, "Cyberspatial Audio Technology", JAESJ, J. Acoustical Society of Japan (English), Vol. 20, No. 6, pp. 389-395, November.

Butz, A., Hollerer, T., Feiner, S., McIntyre, B., Beshers, C., 1999, "Enveloping users and computers in a collaborative 3D augmented reality", IWAR99, San Francisco, pp. 35-44.

Kanda, H., Yogi, T., Ito, Y., Tanaka, S., Watanabe, M and Uchikawa, Y., 1999, "Efficient stimulation inducing neural activity in a retinal implant", Proc. IEEE International Conference on Systems, Man and Cybernetics, Vol 4, pp. 409-413.

Kennedy, P., Bakay, R., Moore, M., Adams, K. and Goldwaithe, J., 2000, "Direct control of a computer from the human central nervous system", IEEE Transactions on Rehabilitation Engineering, Vol. 8, pp. 198-202.

Nam, Y., Chang, J.C.., Wheeler, B.C. and Brewer, G.J., 2004, "Gold-coated microelectrode array with Thiol linked self-assembled monolayers for engineering neuronal cultures", IEEE Transactions on Biomedical Engineering, Vol. 51, No. 1, pp. 158-165.

Gasson, M.., Hutt, B., Goodhew, I., Kyberd, P. and Warwick, K; 2002, "Bi-directional human machine interface via direct neural connection", Proc. IEEE Workshop on Robot and Human Interactive Communication, Berlin, German, pp. 265-270.

Branner, A., Stein, R. B. and Normann, E.A., 2001, "Selective "Stimulation of a Cat Sciatic Nerve Using an Array of Varying-Length Micro electrodes", Journal of Neurophysiology, Vol. 54, No. 4, pp. 1585-1594.

Warwick, K., Gasson, M., Hutt, B., Goodhew, I., Kyberd, P., Andrews, B, Teddy, P and Shad. A, 2003, "The Application of Implant Technology for Cybernetic Systems", Archives of Neurology, Vol. 60, No.10, pp. 1369-1373.

Warwick, K., Gasson, M., Hutt, B., Goodhew, I., Kyberd, K., Schulzrinne, H. and Wu, X., 2004, "Thought Communication and Control: A First Step using Radiotelegraphy", IEE Proceedings-Communications, Vol. 151, No. 3, pp. 185-189.

Gasson, M., Hutt, B., Goodhew, I., Kyberd, P. and Warwick, K., 2005, "Invasive Neural Prosthesis for Neural Signal detection and Nerve Stimulation", International Journal of Adaptive Control and Signal Processing, Vol. 19.

INDUSTRIAL AND REAL WORLD APPLICATIONS OF ARTIFICIAL NEURAL NETWORKS
Illusion or reality?

Kurosh Madani

Intelligence in Instrumentation and Systems Lab. (I^{2}S Lab.),
PARIS XII University, Senart Institute of Technology, Pierre Point avenue, F-77127 Lieusaint, France
Email: madani@univ-paris12.fr

Keywords: Artificial Neural Networks (ANN), Industrial applications, Real-world applications.

Abstract: Inspired from biological nervous systems and brain structure, Artificial Neural Networks (ANN) could be seen as information processing systems, which allow elaboration of many original techniques covering a large field of applications. Among their most appealing properties, one can quote their learning and generalization capabilities. If a large number of works have concerned theoretical and implementation aspects of ANN, only a few are available with reference to their real world industrial application capabilities. In fact, applicability of an available academic solution in industrial environment requires additional conditions due to industrial specificities, which could sometimes appear antagonistic with theoretical (academic) considerations. The main goal of this paper is to present, through some of main ANN models and based techniques, their real application capability in real industrial dilemmas. Several examples dealing with industrial and real world applications have been presented and discussed covering "intelligent adaptive control", "fault detection and diagnosis", "decision support", "complex systems identification" and "image processing".

1 INTRODUCTION

Real world dilemmas, and especially industry related ones, are set apart from academic ones from several basic points of views. The difference appears since definition of the "problem's solution" notion. In fact, academic (called also sometime theoretical) approach to solve a given problem often begins by problem's constraints simplification in order to obtain a "solvable" model (here, solvable model means a set of mathematically solvable relations or equations describing a behavior, phenomena, etc…). If the theoretical consideration is an indispensable step to study a given problem's solvability, in the case of a very large number of real world dilemmas, it doesn't lead to a solvable or realistic solution. A significant example is the modeling of complex behavior, where conventional theoretical approaches show very soon their limitations. Difficulty could be related to several issues among which:

- large number of parameters to be taken into account (influencing the behavior) making conventional mathematical tools inefficient,
- strong nonlinearity of the system (or behavior), leading to unsolvable equations,
- partial or total inaccessibility of system's relevant features, making the model insignificant,
- subjective nature of relevant features, parameters or data, making the processing of such data or parameters difficult in the frame of conventional quantification,
- necessity of expert's knowledge, or heuristic information consideration,
- imprecise information or data leakage.

Examples illustrating the above-mentioned difficulties are numerous and may concern various areas of real world or industrial applications. As first example, one can emphasize difficulties related to economical and financial modeling and prediction, where the large number of parameters, on the one hand, and human related factors, on the other hand, make related real world problems among the most difficult to solve. Another example could be given in the frame of the industrial processes and manufacturing where strong nonlinearities related to complex nature of manufactured products affect controllability and stability of production plants and processes. Finally, one can note the difficult dilemma of complex pattern and signal recognition and analysis, especially when processed patterns or

J. Braz et al. (eds.), Informatics in Control, Automation and Robotics I, 11–26.
© 2006 *Springer. Printed in the Netherlands.*

signals are strongly noisy or deal with incomplete data.

Over the past decades, Artificial Neural Networks (ANN) and issued approaches have allowed the elaboration of many original techniques (covering a large field of applications) overcoming some of mentioned difficulties (Nelles, 1995) (Faller, 1995) (Maidon, 1996), (Madani, 1997) (Sachenco, 2000). Their learning and generalization capabilities make them potentially promising for industrial applications for which conventional approaches show their failure. However, even if ANN and issued approaches offer an attractive potential for industrial world, their usage should always satisfy industrial "specificities". In the context of the present paper, the word "specificity" intends characteristic or criterion channelling industrial preference for a strategy, option or solution as an alternative to the others.

In fact, several specificities distinguish the industrial world and related constraints from the others. Of course, here the goal is not to analyse all those specificities but to overview briefly the most pertinent ones. As a first specificity one could mention the "reproducibility". That means that an industrial solution (process, product, etc…) should be reproducible. This property is also called solution stability. A second industrial specificity is "viability", which means implementation (realization) possibility. That signifies that an industrial solution should be adequate to available technology and achievable in reasonable delay (designable, realizable). Another industrial specificity is "saleability", which means that an industrial solution should recover a well identified field of needs. Finally, an additional important specificity is "marketability" making a proposed industrial solution attractive and concurrent (from the point of view of cost, price-quality ratio, etc…) to other available products (or solutions) concerning the same area.

Another key point to emphasize is related to the real world constraints consideration. In fact, dealing with real world environment and related realities, it is not always possible to put away the lower degree phenomena's influence or to neglect secondary parameters. That's why a well known solved academic problem could sometime appear as an unachieved (unbearable) solution in the case of an industry related dilemma. In the same way a viable and marketable industrial solution may appear as primitive from academic point of view.

The main goal of this paper is to present, through main ANN models and based techniques, the effectiveness of such approaches in real world industrial problems solution. Several examples through real world industrial applications have been shown and discussed. The present paper has been organized as follows: the next section will present the general principle of Artificial Neural Networks relating it to biological considerations. In the same section two classes of neural models will be introduced and discussed: Multi-layer Perceptron and Kernel Functions based Neural Networks. The section 3 and related sub-sections will illustrate real world examples of application of such techniques. Finally, the last section will conclude the paper.

2 FROM NATURAL TO ARTIFICIAL

As mentions Andersen (Anderson, 1995): "*It is not absolutely necessary to believe that neural network models have anything to do with the nervous system, but it helps. Because, if they do, we are able to use a large body of ideas, experiments, and facts from cognitive science and neuroscience to design, construct, and test networks. Otherwise, we would have to suggest functions and mechanism for intelligent behavior without any examples of successful operation*".

Much is still unknown about how the brain trains itself to process information, so theories abound. It is admitted that in the biological systems (human or animal brain), a typical neuron collects signals from others through a host of fine structures called *dendrites*. Figure 1 shows a simplified bloc diagram of biological neural system comparing it to the artificial neuron. The neuron sends out spikes of electrical activity through a long, thin stand known as an *axon*, which splits into thousands of branches. At the end of each branch, a structure called a *synapse* converts the activity from the axon into electrical effects that inhibit or excite activity from the axon into electrical effects that inhibit or excite activity in the connected neurones. When a neuron receives excitatory input that is sufficiently large compared with its inhibitory input, it sends a spike of electrical activity down its axon. Learning occurs by changing the effectiveness of the synapses so that the influence of one neuron on another changes.

Inspired from biological neuron, artificial neuron reproduces a simplified functionality of that complex biological neuron. The neuron's operation could be seen as following: a neuron updates its output from weighted inputs received from all neurons connected to that neuron. The decision to update or not the actual state of the neuron is performed thank to the "decision function" depending to activity of those connected neurons. Let us consider a neuron with its state denoted by x_i (as it is shown in figure 1) connected to M other

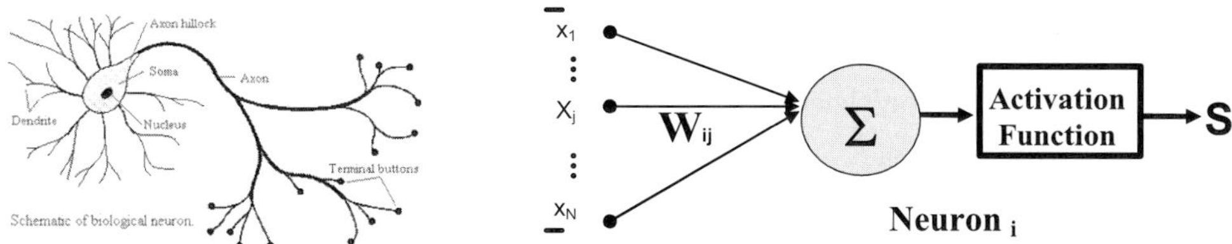

Figure 1: Biological (left) and artificial (right) neurons simplified bloc-diagrams.

neurons, and let x_j represent the state (response) of the j-th neuron interconnected to that neuron with $j \in \{1, \cdots, M\}$. Let W_{ij} be the weight (called also, synaptic weight) between j-th and i-th neurons. In this case, the activity of all connected neurons to the i-th neuron, formalized through the "synaptic potential" of that neuron, is defined by relation (1). Fall back on its synaptic potential (and sometimes to other control parameters), the neuron's decision function will putout (decide) the new state of the neuron according to the relation (2). One of the most commonly used decision functions is the "sigmoidal" function given by relation (3) where η is a control parameter acting on decision strictness or softness, called also "learning rate".

$$V_i = \sum_{j=1}^{j=M} W_{ij} \cdot x_j \qquad (1)$$

$$S_i = F(x_i, V_i) = F\left(x_i, \sum_{j=1}^{j=M} W_{ij} \cdot x_j\right) \qquad (2)$$

$$F(x) = \frac{1}{1 + e^{-\frac{x}{\eta}}} \qquad (3)$$

Also referred to as connectionist architectures, parallel distributed processing, and neuromorphic systems, an artificial neural network (ANN) is an information-processing paradigm inspired by the densely interconnected, parallel structure of the mammalian brain information processes. Artificial neural networks are collections of mathematical models that emulate some of the observed properties of biological nervous systems and draw on the analogies of adaptive biological learning mechanisms. The key element of the ANN paradigm is the novel structure of the information processing system. It is supposed to be composed of a large number of highly interconnected processing elements that are analogous to neurons and are tied together with weighted connections that are analogous to synapses. However, a large number of proposed architectures involve a limited number of neurones.

Biologically, neural networks are constructed in a three dimensional way from microscopic components. These neurons seem capable of nearly unrestricted interconnections. This is not true in any artificial network. Artificial neural networks are the simple clustering of the primitive artificial neurons. This clustering occurs by creating layers, which are then connected to one another. How these layers connect may also vary. Basically, all artificial neural networks have a similar structure or topology. Some of their neurons interface the real world to receive its inputs and other neurons provide the real world with the network's outputs. All the rest of the neurons are hidden form view. Figure 2 shows an artificial neural network's general bloc-diagram.

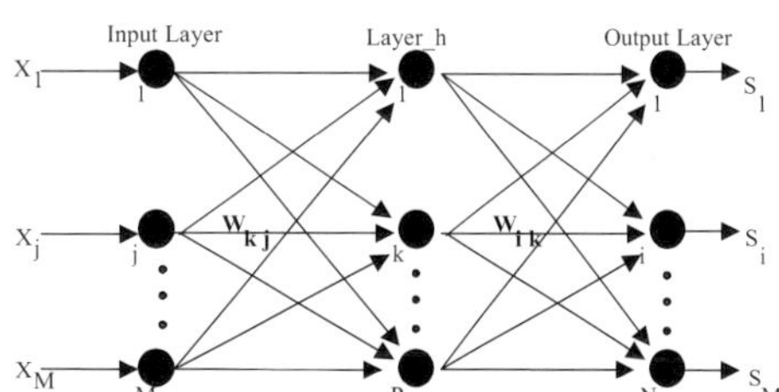

Figure 2: Artificial neural network simplified bloc-diagrams.

In general, the input layer consists of neurons that receive input form the external environment. The output layer consists of neurons that communicate the output of the system to the user or external environment. There are usually a number of hidden layers between these two layers. When the input layer receives the input its neurons produce output, which becomes input to the other layers of the system. The process continues until a certain condition is satisfied or until the output layer is invoked and fires their output to the external environment.

Let us consider a 3 layers standard neural network, including an input layer, a hidden layer and an output layer, conformably to the figure 2. Let us suppose that the input layer includes M neurons, the hidden layer includes P neurons and the output layer includes N neurons. Let

$\mathbf{X} = \left(X_1, ..., X_j, ..., X_M\right)^T$ represents the input vectors, with $j \in \{1, \cdots, M\}$, $\mathbf{H} = \left(H_1, ..., H_k, ..., H_P\right)^T$ represents the hidden layer's output with $k \in \{1, \cdots, P\}$ and $\mathbf{S} = \left(S_1, ..., S_i, ..., S_N\right)^T$ the output vector with $i \in \{1, \cdots, N\}$. Let us note W_{kj}^H and W_{ik}^S synaptic matrixes elements, corresponding to input-hidden layers and hidden-output layers respectively. Neurons are supposed to have a non-linear decision function (activation function) $F(.)$. V_k^H and V_i^S, defined by relation (4), will represent the synaptic potential vectors components of hidden and output neurons, respectively (e.g. vectors V^H and V^S components). Taking into account such considerations, the k-th hidden and the i-th output neurons outputs will be given by relations (5).

$$V_k^H = \sum_{j=1}^{j=M} W_{kj}^H \cdot x_j \text{ and } V_i^S = \sum_{k=1}^{k=P} W_{ik}^S \cdot h_k \qquad (4)$$

$$H_k = F\left(V_k^H\right) \text{ and } S_i = F\left(V_i^S\right) \qquad (5)$$

As it has been mentioned above, learning in biological systems involves adjustments to the synaptic connections that exist between the neurons. This is valid for ANNs as well. Learning typically occurs by example through training, or exposure to a set of input/output data (called also, learning database) where the training algorithm iteratively adjusts the connection weights (synapses). These connection weights store the knowledge necessary to solve specific problems. The strength of connection between the neurons is stored as a weight-value for the specific connection. The system learns new knowledge by adjusting these connection weights. The learning process could be performed in "on-line" or in "off-line" mode. In the off-line learning methods, once the systems enters into the operation mode, its weights are fixed and do not change any more. Most of the networks are of the off-line learning type. In on-line or real time learning, when the system is in operating mode (recall), it continues to learn while being used as a decision tool. This type of learning needs a more complex design structure.

The learning ability of a neural network is determined by its architecture (network's topology, artificial neurons nature) and by the algorithmic method chosen for training (called also, "learning rule"). In a general way, learning mechanisms (learning processes) could be categorized in two classes: "supervised learning" (Arbib, 2003) (Hebb, 1949) (Rumelhart, 1986) and "unsupervised learning" (Kohonen, 1984) (Arbib, 2003). The supervised learning works on reinforcement from the outside. The connections among the neurons in the hidden layer are randomly arranged, then reshuffled according to the used learning rule in order to solving the problem. In general an "error" (or "cost") based criterion is used to determine when stop the learning process: the goal is to minimize that error. It is called supervised learning, because it requires a teacher. The teacher may be a training set of data or an observer who grades the performance of the network results (from which the network's output error is obtained). In the case where the unsupervised learning procedure is applied to adjust the ANN's behaviour, the hidden neurons must find a way to organize themselves without help from the outside. In this approach, no sample outputs are provided to the network against which it can measure its predictive performance for a given vector of inputs. In general, a "distance" based criterion is used assembling the most resembling data. After a learning process, the neural network acts as some non-linear function identifier minimizing the output errors.

ANNs learning dilemma have been the central interest of a large number research investigations during the two past decades, leading to a large number of learning rules (learning processes). The next sub-sections will give a brief overview of the most usual of them: "Back-Propagation" (BP) based learning rule neural network, known also as "Multi-Layer Perceptron and "Kernel Functions" based learning rule based neural networks trough one of their particular cases which are "Radial Basis Functions" (RBF-like neural networks).

2.1 Back-Propagation Learning Rule and Multi-Layer Perceptron

Back-Propagation (Bigot, 1993) (Rumelhart, 1986) (Bogdan, 1994) based neural models, called also Back-Propagation based "Multi-Layer Perceptron" (MLP) are multi-layer neural network (conformably to the general bloc-diagram shown in figure 2). A neuron in this kind of neural network operates conformably to the general ANN's operation frame e.g. according to equations (1), (2) and (3). The specificity of this class of neural network appears in the learning procedure, called "Back-Propagation of error gradient".

The principle of the BP learning rule is based on adjusting synaptic weights proportionally to the neural network's output error. Examples (patterns from learning database) are presented to the neural network, then, for each of learning patterns, the neural network's output is compared to the desired one and an "error vector" is evaluated. Then all synaptic weights are corrected (adjusted) proportionally to the evaluated output error. Synaptic weights correction is performed layer by layer from the output layer to the input layer. So, output error is back-propagated in order to correct synaptic weights. Generally, a quadratic error criterion, given by equation (6), is used. In this relation S_i represents the i-th output vector's component and S_i^d represents the desired value of this component. Synaptic weights are modified according to relation (7), where $dW_{i,j}^h$ represents the synaptic variation (modification) of the synaptic weight connecting the j-th neurone and i-th neuron between two adjacent layers (layer h and layer h-1). η is a real coefficient called also "learning rate".

$$\varepsilon_i = \frac{1}{2}\left(S_i - S_i^d\right)^2 \qquad (6)$$

$$dW_{i,j}^h = -\eta \bullet \mathbf{grad}_W(\varepsilon) \qquad (7)$$

The learning rate parameter is decreased progressively during the learning process. The learning process stops when the output error reaches some acceptable value.

2.2 Kernel Functions Based Neural Models

This kind of neural models belong to the class of "evolutionary" learning strategy based ANN (Reyneri, 1995) (Arbib, 2003) (Tremiolles, 1996). That means that the neural network's structure is completed during the learning process. Generally, such kind of ANNs includes three layers: an input layer, a hidden layer and an output layer. Figure 3 represents the bloc-diagram of such neural net. The

number of neurons in input layer corresponds to the processed patterns dimensionality e.g. to the problem's feature space dimension.

The output layer represents a set of categories associated to the input data. Connections between hidden and output layers are established dynamically during the learning phase. It is the hidden layer which is modified during the learning phase. A neuron from hidden layer is characterized by its "centre" representing a point in an N dimensional space (if the input vector is an N-D vector) and some decision function, called also neuron's "Region Of Influence" (ROI). ROI is a kernel function, defining some "action shape" for neurons in treated problem's feature space. In this way, a new learning pattern is characterized by a point and an influence field (shape) in the problem's N-D feature space. In the other words, the solution is mapped thank to learning examples in problem's N-D feature space. The goal of the learning phase is to partition the input space associating prototypes with a categories and an influence field, a part of the input space around the prototype where generalization is possible. When a prototype is memorized, ROI of neighbouring neurons are adjusted to avoid conflict between neurons and related categories. The neural network's response is obtained from relation (8) where C_j represents a "category", $V = \begin{bmatrix} V_1 & V_2 & \dots & V_N \end{bmatrix}^T$ is the input vector, $P^j = \begin{bmatrix} p_1^j & p_2^j & \dots & p_N^j \end{bmatrix}^T$ represents the j-th "prototype" memorized (learned) thanks to creation of the neuron j in the hidden layer, and λ_j the ROI associated to this neuron (neuron j). F(.) is the neuron's activation (decision) function which is a radial basis function (a Gaussian function for example).

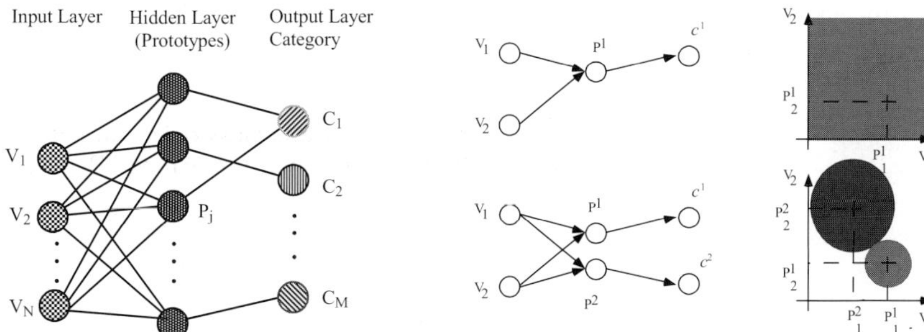

Figure 3: Radial Basis Functions based ANN's bloc-diagram (left). Example of learning process in 2-D feature space (right).

$$C_j = F\big(dist(V, P^j)\big) \quad \text{If} \quad dist(V, P^j) \leq \lambda_j$$
$$C_j = 0 \qquad\qquad\qquad \text{If} \quad dist(V, P^j) > \lambda_j \tag{8}$$

$$dist = \sqrt[n]{\sum_i \left| V_i - p_i^j \right|^n} \tag{9}$$

$$\text{with} \quad \sum_i \left| V_i - p_i^j \right| \leq \left(\sum_i (V_i - p_i^j)^2 \right)^{\frac{1}{2}} \leq \max_i \left| V_i - p_i^j \right| \tag{10}$$

The choice of the distance calculation (choice of the used norm) is one of the main parameters in the case of the RCE-KNN like neural models (and derived approaches). The most usual function used to evaluate the distance between two patterns is the Minkowski function expressed by relation (9), where V_i is the i-th component of the input vector and p_i^j the i-th component of the j-th memorized pattern (learned pattern). Manhattan distance ($n = 1$, called also L1 norm) and Euclidean distance ($n = 2$) are particular cases of the Minkowski function and the most applied distance evaluation criterions. One can write relation (10).

3 ANN BASED SOLUTIONS FOR INDUSTRIAL ENVIRONMENT

If the problem's complexity and the solution consistency, appearing through theoretical tools (modeling or conceptual complexity) needing to solve it, are of central challenges for applicability of a proposed concepts, another key points characterizing application design, especially in industrial environment, is related to implementation requirements. In fact, constraints related to production conditions, quality, etc. set the above-mentioned point as a chief purpose to earn solution's viability. That is why in the next subsections, dealing with real-world, and industrial applications of above-presented ANN models, the implementation issues will be of central considerations. Moreover, progress accomplished during the lasts decades concerning electrical engineering, especially in the microprocessors area, offers new perspectives for real time execution capabilities and enlarges the field for solution implementation ability.

3.1 MLP Based Adaptive Controller

Two meaningful difficulties characterize the controller dilemma, making controllers design one of the most challenging tasks: the first one is the plant parameters identification, and the second one is related to the consideration of interactions between real world (environment) and control system, especially in the case of real-world applications where controlled phenomena and related parameters deal with strong nonlinearities. Neural models and issued approaches offer original perspectives to overcome these two difficulties. However, beside these two difficulties, another chief condition for conventional or unconventional control is related to the controller's implementation which deals with real-time execution capability. Recent progresses accomplished on the one hand, in the microprocessor design and architecture, and on the other hand, in microelectronics technology and manufacturing, leaded to availability of powerful microprocessors, offering new perspectives for software or hardware implementation, enlarging the field in real time execution capability.

Finally, it should always be taken into account that proposed solution to a control dilemma (and so, the issued controller) emerges on the basis of former available equipments (plants, processes, factory, etc.). That's why, with respect to above-discussed industrial specificities, preferentially it should not lead to a substantial modification of the still existent materials. In fact, in the most of real industrial control problems, the solution should enhance existent equipments and be adaptable to an earlier technological environment.

3.1.1 General Frame and Formalization of Control Dilemma

The two usual strategies in conventional control are open-loop and feed-back loop (known also as feedback loop regulation) controllers. Figure 4 gives the general bloc-diagram of these two controllers, were E_k is the "input vector" (called also "order" vector), $Y_k = \begin{pmatrix} y_k & y_{k-1} & \cdots & y_{k-m} \end{pmatrix}^T$ is the "output vector" (plant's or system's state or response) and $U_k = \begin{pmatrix} u_k & u_{k-1} & \cdots & u_{k-n} \end{pmatrix}^T$ is the "command vector". k represents the discrete time variable. The output vector is defined as a vector which components are the m last system's outputs. In the same way, the command vector is defined as a vector which components are the n last commands. Such vectors define output and command feature spaces of the system. Taking into account the general control bloc diagram (figure 4), the goal of the command is to make converge the system's output with respect to some "desired output" noted Y_d. If the command vector is a subject to some modifications, then the output vector will be modified. The output modification will be performed with respect to the system's (plant, process or system under control) characteristics according to equation (11), where **J** represents the Jacobean matrix of the system.

$$dY_k = \mathbf{J}\, dU_k \qquad (11)$$

So, considering that the actual instant is k, it appears that to have an appropriated output ($Y_{k+1} = Y_d$), the output should be corrected according to the output error defined by: $dY_k = Y_k - Y_d$. In the frame of such formulation, supposing that one could compute the system's reverse Jacobean the command correction making system's output to converge to the desired state (or response) will be conform to relation (12).

$$dU_k = \mathbf{J}^{-1}\, dY_k \qquad (12)$$

System's Jacobean is related to plant's features (parameters) involving difficulties mentioned before.

Moreover, system's reverse Jacobean computation is not a trivial task. In the real world applications, only in very few cases (as linear transfer functions) the system's reverse Jacobean is available. So, typically a rough approximation of this matrix is obtained.

3.1.2 ANN Based Controller

Let us consider a neural network approximating (learning) a given system (process or plant). Let Y be the system's output, U be the system's command (U becomes also the neural network's output), W_{ij} be synaptic weights of the neural network and ε be the output error representing some perturbation occurring on output. The part of output perturbation (output error) due to the variation of a given synaptic weight (W_{ij}) of the neural network noted as $\partial\varepsilon / \partial W_{ij}$ could be written conformably to relation (13).

$$\frac{\partial\varepsilon}{\partial W_{ij}} = \frac{\partial\varepsilon}{\partial y}\frac{\partial y}{\partial u}\frac{\partial u}{\partial W_{ij}} \qquad (13)$$

One can remark that $\partial y / \partial u$ is the system's Jacobean element and $\partial u / \partial W_{ij}$ could be interpreted as the "neural network's Jacobean" element. As the output error is related to the system's controller characteristics (represented by system's Jacobean), so the modification of synaptic weights with respect to the measured error (e.g. the neural network appropriated training) will lead to the correction of the command (dU) minimizing the output error.

Several Neural Network based adaptive control architectures have still been proposed. However, taking into account the above-discussed industrial specificities, the most effective scheme is the hybrid neuro-controller (Hormel, 1992) (Miller, 1987) (Madani, 1996) (Albus, 1975) (Comoglio, 1992). This solution operates according to a Neural Network based correction of a conventional controller. Figure 5 shows the bloc diagram of such approach.

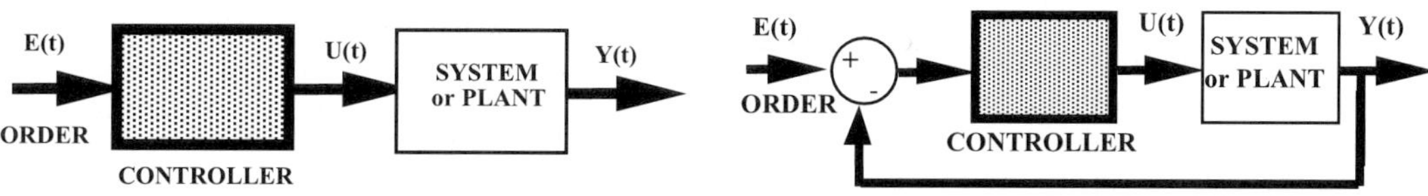

Figure 4: General bloc-diagrams of control strategies showing open-loop controller (left) and feed-back loop controller (Right) principles.

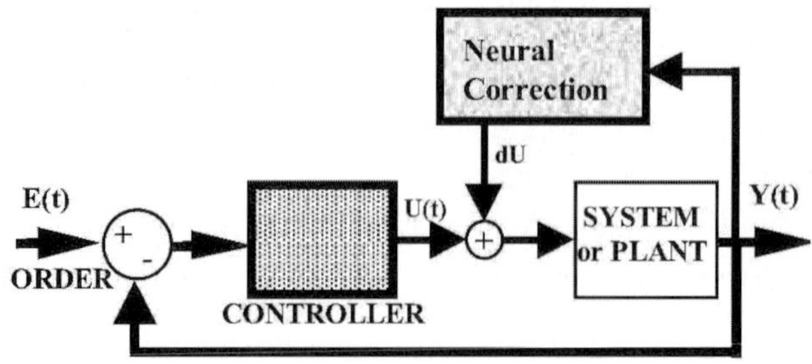

Figure 5: General bloc-diagrams hybrid neuro-controller.

As one can see in our ANN based control strategy, the command U(t) is corrected thanks to the additional correction dU, generated by neural device and added to the conventional command component. The Neural Network's learning could be performed on-line or off-line.

Several advantages characterize the proposed strategy. The first one is related to the control system stability. En fact, in the worst case the controlled plant will operate according to the conventional control loop performances and so, will ensure the control system's stability. The second advantage of such strategy is related to the fact that the proposed architecture acts as a hybrid control system where usual tasks are performed by a conventional operator and unusual operations (such as highly non linear operations or those which are difficult to be modelled by conventional approaches) are realized by neural network based component. This second advantage leads to another main welfare which is the implementation facility and so, the real-time execution capability. Finally, the presented solution takes into account industrial environment reality where most of control problems are related to existent plants behaviours enhancement dealing with an available (still implemented) conventional controller. This last advantage of the proposed solution makes it a viable option for industrial environment.

3.1.3 MLP Based Adaptive Controller Driving Turning Machine

The above-exposed neural based hybrid controller has been used to enhance the conventional vector-control driving a synchronous 3-phased alternative motor. The goal of a vector control or field-oriented control is to drive a 3-phased alternative motor like an independent excitation D.C motor. This consists to control the field excitation current and the torque generating current separately (Madani, 1999). The input currents of the motor should provide an electromagnetic torque corresponding to the command specified by the velocity regulator. For synchronous motor, the secondary magnetic flux (rotor) rotates at the same speed and in the same direction as the primary flux (stator). To achieve the above-mentioned goal, the three phases must be transformed into two equivalent perpendicular phases by using the Park transformation which needs the rotor position, determined by a transducer or a tachometer. In synchronous machine, the main parameters are Ld (inductance of d-phase), Lq (inductance of q-phase), and Rs (statoric resistor), which vary in relation with currents (Id and Iq), voltages (Vd and Vq), mechanical torque and speed (of such machine). The relations between voltages or currents depend on these three parameters defining the motor's model. However, these parameters are not easily available because of their strongly nonlinear dependence to the environment conditions and high number of influent conditions.

The neural network is able to identify these parameters and to correct the machine's reference model, feeding back their real values through the control loop. Parameters are related to voltages, currents, speed and position. The command error (measured as voltage error) could be linked to the plant's parameters values error. In the first step, the command is computed using nominal theoretical plant parameters. The neural network learns the plant's behaviour comparing outputs voltages (Vd ,Vq), extracted from an impedance reference model, with measured voltages (Vdm,Vqm). In the second step when the system is learned, the neural network gives the estimated plant's parameters to the controller (Madani, 1999).

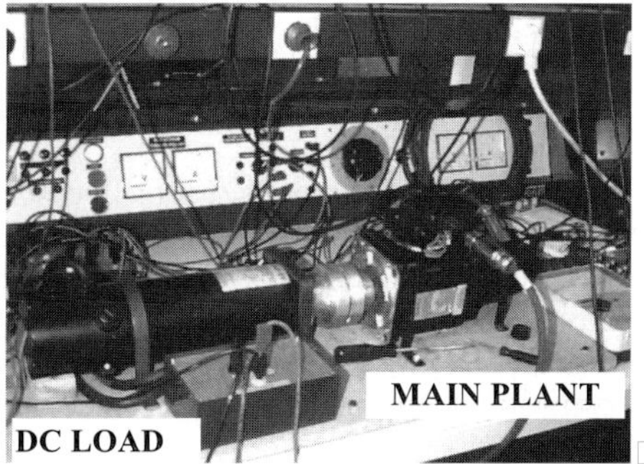

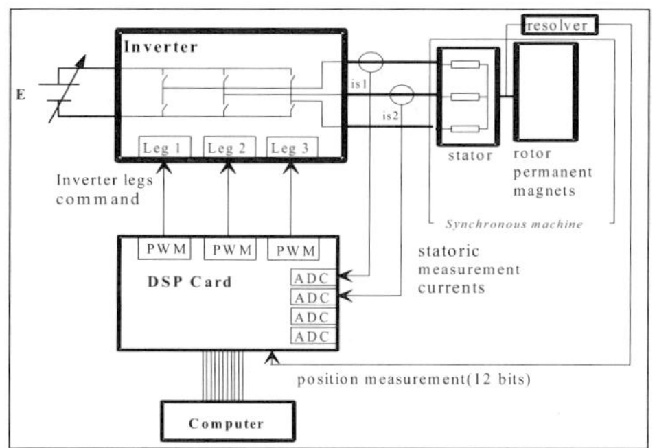

Figure 6: View of the main plant and the load coupled to the main motor (left). Implementation block diagram (right).

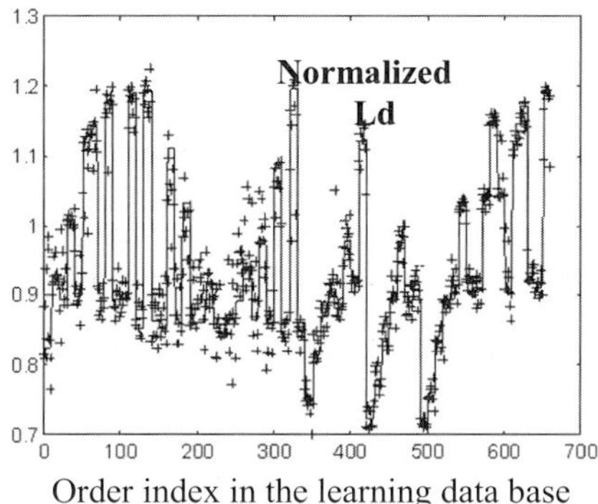

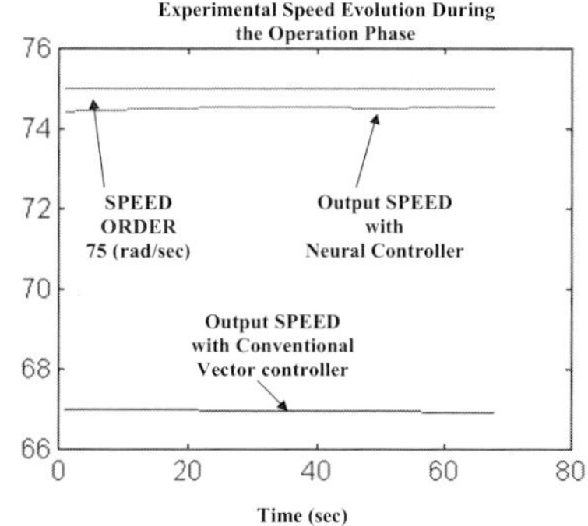

Figure 7: Experimental plant parameters identification by neural net (left). Experimental measured speed when the plant is unloaded (right).

The complete system, including the intelligent neuro-controller, a power interface and a permanent synchronous magnet motor (plant), has been implemented according to the bloc diagram of figure 6. Our intelligent neuro-controller has been implemented on a DSP based board. In this board, the main processor is the TMS C330 DSP from Texas Instruments. The learning data base includes 675 different values of measurement extracted motor's parameters (Ld and Lq). Different values of measurable parameters (currents, voltages, speed and position), leading to motor's parameters extraction, have been obtained for different operation modes of the experimental plant, used to validate our concepts. The ANN learning is shifted for 4 seconds after power supply application to avoid unstable data in the starting phase of the motor.

Figures 7 gives experimental results relative to the motor's internal parameter evolution and the plant's measured speed, respectively. One can remark from those figures that:
- Internal plant model's parameters are identified by the neural network.
- Such neural based controller compensates the

inefficiency of the conventional control loop (achieving a 74 rad/sec angular speed).

3.2 Kernel Functions ANN Based Image Processing for Industrial Applications

Characterization by a point and an influence field (shape) in the problem's N-D feature space of a learned becomes particularly attractive when the problem's feature space could be reduced to a 2-D space. In fact, in this case, the learning process, in the frame of kernel functions ANN, could be interpreted by a simple mapping model. In the case of images the bi-dimensionality (2-D nature) is a natural property. That's why, such kind of neural models and issued techniques become very attractive for image processing issues. Moreover, their relative implementation facility makes them powerful candidates to overcome a large class of industrial requirements dealing with image processing and image analysis.

Before presenting the related industrial applications, let focus the next sub-section on a brief description of ZISC-036 neuro-processor from IBM, which implements some of kernel functions ANN based models.

3.2.1 IBM ZISC-036 Neuro-Processor

The IBM ZISC-036 (Tremiolles, 1996) (Tremiolles, 1997) is a parallel neural processor based on the RCE and KNN algorithms. Each chip is capable of performing up to 250 000 recognitions per second. Thanks to the integration of an incremental learning algorithm, this circuit is very easy to program in order to develop applications; a very few number of functions (about ten functions) are necessary to control it. Each ZISC-036 like neuron implements two kinds of distance metrics called L1 and LSUP respectively. Relations (14) and (15) define the above-mentioned distance metrics were P_i represents the memorized prototype and V_i is the input pattern. The first one (L1) corresponds to a polyhedral volume influence field and the second (LSUP) to a hyper-cubical influence field.

$$\text{L1:}\ dist = \sum_{i=0}^{n} \left| V_i - P_i \right| \qquad (14)$$

$$\text{LSUP:}\ dist = \max_{i=0...n} \left| V_i - P_i \right| \qquad (15)$$

Figure 8 gives the ZISC-036 chip's bloc diagram and an example of input feature space mapping in a 2-D space. A 16 bit data bus handles input vectors as well as other data transfers (such as category and distance), and chip controls. Within the chip, controlled access to various data in the network is performed through a 6-bit address bus. ZISC-036 is composed of 36 neurons. This chip is fully cascadable which allows the use of as many neurons as the user needs (a PCI board is available with a 684 neurons). A neuron is an element, which is able to:

- memorize a prototype (64 components coded on 8 bits), the associated category (14 bits), an influence field (14 bits) and a context (7 bits),
- compute the distance, based on the selected norm (norm L1 given by relation or LSUP) between its memorized prototype and the input vector (the distance is coded on fourteen bits),
- compare the computed distance with the influence fields,
- communicate with other neurons (in order to find the minimum distance, category, etc.),
- adjust its influence field (during learning phase).

Two kinds of registers hold information in ZISC-O36 architecture: global registers and neuron registers. Global registers hold information for the device or for the full network (when several devices are cascaded). There are four global registers implemented in ZISC-036: a 16-bits Control & Status Register (CSR), a 8-bits Global Context Register (GCR), a 14-bits Min. Influence Field register (MIF) and a 14-bits Max. Influence Field register (MAF). Neuron registers hold local data for each neuron. Each neuron includes five neuron registers: Neuron Weight Register (NWR), which is a 64-by-8 bytes register, a 8-bits Neuron Context Register (NCR), Category register (CAT), Distance register (DIST) and Neuron Actual Influence Field register (NAIF). The last three registers are both 14-bites registers. Association of a context to neurons is an interesting concept, which allows the network to be divided in several subsets of neurons. Global Context Register (GCR) and Neuron Context Register (NCR) hold information relative to such subdivision at network and neuron levels respectively. Up to 127 contexts can be defined.

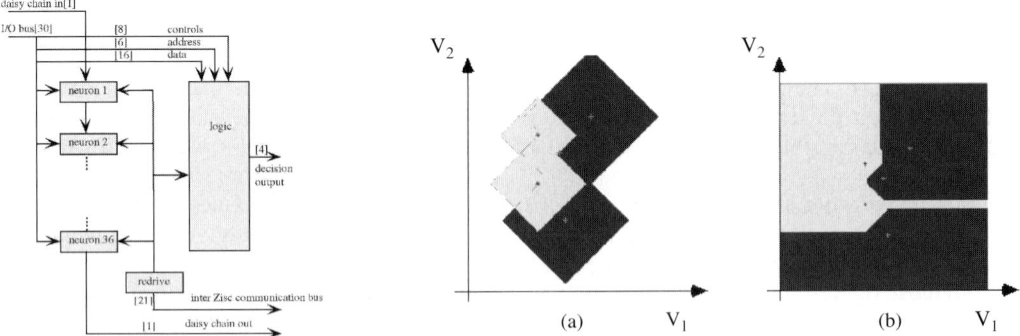

Figure 8: IBM ZISC-036 chip's bloc diagram (left) and an example of input feature space mapping in a 2-D space using ROI and 1-NN modes, using norm L1 (right).

3.2.2 Application in Media and Movie Production Industry

The first class of applications concerns image enhancement in order to: restore old movies (noise reduction, focus correction, etc.), improve digital television, or handle images which require adaptive processing (medical images, spatial images, special effects, etc.).

The used principle is based on an image's physics phenomenon which states that when looking at an image through a small window, there exist several kinds of shapes that no one can ever see due to their proximity and high gradient (because, the number of existing shapes that can be seen with the human eye is limited). ZISC-036 is used to learn as many shapes as possible that could exist in an image, and then to replace inconsistent points by the value of the closest memorized example. The learning phase consists of memorizing small blocks of an image (as an example 5x5) and associating to each the middle pixel's value as a category. These blocks must be chosen in such a way that they represent the maximum number of possible configurations in an image. To determine them, the proposed solution consists of computing the distances between all the blocks and keeping only the most different.

The learning algorithm used here incorporates a threshold and learning criteria (Learn_Crit (V)). The learning criteria is the criteria given by relation (16) where V_l^k represents the l-th component of the input vector V^k, P_l^j represents the l-th component of the j-th memorized prototype, C^k represents the category value associated to the input vector V^k, C^j is the category value associated to the memorized prototype P^j and, α and β are real coefficients adjusted empirically.

$$Learn\_Crit(V^k) = \alpha \sum_l \left| V_l^k - P_l^j \right| + \beta \left| C^k - C^j \right| \quad (16)$$

An example (pattern) from the learning base is chosen and the learning criterion for that example is calculated. If the value of the learning criteria is greater than the threshold, then a neuron is engaged (added). If the learning criteria's value is less than the threshold, no neuron is engaged. The aforementioned threshold is decreased progressively. Once learning database is learned the training phase is stopped. Figure 9 shows a pattern-to-category association learning example and the generalization (application) phase for the case of an image enhancement process.

The image enhancement or noise reduction principles are the same as described above. The main difference lies in the pixel value associated to each memorized example. In noise reduction, the learned input of the neural network is a noisy form of the original image associated with the correct value (or form). For example, in the figure 9 learning process example, for each learned pattern (a block of 5x5) from the input image (degraded one), the middle pixel of the corresponding block from the output image (correct one) is used as the "corrected pixel value" and is memorized as the associated category. After having learned about one thousand five hundred examples, the ZISC-036 based system is able to enhance an unlearned image.

Figure 10 gives results corresponding to movie sequences coloration. In this application unlearned scenes of a same sequence are collared (restored) by learning a representative (sample) scene of the same sequence. For both cases of image restoration and coloration it has been shown (Tremiolles, 1998) (Madani, 2003) that the same neural concept could perform different tasks as noise reduction, image enhancement and image coloration which are necessary to restore a degraded movie. Quantitative comparative studies established and analysed in above-mentioned references show pertinence of such techniques.

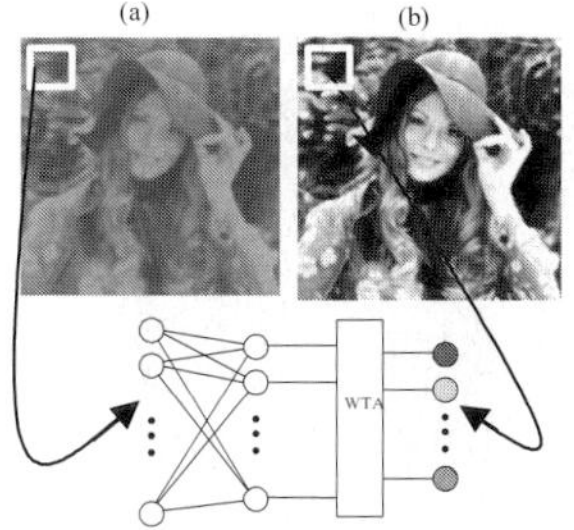

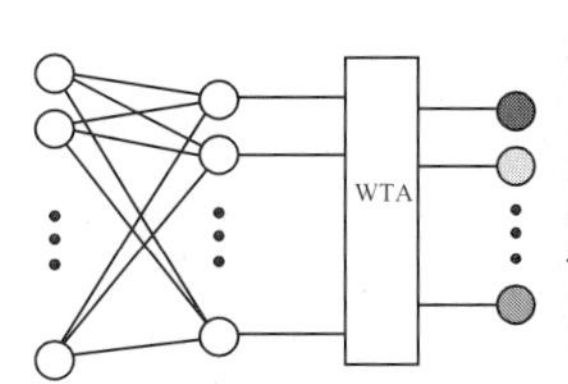

Figure 9: Image enhancement learning process using association of regions from the degraded and correct images (left) and image enhancement operation in generalization phase using an unlearned image (right).

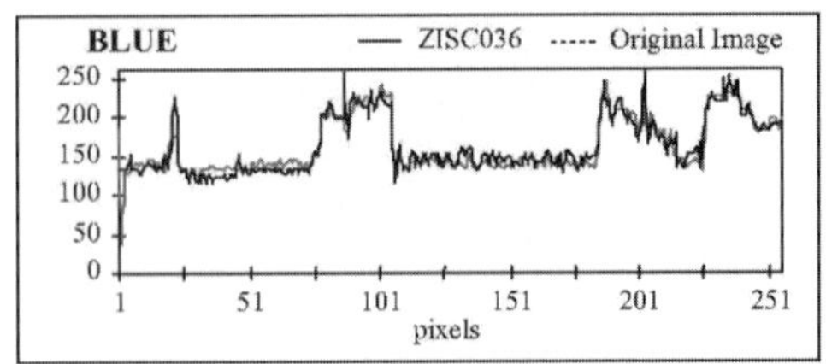

Figure 10: Result concerning movie coloration showing from left to right the image used to train the neural network and the result obtained on coloration of an unlearned scene (left). Blue component cross sections comparison between the coloured (reconstructed) and the original images in generalization phase (right).

3.2.3 Probe Mark Inspection in VLSI Chips Production

One of the main steps in VLSI circuit production is the testing step. This step verifies if the final product (VLSI circuit) operates correctly or not. The verification is performed thank to a set of characteristic input signals (stimulus) and associated responses obtained from the circuit under test. A set of such stimulus signals and associated circuit's responses are called test vectors. Test vectors are delivered to the circuit and the circuit's responses to those inputs are catch through standard or test dedicated Input-Output pads (I/O pads) called also vias. As in the testing step, the circuit is not yet packaged the test task is performed by units, which are called probers including a set of probes performing the communication with the circuit. Figure 11 shows a picture of probes relative to such probers. The problem is related to the fact that the probes of the prober may damage the circuit under test. So, an additional step consists of inspecting the circuit's area to verify vias (I/O pads) status after circuit's testing: this operation is called developed Probe Mark Inspection (PMI). Figure 11 shows a view of an industrial prober and examples of faulty and correct vias.

Many prober constructors had already developed PMI software based on conventional pattern recognition algorithms with little success]. The difficulty is related to the compromise between real time execution (production constraints) and methods reliability. In fact, even sophisticated hardware implementations using DSPs and ASICs specialized in image processing are not able to perform sufficiently well to convince industrials to switch from human operator (expert) defects recognition to electronically automatic PMI. That's why a neural network based solution has been developed and implemented on ZISC-036 neuro-processor, for the IBM Essonnes plant. The main advantages of developed solutions are real-time control and high reliability in fault detection and classification tasks. Our automatic intelligent PMI application, detailed in (Tremiolles, 1997) and (Madani, 2003, a), consists of software and a PC equipped with this neural board, a video acquisition board connected to a camera and a GPIB control board connected to a wafer prober system. Its goal is image analysis and prober control.

The process of analyzing a probe mark can be described as following: the PC controls the prober to move the chuck so that the via to inspect is precisely located under the camera; an image of the via is taken through the video acquisition board, then, the ZISC-036 based PMI:

- finds the via on the image,
- checks the integrity of the border (for damage) of via,
- locates the impact in the via and estimates its surface for statistics.

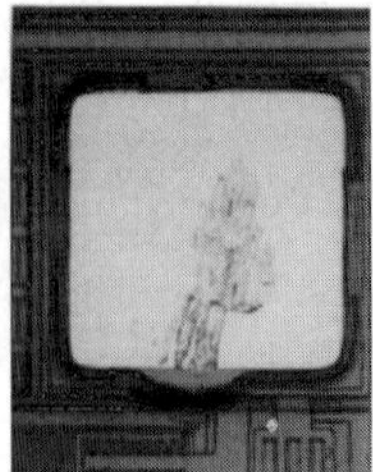

Figure 11: Photograph giving an example of probes in industrial prober (left). Example of probe impact: correct and faulty (right).

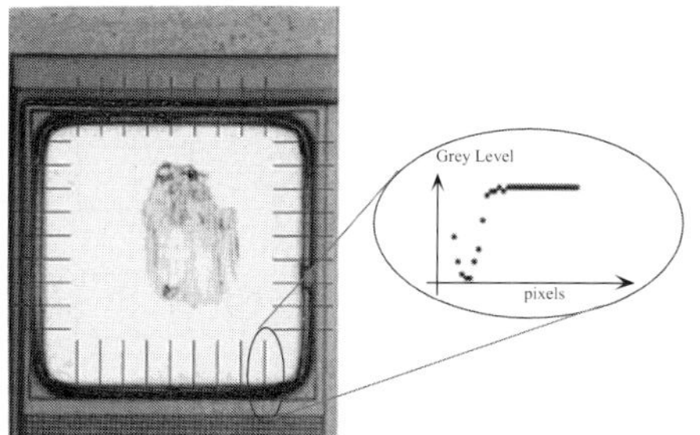

Figure 12: Example of profiles extraction after via centring process (left). Example of profiles to category association during the learning phase (right).

Figure 13: Profiles extraction for size and localization of the probe mark (left). Experimental result showing a fault detection and its localization in the via (right).

All vias of a tested wafer are inspected and analysed. At the end of the process, the system shows a wafer map which presents the results and statistics on the probe quality and its alignment with the wafer. All the defects are memorized in a log file. In summary, the detection and classification tasks of our PMI application are done in three steps: via localization in the acquired image, mark size estimation and probe impact classification (good, bad or none).

The method, which was retained, is based on profiles analysis using kennel functions based ANN. Each extracted profile of the image (using a square shape, figures 12 and 13) is compared to a reference learned database in which each profile is associated with its appropriated category. Different categories, related to different needed features (as: size, functional signature, etc).

Experiments on different kinds of chips and on various probe defects have proven the efficiency of the neural approach to this kind of perception problem. The developed intelligent PMI system outperformed the best solutions offered by competitors by 30%: the best response time per via obtained using other wafer probers was about 600 ms and our neural based system analyzes one via every 400 ms, 300 of which were taken for mechanical movements. Measures showed that the defect recognition neural module's execution time was negligible compared to the time spent for mechanical movements, as well as for the image acquisition (a ratio of 12 to 1 on any via). This application is presently inserted on a high throughput production line.

3.3 Bio-inspired Multiple Neural Networks Based Process Identification

The identification task involves two essential steps: structure selection and parameter estimation. These two steps are linked and generally have to be realized in order to achieve the best compromise between error minimization and the total number of parameters in the final global model. In real world applications (situations), strong linearity, large number of related parameters and data nature complexity make the realization of those steps challenging, and so, the identification task difficult.

To overcome mentioned difficulties, one of the key points on which one can act is the complexity reduction. It may concern not only the problem representation level (data) but also may appear at processing procedure level. An issue could be model complexity reduction by splitting a complex problem into a set of simpler problems: multi-modelling where a set of simple models is used to

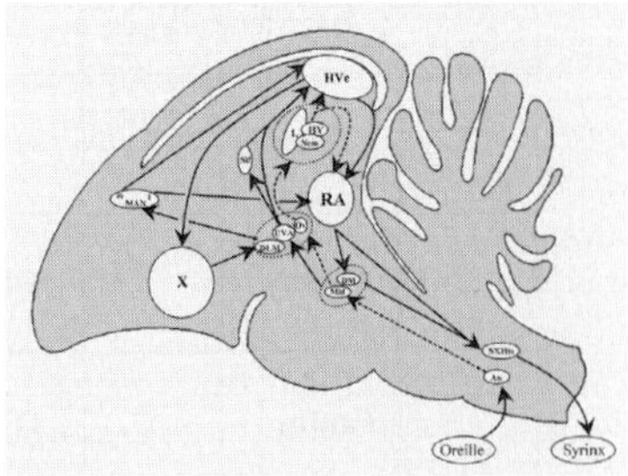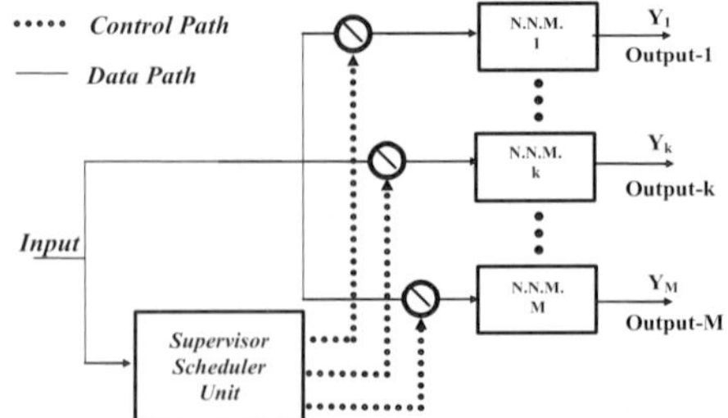

Figure 14: Left side of a bird's brain scheme and it's auditory and motor pathways involved in the recognition and the production of song (left) and inspired T-DTS multiple model generator's general bloc-diagram (right).

sculpt a complex behaviour (Murray, 1997). On this basis and inspired from animal brain structure (left picture of figure 14, showing the left side of a bird's brain scheme and it's auditory and motor pathways involved in the recognition and the production of song), we have designed an ANN based data driven treelike Multiple Model generator, that we called T-DTS (Treelike Divide To Simplify).This data driven neural networks based Multiple Processing (multiple model) structure is able to reduce complexity on both data and processing levels (Madani, 2003, b) (right picture of figure 14). T-DTS and associated algorithm construct a treelike evolutionary neural architecture automatically where nodes, called also "Supervisor/Scheduler Units" (SU) are decision units and leafs called also "Neural Network based Models" (NNM) correspond to neural based processing units.

The T-DTS includes two main operation modes. The first is the learning phase, when T-DTS system decomposes the input data and provides processing sub-structures and tools for decomposed sets of data. The second phase is the operation phase (usage the system to process unlearned data). There could be also a pre-processing phase at the beginning, which arranges (prepare) data to be processed. Pre-processing phase could include several steps (conventional or neural stages). The learning phase is an important phase during which T-DTS performs several key operations: splitting the learning database into several sub-databases, constructing (dynamically) a treelike Supervision/Scheduling Unit (SSU) and building a set of sub-models (NNM) corresponding to each sub-database. Figure 15 represents the division and NNM construction bloc diagrams. As shown in this figure, if a neural based model cannot be built for an obtained sub-database, then, a new decomposition will be performed dividing the concerned sub-space into several other sub-spaces.

The second operation mode corresponds to the use of the constructed neural based Multi-model system for processing unlearned (work) data. The SSU, constructed during the learning phase, receives data and classifies that data (pattern). Then, the most appropriated NNM is authorized (activated) to process that pattern.

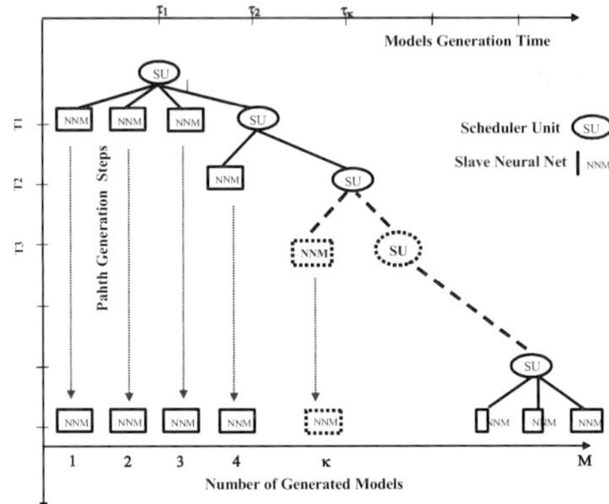

Figure 15: General bloc diagram of T-DTS learning phase, splitting and NNM generation process.

We have applied T-DTS based Identifier to a real world industrial process identification and control problem. The process is a drilling rubber process used in plastic manufacturing industry. Several non-linear parameters influence the manufacturing process. To perform an efficient control of the manufacturing quality (process quality), one should identify the global process (Chebira, 2003).

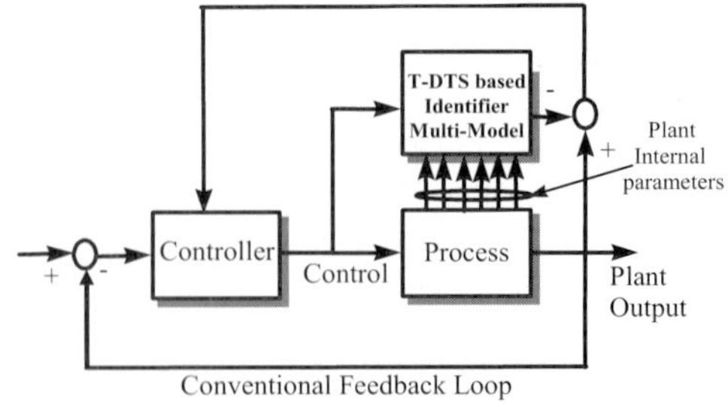

Figure 16: Industrial processing loop bloc diagram.

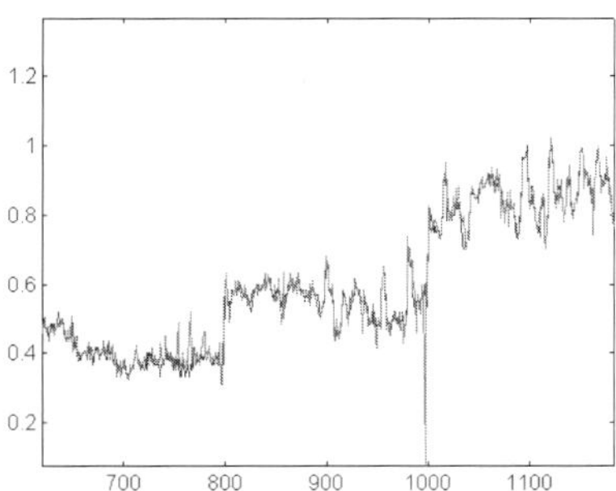

Figure 17: Process identification results showing the plant's output prediction in the working phase.

A Kohonen SOM based SU with a 4x3 grid generates and supervises 12 NNM trained from learning database. Figures 16 and 17 show the bloc diagram of industrial processing loop and the identification result in the working phase, respectively. One can conclude that the predicted output is in accord with the measured one, obtained from the real plant.

4 CONCLUSION

Advances accomplished during last decades in Artificial Neural Networks area and issued techniques made possible to approach solution of a large number of difficult problems related to optimization, modeling, decision making, classification, data mining or nonlinear functions (behavior) approximation. Inspired from biological nervous systems and brain structure, these models take advantage from their learning and generalization capabilities, overcoming difficulties and limitations related to conventional techniques. Today, conjunction of these new techniques with recent computational technologies offers attractive potential for designing and implementation of real-time intelligent industrial solutions. The main goal of the present paper was focused on ANN based techniques and their application to solve real-world and industrial problems. Of course, the presented models and applications don't give an exhaustive state of art concerning huge potential offered by such approaches, but they could give, through above-presented ANN based applications, a good idea of promising capabilities of ANN based solutions to solve difficult future industrial changes.

ACKNOWLEDGEMENTS

Reported works were sponsored and supported by several research projects and industrial partners among which, French Ministry of Education, French Ministry of Research and IBM-France Company. Author whish thank especially Dr. P. Tanhoff, and Dr. G. Detremiolles from IBM-France for their partnership and joint collaboration. I would also acknowledge Dr. V. Amarger, Dr. A. Chebira and Dr. A. Chohra from my research team (I^2S Lab.) involved in presented works. I would thank Dr. G. Mercier from PARIS XII University, who worked with me, during his Ph.D. and during several years in my lab, on intelligent adaptive control dilemma. Finally, I would express my gratitude to Mr. M. Rybnik, my Ph.D. student, working on aspects related to T-DTS, for useful discussions concerning the last part of this paper.

REFERENCES

Nelles, O., 1995. On the identification with neural networks as series-parallel and parallel models. In *ICANN'95, International Conference on Artificial Neural Networks, Paris, France*.

Faller W., Schreck S., 1995. Real-Time Prediction Of Unsteady Aerodynamics : Application for Aircraft Control and Manoeuvrability Enhancement. In *IEEE Transac. on Neural Networks, Vol. 6, N¡ 6, Nov. 95*.

Maidon Y., Jervis B. W., Dutton N., Lesage S., 1996. Multifault Diagnosis of Analogue Circuits Using Multilayer Perceptrons. In *IEEE European Test Workshop 96, Montpellier, June 12-14, 1996*.

Anderson C.W., Devulapalli S.V., Stolz E.A., 1995. Determining Mental State from EEG Signals Using Parallel Implementations of Neural Networks. In *Scientific Programming, Special Issue on Applications Analysis, 4, 3, Fall, pp. 171-183*.

Madani K., Bengharbi A., Amarger V., 1997. Neural Fault Diagnosis Techniques for Nonlinear Analog Circuit. In *SPIE Vol. 3077, pp 491-502, Orlando, Florida, U.S.A., April 21 to 24, 1997*.

Sachenko A., Kochan V., Turchenko V., Golovko V., Savitsky J., Dunets A., Laopoulos T., 2000. Sensor errors prediction using neural networks. In *Proceedings IJCNN'2000, Jul 24-Jul 27 2000, Como, Italy, pp. 441-446*.

Arbib M.A., 2003. Handbook of Brain Theory and Neural Networks" 2ed. M.I.T. Press. 2003.

Hebb S., 1949. The Organization of Behaviour, *Wiley and Sons, New-York, U.S.A., 1949.*

Kohonen T., 1984. Self-Organization and Associative Memory, *Springer-Verlag, Germany, 1984.*

Rumelhart D., Hinton G., Williams R., 1986. Learning Internal Representations by Error Propagation". *Parallel Distributed Processing: Explorations in the Microstructure of Cognition, MIT Press, MA, 1986.*

Bigot P., Cosnard M., 1993. Probabilistic Decision Trees and Multilayered Perceptrons. In *Proc. of the Europ. Symp. on A. N. N., ESANN'93, pp. 91-96, 1993.*

Bogdan M., Speakman H., Rosenstiel W., 1994. Kobold: A neural Coprocessor for Back-propagation with on-line learning. In *Proc. NeuroMicro 94, Torino, Italy, pp. 110-117.*

Reyneri L.M., 1995. Weighted Radial Basis Functions for Improved Pattern Recognition and Signal Processing. *Neural Processing Letters, Vol. 2, No. 3, pp. 2-6, May 1995.*

Trémiolles G., Madani K., Tannhof P., 1996. A New Approach to Radial Basis Function's like Artificial Neural Networks. In *NeuroFuzzy'96, IEEE European Workshop, Vol. 6 N° 2, pp. 735-745, April 16 to 18, Prague, Czech Republic, 1996.*

Hormel M.A., 1992. Self-organizing associative memory system for control applications. *Advances Neural Information processing Systems 2,Touretzky Ed. Los Altos ,CA: Morgan Kaufmann,1992 pp. 332-339.*

Miller W.T, 1987. Sensor based control of robotic manipulators using a general learning algorithm, *IEEE J. of Robotics and Automation, pp. 157-165.*

Madani K., Mercier G., Chebira A., Duchesne S., 1996. Implementation of Neural Network Based Real Time Process Control on IBM Zero Instruction Set Computer. In *SPIE AeroSense'96, Applications and Science of Artificial Neural Networks , Orlando, USA, SPIE Proc. Vol. 2760, pp. 250-261.*

Albus J.S., 1975. A new Approach to Manipulator Control, *Transa of ASME, Sept. 1975 pp. 220-227.*

Comoglio R.F, 1992. CMAC Neural network Architecture of Autonomous Undersea Vehicle. In *SPIE vol 1709. Applications of artificial Neural Networks III, pp. 517-527, 1992.*

Madani K., Mercier G., Dinarvand M., Depecker JC, 1999. A Neuro-Vector based electrical machines driver combining a neural plant identifier and a conventional vector controller. In *SPIE Intern. Symp. AeroSense'99, Orlando, Florida, USA, April 1999.*

Trémiolles G., Tannhof P., Plougonven B., Demarigny C., Madani K., 1997. Visual Probe Mark Inspection, using Hardware Implementation of Artificial Neural Networks, in VLSI Production. In *LNCS. Biological and Artificial Computation:From Neuroscience to Technology, Ed.: J. Mira, R.M. Diaz and J. Cabestany - Springer Verlag Berlin Heidelberg, pp. 1374-1383, 1997.*

Tremiolles G, 1998. Contribution to the theoretical study of neuromimetic models and to their experimental validation: use in industrial applications. *Ph.D. Report, University of PARIS XII – Val de Marne.*

Madani K., Tremiolles G., Tanhoff P., 2003 - a. Image processing using RBF like neural networks: A ZISC-036 based fully parallel implementation solving real world and real complexity industrial problems. In *Journal of Applied Intelligence N°18, 2003, Kluwer Academic Publishers, pp. 195-231.*

Murray-Smith R., Johansen T.A., 1997. Multiple Model Approaches to Modeling and Control. Taylor & Francis Publishers, 1997, ISBN 0-7484-0595-X.

Madani K., Chebira A., Rybnik M., 2003 - b. Data Driven Multiple Neural Network Models Generator Based on a Tree-like Scheduler, LNCS "Computational Methods in Neural Modeling", Ed. J. Mira, J.R. Alvarez - Springer Verlag 2003, ISBN 3-540-40210-1, pp. 382-389.

THE DIGITAL FACTORY
Planning and simulation of production in automotive industry

F. Wolfgang Arndt

Fachhochschule Konstanz, Fachbereich Informatik, Brauneggerstr. 55, 76xxx Konstanz, Germany
Email: arndt@fh-konstanz,de; wolfgang@ppgia.pucpr.bru

Keywords: simulation, digital factory, automotive industry.

Abstract: To improve and to shorten product and production planning as well as to get a higher planning and product quality the idea of the Digital Factory was invented. Digital Factory means   general digital support of planning following the process chain from **product** development over process and product planning to production by using virtual working techniques. It follows that the whole process of developing a new product with its associated production equipment has to be completely simulated before starting any realisation. That calls for the integration of heterogeneous processes and a reorganisation of the whole factory. The Digital Factory will be one of the core future technologies and revolutionize all research, development and production activities in mechanical and electrical industries. International enterprises like DaimlerChrysler and Toyota are making a considerable effort to introduce this technique as soon as possible in their factories. It will give them a significant advance in time and costs.

1 INTRODUCTION

Automotive industry is in many areas of automation a forerunner. This is due to some special characteristics of this industrial area. To make profit each type of a car must be produced in a great number of pieces, which makes it worth to automate production as much as possible. The competition on the automotive market is very hard and forces low retail prices. Larger design changes of one type of a car necessitate a complete rebuilt of the production line in the body shop and partially in the assembly area.

But changes of a production line are expensive, because a lot of special equipment is needed and the installation of new equipment is very labour intensive.   Investments available to rebuild or to change production lines are limited. Therefore any modification of a production line must be planed very carefully. The planning procedure involves a lot of different departments and is a relatively time consuming task.

On the other hand at the beginning of the planning activities the day, when the new production has to be started, the date of SOP (start of production) is already fixed. All planning suffers therefore from a limited amount of investments and a lack of time to do planning in detail. To overcome these problems intensive engineering should be done using simulation tools. A large variety of tools is today available on the market. Nearly every activity can be simulated by a special tool as digital mock-up (car assembly), work place optimisation and workflow management.

But the traditional use of simulation tools deals only with isolated, limited problems. A conveyor systems can be optimised, the optimal way to mount front windows investigated or the optimal distribution of work among several robots determined. But the final goal is the *digital factory*.

2 DEFINITION

The term digital factory means simulation of all activities, which are taking place in the area of R&D as well as in the production area of a factory. The digital factory involves the use of digital simulation along the entire process chain, from developing of new product, planning the associated production equipment and organizing mass production.

The digital factory involves therefore much more than only the use of simulation tools. It imposes new types of organisation of the factory and an intensive collaboration between the car manufacturer and his subcontractors. All activities in the plant – that means the whole workflow - have to be standar-

J. Braz et al. (eds.), Informatics in Control, Automation and Robotics I, 27–29.
© 2006 *Springer. Printed in the Netherlands.*

dized. The data outcome of every step of the workflow has to be specified and measures have to be taken, that the data of the workflow, when a step is finished, are immediately stored into a global factory wide database. The final target is to start development and production -that means- any realisation only, if the simulation shows, that product and production will met the given investments, the predefined time schedule and the necessary quality.

3 GLOBAL DATA BASE

The success of simulation depends on the quality and actuality of the available data. At any moment simulation must have access to the actual data of development and planning. The quality of the simulation results depends on the quality and actuality of the available data.

There are in every factory two different types of data, the geometric or engineering data, which are produced, when a new product is developed and the associated production line is planed, and the commercial and administrative data, which are used by the purchase, sales and controlling departments.

First of all these both areas have to be to integrated. To optimise e.g. the planning of a production line engineering and commercial data are needed. Special data structures have to be implemented to enable a data flow between these two areas. The data structures must be able to stores both types of data and to store all data, which the engineering and commercial departments need.

The design of theses structures is of paramount importance for the well working of the digital factory. They must de designed to store all data and to permit an effective and fast access to the data. When the data structures are defined, it must be assured that they always contain actual data.

4 WORKFLOW MANAGEMENT

In an automotive factory at any moment a large number of activities are taking place. These activities are embedded in different workflows. A workflow consists of a sequence of different process steps. The activities of each step and the sequence of the different activities must be defined. Every employee must respect the factory specific workflow directions.

For every process step the data outcome has to be defined. As soon as a step is executed the generated data are stored in the global database. That assures, that the global database always contains the actual data.

There are standardized workflows and order or client specific ones. The data outcome of the special workflow cannot be defined in advance, because the activities or the workflow respectively depends on the work to do. Therefore it is very important to reduce the number of special workflows. They must be converted to standard workflows or integrated into existing standard ones.

Every activity will be computer based. Each process step has to be supported by a computer system. There are different software systems available on the market like *Delmia Process Engineer* of *Dassault Systémes*. That assures that the defined sequence of steps of a workflow is respected. A new step can only be initiated, if the previous one has been completed. After the definition of the workflows and the outcome of each step, the management of the workflows is done by computer.

5 SUBCONTRACTORS AND COLLABORATIVE ENDINEERING

Generally the production of a car manufacturer is limited to the car body, the engine, the gearbox and the assembly of the whole car. Therefore a large part of the car components has to be developed and produced by subcontractors. But the concept of a digital factory requires the simulation of the whole vehicle and consequently the integration of the subcontractors.

The exchange of information between car manufacturer and subcontractors has to be started, before a component is completely developed. Both sides have to inform each other of the daily progress of work. That calls for collaborative engineering. Something like a virtual team has to be created. The team members are working in different places, but there is a permanent information exchange between them. If the development of the car body and the engine is terminated, all components will be also available.

This gives to the subcontractors tremendous problems. They have to dispose of the same simulation tools as the car manufacturer, which will cause high investments. They have to standardize their workflows, to deliver to the car manufacturer continuously information about the work in progress. They need to use the similar data structures to store engineering and commercial data. Every partner must give to access to his engineering data to the other participants. As subcontractors normally are working for different car manufacturers, who are using different simulation tools and have different data structures, they will face considerable problems in the future.

6 DANGER AND UNRESOLVED PROBLEMS

Such a close collaboration needs a very intensive data interconnection, which brings with out doubts considerable dangers. There are still today no measures to protect data networks against foreigners with an absolute security.

There are unresolved problems like how the own data can be protected against competitors, how a furnisher can be hindered from passing secret information to another car manufacturer, how to secret services can be prevented to seize data or hackers to destroy them.

7 SUMMARY

Concept and realisation of the digital factory form a key technology and will revolutionize significantly the way development and planning is done in mechanical and electrical industry. It will help to reduce significantly the production costs and speed up the product live cycle. But it will also increase the interdependence between the automotive industry or leading companies respectively and their subcontractors and make companies more vulnerable.

REFERENCES

Schiller, E.; Seuffert, W.-P. Digitale Fabrik bei DaimlerChrysler. In *Sonderdruck Automobil Produktion 2/2002, 2002.*

WHAT'S REAL IN "REAL-TIME CONTROL SYSTEMS"?
Applying formal verification methods and real-time rule-based systems to control systems and robotics

Albert M. K. Cheng

cheng@cs.uh.edu

University of Houston, Texas

USA

1 INTRODUCTION

Engineers focus on the dynamics of control systems and robotics, addressing issues such as controllability, safety, and stability. To facilitate the control of increasingly complex physical systems such as drive-by-wire automobiles and fly-by-wire airplanes, high-performance networked computer systems with numerous hardware and software components are increasingly required. However, this complexity also leads to more potential errors and faults, during both the design/implementation phase and the deployment/runtime phase. It is therefore essential to manage the control system's complexity with the help of smart information systems and to increase its reliability with the aid of mechanical verification tools. Software control programs provide greater flexibility, higher precision, and better complexity management. However, these safety-critical real-time software must themselves be formally analyzed and verified to meet logical and timing correctness specifications.

This keynote explores the use of rule-based systems in control systems and robotics, and describes the latest computer-aided verification tools for checking their correctness and safety.

2 MODEL CHECKING

To verify the logical and timing correctness of a control program or system, we need to show that it meets the designer's specification. One way is to manually construct a proof using axioms and inference rules in a deductive system such as temporal logic, a first-order logic capable of expressing relative ordering of events.

This traditional approach toconcurrent program verification is tedious and error-prone even for small programs. For finite-state systems and restricted classes of infinite-state systems, we can use model checking (first developed by Clarke, Emerson, and Sistla in the 1980s) instead of proof construction to check their correctness relative to their specifications.

We represent the control system as a finite-state graph. The specification or safety assertion is expressed in propositional temporal logic formulas. We can then check whether the system meets its specification using an algorithm called a model checker, which determines whether the finite-state graph is a model of the formula(s). Several model checkers are available and they vary in code and runtime complexity, and performance:
(1) explicit-state, [Clarke, Emerson, and Sistla 1986], (2) symbolic (using Binary Decision Diagrams or BDDs) [Burch, Clarke, McMillan, Dill, and Hwang 1990], and (3) model checkers with real-time extensions.

In Clarke, Emerson, and Sistla's approach, the system to be checked is represented by a labeled finite-state graph and the specification is written in a propositional, branching-time temporal logic called computation tree logic (CTL).

The use of linear-time temporal logic, which can express fairness properties, is ruled out since a model checker for such as logic has high complexity.

Instead, fairness requirements are moved into the semantics of CTL.

3 VISUAL FORMALISM, STATECHARTS, AND STATEMATE

Model checking uses finite state machines (FSMs) to represent the control system's specification, as is the case in the specification and analysis of many computer-based as well as non-computer-based

31

J. Braz et al. (eds.), Informatics in Control, Automation and Robotics I, 31–35.
© 2006 *Springer. Printed in the Netherlands.*

systems, ranging from electronic circuits to enterprise models. They can model in detail the behavior of a system and several algorithms exist to perform the analysis. Unfortunately, classical state machines as those employed in the standard, explicit-state CTL model-checking approach lack support for modularity and suffer from exponential state explosion. The first problem often arises when FSMs are used to model complex systems which contain similar subsystems. The second problem is evident in systems where the addition of a few variables or components can substantially increase the number of states and transitions, and hence the size of the FSM. Furthermore, the inability to specify absolute time and time intervals limit the usability of classical FSMs for the specification of real-time systems.

We can introduce modular and hierarchical features to classical FSMs to solve the first two problems. Harel et al., in 1987 developed a visual formalism called Statecharts to solve these two problems as well as the problem of specifying reactive systems. Reactive systems are complex control-driven mechanisms that interact with discrete occurrences in the environment in which they are embedded. They include real-time computer systems, communication devices, control plants, VLSI circuits, robotics, and airplane avionics. The reactive behavior of these systems cannot be captured by specifying the corresponding outputs resulting from every possible set of inputs. Instead, this behavior has to be described by specifying the relationship of inputs, outputs, and system state over time under a set of system- and environment-dependent timing and communication constraints.

Graphic features (labeled boxes) in the Statecharts language are used to denote states (or sets of states) and transitions between states. A transition from one state to another state takes place when the associated event(s) and condition(s) are enabled. A state can be decomposed into lower-level states via refinement, and a set of states can be combined into a higher-level state via clustering. This hierarchical specification approach makes it possible for the specifier to zoom-in and zoom-out of a section of the Statecharts specification, thus partially remedying the exponential state explosion problem in classical FSMs. Furthermore, AND and OR clustering relations, together with the notions of state exclusivity and orthogonality, can readily support concurrency and independence in system specification. These features dramatically reduce the state explosion problem by not considering all states in a classical FSM at once. The entire specification and development environment is known as STATEMATE.

To develop a comprehensive tool capable of not only system specification, Harel et al., in 1990 extended the work on Statecharts, which is capable of behavioral description, to derive high-level languages for structural and functional specifications. The language module-charts is used to describe a structural view with a graphical display of the components of the system. The language activity-charts is used to describe a functional view with a graphical display of the functions of the system. They also added mechanisms that provide a friendly user interface, simulated system executions, dynamic analysis, code generation, and rapid prototyping.

4 TIMED AUTOMATA

Finite automata and temporal logics have been used extensively to formally verify qualitative properties of concurrent systems. The properties include deadlock or livelock-freedom, the eventual occurrence of an event, and the satisfaction of a predicate. The need to reason with absolute time is unnecessary in these applications, whose correctness depends only on the relative ordering of the associated events and actions. These automata-theoretic and temporal logic techniques using finite-state graphs are practical in a variety of verification problems in network protocols, electronic circuits, and concurrent programs. More recently, several researchers have extended these techniques to timed or real-time systems while retaining many of the desirable features of their untimed counterparts.

Two popular automata-theoretic techniques based on timed automata are the Lynch-Vaandrager approach and the Alur-Dill approach. The Lynch-Vaandrager approach (1991, 1994) is more general and can handle finite and infinite state systems, but it lack an automaticverification mechanism. Its specification can be difficult to write and understand even for relatively small systems. The Alur-Dill approach (1994) is less ambitious and is based on finite automata, but it offers an automated tool for verification of desirable properties. Its dense-time model can handle time values selected from the set of real numbers whereas discrete-time models such as those in Statecharts and Modecharts use only integer time values.

5 REAL-TIME LOGIC, GRAPH-THEORETIC ANALYSIS, AND MODECHART

A real-time control system can be specified in one of two ways. The first is to structurally and functionally describe the system by specifying its mechanical, electrical, and electronic components. This type of specification shows how the components of the system work as well as their functions and operations. The second is to describe the behavior of the system in response to actions and events. Here, the specification tells sequences of events allowed by the system. For instance, a structural-functional specification of the real-time anti-lock braking system in an automobile describes the braking system components and sensors, how they are interconnected, and how the actions of each component affects each other. This specification shows, for example, how to connect the wheel sensors to the central decision-making computer which controls the brake mechanism.

In contrast, a behavioral specification only shows the response of each braking system component in response to an internal or external event, but does not describe how one can build such a system. For instance, this specification shows that when the wheel sensors detect wet road condition, the decision-making computer will instruct the brake mechanism to pump the brakes at a higher frequency within 80 milliseconds. Since we are interested in the timing properties of the system, a behavioral specification without the complexity of the structural specification often suffices for verifying the satisfaction of many timing constraints. ferthermore, to reduce specification and analysis complexity, we restrict the specification language to handle only timing relations. This is a departure from techniques which employ specification languages capable of describing logical as well as timing relations such as real-time CTL.

To show that a system or program meets certain safety properties, we can relate the specification of the system to the safety assertion representing the desired safety properties. This assumes that the actual implementation of the system is faithful to the specification. Note that even though a behavioral specification does not show how one can build the specified system, one can certainly show that the implemented system, built from the structural-functional specification, satisfy the behavioral specification.

One of the following three cases may result from the analysis relating the specification and the safety assertion. (1) The safety assertion is a theorem derivable from the specification, thus the system is safe with respect to the behavior denoted by the safety assertion. (2) The safety assertion is unsatisfiable with respect to the specification, so the system is inherently unsafe since the specification will cause the safety assertion to be violated. (3) The negation of the safety assertion is satisfiable under certain conditions, meaning that additional constraints must be added to the system to ensure its safety.

The specification and the safety assertion can be written in one of several real-time specification languages. The choice of language would in turn determine the algorithms that can be used for the analysis and verification. The first-order logic called Real-Time Logic (RTL) [Jahanian and Mok 1986] has been developed for writing specifications and safety assertions. For a restricted but practical class of RTL formulas, an efficient graph-theoretic analysis can be applied to perform verification. Modechart is a graphical tool based on RTL for specifying real-time control systems.

Timed Petri Nets:

Petri nets provide an operational formalism for specifying untimed concurrent systems. They can show concurrent activities by depicting control and data flows in different parts of the modeled system. A Petri net gives a dynamic representation of the state of a system through the use of moving tokens. The original, classical, untimed Petri nets have been used successfully to model a variety of industrial systems. More recently, time extensions of Petri nets have been developed to model and analyze time-dependent or real-time systems. The fact that Petri nets can show the different active components of the modeled system at different stages of execution or at different instants of time makes this formalism especially attractive for modeling embedded systems that interact with the external environment.

6 PROCESS ALGEBRA

A computer process is a program or section of a program (such as a function) in execution. It may be in one of the following states: ready, running, waiting, or terminated. A process algebra is a concise language for describing the possible execution steps of computer processes. It has a set of operators and syntactic rules for specifying a process using simple, atomic components. It is usually not a logic-based language.

Central to process algebras is the notion of equivalence, which is used to show that two processes have the same behavior. Well-established process algebras such as Hoare's Communicating Sequential Processes (CSP) (1978, 1985), Milner's Calculus of Communicating Systems (CCS) (1980, 1989), and Bergstra and Klop's Algebra of Communicating Processes (ACP) (1985) have been used to specify and analyze concurrent processes with interprocess communication. These are untimed algebras since they only allow one to reason about the relative ordering of the execution steps and events.

To use a process algebra or a process-algebraic approach to specify and analyze a system, we write the requirements specification of the system as an abstract process and the design specification as a detailed process. We then show that these two processes are equivalent, thus showing the design specification is correct with respect to the requirements specification. Here, the requirements specification may include the desired safety properties.

7 REAL-TIME RULE-BASED DECISION SYSTEMS

As the complexity of control systems increases, it is obvious that more intelligent decision/monitoring/control software must be developed and installed to monitor and control these control systems. Real-time decision systems must react torrr events in the external environment by making decisions based on sensor inputs and state information sufficiently fast to meet environment-imposed timing constraints. They are used in applications that would require human expertise if such decision systems are not available. Human beings tend to be overwhelmed by a transient information overload resulting from an emergency situation, thus expert systems are increasingly used under many circumstances to assist human operators.

These embedded control systems include airplane avionics navigation systems, automatic vehicle control systems fly-by-wire Airbus 330/340/380 and Boeing 777/7E7, smart robots (e.g., the Autonomous Land Vehicle), space vehicles (e.g., unmanned spacecrafts), the Space Shuttle and satellites, and the International Space Station), electric and communication grid monitoring centers, portable wireless devices and hospital patient-monitoring devices.

Since the solutions to many of these decision problems are often nondeterministic or cannot be easily expressed in algorithmic form, these applications increasingly employ rule-based (or knowledge-based) expert systems. In recent years, such systems are also increasingly used to monitor and control the operations of complex safety-critical real-time systems.

8 REAL-TIME DECISION SYSTEMS

A real-time decision system interacts with the external environment by taking sensor readings and computing control decisions based on sensor readings and stored state information. We can characterize a real-time decision system by the following model with 7 components:

(1) a sensor vector x in X,
(2) a decision vector y in Y,
(3) a system state vector s in S,
(4) a set of environmental constraints A,
(5) a decision map D, D: S * X -> S * Y,
(6) a set of timing constraints T, and
(7) a set of integrity constraints I.

In this model, X is the space of sensor input values, Y is the space of decision values, and S is the space of system state values. (We shall use x(t) to denote the value of the sensor input x at time t, etc.)

The environmental constraints A are relations over X, Y, S and are assertions about the effect of a control decision on the external world which in turn affect future sensor input values. Environmental constraints are usually imposed by the physical environment in which the real-time decision system functions.

The decision map D relates y(t+1), s(t+1) to x(t), s(t), i.e., given the current system state and sensor input, D determines the next decisions and system state values. For our purpose, decision maps are implemented by rule-based programs.

The decisions specified by D must conform to a set of integrity constraints I. Integrity constraints are relations over X, S, Y and are assertions that the decision map D must satisfy in order to ensure safe operation of the physical system under control. The implementation of the decision map D is subject to a set of timing constraints T which are assertions about how fast the map D has to be performed. Let us consider a simple example of a real-time decision system. Suppose we want to automate a toy race car so that it will drive itself around a track as fast as possible. The sensor vector consists of variables denoting the position of the car and the distance of the next obstacle ahead. The decision vector consists of two variables: one variable to indicate whether to accelerate, decelerate or maintain the same speed, another variable to indicate whether to turn left, right or keep the same heading. The system state vector consists of variables denoting the current speed and heading of the car. The set of environmental constraints consists of assertions that express the physical laws governing where the next position of the car will be, given its current position, velocity, and acceleration. The integrity constraints are assertions restricting the acceleration and heading of the car so that it will stay on the race track and not to run into an obstacle. The decision map may be implemented by some equational rule-based program. The input and decision variables of this program are respectively the sensor vector and decision vectors. The timing constraint consists of a bound on the length of the monitor-decide cycle of the program, i.e., the maximum number of rule firings before a fixed point is reached.

There are two practical problems of interest with respect to this model: Analysis problem: Does a given rule-based decision program satisfy   the integrity and timing constraints of the real-time control system?

Synthesis problem: Given a rule-based decision program that
satisfies the integrity constraints but is not fast enough to meet
the timing constraints, can we transform the given program into one
which meets both the integrity and timing constraints?

In view of the safety-critical functions that computers and software are beginning to be relied upon to perform in real time, it is incumbent upon us to ensure that some acceptable performance level can be provided by a rule-based program, subject to reasonable assumptions about the quality of the input.

SUFFICIENT CONDITIONS FOR THE STABILIZABILITY OF MULTI-STATE UNCERTAIN SYSTEMS, UNDER INFORMATION CONSTRAINTS

Nuno C. Martins, Munther A. Dahleh

Massashusetts Institute of Technology
77, Massachusetts Avenue, Cambridge - MA
Email: nmartins@mit.edu

Nicola Elia
Iowa State University

Keywords: Information Theory, Control, Stabilization, Uncertainty

Abstract: We study the stabilizability of uncertain stochastic systems in the presence of finite capacity feedback. Motivated by the structure of communication networks, we consider a stochastic digital link that sends words whose size is governed by a random process. Such link is used to transmit state measurements between the plant and the controller. We extend previous results by deriving sufficient conditions for internal and external stabilizability of multi-state linear and time-invariant plants. In accordance with previous publications, stabilizability of unstable plants is possible if the link's average transmission rate is above a positive critical value. In our formulation the plant and the link can be stochastic. In addition, stability in the presence of uncertainty in the plant is studied using a small-gain argument.

1 INTRODUCTION

Various publications in this field have introduced necessary and sufficient conditions for the stabilizability of unstable plants in the presence of data-rate constraints. The construction of a stabilizing controller requires that the data-rate of the feedback loop is above a non-zero critical value (Tatikonda, 2000b; Tatikonda, 2000a; G. Nair, 2000; Sahai, 1998; Liberzon, 2003). Different notions of stability have been investigated, such as containability (W. S. Wong, 1997; W. S. Wong, 1999), moment stability (Sahai, 1998) and stability in the almost sure sense (Tatikonda, 2000b). The last two are different when the state is a random variable. That happens when disturbances are random or if the communication link is stochastic. In (Tatikonda, 2000b) it is shown that the necessary and sufficient condition for almost sure stabilizability of finite dimensional linear and time-invariant systems is given by an inequality of the type $C > \mathcal{R}$. The parameter C represents the average data-rate of the feedback loop and $\mathcal{R}$ is a quantity that depends on the eigenvalues of A, the dynamic matrix of the system. If a well defined channel is present in the feedback loop then C may be taken as the Shannon Capacity. If it is a digital link then C is the average transmission rate. Different notions of stability may lead to distinct requirements for stabiliza-

tion. For tighter notions of stability, such as in the m-th moment sense, the knowledge of C may not suffice. More informative notions, such as higher moments or any-time capacity (Sahai, 1998), are necessary. Results for the problem of state estimation in the presence of information constraints can be found in (W. S. Wong, 1997), (Sahai, 2001) and (X. Li, 1996). The design problem was investigated in (Borkar and Mitter, 1997).

1.1 Main Contributions of the Paper

In this paper we study the moment stabilizability of uncertain time-varying stochastic systems in the presence of a stochastic digital link. In contrast with (G. Nair, 2003), we consider systems whose time-variation is governed by an identically and independently distributed (i.i.d.) process which may be defined over a continuous and unbounded alphabet. We also provide complementary results to (G. Nair, 2003; Elia, 2002; Elia, 2003; R. Jain, 2002) because we use a different problem formulation where we consider external disturbances and uncertainty on the plant and a stochastic digital link.

In order to focus on the fundamental issues and keep clarity, we start by deriving our results for first order linear systems (N. Martins and Dahleh, 2004). Subsequently, we provide an extension to a class of

J. Braz et al. (eds.), Informatics in Control, Automation and Robotics I, 37–50.
© 2006 *Springer. Printed in the Netherlands.*

multi-state linear systems. Necessary conditions for the multi-state case can also be found in (N. Martins and Dahleh, 2004). As pointed out in (G. Nair, 2003), non-commutativity creates difficulties in the study of arbitrary time-varying stochastic systems. Results for the fully-observed Markovian case over finite alphabets, in the presence of a deterministic link, can be found in (G. Nair, 2003).

Besides the introduction, the paper has 3 sections: section 2 comprises the problem formulation and preliminary definitions; in section 3 we prove sufficiency conditions by constructing a stabilizing feedback scheme for first order systems and section 4 extends the sufficient conditions to a class of multi-state linear systems.

The following notation is adopted:

- Whenever that is clear from the context we refer to a sequence of real numbers $x(k)$ simply as x. In such cases we may add that $x \in \mathbb{R}^\infty$.

- Random variables are represented using boldface letters, such as $\mathbf{w}$

- if $\mathbf{w}(k)$ is a stochastic process, then we use $w(k)$ to indicate a specific realization. According to the convention used for sequences, we may denote $\mathbf{w}(k)$ just as $\mathbf{w}$ and $w(k)$ as w.

- the expectation operator over $\mathbf{w}$ is written as $\mathcal{E}[\mathbf{w}]$

- if E is a probabilistic event, then its probability is indicated as $\mathcal{P}(E)$

- we write $\log_2(.)$ simply as $\log(.)$

- if $x \in \mathbb{R}^\infty$, then

$$\|x\|_1 = \sum_{i=0}^{\infty} |x(i)|$$

$$\|x\|_\infty = \sup_{i \in \mathbb{N}} |x(i)|$$

Definition 1.1 *Let* $\varrho \in \mathbb{N}_+ \bigcup \{\infty\}$ *be an upperbound for the memory horizon of an operator. If* $G_f : \mathbb{R}^\infty \to \mathbb{R}^\infty$ *is a causal operator then we define* $\|G_f\|_{\infty(\varrho)}$ *as:*

$$\|G_f\|_{\infty(\varrho)} = \sup_{k \geq 0, x \neq 0} \frac{|G_f(x)(k)|}{\max_{j \in \{k-\varrho+1,\dots,k\}} |x(j)|} \tag{1}$$

Note that, since G_f *is causal,* $\|G_f\|_{\infty(\infty)}$ *is just the infinity induced norm of* G_f:

$$\|G_f\|_{\infty(\infty)} = \|G_f\|_\infty = sup_{x \neq 0} \frac{\|G_f(x)\|_\infty}{\|x\|_\infty}$$

2   PROBLEM FORMULATION

We study the stabilizability of uncertain stochastic systems under communication constraints. Motivated by the type of constraints that arise in most computer networks, we consider the following class of stochastic links:

Definition 2.1 *(Stochastic Link) Consider a link that, at every instant k, transmits $\mathbf{r}(k)$ bits. We define it to be a stochastic link, provided that $\mathbf{r}(k) \in \{0, \dots, \bar{r}\}$ is an independent and identically distributed (i.i.d.) random process satisfying:*

$$\mathbf{r}(k) = C - \mathbf{r}^\delta(k) \tag{2}$$

where $\mathcal{E}[\mathbf{r}^\delta(k)] = 0$ and $C \geq 0$. The term $\mathbf{r}^\delta(k)$ represents a fluctuation in the transfer rate of the link.

More specifically, the link is a stochastic truncation operator $\mathcal{F}_k^l : \{0,1\}^{\bar{r}} \to \bigcup_{i=0}^{\bar{r}} \{0,1\}^i$ defined as:

$$\mathcal{F}_k^l(b_1, \dots, b_{\bar{r}}) = (b_1, \dots, b_{\mathbf{r}(k)}) \tag{3}$$

where $b_i \in \{0,1\}$.

Given $x(0) \in [-\frac{1}{2}, \frac{1}{2}]$ and $\bar{d} \geq 0$, we consider nominal systems of the form:

$$\mathbf{x}(k+1) = \mathbf{a}(k)\mathbf{x}(k) + \mathbf{u}(k) + \mathbf{d}(k) \tag{4}$$

with $|\mathbf{d}(k)| \leq \bar{d}$ and $\mathbf{x}(i) = 0$ for $i < 0$.

2.1   Description of Uncertainty in the Plant

Let $\varrho \in \mathbb{N}_+ \bigcup \{\infty\}$, $\bar{z}_f \in [0,1)$ and $\bar{z}_a \in [0,1)$ be given constants, along with the stochastic process $\mathbf{z}_a$ and the operator $G_f : \mathbb{R}^\infty \to \mathbb{R}^\infty$ satisfying:

$$|\mathbf{z}_a(k)| \leq \bar{z}_a \tag{5}$$

$$G_f \text{ causal and } \|G_f\|_{\infty(\varrho)} \leq \bar{z}_f \tag{6}$$

Given $x(0) \in [-\frac{1}{2}, \frac{1}{2}]$ and $\bar{d} \geq 0$, we study the existence of stabilizing feedback schemes for the following perturbed plant:

$$\mathbf{x}(k+1) = \mathbf{a}(k)\left(1 + \mathbf{z}_a(k)\right)\mathbf{x}(k) + \mathbf{u}(k)$$
$$+ G_f(\mathbf{x})(k) + \mathbf{d}(k) \tag{7}$$

where the perturbation processes $\mathbf{z}_a$ and $G_f(\mathbf{x})$ satisfy (5)-(6). Notice that $\mathbf{z}_a(k)$ may represent uncertainty in the knowledge of $\mathbf{a}(k)$, while $G_f(\mathbf{x})(k)$ is the output of the feedback uncertainty block G_f. We chose this structure because it allows the representation of a wide class of model uncertainty. It is also allows the construction of a suitable stabilizing scheme.

Example 2.1 *If $G_f(\mathbf{x})(k) = \mu_0 \mathbf{x}(k) + \dots + \mu_{n-1}\mathbf{x}(k - n + 1)$ then $\|G_f\|_{\infty(\varrho)} = \sum |\mu_i|$ for $\varrho \geq n$.*

In general, the operator G_f may be nonlinear and time-varying.

2.2 Statistical Description of $\mathbf{a}(k)$

The process $\mathbf{a}(k)$ is i.i.d. and independent of $\mathbf{r}(k)$ and $x(0)$, meaning that it carries no information about the link nor the initial state. In addition, for convenience, we use the same representation as in (2) and write:

$$\log(|\mathbf{a}(k)|) = \mathcal{R} + \mathbf{l}_a^\delta(k) \qquad (8)$$

where $\mathcal{E}[\mathbf{l}_a^\delta(k)] = 0$. Notice that $\mathbf{l}_a^\delta(k)$ is responsible for the stochastic behavior, if any, of the plant. Since $\mathbf{a}(k)$ is ergodic, we also assume that $\mathcal{P}\,(a(k) = 0) = 0$, otherwise the system is trivially stable. Such assumption is also realistic if we assume that (7) comes from the discretization of a continuous-time system.

2.3 Functional Structure of the Feedback Interconnection

In the subsequent text we describe the feedback loop structure, which might also be designated as information pattern (Witsenhausen, 1971). Besides the plant, there are two blocks denoted as encoder and controller which are stochastic operators. At any given time k, we assume that both the encoder and the controller have access to $a(0), \ldots, a(k)$ and $r(k-1)$ as well as the constants ϱ, $\bar{z}_f$, $\bar{z}_a$ and $\bar{d}$. The encoder and controller are described as:

- The encoder is a function $\mathcal{F}_k^e : \mathbb{R}^{k+1} \to \{0,1\}^{\bar{r}}$ that has the following dependence on observations:

$$\mathcal{F}_k^e(x(0), \ldots, \mathbf{x}(k)) = (\mathbf{b}_1, \ldots, \mathbf{b}_{\bar{r}}) \qquad (9)$$

- The control action results from a map, not necessarily memoryless, $\mathcal{F}_k^c : \bigcup_{i=0}^{\bar{r}} \{0,1\}^i \to \mathbb{R}$ exhibiting the following functional dependence:

$$\mathbf{u}(k) = \mathcal{F}_k^c(\vec{\mathbf{b}}(k)) \qquad (10)$$

where $\vec{b}(k)$ are the bits successfully transmitted through the link, i.e.:

$$\vec{\mathbf{b}}(k) = \mathcal{F}_k^l\,(\mathbf{b}_1, \ldots, \mathbf{b}_{\bar{r}}) = (\mathbf{b}_1, \ldots, \mathbf{b}_{\mathbf{r}(k)}) \qquad (11)$$

As such, $\mathbf{u}(k)$ can be equivalently expressed as

$$\mathbf{u}(k) = (\mathcal{F}_k^c \circ \mathcal{F}_k^l \circ \mathcal{F}_k^e)(x(0), \ldots, \mathbf{x}(k))$$

Definition 2.2 *(Feedback Scheme) We define a feedback scheme as the collection of a controller $\mathcal{F}_k^c$ and an encoder $\mathcal{F}_k^e$.*

2.4 Problem Statement and M-th Moment Stability

Definition 2.3 *(Worst Case Envelope) Let $\mathbf{x}(k)$ be the solution to (7) under a given feedback scheme. Given any realization of the random variables $\mathbf{r}(k)$,* $\mathbf{a}(k)$, $G_f(\mathbf{x})(k)$, $\mathbf{z}_a(k)$ *and* $\mathbf{d}(k)$, *the worst case envelope $\bar{\mathbf{x}}(k)$ is the random variable whose realization is defined by:*

$$\bar{x}(k) = \sup_{x(0) \in [-\frac{1}{2}, \frac{1}{2}]} |x(k)| \qquad (12)$$

Consequently, $\bar{\mathbf{x}}(k)$ is the smallest envelope that contains every trajectory generated by an initial condition in the interval $x(0) \in [-\frac{1}{2}, \frac{1}{2}]$. We adopted the interval $[-\frac{1}{2}, \frac{1}{2}]$ to make the paper more readable. All results are valid if it is replaced by any other symmetric bounded interval.

Our problem consists in determining sufficient conditions that guarantee the existence of a stabilizing feedback scheme. The results must be derived for the following notion of stability.

Definition 2.4 *(m-th Moment Robust Stability) Let $m > 0$, $\varrho \in \mathbb{N}_+ \bigcup\{\infty\}$, $\bar{z}_f \in [0,1)$, $\bar{z}_a \in [0,1)$ and $\bar{d} \geq 0$ be given. The system (7), under a given feedback scheme, is m-th moment (robustly) stable provided that the following holds:*

$$\begin{cases} \lim_{k \to \infty} \mathcal{E}\left[\bar{\mathbf{x}}(k)^m\right] = 0 & \bar{z}_f = \bar{d} = 0 \\ \exists b > 0, \limsup_{k \to \infty} \mathcal{E}\left[\bar{\mathbf{x}}(k)^m\right] < b & otherwise \end{cases} \qquad (13)$$

The first limit in (13) is an internal stability condition while the second establishes external stability. The constant b must be such that $\limsup_{k \to \infty} \mathcal{E}\left[\bar{\mathbf{x}}(k)^m\right] < b$ holds for all allowable perturbations $\mathbf{z}_a$ and $G_f(\mathbf{x})$ satisfying (5)-(6).

2.5 Motivation for our Definition of Stochastic Link and Further Comments on the Information Pattern

The purpose of this section[1] is to motivate the truncation operator of definition 2.1. In addition, details of the synchronization between the encoder and the decoder are discussed in subsection 2.5.1.

Consider that we want to use a wireless medium to transmit information between nodes A and B. In our formulation, node A represents a central station, which measures the state of the plant. The goal of the transmission system is to send information, about the state, from node A to node B, which represents a controller that has access to the plant. Notice that node A maybe a communication center which may communicate to several other nodes, but we assume that node B only communicates with node A. Accordingly, we will concentrate on the communication problem between nodes A and B only, without loss of generality.

---

[1]This section is not essential for understanding the sufficiency theorems of sections 3 and 4.

Definition 2.5 *(Basic Communication Scheme) We assume the existence of an external time-synchronization variable, denoted as k. The interval between k and $k+1$ is of T seconds, of which $T_T < T$ is reserved for transmission. We also denote the number of bits in each packet as Π, excluding headers. In order to submit an ordered set of packets for transmission, we consider the following basic communication protocol, at the media access control level:*

*(**Initialization**) A variable denoted by $c(k)$ is used to count how many packets are sent in the interval $t \in [kT, kT + T_T]$. We consider yet another counter p, which is used to count the number of periods for which no packet is sent. The variables are initialized as $k = 0$, $p = 0$ and $c(0) = 0$.*

*(**For node A**)*

*(**Synchronization**) If k changes to $k := k + 1$ then step 1 is activated.*

- ***Step1*** *The packets to be submitted for transmission are numbered according to their priority; 0 is the highest priority. The order of each packet is included in the header of the packet. The variable $c(k)$ is initialized to $c(k) = 0$ and p is incremented to $p := p + 1$. The first packet (packet number 0) has an extra header, comprising the pair $(c(k - p - 1), p)$. Move to step 2.*

- ***Step 2:*** *Stands by until it can send packet number $c(k)$. If such opportunity occurs, move to step 3.*

- ***Step 3:*** *Node A sends packet number $c(k)$ to node B and waits for an ACK signal from node B. If node A receives an ACK signal then $c(k) := c(k) + 1$, $p = 0$ and move back to step 2. If time-out then go back to Step 2.*

The time-out *decision may be derived from several events: a fixed waiting time; a random timer or a new opportunity to send a packet.*

*(**For node B**)*

- ***Step 1:*** *Node B stands by until it receives a packet from node A. Once a packet is received, check if it is a first packet: if so, extract $(c(k - p - 1), p)$ and construct $\mathbf{r}_{dec}(i)$, with $i \in \{k - p - 1, \ldots, k - 1\}$, according to:*

*(**If** $p \geq 1$)*

$$\mathbf{r}_{dec}(k - p - 1) = \mathbf{c}(k - p - 1)\Pi \qquad (14)$$

$$\mathbf{r}_{dec}(i) = 0, i \in \{k - p, \ldots, k - 1\} \qquad (15)$$

*(**If** $p = 0$)*

$$\mathbf{r}_{dec}(k - 1) = \mathbf{c}(k - 1)\Pi \qquad (16)$$

where Π is the size of the packets, excluding the header. If the packet is not duplicated then make the packet available to the controller. Move to step 2.

- ***Step 2:*** *Wait until it can send an ACK signal to node A. Once ACK is sent, go to step 1.*

The scheme of definition 2.5 is the simplest version of a class of media access control (MAC) protocols, denoted as Carrier Sense Multiple Access (*CSMA*). A recent discussion and source of references about CSMA is (M. Heusse, 2003). Such scheme also describes the MAC operation for a wireless communication network between two nodes. Also, we adopt the following strong assumptions:

- Every time node A sends a packet to node B: either it is sent without error or it is lost. This assumption means that we are not dealing with, what is commonly referred to as, a *noisy channel*.

- Every ACK signal sent by node B will reach node A before k changes to $k + 1$. This assumption is critical to guarantee that no packets are wasted. Notice that node B can use the whole interval $t \in (kT + T_T, (k + 1)T)$ to send the last ACK. During this period, the controller is not expecting new packets. The controller will generate $\mathbf{u}(k)$ using the packets that were sent in the interval $t \in [kT, kT + T_T]$. Consequently, such ACK is not important in the generation of $\mathbf{u}(k)$. It will be critical only for $\mathbf{u}(i)$ for $i > k$.

We adopt k, the discrete-time unit, as a reference. According to the usual framework of digital control, k will correspond to the discrete time unit obtained by partitioning the continuous-time in periods of duration T. Denote by $T_T < T$ the period allocated for transmission. Now, consider that the aim of a discrete-time controller is to control a continuous-time linear system, which admits[2] a discretization of the form $\mathbf{x}_c((k + 1)T) = \mathbf{A}(k)\mathbf{x}_c(kT) + \mathbf{u}(k)$. The discretization is such that $u(k)$ represents the effect of the control action over $t \in (kT + T_T, (k + 1)T)$. Information about $\mathbf{x}(k) = \mathbf{x}_c(kT)$, the state of the plant at the sampling instant $t = kT$, is transmitted during $t \in [kT, kT + T_T]$. Whenever k changes, we construct a new queue and assume that the cycle of definition 2.5 is reset to step 1.

2.5.1 Synchronization Between the Encoder and the Decoder

Denote by $\mathbf{r}_{enc}(k)$ the total number of bits that the encoder has successfully sent between k and $k + 1$, i.e., the number of bits for which the encoder has received an ACK. The variable $\mathbf{r}_{enc}(k)$ is used by the encoder to keep track of how many bits were sent. The corresponding variable at the decoder is represented as $\mathbf{r}_{dec}(k)$. From definition 2.5, we infer that $\mathbf{r}_{dec}(k - 1)$

---

[2]A controllable linear and time-invariant system admits a discretization of the required form. If the system is stochastic an equivalent condition has to be imposed

may not be available at all times. On the other hand, we emphasize that the following holds:

$$\mathbf{c}(k) \neq 0 \implies \mathbf{r}_{dec}(i) = \mathbf{r}_{enc}(i), i \in \{0, \ldots, k-1\} \quad (17)$$

In section 3, the stabilizing control is constructed in a way that: if no packet goes through between k and $k+1$, i.e., $c(k) = 0$ then $\mathbf{u}(k) = 0$. That shows that $\mathbf{r}_{dec}(k-1)$ is not available only when it is not needed. That motivated us to adopt the simplifying assumption that $\mathbf{r}(k-1) = \mathbf{r}_{enc}(k-1) = \mathbf{r}_{dec}(k-1)$. We denote by $\mathbf{r}(k)$ the random variable which represents the total number of bits that are transmitted in the time interval $t \in [kT, kT + T_T]$. The $\mathbf{r}(k)$ transmitted bits are used by the controller to generate $\mathbf{u}(k)$. Notice that our scheme does not pressupose an extra delay, because the control action will act, in continuous time, in the interval $t \in (kT + T_T, (k+1)T)$.

2.5.2 Encoding and Decoding for First Order Systems

Given the transmission scheme described above, the only remaining degrees of freedom are how to encode the measurement of the state and how to construct the queue. From the proofs of theorems 3.2 and 3.4, we infer that a sufficient condition for stabilization is the ability to transmit, between nodes A and B, an estimate of the state $\hat{\mathbf{x}}(k)$ with an accuracy lower-bounded[3] by $\mathcal{E}[|\hat{\mathbf{x}}(k) - \mathbf{x}(k)|^m] < 2^{-R}$, where $R > 0$ is a given constant that depends on the state-space representation of the plant. Since the received packets preserve the original order of the queue, we infer that the *best* way to construct the queues, at each k, is to compute the binary expansion of $\mathbf{x}(k)$ and position the packets so that the bits corresponding to higher powers of 2 are sent first. The lost packets will always[4] be the *less important*. The abstraction of such procedure is given by the truncation operator of definition 2.1. The random behavior of $\mathbf{r}(k)$ arises from random time-out, the existence of collisions generated by other nodes trying to communicate with node A or from the fading that occurs if node B is moving. The fading fenomena may also occur from interference.

---

[3]This observation was already reported in (Tatikonda, 2000a)

[4]The situation were the packets lost are in random positions is characteristic of large networks where packets travel through different routers.

3 SUFFICIENCY CONDITIONS FOR THE ROBUST STABILIZATION OF FIRST ORDER LINEAR SYSTEMS

In this section, we derive constructive sufficient conditions for the existence of a stabilizing feedback scheme. We start with the deterministic case in subsection 3.1, while 3.2 deals with random $\mathbf{r}$ and $\mathbf{a}$. We stress that our proofs hold under the framework of section 2. The strength of our assumptions can be accessed from the discussion in section 2.5.

The following definition introduces the main idea behind the construction of a stabilizing feedback scheme.

Definition 3.1 *(Upper-bound Sequence) Let $\bar{z}_f \in [0, 1)$, $\bar{z}_a \in [0, 1)$, $\bar{d} \geq 0$ and $\varrho \in \mathbb{N}_+ \bigcup \{\infty\}$ be given. Define the upper-bound sequence as:*

$$\mathbf{v}(k+1) = |\mathbf{a}(k)| 2^{-\mathbf{r}_e(k)} \mathbf{v}(k)$$
$$+ \bar{z}_f \max\{\mathbf{v}(k-\varrho+1), \ldots, \mathbf{v}(k)\} + \bar{d}, \quad (18)$$

where $v(i) = 0$ for $i < 0$, $v(0) = \frac{1}{2}$ and $\mathbf{r}_e(k)$ is an effective rate given by:

$$\mathbf{r}_e(k) = -\log(2^{-\mathbf{r}(k)} + \bar{z}_a) \quad (19)$$

Definition 3.2 *Following the representation for $\mathbf{r}(k)$ we also define C_e and $\mathbf{r}_e^\delta(k)$ such that:*

$$\mathbf{r}_e(k) = C_e - \mathbf{r}_e^\delta(k) \quad (20)$$

where $\mathcal{E}[\mathbf{r}_e^\delta(k)] = 0$.

We adopt $v(0) = \frac{1}{2}$ to guarantee that $|x(0)| \leq v(0)$. If $x(0) = 0$ then we can select $v(0) = 0$. Notice that the multiplicative uncertainty $\bar{z}_a$ acts by reducing the effective rate $\mathbf{r}_e(k)$. After inspecting (19), we find that $\mathbf{r}_e(k) \leq \min\{\mathbf{r}(k), -\log(\bar{z}_a)\}$. Also, notice that:

$$\bar{z}_a = 0 \implies (\mathbf{r}_e(k) = \mathbf{r}(k), \mathbf{r}_e^\delta(k) = \mathbf{r}^\delta(k), C = C_e) \quad (21)$$

Definition 3.3 *(Stabilizing feedback scheme) We make use of the sequence specified in definition 3.1. Notice that $\mathbf{v}(k)$ can be constructed at the controller and the encoder because both have access to ϱ, $\bar{z}_f$, $\bar{z}_a$, $\bar{d}$, $\mathbf{r}(k-1)$ and $\mathbf{a}(k-1)$.*

The feedback scheme is defined as:

- *Encoder: Measures $x(k)$ and computes $b_i \in \{0, 1\}$ such that:*

$$(b_1, \ldots, b_{\bar{r}}) = arg \max_{\sum_{i=1}^{\bar{r}} b_i \frac{1}{2^i} \leq \left(\frac{\mathbf{x}(k)}{2\mathbf{v}(k)} + \frac{1}{2}\right)} \sum_{i=1}^{\bar{r}} b_i \frac{1}{2^i} \quad (22)$$

Place $(b_1, \ldots, b_{\bar{r}})$ for transmission. For any $r(k) \in \{0, \ldots, \bar{r}\}$, the above construction provides the following centroid approximation $\hat{x}(k)$ for $x(k) \in [-v(k), v(k)]$:

$$\hat{x}(k) = 2v(k)\left(\sum_{i=1}^{r(k)} b_i \frac{1}{2^i} + \frac{1}{2^{r(k)+1}} - \frac{1}{2}\right) \quad (23)$$

which satisfies $|x(k) - \hat{x}(k)| \leq 2^{-r(k)}v(k)$.

- *Controller: From the $\bar{r}$ bits placed for transmission in the stochastic link, only $\mathbf{r}(k)$ bits go through. Compute $\mathbf{u}(k)$ as:*

$$\mathbf{u}(k) = -\mathbf{a}(k)\hat{\mathbf{x}}(k) \quad (24)$$

As expected, the transmission of state information through a finite capacity medium requires quantization. The encoding scheme of definition 3.3 is not an exception and is structurally identical to the ones used by (R. W. Brocket, 2000; Tatikonda, 2000b), where sequences were already used to upper-bound the state of the plant.

The following lemma suggests that, in the construction of stabilizing controllers, we may choose to focus on the dynamics of the sequence $v(k)$. That simplifies the analysis in the presence of uncertainty because the dynamics of $v(k)$ is described by a first-order difference equation.

Lemma 3.1 *Let $\bar{z}_f \in [0, 1)$, $\bar{z}_a \in [0, 1)$ and $\bar{d} \geq 0$ be given. If $\mathbf{x}(k)$ is the solution of (7) under the feedback scheme of definition 3.3, then the following holds:*

$$\bar{\mathbf{x}}(k) \leq \mathbf{v}(k)$$

for all $\varrho \in \mathbb{N}_+ \bigcup\{\infty\}$, every choice $G_f \in \Delta_{f,\varrho}$ and $|\mathbf{z_a}(k)| \leq \bar{z}_a$, where

$$\Delta_{f,\varrho} = \{G_f : \mathbb{R}^\infty \to \mathbb{R}^\infty : \|G_f\|_{\infty(\varrho)} \leq \bar{z}_f\} \quad (25)$$

Proof: We proceed by induction, assuming that $\bar{x}(i) \leq v(i)$ for $i \in \{0, \ldots, k\}$ and proving that $\bar{x}(k+1) \leq v(k+1)$. From (7), we get:

$$|x(k+1)| \leq |a(k)||x(k) + \frac{u(k)}{a(k)}|$$

$$+ |z_a(k)||a(k)||x(k)| + |G_f(x)(k)| + |d(k)| \quad (26)$$

The way the encoder constructs the binary expansion of the state, as well as (24), allow us to conclude that

$$|x(k) + \frac{u(k)}{a(k)}| \leq 2^{-r(k)}v(k)$$

Now we recall that $|z_a(k)| \leq \bar{z}_a$, $|G_f(x)(k)| \leq \bar{z}_f \max\{v(k-\varrho+1), \ldots, v(k)\}$ and that $|d(k)| \leq \bar{d}$, so that (26) implies:

$$|x(k+1)| \leq |a(k)|(2^{-r(k)} + \bar{z}_a)v(k)$$

$$+ \bar{z}_f \max\{v(k-\varrho+1), \ldots, \mathbf{v}(k)\} + \bar{d} \quad (27)$$

The proof is concluded once we realize that $|x(0)| \leq v(0)$.

$$\square$$

3.1 The Deterministic Case

We start by deriving a sufficient condition for the existence of a stabilizing feedback scheme in the deterministic case, i.e., $\mathbf{r}(k) = \mathcal{C}$ and $\log(|\mathbf{a}(k)|) = \mathcal{R}$. Subsequently, we move for the stochastic case where we derive a sufficient condition for stabilizability.

Theorem 3.2 (Sufficiency conditions for Robust Stability) *Let $\varrho \in \mathbb{N}_+ \bigcup\{\infty\}$, $\bar{z}_f \in [0, 1)$, $\bar{z}_a \in [0, 1)$ and $\bar{d} \geq 0$ be given and $h(k)$ be defined as*

$$h(k) = 2^{k(\mathcal{R}-\mathcal{C}_e)}, \; k \geq 0$$

where $\mathcal{C}_e = r_e = -\log(2^{-\mathcal{C}} + \bar{z}_a)$.

Consider that $\mathbf{x}(k)$ is the solution of (7) under the feedback scheme of definition 3.3 as well as the following conditions:

- *(C 1) $\mathcal{C}_e > \mathcal{R}$*
- *(C 2) $\bar{z}_f \|h\|_1 < 1$*

If conditions (C 1) and (C 2) are satisfied then the following holds for all $|\mathbf{d}(t)| \leq \bar{d}$, $G_f \in \Delta_{f,\varrho}$ and $|\mathbf{z_a}(k)| \leq \bar{z}_a$:

$$\bar{x}(k) \leq \|h\|_1 \left(\bar{z}_f \frac{\|h\|_1 \bar{d} + \frac{1}{2}}{1 - \|h\|_1 \bar{z}_f} + \bar{d}\right) + h(k)\frac{1}{2} \quad (28)$$

where $\Delta_{f,\varrho}$ is given by:

$$\Delta_{f,\varrho} = \{G_f : \mathbb{R}^\infty \to \mathbb{R}^\infty : \|G_f\|_{\infty(\varrho)} \leq \bar{z}_f\} \quad (29)$$

Proof: From definition 3.1, we know that, for arbitrary $\varrho \in \mathbb{N}_+ \bigcup\{\infty\}$, the following is true:

$$v(k+1) = 2^{\mathcal{R}-\mathcal{C}_e}v(k)$$

$$+ \bar{z}_f \max\{v(k-\varrho+1), \ldots, v(k)\} + \bar{d} \quad (30)$$

Solving the difference equation gives:

$$v(k) = 2^{k(\mathcal{R}-\mathcal{C}_e)}v(0)$$

$$+ \sum_{i=0}^{k-1} 2^{(k-i-1)(\mathcal{R}-\mathcal{C}_e)}$$

$$\times \left(\bar{z}_f \max\{v(i-\varrho+1), \ldots, v(i)\} + \bar{d}\right) \quad (31)$$

which, using $\|\Pi_k v\|_\infty = \max\{v(0), \ldots, v(k)\}$, leads to:

$$v(k) \leq \|h\|_1(\bar{z}_f \|\Pi_k v\|_\infty + \bar{d}) + 2^{k(\mathcal{R}-\mathcal{C}_e)}v(0) \quad (32)$$

But we also know that $2^{k(\mathcal{R}-\mathcal{C}_e)}$ is a decreasing function of k, so that:

$$\|\Pi_k v\|_\infty \leq \|h\|_1(\bar{z}_f \|\Pi_k v\|_\infty + \bar{d}) + v(0) \quad (33)$$

which implies:

$$\|\Pi_k v\|_\infty \leq \frac{\|h\|_1 \bar{d} + v(0)}{1 - \|h\|_1 \bar{z}_f} \quad (34)$$

Direct substitution of (34) in (32) leads to:

$$v(k) \leq \|h\|_1 \left(\bar{z}_f \frac{\|h\|_1 \bar{d} + v(0)}{1 - \|h\|_1 \bar{z}_f} + \bar{d}\right) + 2^{k(\mathcal{R}-\mathcal{C}_e)}v(0) \quad (35)$$

The proof is complete once we make $v(0) = \frac{1}{2}$ and use lemma 3.1 to conclude that $\bar{x}(k) \leq v(k)$.

$$\square$$

3.2 Sufficient Condition for the Stochastic Case

The following lemma provides a sequence, denoted by $v_m(k)$, which is an upper-bound for the m-th moment of $\bar{\mathbf{x}}(k)$. We show that v_m is propagated according to a first-order difference equation that is suitable for the analysis in the presence of uncertainty.

Lemma 3.3 *(M-th moment boundedness) Let $\varrho \in \mathbb{N}_+$, $\bar{z}_f \in [0,1)$, $\bar{z}_a \in [0,1)$ and $\bar{d} \geq 0$ be given along with the following set:*

$$\Delta_{f,\varrho} = \{G_f : \mathbb{R}^\infty \to \mathbb{R}^\infty : \|G_f\|_{\infty(\varrho)} \leq \bar{z}_f\} \quad (36)$$

Given m, consider the following sequence:

$$v_m(k) = h_m(k)v_m(0)$$
$$+ \sum_{i=0}^{k-1} h_m(k-i-1)$$
$$\times \left(\varrho^{\frac{1}{m}} \bar{z}_f \max\{v_m(i-\varrho+1), \ldots, v_m(i)\} + \bar{d} \right)$$
$$(37)$$

where $v_m(i) = 0$ for $i < 0$, $v_m(0) = \frac{1}{2}$, $h_m(k)$ is the impulse response given by:

$$h_m(k) = \left(\mathcal{E}[2^{m(\log(|a(k)|)-r_e(k))}] \right)^{\frac{k}{m}}, \ k \geq 0 \quad (38)$$

and $r_e(k) = -\log\left(2^{-r(k)} + \bar{z}_a\right)$. If $\mathbf{x}(k)$ is the solution of (7) under the feedback scheme of definition 3.3, then the following holds

$$\mathcal{E}[\bar{\mathbf{x}}(k)^m] \leq v_m(k)^m$$

for all $|d(t)| \leq \bar{d}$, $G_f \in \Delta_{f,\varrho}$ and $|\mathbf{z}_a(k)| \leq \bar{z}_a$.

Proof: Since lemma 3.1 guarantees that $\bar{x}(k+1) \leq v(k+1)$, we only need to show that $\mathcal{E}[\mathbf{v}(k+1)^m]^{\frac{1}{m}} \leq v_m(k+1)$. Again, we proceed by induction by noticing that $v(0) = v_m(0)$ and by assuming that $\mathcal{E}[\mathbf{v}(i)^m]^{\frac{1}{m}} \leq v_m(i)$ for $i \in \{1, \ldots, k\}$. The induction hypothesis is proven once we establish that $\mathcal{E}[\mathbf{v}(k+1)^m]^{\frac{1}{m}} \leq v_m(k+1)$. From definition 3.1, we know that:

$$\mathcal{E}[\mathbf{v}(k+1)^m]^{\frac{1}{m}} = \mathcal{E}[\left(2^{\log(|\mathbf{a}(k)|)-\mathbf{r}_e(k)}\mathbf{v}(k) \right.$$
$$\left. + \bar{z}_f \max\{\mathbf{v}(k-\varrho+1), \ldots, \mathbf{v}(k)\} + \bar{d}\right)^m]^{\frac{1}{m}} \quad (39)$$

Using Minkowsky's inequality (Halmos, 1974) as well as the fact that $\mathbf{v}(i)$ is independent of $\mathbf{a}(j)$ and $\mathbf{r}_e(j)$ for $j \geq i$, we get:

$$\mathcal{E}[\mathbf{v}(k+1)^m]^{\frac{1}{m}}$$
$$\leq \mathcal{E}[2^{m(\log(|\mathbf{a}(k)|)-\mathbf{r}_e(k))}]^{\frac{1}{m}} \mathcal{E}[\mathbf{v}(k)^m]^{\frac{1}{m}}$$
$$+ \bar{z}_f \mathcal{E}[\max\{\mathbf{v}(k-\varrho+1), \ldots, \mathbf{v}(k)\}^m]^{\frac{1}{m}} + \bar{d} \quad (40)$$

which, using the inductive assumption, implies the following inequality:

$$\mathcal{E}[\mathbf{v}(k+1)^m]^{\frac{1}{m}}$$
$$\leq \mathcal{E}[2^{m(\log(|\mathbf{a}(k)|)-\mathbf{r}_e(k))}]^{\frac{1}{m}} v_m(k)$$
$$+ \varrho^{\frac{1}{m}} \bar{z}_f \max\{v_m(k-\varrho+1), \ldots, v_m(k)\} + \bar{d} \quad (41)$$

where we used the fact that, for arbitrary random variables $\mathbf{s}_1, \ldots, \mathbf{s}_n$, the following holds:

$$\mathcal{E}[\max\{|\mathbf{s}_1|, \ldots, |\mathbf{s}_n|\}^m] \leq \mathcal{E}[\sum_{i=1}^{n} |\mathbf{s}_i|^m]$$
$$\leq n \max\{\mathcal{E}[|\mathbf{s}_1|^m], \ldots, \mathcal{E}[|\mathbf{s}_n|^m]\} \quad (42)$$

The proof follows once we notice that the right hand side of (41) is just $v_m(k+1)$.

$\square$

Theorem 3.4 *(Sufficient Condition) Let m, $\varrho \in \mathbb{N}_+$, $\bar{z}_f \in [0,1)$, $\bar{z}_a \in [0,1)$ and $\bar{d} \geq 0$ be given along with the quantities bellow:*

$$\beta(m) = \frac{1}{m} \log \mathcal{E} \left[2^{m\mathbf{1}_a^\delta(k)} \right]$$
$$\alpha_e(m) = \frac{1}{m} \log \mathcal{E} \left[2^{m\mathbf{r}_e^\delta(k)} \right]$$
$$h_m(k) = 2^{k(\mathcal{R}+\beta(m)+\alpha_e(m)-\bar{C}_e)}, \ k \geq 0$$

where $\mathbf{r}_e^\delta$ comes from (20). Consider that $\mathbf{x}(k)$ is the solution of (7) under the feedback scheme of definition 3.3 as well as the following conditions:

- **(C 3)** $C_e > \mathcal{R} + \beta(m) + \alpha_e(m)$
- **(C 4)** $\varrho^{\frac{1}{m}} \bar{z}_f \|h_m\|_1 < 1$

If conditions (C 3) and (C 4) are satisfied, then the following holds for all $|d(t)| \leq \bar{d}$, $G_f \in \Delta_{f,\varrho}$ and $|\mathbf{z}_a(k)| \leq \bar{z}_a$:

$$\mathcal{E}[\bar{\mathbf{x}}(k)^m]^{\frac{1}{m}}$$
$$\leq \|h_m\|_1 \left(\varrho^{\frac{1}{m}} \bar{z}_f \frac{\|h_m\|_1 \bar{d} + \frac{1}{2}}{1 - \varrho^{\frac{1}{m}} \bar{z}_f \|h_m\|_1} + \bar{d} \right) + h_m(k)\frac{1}{2}$$
$$(43)$$

where $\Delta_{f,\varrho}$ is given by:

$$\Delta_{f,\varrho} = \{G_f : \mathbb{R}^\infty \to \mathbb{R}^\infty : \|G_f\|_{\infty(\varrho)} \leq \bar{z}_f\} \quad (44)$$

Proof: Using v_m from lemma 3.3, we arrive at:

$$v_m(k) \leq h_m(k)v_m(0)$$
$$+ \|h_m\|_1 \left(\varrho^{\frac{1}{m}} \bar{z}_f \|\Pi_k v_m\|_\infty + \bar{d} \right) \quad (45)$$

where we use

$$\|\Pi_k v_m\|_\infty = \max\{v_m(0), \ldots, v_m(k)\}$$

But from (45), we conclude that:

$$\|\Pi_k v_m\|_\infty \leq v_m(0) + \|h_m\|_1 \left(\varrho^{\frac{1}{m}} \bar{z}_f \|\Pi_k v_m\|_\infty + \bar{d} \right) \tag{46}$$

or equivalently:

$$\|\Pi_k v_m\|_\infty \leq \frac{v_m(0) + \|h_m\|_1 \bar{d}}{1 - \|h_m\|_1 \varrho^{\frac{1}{m}} \bar{z}_f} \tag{47}$$

Substituting (47) in (45), gives:

$$v_m(k) \leq h_m(k) v_m(0)$$
$$+ \|h_m\|_1 \left(\varrho^{\frac{1}{m}} \bar{z}_f \frac{v_m(0) + \|h_m\|_1 \bar{d}}{1 - \|h_m\|_1 \varrho^{\frac{1}{m}} \bar{z}_f} + \bar{d} \right) \tag{48}$$

The proof follows from lemma 3.3 and by noticing that $h_m(k)$ can be rewritten as:

$$h_m(k) = \left(\mathcal{E}[2^{m(\log(|\mathbf{a}(k)|) - \mathbf{r}_e(k))}] \right)^{\frac{k}{m}}$$
$$= 2^{k(\mathcal{R} + \beta(m) + \alpha_e(m) - C_e)}, \ k \geq 0 \tag{49}$$

$\square$

4 SUFFICIENT CONDITIONS FOR A CLASS OF SYSTEMS OF ORDER HIGHER THAN ONE

The results, derived in section 3, can be extended, in specific cases, to systems of order higher than one (see section 4.1). In the subsequent analysis, we outline how and suggest a few cases when such extension can be attained. Our results do not generalize to arbitrary stochastic systems of order $n > 1$. We emphasize that the proofs in this section are brief as they follow the same structure of the proofs of section 3 [5].

We use n as the order of a linear system whose state is indicated by $x(k) \in \mathbb{R}^n$. The following is a list of the adaptations, of the notation and definitions of section 1, to the multi-state case:

- if $x \in \mathbb{R}^n$ then we indicate its components by $[x]_i$, with $i \in \{1, \ldots, n\}$. In a similar way, if M is a matrix then we represent the element located in the i-th row and j-th column as $[M]_{ij}$. We also use $|M|$ to indicate the matrix whose elements are obtained as $[|M|]_{ij} = |[M]_{ij}|$.

- $\mathbb{R}^{n \times \infty}$ is used to represent the set of sequences of n-dimensional vectors, i.e., $x \in \mathbb{R}^{n \times \infty} \implies x(k) \in \mathbb{R}^n, k \in \mathbb{N}$.

- the infinity norm in $\mathbb{R}^{n \times \infty}$ is defined as:

$$\|x\|_\infty = \sup_i \max_j |[x(i)]_j|$$

---

[5]The authors suggest the reading of section 3 first.

It follows that if $x \in \mathbb{R}^n$ then $\|x\|_\infty = \max_{j \in \{1,\ldots,n\}} |[x]_j|$. Accordingly, if $x \in \mathbb{R}^{n \times \infty}$ we use $\|x(k)\|_\infty = \max_{j \in \{1,\ldots,n\}} |[x(k)]_j|$ to indicate the norm of a single vector, at time k, in contrast with $\|x\|_\infty = \sup_i \max_j |[x(i)]_j|$.

- the convention for random variables remains unchanged, e.g., $[\mathbf{x}(k)]_j$ is the jth component of a n-dimensional random sequence whose realizations lie on $\mathbb{R}^{n \times \infty}$

- If H is a sequence of matrices, with $H(k) \in \mathbb{R}^{n \times n}$, then

$$\|H\|_1 = \max_i \sum_{k=0}^{\infty} \sum_{j=1}^{n} |[H(k)]_{ij}|$$

For an arbitrary vector $x \in \mathbb{R}^n$ we use Hx to represent the sequence $H(k)x$. For a particular matrix $H(k)$, we also use $\|H(k)\|_1 = \max_i \sum_{j=1}^{n} |[H(k)]_{ij}|$.

- we use $\vec{1} \in \mathbb{R}^n$ to indicate a vector of ones, i.e., $[\vec{1}]_j = 1$ for $j \in \{1, \ldots, n\}$.

4.1 Description of the Nominal Plant and Equivalent Representations

In this section, we introduce the state-space representation of the nominal discrete-time plant, for which we want to determine robust stabilizability. We also provide a condition, under which the stabilizability, of such state-space representation, can be inferred from the stabilizability of another representation which is more convenient. The condition is stated in Proposition 4.1 and a few examples are listed in remark 4.2. Such equivalent representation is used in section 4.2 as way to obtain stability conditions that depend explicitly on the eigenvalues of the dynamic matrix.

Consider the following nominal state-space realization:

$$\tilde{\mathbf{x}}(k+1) = \tilde{\mathbf{A}}(k)\tilde{\mathbf{x}}(k) + \tilde{\mathbf{u}}(k) + \tilde{\mathbf{d}}(k) \tag{50}$$

where $\|\tilde{\mathbf{d}}\|_\infty \leq \bar{\bar{d}}$ and $\bar{\bar{d}}$ is a pre-specified constant.

We also consider that $\tilde{\mathbf{A}}$ is a real Jordan form with a structure given by:

$$\tilde{\mathbf{A}}(k) = diag(\mathbf{J}_1(k), \ldots, \mathbf{J}_{\mathbf{q}(k)}(k)) \tag{51}$$

where $J_i(k)$ are real Jordan blocks(R. Horn, 1985) with multiplicity q_i satisfying $\sum_i q_i = n$.

The state-space representation of a linear and time-invariant system can always be transformed in a way that $\tilde{A}$ is in real Jordan form. The discretization of a controllable continuous-time and time-invariant linear system can always be expressed in the form (50), i.e., with $B = I$. If the system is not controllable, but stabilizable, then we can ignore the stable dynamics and consider only the unstable part which can be written in the form (50).

If the system is stochastic then, in general, there is no state transformation leading to $\mathbf{A}(k)$ in real Jordan form. The following is a list of conditions, under which a state-space representation of the form $\mathbf{x}(k+1) = \mathbf{A}(k)\mathbf{x}(k) + \mathbf{u}(k)$ can be transformed in a new one, for which $\tilde{\mathbf{A}}$ is in real Jordan form:

- When the original dynamic matrices are already in real Jordan form. A particular instance of that are the second order stochastic systems with complex poles.

- A collection of systems with a state-space realization of the type $\mathbf{x}(k+1) = \mathbf{J}(k)\mathbf{x}(k) + \mathbf{u}(k)$ which connected in a shift-invariant topology. Here we used the fact that if several *copies* of the same system are connected in a shift-invariant topology then they can be decoupled by means of a time-invariant transformation (N. C. Martins, 2001).

Still, the representation (50)-(51) cannot be studied directly due to the fact that it may have complex eigenvalues. We will use the idea in (Tatikonda, 2000a) and show that, under certain conditions, there exists a transformation which leads to a more convenient state-space representation. Such representation has a dynamic matrix which is upper-triangular and has a diagonal with elements given by $|\lambda_i(\tilde{\mathbf{A}}(k))|$.

If we denote $R(\theta)$ as the following rotation:

$$R(\theta) = \begin{bmatrix} cos(\theta) & sin(\theta) \\ -sin(\theta) & cos(\theta) \end{bmatrix} \quad (52)$$

then the general structure of $J_i(k) \in \mathbb{R}^{q_i}$ is:
(**If $\eta_i(k)$ is real**)

$$J_i(k) = \begin{bmatrix} \eta_i(k) & 1 & \cdots & \\ 0 & \eta_i(k) & 1 & \ddots \\ \vdots & & & \ddots \\ 0 & 0 & 0 & \end{bmatrix} \quad (53)$$

(**otherwise**)

$$J_i(k)$$
$$= \begin{bmatrix} |\eta_i(k)|R(\theta_i(k)) & I & \cdots & 0 \\ 0 & |\eta_i(k)|R(\theta_i(k)) & I & 0 \\ \vdots & & & \ddots \end{bmatrix} \quad (54)$$

We start by following (Tatikonda, 2000a) and defining the following matrix:

Definition 4.1 *(Rotation dynamics) Let the real Jordan form $\tilde{\mathbf{A}}(k)$ of (50) be given by (51). We define the rotation dynamics $\mathbf{R}_{\tilde{A}}(k)$ as the following matrix:*

$$\mathbf{R}_{\tilde{A}}(k) = diag(\mathbf{J}_1^R(k), \ldots, \mathbf{J}_q^R(k)) \quad (55)$$

where $J_i^R(k) \in \mathbb{R}^{q_i}$ are given by:

$$J_i^R(k) = \begin{cases} sgn(\eta_i(k))I & \eta_i(k) \in \mathbb{R} \\ diag(R(\theta_i(k)), \ldots, R(\theta_i(k))) & otherwise \end{cases} \quad (56)$$

For technical reasons, we use the idea of (Tatikonda, 2000a) and study the stability of $\mathbf{x}$ given by:

$$\mathbf{x}(k) = \mathbf{R}_{\tilde{A}}(k-1)^{-1} \cdots \mathbf{R}_{\tilde{A}}(0)^{-1}\tilde{\mathbf{x}}(k) \quad (57)$$

Remark 4.1 *The motivation for such time-varying transformation is that, by multiplying (50) on the left by $\mathbf{R}_{\tilde{A}}(k)^{-1} \cdots \mathbf{R}_{\tilde{A}}(0)^{-1}$, the nominal dynamics of $\mathbf{x}$ is given by:*

$$\mathbf{x}(k+1) = \mathbf{A}(k)\mathbf{x}(k) + \mathbf{d}(k) + \mathbf{u}(k) \quad (58)$$

where $\mathbf{d}(k) = \mathbf{R}_{\tilde{A}}(k-1)^{-1} \cdots \mathbf{R}_{\tilde{A}}(0)^{-1}\tilde{\mathbf{d}}(k)$, $\mathbf{u}(k) = \mathbf{R}_{\tilde{A}}(k-1)^{-1} \cdots \mathbf{R}_{\tilde{A}}(0)^{-1}\tilde{\mathbf{u}}(k)$ and $\mathbf{A}(k)$ is the following upper-triangular matrix[6]:

$$\mathbf{A}(k) = \mathbf{R}_{\tilde{A}}(k)^{-1}\tilde{\mathbf{A}}(k)$$
$$= \begin{bmatrix} |\lambda_1(\tilde{\mathbf{A}}(k))| & \cdots & \\ 0 & \ddots & \vdots \\ 0 & 0 & |\lambda_n(\tilde{\mathbf{A}}(k))| \end{bmatrix} \quad (59)$$

The following proposition is a direct consequence of the previous discussion:

Proposition 4.1 *(Condition for equivalence of representations) Let $\tilde{\mathbf{A}}(k)$ be such that $\mathbf{R}_{\tilde{A}}(k)$ satisfies:*

$$\sup_k \|\mathbf{R}_{\tilde{A}}(k)^{-1} \cdots \mathbf{R}_{\tilde{A}}(0)^{-1}\|_1 \leq \Gamma_1 < \infty \quad (60)$$

$$\sup_k \|\left(\mathbf{R}_{\tilde{A}}(k)^{-1} \cdots \mathbf{R}_{\tilde{A}}(0)^{-1}\right)^{-1}\|_1 \leq \Gamma_2 < \infty \quad (61)$$

Under the above conditions, the stabilization of (58)-(59) and the stabilization of (50)-(51) are equivalent in the sense of (62)-(63).

$$\limsup_{k \to \infty} \|\mathbf{x}(k)\|_\infty \leq \Gamma_1 \limsup_{k \to \infty} \|\tilde{\mathbf{x}}(k)\|_\infty$$
$$\leq \Gamma_1\Gamma_2 \limsup_{k \to \infty} \|\mathbf{x}(k)\|_\infty \quad (62)$$

$$\limsup_{k \to \infty} \mathcal{E}[\|\mathbf{x}(k)\|_\infty^m] \leq \Gamma_1^m \limsup_{k \to \infty} \mathcal{E}[\|\tilde{\mathbf{x}}(k)\|_\infty^m]$$
$$\leq \Gamma_1^m\Gamma_2^m \limsup_{k \to \infty} \mathcal{E}[\|\mathbf{x}(k)\|_\infty^m] \quad (63)$$

---

[6]Here we use an immediate modification of lemma 3.4.1, from (Tatikonda, 2000a). It can be shown that, if $\tilde{\mathbf{A}}(k)$ is a real Jordan form then $\mathbf{R}_{\tilde{A}}(j)^{-1}\tilde{\mathbf{A}}(k) = \tilde{\mathbf{A}}(k)\mathbf{R}_{\tilde{A}}(j)^{-1}$ holds for any j, k. This follows from the fact that $R(\theta_1)R(\theta_2) = R(\theta_1 + \theta_2) = R(\theta_2)R(\theta_1)$ holds for arbitrary θ_1 and θ_2.

Remark 4.2 *Examples of $\tilde{\mathbf{A}}(k)$ for which (60)-(61) hold are:*

- *all time-invariant $\tilde{A}$*
- *$\tilde{\mathbf{A}}(k) = diag(\mathbf{J}_1(k), \ldots, \mathbf{J}_q(k))$ where q_i are invariant.*
- *all 2-dimensional $\mathbf{A}(k)$. In this case $\mathbf{R}_{\tilde{A}}(k)$ is always a rotation matrix, which includes the identity as a special case. Under such condition, the bounds in (60)-(61) are given by*

$$\sup_k \|\mathbf{R}_{\tilde{A}}(k)^{-1} \cdots \mathbf{R}_{\tilde{A}}(0)^{-1}\|_1 \leq 2$$

$$\sup_k \| \left(\mathbf{R}_{\tilde{A}}(k)^{-1} \cdots \mathbf{R}_{\tilde{A}}(0)^{-1}\right)^{-1} \|_1 \leq 2$$

4.2 Description of Uncertainty and Robust Stability

Let $\varrho \in \mathbb{N}_+ \bigcup \{\infty\}$, $\bar{d} \geq 0$, $\bar{z}_f \in [0,1)$ and $\bar{z}_a > 0$ be given constants then we study the stabilizability of the following uncertain system:

$$\mathbf{x}(k+1) = \mathbf{A}(k)(I + \mathbf{Z}_a(k))\mathbf{x}(k) + \mathbf{u}(k) + \mathbf{d}(k)$$
$$+ G_f(\mathbf{x})(k) \quad (64)$$

$$\mathbf{A}(k) = \begin{bmatrix} |\lambda_1(\mathbf{A}(k))| & \cdots & \\ 0 & \ddots & \vdots \\ 0 & 0 & |\lambda_n(\mathbf{A}(k))| \end{bmatrix} \quad (65)$$

where $\|\mathbf{d}\|_\infty \leq \bar{d}$, $|[\mathbf{Z}_a(k)]_{ij}| \leq \bar{z}_a$ and $\|G_f\|_{\infty(\varrho)} \leq \bar{z}_f$. Recall, from Proposition 4.1, that the stabilizability of a state-space representation of the form (50), satisfying (60)-(61), can be studied in the equivalent form where (65) holds.

A given feedback scheme is robustly stabilizing if it satisfies the following definition.

Definition 4.2 *(m-th Moment Robust Stability)* *Let $m > 0$, $\varrho \in \mathbb{N}_+ \bigcup \{\infty\}$, $\bar{z}_f \in [0,1)$, $\bar{z}_a \in [0,1)$ and $\bar{d} \geq 0$ be given. The system (64), under a given feedback scheme, is m-th moment (robustly) stable provided that the following holds:*
*(**if** $\bar{z}_f = \bar{d} = 0$)*

$$\lim_{k \to \infty} \mathcal{E}\left[\|\bar{\mathbf{x}}(k)\|_\infty^m\right] = 0 \quad (66)$$

*(**Otherwise**)*

$$\exists b \in \mathbb{R}_+ \ s.t. \ \limsup_{k \to \infty} \mathcal{E}\left[\|\bar{\mathbf{x}}(k)\|_\infty^m\right] < b \quad (67)$$

where $\bar{\mathbf{x}}(k)$ is given by:

$$[\bar{\mathbf{x}}(k)]_i = \sup_{x(0) \in [-1/2, 1/2]^n} |[\mathbf{x}(k)]_i|$$

4.3 Feedback Structure and Channel Usage Assumptions

In order to study the stabilization of systems of order higher than one, we assume the existence of a channel allocation $\vec{\mathbf{r}}(k) \in \{0, \ldots, \bar{r}\}^n$ satisfying:

$$\sum_{j=1}^n [\vec{\mathbf{r}}(k)]_j = \mathbf{r}(k) \quad (68)$$

where $\mathbf{r}(k)$ is the instantaneous rate sequence as described in section 2.5.1. We also emphasize that $\mathbf{A}(k)$ and $\vec{\mathbf{r}}(k)$ are i.i.d and independent of each other.

Using the same notation of section 1, we define C_j and $[\vec{\mathbf{r}}^\delta(k)]_j$ as:

$$[\vec{\mathbf{r}}(k)]_j = C_j - [\vec{\mathbf{r}}^\delta(k)]_j \quad (69)$$

Similarly, we also define $\alpha_i(m)$ as:

$$\alpha_i(m) = \frac{1}{m} \log \mathcal{E}[2^{m[\vec{\mathbf{r}}^\delta(k)]_i}] \quad (70)$$

In the general case, the allocation problem is difficult because it entails a change of the encoding process described in section 2.5.2. The encoding must be such that each $[\vec{\mathbf{r}}^\delta(k)]_i$ corresponds to the instantaneous rate of a truncation operator. In section 4.8 we solve the allocation problem for a class of stochastic systems in the presence of a stochastic link.

As in the one dimensional case, we assume that both the encoder and the controller have access to $A(0), \ldots, A(k)$ and $\vec{r}(k-1)$ as well as the constants ϱ, $\bar{z}_f$, $\bar{z}_a$ and $\bar{d}$. The encoder and the controller are described as:

- The encoder is a function $\mathcal{F}_k^e : \mathbb{R}^{n \times (k+1)} \to \{0,1\}^{n \times \bar{r}}$ that has the following dependence on observations:

$$\mathcal{F}_k^e(x(0), \ldots, \mathbf{x}(k)) = (\mathbf{b}_1, \ldots, \mathbf{b}_{\bar{r}}) \quad (71)$$

where $\mathbf{b}_i \in \{0,1\}^n$.

- The control action results from a map, not necessarily memoryless, $\mathcal{F}_k^c : \bigcup_{i=0}^{\bar{r}} \{0,1\}^{n \times i} \to \mathbb{R}$ exhibiting the following functional dependence:

$$\mathbf{u}(k) = \mathcal{F}_k^c(\vec{\mathbf{b}}(k)) \quad (72)$$

where $\vec{\mathbf{b}}(k)$ is the vector for which, each component $[\vec{\mathbf{b}}(k)]_j$, comprises a string of $[\vec{\mathbf{r}}(k)]_j$ bits successfully transmitted through the link, i.e.:

$$[\vec{\mathbf{b}}(k)]_j = [\mathcal{F}_k^l(\mathbf{b}_1, \ldots, \mathbf{b}_{\bar{r}})]_j$$
$$= ([\mathbf{b}_1]_j, \ldots, [\mathbf{b}_{[\vec{\mathbf{r}}(k)]_j}]_j) \quad (73)$$

As such, $\mathbf{u}(k)$ can be equivalently expressed as

$$\mathbf{u}(k) = (\mathcal{F}_k^c \circ \mathcal{F}_k^l \circ \mathcal{F}_k^e)(x(0), \ldots, \mathbf{x}(k))$$

4.4 Construction of a Stabilizing Feedback Scheme

The construction of a stabilizing scheme follows the same steps used in section 3. The following is the definition of a multidimensional upper-bound sequence.

Definition 4.3 (Upper-bound Sequence) *Let* $\bar{z}_f^x \in [0,1)$, $\bar{z}_a \in [0,1)$, $\bar{d} \geq 0$ *and* $\varrho \in \mathbb{N}_+ \bigcup\{\infty\}$ *be given. Define the upper-bound sequence* $\mathbf{v}(k)$, *with* $v(k) \in \mathbb{R}^n$, *as:*

$$\mathbf{v}(k+1) = \mathbf{A}_{cl}(k)\mathbf{v}(k)$$
$$+ \bar{z}_f \max\{\|\mathbf{v}(k-\varrho+1)\|_\infty,\ldots,\|\mathbf{v}(k)\|_\infty\}\vec{1} + \bar{d}\vec{1}, \tag{74}$$

where $\|A(k)\|_{ij} = |[A(k)]_{ij}|$, $v(i) = 0$ *for* $i < 0$, $[v(0)]_j = \frac{1}{2}$ *and* $A_{cl}(k)$ *is given by:*

$$A_{cl}(k)$$
$$= |A(k)|\left(diag(2^{-[\vec{r}(k)]_1},\ldots,2^{-[\vec{r}(k)]_n}) + \bar{z}_a\vec{1}\vec{1}^T\right) \tag{75}$$

Adopt the feedback scheme of definition 3.3, mutatis mutandis, for the multi-dimensional case. By measuring the state $x(k)$ and using $[\vec{r}(k)]_j$ bits, at time k, to encode each component $[x(k)]_j$, we construct $[\hat{x}(k)]_j$ such that

$$|[x(k)]_j - [\hat{x}(k)]_j| \leq 2^{-[\vec{r}(k)]_j}[v(k)]_j \tag{76}$$

Accordingly, $\mathbf{u}(k)$ is defined as:

$$\mathbf{u}(k) = -\mathbf{A}(k)\hat{\mathbf{x}}(k) \tag{77}$$

The following lemma establishes that the stabilization of $\mathbf{v}(k)$ is sufficient for the stabilization of $\mathbf{x}(k)$.

Lemma 4.2 *Let* $\bar{z}_f \in [0,1)$, $\bar{z}_a \in [0,1)$ *and* $\bar{d} \geq 0$ *be given. If* $\mathbf{x}(k)$ *is the solution of (64) under the feedback scheme given by (76)-(77), then the following holds:*

$$[\bar{\mathbf{x}}(k)]_j \leq [\mathbf{v}(k)]_j$$

for all $\varrho \in \mathbb{N}_+ \bigcup\{\infty\}$, $\|\mathbf{d}(k)\|_\infty \leq \bar{d}$, *every choice* $|[Z_a]_{ij}| \leq \bar{z}_a$ *and* $\|G_f\|_{\infty(\varrho)} \leq \bar{z}_f$.

Proof: The proof follows the same steps as in lemma 3.1. We start by assuming that $[\bar{x}(i)]_j \leq [v(i)]_j$ for $i \in \{0,\ldots,k\}$ and proceed to prove that $[\bar{x}(k+1)]_j \leq [v(k+1)]_j$. From (64) and the feedback scheme (76)-

(77), we find that:

$$\begin{bmatrix} |[x(k+1)]_1| \\ \vdots \\ |[x(k+1)]_n| \end{bmatrix} \underset{element-wise}{\leq}$$

$$|A(k)|\begin{bmatrix} |[v(k)]_1|2^{-\vec{r}_1(k)} \\ \vdots \\ |[v(k)]_n|2^{-\vec{r}_n(k)} \end{bmatrix}$$

$$+|A(k)|\,|Z_a(k)|\begin{bmatrix} |[v(k)]_1| \\ \vdots \\ |[v(k)]_n| \end{bmatrix} + \bar{d}\vec{1}$$

$$+ \bar{z}_f\vec{1}\max\{\|v(k-\varrho+1)\|_\infty,\ldots,\|v(k)\|_\infty\} \tag{78}$$

$\square$

4.5 Sufficiency for the Deterministic/Time-Invariant Case

Accordingly, the following theorem establishes the multi-dimensional analog to theorem 3.2.

Theorem 4.3 (Sufficiency conditions for Robust Stability) *Let* A *be the dynamic matrix of (64),* $\varrho \in \mathbb{N}_+\bigcup\{\infty\}$, $\bar{z}_f \in [0,1)$, $\bar{z}_a \in [0,1)$ *and* $\bar{d} \geq 0$ *be given and* $H(k)$ *be defined as*

$$H(k) = A_{cl}^k, \ k \geq 0$$

where $A_{cl} = |A|\left(diag(2^{-C_1},\ldots,2^{-C_n}) + \bar{z}_a\vec{1}\vec{1}^T\right)$ *and* $\|A\|_{ij} = |[A]_{ij}|$. *Consider that* $\mathbf{x}(k)$ *is the solution of (64) under the feedback scheme of (76)-(77) as well as the following conditions:*

- **(C 1)** $\max_i \lambda_i(A_{cl}) < 1$
- **(C 2)** $\bar{z}_f\|H\|_1 < 1$

If conditions **(C 1)** *and* **(C 2)** *are satisfied then the following holds:*

$$\|\bar{x}(k)\|_\infty$$
$$\leq \|H\|_1\left(\bar{z}_f\frac{\|H\|_1\bar{d} + \|\tilde{g}\|_\infty\frac{1}{2}}{1 - \bar{z}_f\|H\|_1} + \bar{d}\right) + \|\tilde{g}(k)\|_\infty\frac{1}{2} \tag{79}$$

where $\tilde{g}(k) = A_{cl}^k\vec{1}$.

Proof: We start by noticing that the condition (C1) is necessary and sufficient to guarantee that $\|H\|_1$ is finite. Following the same steps, used in the proof of theorem 3.2, from definition 4.3, we have that:

$$\mathbf{v}(k) = A_{cl}^k v(0) + \sum_{i=0}^{k-1} A_{cl}^{k-i-1}$$
$$\times\left(\bar{z}_f \max\{\|\mathbf{v}(i-\varrho+1)\|_\infty,\ldots,\|\mathbf{v}(i)\|_\infty\}\vec{1} + \bar{d}\vec{1}\right) \tag{80}$$

$$\|\pi_k \mathbf{v}\|_\infty \leq \|\tilde{g}\|_\infty \frac{1}{2} + \|H\|_\infty \left(\bar{z}_f \|\pi_k \mathbf{v}\|_\infty + \bar{d}\right) \quad (81)$$

where we use $\|\pi_k \mathbf{v}\|_\infty = \max\{\|\mathbf{v}(0)\|_\infty, \ldots, \|\mathbf{v}(k)\|_\infty\}$ and $\tilde{g}(k) = A_{cl}^k \vec{1}$. By means of lemma 4.2, the formula (79) is obtained by isolating $\|\pi_k \mathbf{v}\|_\infty$ in (81) and substituting it back in (80).$\square$

4.5.1 Interpretation for $\bar{z}_a = 0$

If $\bar{z}_a = 0$ then A_{cl}, of Theorem 4.3, can be written as:

$$A_{cl} = \begin{bmatrix} |\lambda_1(A)|2^{-C_1} & \cdots & \\ 0 & \ddots & \vdots \\ 0 & 0 & |\lambda_n(A)|2^{-C_n} \end{bmatrix} \quad (82)$$

The increase of C_i causes the decrease of $\lambda_i(A_{cl})$ and $\|H\|_1$. Accordingly, conditions (C1) and (C2), from theorem 4.3, lead to the conclusion that the increase in C_i gives a guarantee that the feedback loop is stable under larger uncertainty, as measured by $\bar{z}_f$. In addition, if we denote $R_i = \log(|\lambda_i(A)|)$ then we can cast condition (C1) as:

$$C_i > R_i \quad (83)$$

4.6 Sufficiency for the Stochastic Case

We derive the multi-dimensional version of the sufficiency results in section 3.2. The results presented below are the direct generalizations of lemma 3.3 and theorem 3.4.

Definition 4.4 (*Upper-bound sequence for the stochastic case*) *Let* $\varrho \in \mathbb{N}_+$, $\bar{z}_f \in [0,1)$, $\bar{z}_a \in [0,1)$ *and* $\bar{d} \geq 0$ *be given. Given* m, *consider the following sequence:*

$$v_m(k+1) = A_{cl,m} v_m(k) + \vec{1}\left(d + (n\varrho)^{\frac{1}{m}} \bar{z}_f\right)$$

$$\times \max\{\|v_m(k-\varrho+1)\|_\infty, \ldots, \|v_m(k)\|_\infty\} \quad (84)$$

where $v_m(i) = 0$ *for* $i < 0$, $v_m(0) = \frac{1}{2}\vec{1}$ *and* $A_{cl,m}$ *is defined as*

$$A_{cl,m} = A_m \left(\bar{z}_a \vec{1}\vec{1}^T \right.$$

$$\left. + diag(2^{-C_1+\alpha_1(m)}, \ldots, 2^{-C_n+\alpha_n(m)})\right) \quad (85)$$

and

$$[A_m]_{ij} = \mathcal{E}[|[\mathbf{A}(k)]_{ij}|^m]^{\frac{1}{m}}$$

Lemma 4.4 (*M-th moment boundedness*)

If $\mathbf{x}(k)$ *is the solution of (64) under the feedback scheme of (76)-(77), then the following holds*

$$\mathcal{E}[|\bar{\mathbf{x}}(k)]_i^m]^{\frac{1}{m}} \leq [v_m(k)]_i$$

Proof: We start by showing that $\mathcal{E}[|\mathbf{v}(k)]_i^m]^{\frac{1}{m}} \leq [v_m(k)]_i$. We proceed by induction, by assuming that $\mathcal{E}[|\mathbf{v}(j)]_i^m]^{\frac{1}{m}} \leq [v_m(j)]_i$ holds for $j \in \{0, \ldots, k\}$ and proving that $\mathcal{E}[|\mathbf{v}(k+1)]_i^m]^{\frac{1}{m}} \leq [v_m(k+1)]_i$.

Let $\mathbf{z}$, $\mathbf{s}$ and $\mathbf{g}$ be random variables with $\mathbf{z}$ independent of $\mathbf{s}$. By means of the Minkovsky inequality, we know that $\mathcal{E}[|\mathbf{zs}+\mathbf{g}|^m]^{\frac{1}{m}} \leq \mathcal{E}[|\mathbf{z}|^m]^{\frac{1}{m}} \mathcal{E}[|\mathbf{s}|^m]^{\frac{1}{m}} + \mathcal{E}[|\mathbf{g}|^m]^{\frac{1}{m}}$. Using such property, the following inequality is a consequence of (74):

$$\begin{bmatrix} \mathcal{E}[|\mathbf{v}(k+1)]_1^m]^{\frac{1}{m}} \\ \vdots \\ \mathcal{E}[|\mathbf{v}(k+1)]_n^m]^{\frac{1}{m}} \end{bmatrix} \underset{element-wise}{\leq}$$

$$A_{cl,m} \begin{bmatrix} \mathcal{E}[|\mathbf{v}(k)]_1^m]^{\frac{1}{m}} \\ \vdots \\ \mathcal{E}[|\mathbf{v}(k)]_n^m]^{\frac{1}{m}} \end{bmatrix} + \vec{d1}$$

$$+ \bar{z}_f \mathcal{E}[\max\{\|\mathbf{v}(k-\varrho+1)\|_\infty, \ldots, \|\mathbf{v}(k)\|_\infty\}^m]^{\frac{1}{m}} \vec{1} \quad (86)$$

But using the inductive assumption that $\mathcal{E}[|\mathbf{v}(j)]_i^m]^{\frac{1}{m}} \leq [v_m(j)]_i$ holds for $j \in \{0, \ldots, k\}$ and that:

$$\mathcal{E}[\max\{\|\mathbf{v}(k-\varrho+1)\|_\infty, \ldots, \|\mathbf{v}(k)\|_\infty\}^m]^{\frac{1}{m}}$$

$$\leq (n\varrho)^{\frac{1}{m}} \max_{j \in [k-\varrho+1,k]} \max_{i \in [1,n]} \mathcal{E}[|\mathbf{v}(j)]_i^m]^{\frac{1}{m}} \quad (87)$$

we can rewrite (86) as:

$$\begin{bmatrix} \mathcal{E}[|\mathbf{v}(k+1)]_1^m]^{\frac{1}{m}} \\ \vdots \\ \mathcal{E}[|\mathbf{v}(k+1)]_n^m]^{\frac{1}{m}} \end{bmatrix} \underset{element-wise}{\leq} A_{cl,m} v_m(k)$$

$$+ \vec{1}\left(\bar{d} + (n\varrho)^{\frac{1}{m}} \bar{z}_f\right)$$

$$\times \max\{\|v_m(k-\varrho+1)\|_\infty, \ldots, \|v_m(k)\|_\infty\} \quad (88)$$

Since the induction hypothesis is verified, we can use lemma 4.2 to finalize the proof. $\square$

Theorem 4.5 (*Sufficiency conditions for Robust m-th moment Stability*) *Let* A *be the dynamic matrix of (64),* $\varrho \in \mathbb{N}_+$, $\bar{z}_f \in [0,1)$, $\bar{z}_a \in [0,1)$ *and* $\bar{d} \geq 0$ *be given and* $H_m(k)$ *be defined as*

$$H_m(k) = A_{cl,m}^k, \ k \geq 0$$

where

$$A_{cl,m} = A_m \left(\bar{z}_a \vec{1}\vec{1}^T \right.$$

$$\left. + diag(2^{-C_1+\alpha_1(m)}, \ldots, 2^{-C_n+\alpha_n(m)})\right) \quad (89)$$

$[A_m]_{ij} = \mathcal{E}[|[A(k)]_{ij}|^m]^{\frac{1}{m}}$ *and* $[|A|]_{ij} = |[A]_{ij}|$. *Consider that* $\mathbf{x}(k)$ *is the solution of (64) under the feedback scheme of (76)-(77) as well as the following conditions:*

- **(C 1)** $\max_i \lambda_i(A_{cl,m}) < 1$
- **(C 2)** $(n\varrho)^{\frac{1}{m}} \bar{z}_f \|H_m\|_1 < 1$

*If conditions **(C 1)** and **(C 2)** are satisfied then the following holds:*

$$\mathcal{E}[[\bar{\mathbf{x}}(k)]_i^m]^{\frac{1}{m}} \le \|\tilde{g}_m(k)\|_\infty \frac{1}{2}$$

$$\|H_m\|_1 \left((n\varrho)^{\frac{1}{m}} \bar{z}_f \frac{\|H_m\|_1 \bar{d} + \|\tilde{g}_m\|_\infty \frac{1}{2}}{1 - (n\varrho)^{\frac{1}{m}} \bar{z}_f \|H_m\|_1} + \bar{d}\right)$$

$$(90)$$

where $\tilde{g}_m(k) = A_{cl,m}^k \vec{1}$.

Proof: We start by noticing that the condition (C1) is necessary and sufficient to guarantee that $\|H_m\|_1$ is finite. From definition 4.4, we have that:

$$v_m(k) = A_{cl,m}^k v(0) + \vec{1} \sum_{i=0}^{k-1} A_{cl,m}^{k-i-1} \left(\bar{d} + (n\varrho)^{\frac{1}{m}}\right)$$

$$\times \bar{z}_f \max\{\|v_m(i - \varrho + 1)\|_\infty, \ldots, \|v_m(i)\|_\infty\}$$

$$(91)$$

$$\|\pi_k \mathbf{v}_m\|_\infty$$

$$\le \|\tilde{g}_m\|_\infty \frac{1}{2} + \|H_m\|_1 \left((n\varrho)^{\frac{1}{m}} \bar{z}_f \|\pi_k \mathbf{v}_m\|_\infty + \bar{d}\right)$$

$$(92)$$

where we use $\|\pi_k \mathbf{v}_m\|_\infty = \max\{\|\mathbf{v}_m(0)\|_\infty, \ldots, \|\mathbf{v}_m(k)\|_\infty\}$ and $\tilde{g}_m(k) = A_{cl,m}^k \vec{1}$. By means of lemma 4.4, the formula (90) is obtained by isolating $\|\pi_k \mathbf{v}_m\|_\infty$ in (92) and substituting it back in (91). $\square$

4.7 Sufficiency for the case $\bar{z}_a = 0$

If we define

$$\beta_i(m) = \frac{1}{m} \log(\mathcal{E}[2^{m(\log(|\lambda_i(\mathbf{A}(k))|) - R_i)}])$$

where $R_i = \mathcal{E}[\log(|\lambda_i(\mathbf{A}(k))|)]$, then, after a few algebraic manipulations, condition (C1) of Theorem 4.5 can be written as:

$$C_i > R_i + \alpha_i(m) + \beta_i(m) \qquad (93)$$

Also, from condition (C2), we find that by increasing the difference $C_i - (R_i + \alpha_i(m) + \beta_i(m))$ we reduce $\|H_m\|_1$ and that improves robustness to uncertainty as measured by $\bar{z}_f$.

4.8 Solving the Allocation Problem for a Class of Stochastic Systems

Given $\bar{d} > 0$, $\varrho \in \mathbb{N}_+$ and $\bar{z}_f \in [0, 1)$, consider the following n-th order state-space representation:

$$\mathbf{x}(k+1) = \mathbf{J}(k)\mathbf{x}(k) + \mathbf{u}(k) + \mathbf{d}(k) + G_f(\mathbf{x})(k) \qquad (94)$$

where G_f is a causal operator satisfying $\|G_f\|_{\infty(\varrho)} \le \bar{z}_f$, $\|\mathbf{d}\|_\infty \le \bar{d}$ and $\mathbf{J}(k)$ is a real Jordan block of the form:

if $\mathbf{a}(k)$ is real

$$\mathbf{J}(k) = \begin{bmatrix} \mathbf{a}(k) & 1 & \cdots & 0 & 0 \\ 0 & \mathbf{a}(k) & 1 & \ddots & 0 \\ \vdots & & \ddots & & \vdots \\ 0 & 0 & 0 & 0 & \mathbf{a}(k) \end{bmatrix} \qquad (95)$$

otherwise

$$\mathbf{J}(k) = \begin{bmatrix} |\mathbf{a}(k)|\mathbf{Rot}(k) & I & \cdots & 0 \\ 0 & |\mathbf{a}(k)|\mathbf{Rot}(k) & I & \ddots \\ \vdots & & & \ddots \end{bmatrix}$$

$$(96)$$

where $\mathbf{Rot}$ is an i.i.d. sequence of random rotation matrices.

Take the scheme of section 2.5 as a starting point. Assume also that $\mathbf{r}(l)$ is always a multiple of n, i.e., $r(l) \in \{0, n, 2n, \ldots, \bar{r}\}$. In order to satisfy this assumption, we only need to adapt the scheme of section 2.5 by selecting packets whose size is a multiple of n. By doing so, we can modify the encoding/decoding scheme of section 2.5.2 and include in each packet an equal number of bits from each $[\vec{\mathbf{r}}(l)]_i$. By including the most important bits in the highest priority packets, we guarantee that each $[\vec{\mathbf{r}}(l)]_i$ corresponds to the instantaneous rate of a truncation operator. As such, we adopt the following allocation:

$$[\vec{\mathbf{r}}(l)]_i = \frac{\mathbf{r}(l)}{n} \qquad (97)$$

where we also use C_i and define the zero mean i.i.d. random variable $[\vec{\mathbf{r}}^\delta(l)]_i$, satisfying:

$$[\vec{\mathbf{r}}(l)]_i = C_i - [\vec{\mathbf{r}}^\delta(l)]_i \qquad (98)$$

From the definition of $\alpha_i(m)$ and $\alpha(m)$, the parameters characterizing the allocation (97) and $\mathbf{r}(l)$ are related through:

$$\alpha_i(m) = \frac{1}{m} \log \mathcal{E}[2^{-m[\vec{\mathbf{r}}^\delta(l)]_i}]$$

$$= \frac{1}{n} \frac{n}{m} \log \mathcal{E}[2^{-\frac{m}{n}\mathbf{r}^\delta(l)}] = \frac{1}{n}\alpha(\frac{m}{n}) \qquad (99)$$

$$C_i = \frac{1}{n}C \qquad (100)$$

We also adopt $\beta(m)$ according to section 4.7:

$$\beta(m) = \frac{1}{m} \log \mathcal{E}[2^{m\mathbf{1}_a^\delta(k)}] \qquad (101)$$

where $\log |\mathbf{a}(k)| = R + \mathbf{1}_a^\delta(k)$.

The following Proposition shows that, under the previous assumptions, the necessary condition derived in (N. Martins and Dahleh, 2004) is not conservative.

Proposition 4.6 *Let $\varrho \in \mathbb{N}_+$ be a given constant. If $C - \alpha(\frac{m}{n}) > n\beta(m) + nR$ then there exists constants $\bar{d} > 0$ and $\bar{z}_f \in [0, 1)$ such that the state-space representation (94) can be robustly stabilized in the m-th moment sense.*

Proof: From Proposition 4.1, we know that (94) can be written in the form (64)-(65). From section 4.7 we know that we can use Theorem 4.5 to guarantee that the following is a sufficient condition for the existence of $\bar{d} > 0$ and $\bar{z}_f \in [0, 1)$ such that (94) is robustly stabilizable:

$$C_i > \alpha_i(m) + \beta_i(m) + R_i \qquad (102)$$

where, in this case, $R_i = R$ and $\beta_i(m) = \beta(m)$ is given by (101). On the other hand, by means of (99)-(100), the assumption $C - \alpha(\frac{m}{n}) > n\beta(m) + nR$ can be written as (102). $\square$

ACKNOWLEDGEMENTS

The authors would like to thank Prof. Sekhar Tatikonda (Yale University) for providing pre-print papers comprising some of his most recent results. We also would like to thank Prof. Sanjoy Mitter (M.I.T.) for interesting discussions. This work was sponsored by the University of California - Los Angeles, MURI project title: *"Cooperative Control of Distributed Autonomous Vehicles in Adversarial Environments"*, award: 0205-G-CB222. Nuno C. Martins was supported by the Portuguese Foundation for Science and Technology and the European Social Fund, PRAXIS SFRH/BPD/19008/2004/JS74. Nicola Elia has been supported by NSF under the Career Award grant number ECS-0093950.

REFERENCES

Borkar, V. S. and Mitter, S. K. (1997). Lqg control with communication constraints. In *Communications, Computation, Control and Signal Processing:a tribute to Thomas Kailath*. Kluwer Academic Publishers.

Elia, N. (2002). Stabilization of systems with erasure actuation and sensory channels. In *Proceedings of 40th Allerton Conference*. UIUC.

Elia, N. (2003). Stabilization in the presence of fading channels. In *Proceedings of the American Control Conference*. ACC.

G. Nair, R. J. E. (2000). Stabilization with data-rate-limited feedback: Tightest attainable bounds. In *Systems and Control Letters, Vol 41, pp. 49-76*.

G. Nair, S. Dey, R. J. E. (2003). Infimum data rates for stabilizing markov jump linear systems. In *Proc. IEEE Conference on Decision and Control, pp. 1176-81.* IEEE.

Halmos, P. R. (1974). *Measure Theory*. Springer Verlag.

Liberzon, D. (2003). On stabilization of linear systems with limited information. In *IEEE Transactions on Automation and Control, Vol 48(2), pp. 304-7*.

M. Heusse, F. Rousseau, G. B.-S. A. D. (2003). sperformance anomaly of 802.11b. In *Proc. IEEE Infocom*. IEEE.

N. C. Martins, S. Venkatesh, M. A. D. (2001). Controller design and implementation for large scale systems, a block decoupling approach. In *Proceedings of the American Control Conference*. ACC.

N. Martins, N. E. and Dahleh, M. A. (2004). Feedback stabilization of uncertain systems using a stochastic digital link. In *Proceedings of the IEEE Conference on Decision and Control*. IEEE.

R. Horn, C. J. (1985). *Matrix Analysis*. Cambridge University Press, New York, NY:, 2nd edition.

R. Jain, T. Simsek, P. V. (2002). Control under communication constraints. In *Proc. IEEE Conference on Decision and Control, Vol. 5, pp. 3209-15*. IEEE.

R. W. Brocket, D. L. (2000). Quantized feedback stabilization of linear systems. In *IEEE Trans. Automat. Control, vol 45, pp. 1279-1289*. IEEE.

Sahai, A. (1998). Evaluating channels for control: Capacity reconsidered. In *Proceedings of the American Control Conference*. ACC.

Sahai, A. (2001). Anytime information theory. In *Ph.D. Dissertation*. M.I.T.

Tatikonda, S. (2000a). Control under communication constraints. In *Ph.D. Dissertation*. M.I.T.

Tatikonda, S. (2000b). Control under communication constraints: Part i and ii. In *IEEE Trans. Automat. Control (submitted)*. IEEE.

W. S. Wong, R. W. B. (1997). Systems with finite communication bandwidth constraints -i: State estimation problems. In *IEEE Trans. Automat. Control, Vol 42, pp. 1294-1298*. IEEE.

W. S. Wong, R. W. B. (1999). Systems with finite communication bandwidth constraints -ii: Stabilization with limited information feedback. In *IEEE Trans. Automat. Control, Vol 44, No. 5 pp. 1049-1053*. IEEE.

Witsenhausen, H. (November 1971). Separation of estimation and control for discrete-time systems. In *Proceeding of the IEEE, Volume 59, No 11*. IEEE.

X. Li, W. S. W. (1996). State estimation with communication constraints. In *Systems and Control Letters 28, pp. 49-54*.

PART 1

Intelligent Control Systems and Optimization

DEVICE INTEGRATION INTO AUTOMATION SYSTEMS WITH CONFIGURABLE DEVICE HANDLER

Anton Scheibelmasser, Udo Traussnigg

Department of Automation Technology,CAMPUS 02, Körblergasse 111, Graz, Austria
Email: : anton.scheibelmasser@campus02.at, udo.traussnigg@campus02.at

Georg Schindin, Ivo Derado

Test Bed Automation and Control Systems, AVL List GmbH, Hans List Platz 1, Graz, Austria
Email: georg.schindin@avl.com, ivo.derado@avl.com

Keywords: Measurement device, automation system, Configurable Device Handler, test bed.

Abstract: One of the most important topics in the field of automation systems is the integration of sensors, actuators, measurement devices and automation subsystems. Especially automation systems like test beds in the automotive industry impose high requirements regarding flexibility and reduced setup and integration time for new devices and operating modes. The core function of any automation system is the acquisition, evaluation and control of data received by sensors and sent to actuators. Sensors and actuators can be connected directly to the automation systems. In this case they are parameterised using specific software components, which determine the characteristics of every channel. In contrast to this, smart sensors, measurement devices or complex subsystems have to be integrated by means of different physical communication lines and protocols. The challenge for the automation system is to provide an integration platform, which will offer easy and flexible way for the integration of this type of devices. On the one hand, a sophisticated interface to the automation system should trigger, synchronise and evaluate values of different devices. On the other hand, a simple user interface for device integration should facilitate the flexible and straightforward device integration procedure for the customer. Configurable Device Handler is a software layer in the automation system, which offers a best trade-off between the complex functionality of intelligent devices and their integration in a simple, fast and flexible way. Due to a straightforward integration procedure, it is possible to integrate new devices and operation modes in a minimum of time by focusing on device functions and configuring the automation system, rather than writing software for specific device subsystems. This article gives an overview of Configurable Device Handler, which was implemented in the test bed automation system PUMA Open developed at the AVL. It provides an insight into the architecture of the Configurable Device Handler and shows the principles and the ideas behind it. Finally, new aspects and future developments are discussed.

1 INTRODUCTION

The general task of an automation system is to control a system in a defined mode of operation. In order for the automation system to perform this task, a number of sensors and actuators have to be evaluated and controlled. Since the timing of the automation system is critical, its software has to establish real time data acquisition and control. A further requirement is to evaluate and calculate results based on the acquired data and to store them. Particularly, automation systems used in the field of CAT (Computer Aided Testing, e.g. combustion engine development), called test beds, have to store the acquired data during the test run into a database, so that the developer can calculate specific quantities for quality assurance or optimisation. Systems under test are monitored and controlled by various devices connected to the automation systems.

Concerning the complexity of the data acquisition and control, devices can be divided into two categories. The first category of devices can be described as simple actuators (e.g. valves) or sensors (e.g. temperature sensor Pt100), which are connected to an input/output system as part of the automation

53

J. Braz et al. (eds.), Informatics in Control, Automation and Robotics I, 53–60.
© 2006 *Springer. Printed in the Netherlands.*

system. Since the sensors and actuators are under the control of the real time system, they are usually completely integrated into the automation system. Depending on the use of the sensor/actuator, specific parameterisation (e.g. acquisition rate, filtering or buffering) has to be performed by the customer.

The second category could be described as intelligent subsystems (e.g. measurement devices). In contrast to the first category, they are controlled by a local microprocessor, which provides functions comparable to those of the automation system (e.g. real time data acquisition or statistical calculation). Devices of this group could be qualified as finite automata with several internal states and transitions. Usually they are connected to the automation system via physical line (e.g. RS232, Ethernet). Consequently, data acquisition and control is possible only via this physical line and by using diverse communication protocols on it.

These various types of devices have to be uniformly integrated into the automation system, so that data acquisition and device control can be performed in a common way. Thus, automation systems typically contain a software layer, with the main task to make device specific functions available via standard automation interfaces. We refer to this software layer as Device Handler (see figure 1). Specifically, there are two main automation interfaces in PUMA Open (AVL, 2004), which are used by Device Handler, i.e., Platform Adapter and Device Framework.

The former interface offers functions compatible to those of the ASAM-GDI Platform Adapter (ASAM, 2001). There are two main advantages of the Platform Adapter. Firstly, it abstracts the complexity of lower ISO/OSI layers 1-6 (ISO, 1990) and it provides to the client a generic interface where common read/write commands from/to the device are independent from the specific lower ISO/OSI layers (TCP/IP over Ethernet, RS232, CAN, etc.). Secondly, it provides standard OS-services (e.g. memory handling and synchronisation) and thus enables the client to be independent from the specific OS. Consequently, the client, i.e. the Device Handler, deals exclusively with device specific functions and it is therefore robust to changes of lower communication layers and/or OS's.

The later interface must be implemented by the Device Handler and it contains services that are to be used by the automation system, i.e., user. This comprises services, such as: handling of device channels (System Channels), device parameterisation, support of the standard Device Handler's state machine, data storage, etc.

2 DEVICE HANDLER TYPES

One of the most important aspects of the automation system is the synchronisation of all test bed components (software and hardware) in order to perform specific control and/or measurement tasks (e.g. power-up all devices or get ready to start the measurement of the system under test). As mentioned in the previous chapter, all devices are synchronised due to the fact that all Device Handlers behave in a uniform way, which is ensured by supporting the state machine, i.e., states and transitions of the Device Framework interface. For instance, if the automation system wants to perform a measurement, it simply invokes the transition *StartMeasurement* of each Device Handler. The Device Handler interprets this command in a device specific way and accordingly sends one or more protocol frames to the device. Depending on the physical connection (e.g. RS232, CAN), the protocol mode (e.g. bus, peer to peer), the communication mode (e.g. master-slave, burst-mode) and the functionality (e.g. measure, calibrate), one could distinguish between various families of devices, i.e., Device Handlers.

As a result of this, the device is switched to the intended state (e.g. measurement mode) and is able to perform the specific activities. Acquired data are analysed and accordingly the transition is performed, the values of System Channels are updated, etc.

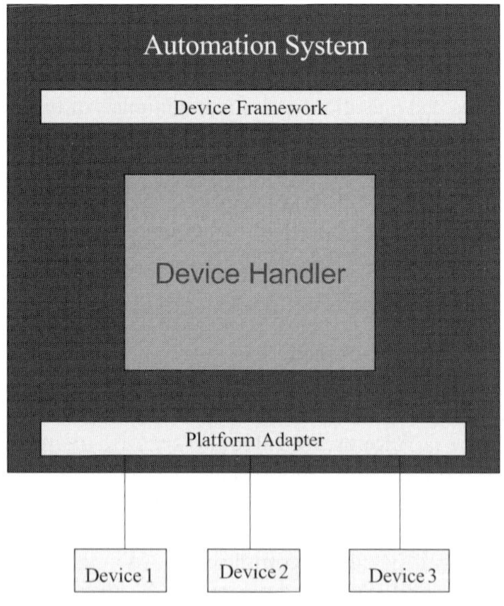

Figure 1: Device Handler.

A vital part of the Device Handler is its visualisation, or graphical user interface (GUI). Typically, it is implemented as a separate component and provides a visualisation for services, such as device parameterisation, maintenance, or visualisation of on-line values.

From the implementation'spoint of view, we can identify two types of Device Handler (see figure 2).

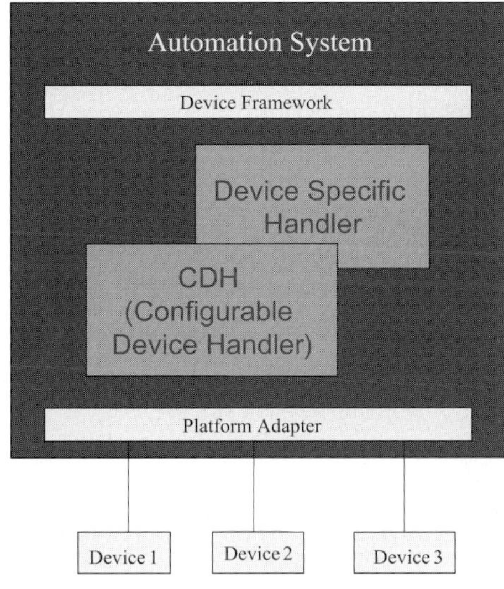

Figure 2: Device Handler types.

2.1 Device Specific Handler

In case of very specific and high complex devices, with sophisticated GUI requirements, it is hardly possible to specify the device functions in a generic manner. For this kind of devices, it is more appropriate and efficient to implement the corresponding Device Handler within traditional software development projects. This implies that all device functions (e.g. protocol, logic, GUI, state-machine) are implemented hard-coded at compile time. Therefore, the functions of the handler are fixed. In case of changing customer requirements or firmware changes on the device, the software has to be modified and compiled again. Hence, there is no flexibility concerning customer or application specific modification. Only device parameters which are handled in the automation system as System Channels, can be customized.

Moreover, a person responsible for the integration of the device into the automation system has to be not only familiar with the device specifics, but also with programming language and the

software object model behind the automation interfaces.

2.2 Configurable Device Handler

The idea behind the Configurable Device Handler (CDH) is to simplify and speed up the integration of devices by configuring the automation systems and thus allowing the responsible person (e.g. customer) to focus on device functions and not on object models behind automation interfaces. In order to achieve this, a generic Device Handler was implemented, which can cover general device functions, such as RS232 or TCP/IP connection, master-slave protocol, ASCII protocol, etc. During the configuration procedure specific device functions are identified. The mapping of device functions to automation interfaces is done automatically with the instantiation of the generic Device Handler including device specific information. Therefore, there is no need for programming or learning a programming language and object models behind automation interfaces. The information gained during the configuration procedure is stored in a so called Measurement Device Description (MDD) file. As the content of this file incorporates only the necessary device specific information, platform independent device integration is achieved. The MDD file can be saved as an ASCII file and it could be used on other operating system platforms, if similar generic handlers are installed.

The generated MDD file is stored together with all the other parameters in the automation system's database.

For each MDD file in the database exists a corresponding instance of the generic Device Handler, which is instantiated at start-up time of the automation system. The automation system does not distinguish between generic and specific Device Handlers.

2.3 Related Work

In the past the integration of devices in automation systems was performed typically by developing the device handler in the specific programming language and development environment. This work was done as a part of the service provided by the developer of the automation system. The cost of the integration was significant, especially if the device was developed by the customer or third party.

In the 90^s the scripting technology (Clark, 1999) (Schneider, 2001) (Wall, 2000) (Hetland, 2002) emerged in most automation systems in the automotive industry. Its popularity was primarily due to the fact that the customisation of automation systems was possible at the customer site. However, although the flexibility has increased, the costs of the integration were lower only at the first sight. The changes done in the automation system, at the customer site, had to be integrated back into the product, i.e., the product development process had to be followed to support the customer also in the maintenance phase. In addition, the changes could only be done by people with the skills in programming and automation system, which is again the personal of the provider of the automation system.

Exactly these problems were the motivation for the development of the CDH. The authors are not aware of any similar concept for the integration of devices in any other automation system, i.e., in test beds, especially in the automotive industry.

The concepts, as introduced in the product LabVIEW (National Instruments, 2003) with the graphical programming language G or TestPoint (Capital Equipment Corporation, 2003), offers an easy-to-use graphical environment to the customer for the development of instrumentation panels, device drivers and monitors of technical processes on the PC-based platform. Nevertheless, these tools are suitable only for the development of simple automation systems and not for the integration in large automation systems. Moreover, the graphical programming languages require more than a moderate amount of programming skills.

Standards, such as ASAM-GDI or IVI (IVI Foundation, 2003), specify interfaces between the Device Handler and the automation system. The art of the development of Device Handlers is out of the scope of these standards. CDH is compatible with these standards, because it identifies similar interfaces and functions in the automation system. In the chapter 6.5 we discuss the possibility of CDH supporting the ASAM-GDI standard.

3 CDH DEVICE INTEGRATION

Figure 3 shows the component view of the CDH. It comprises the following components: Configurable Device Generator (CDG), Configurable Device Engine (CDE), Configurable Device Panel (CDP), and finally the MDD file. The main features of these components are described in the following sections.

In order to achieve a device integration using the CDH, first the configuration procedure has to be performed, i.e., the device functions must be defined and saved in the MDD file. There are two main prerequisites for this task to be fulfilled. Firstly, the

person responsible for the device integration (in the following: device integrator) has to have the knowledge about the device functions. This implies that the device states, the intended device modes and the necessary protocol frames are well known. Secondly, it is necessary to understand the standard Device Handler's state machine (see fig. 5) defined within the Device Framework in order to integrate the device functions into the automation system appropriately. This enables the correct synchronisation of the device by the automation system. With the help of the CDG component, the device integrator can perform the creative part of the device integration in a straightforward manner.

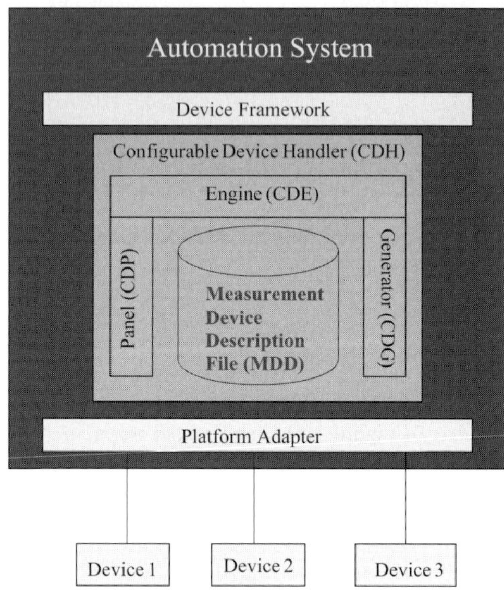

Figure 3: CDH components.

4 OFFLINE CONFIGURATION

The main part of the device integration using the CDH is the configuration procedure, where the device specifics are defined and saved in form of the MDD file. These activities are supported with the CDG, which conducts the device integrator by dividing the integration in several precise and clear defined steps. Hence, the configuration procedure can be mastered after a short time and in an intuitive way, and it is therefore especially suitable for customers with good knowledge of the device, but less of the automation system. At the beginning, the CDG steps are followed in a sequential way. Later as the configuration procedure progresses, it makes sense to use these steps interchangeably. In the following sections the steps are described in more details. However, it is out of the scope of this document to describe each attribute and feature, and therefore only excerpts are shown.

4.1 Physical Line

First, the device name used internally in the automation system is specified, followed by the definition of the parameters for the physical line (e.g. RS232). The following structure shows an example for the RS232 parameter definition.

```
Type: RS232
   Baudrate: 9600
   Bits: 8
   Parity: None
   Stop bit: One
   Port number: COM1
   TimeOut: 1000 [msec]
```

An excerpt from the definition of a physical line

4.2 Device Variables

For every value from the physical device, which should be available in the automation system, a name and several characteristics, such as *Unit, Minimum, Maximum,* data *Type*, and *Initial Value* have to be defined. The following description gives an example:

```
Value: FB_Temperature
   Unit: °C
   I/O-Type: Output
   Type: Float
   Initial Value: 0
   Minimum: -10
   Maximum: 70
```

An excerpt from the definition of a Device Variable

The attribute *I/O-Type* denotes the device variable either as Output (device defines its value) or Input (device needs a value from the automation system, i.e., from some other SW or HW component). This attribute is set automatically by the CDG as described in the following section 4.3.

4.3 Telegrams

Since the access to the physical device occurs exclusively via the communication line, each value and command has to be transmitted by the appropriate communication telegram. Therefore, each telegram, which is used for control or data acquisition has to be defined in this configuration step. The following example shows the definition structure of a command and a data inquiry telegram, using AK protocol commands (Arbeitskreis, 1991):

```
Telegram: SetToRemote
Type: Send and receive
Send: <02> SREM K0<03>
Receive: <02> SREM  #ERROR_Status#
$AK_Error$<03>

Telegram: GetMeasResult
Type: Send and receive
Send: <02> AERG K0<03>
Receive: <02> AERG #ERROR_Status#
#FB_MeasCycle# #ignore#
#FB_MeasTime# #FB_MeasWeight#<03>
```

Simplified example for two telegrams

The telegram definition can contain the definition for two protocol frames (one to send and one to receive), or for only one (only send or only receive). A minimum syntax is used to define a protocol frame. This includes, e.g., the definition of fixed (# #) and optional ($ $) position for a device variable's value and the definition of not-readable ASCII characters (< >). This syntax could be extended with the more powerful pattern-matching techniques for text processing, such as regular expressions (Friedl, 2002).

A device variable is denoted as an Output variable, if it used exclusively in receive protocol frames, otherwise it is an Input variable (see 4.2).

Using telegram definitions, CDH can send, receive and analyse protocol frames at run-time. Failures in terms of transmission timeout or parsing error are handled within the execution model of the CDH (within CDE) and are mapped to an error handler as described later.

4.4 Sequences

Transitions in the Device Framework's state machine that trigger and synchronise complex device activities are usually implemented by the Device Handler with more than one communication telegram. The order of the different telegrams and their correct invocation ensures the right device behaviour. Therefore, there is a need to define the logical order.

From the programming point of view three elementary software constructs are sufficient to support this, i.e., the commands, the conditional branches, and the jump commands or loops.

According to this fact, the CDG offers elements to define telegram invocation orders without the need for classic programming.

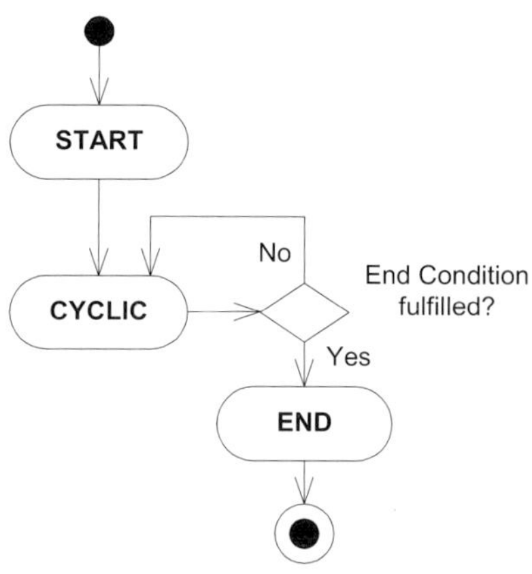

Figure 4: Simplified model of the Sequence execution.

Moreover, to facilitate the typical order patterns in the automation, the telegram invocation order is organised in three blocks, which together constitute the CDH Sequence (see fig. 4).

The first block called *Start* defines a sequence of telegrams, which is executed only once. The second block, called *Cyclic*, allows the execution of telegrams cyclically until the end loop condition is fulfilled. A third, optional *End* block is used to concatenate Sequences, depending on whether a current Sequence ended with success or failure. The *End* block and Global Conditions (see section 4.5), support the error handling in the CDH handler.

At run-time, the invoked transition in the Device Framework triggers the execution of the Sequence by the CDE component. The execution is done according to the execution model for the MDD structures (Sequence, telegram,…). The description of the execution model is out of the scope of this document. Hence, the Sequences specified by the user are actually implementations of the transitions in the Device Framework's state machine. The following description gives an example for a Sequence *Measure,* which is invoked, when the automation system triggers the start of a measurement:

```
Sequence: Measure
  Start Block:
   If Condition:
    IF RequestArgument <= 0 THEN
    Invoke Sequence NotOk AND
    Display INFO Message:
```

```
   "Measurement mode not supported"
  Function:
   SetChannelValue(PARA_MeasTime,
    #RequestArgument)
 Cyclic Block:
  If Condition:
   IF FT_ControlSystem < 0 THEN
    Invoke Telegram ADAT_cascade
  Cycle Time: 1000 msec
  End Condition: TIME = PARA_MeasTime
 End Block:
   Sequence if OK: Online
   Sequence if NOT OK: AnalyseFailure
```

Example of a Sequence

4.5 Global Conditions

A number of protocol frames require the same reactions (e.g. error handling). Therefore, every protocol frame has to be checked whether, e.g., an exception has occurred or not. In order to reduce the effort of writing a number of same conditions, the CDG offers an additional definition step called Global Condition.

Conditions defined in this step are checked automatically at run-time whenever a telegram is executed. If the condition is true the corresponding reaction is invoked (e.g. telegram or Sequence execution).

5 ON-LINE USAGE

At run-time the automation system loads MDD files from the database and generates a CDH instance for each of them. The communication to the device is established according to the parameters of the corresponding physical line. Every state transition in the Device Handler's state machine triggers the execution of the corresponding CDH Sequence. Figure 5 shows the simplified description of the state machine.

The CDE component interprets the content of the corresponding MDD file and executes Sequences according to the specified execution model. Consequently, the defined telegrams, conditions and functions are executed, in order to control the physical device appropriately. Additionally, for every device variable defined in the MDD file, the CDE generates a System Channel if needed, and provides them with values received from the telegram frames.

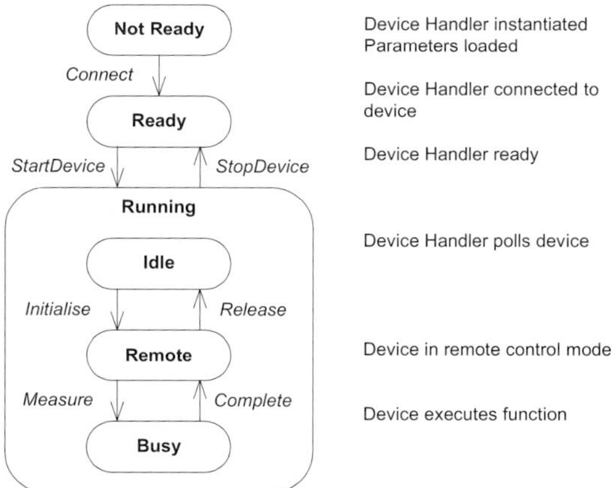

Figure 5: Simplified Device Handler's state machine.

The CDP component provides a graphical view on active System Channels and their values and the possibility to trigger each Sequence manually. As shown in figure 6, the CDP offers a common GUI for all CDH instances, which can be used for service or maintenance reasons.

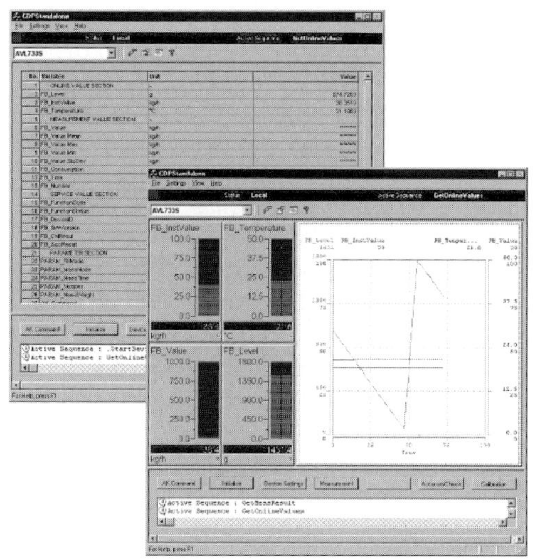

Figure 6: CDP visualisation.

6 FURTHER DEVELOPMENT

The first implementations of the CDH were successfully done for a number of measurement devices and subsystems with RS232 or Ethernet (TCP/IP, UDP/IP) connection. The common protocol of these devices was restricted to ASCII protocol communication. Experience gained with the integration of these devices was the base for the further development issues described in the following sections.

6.1 Calculation Capability

The execution model of the CDE is restricted on simple value extraction or insertion in protocol frames. Since there are a number of protocols, which use checksums or CRC's, because of data security, it is necessary to perform such calculations inside the Device Handler. An additional aspect is the arithmetical calculation of different device values. By introducing a formula interpreter in the CDH, both examples can be solved.

6.2 Automatic Detection of Devices

The idea behind the automatic detection of devices is to automatically detect known devices on different communication lines. This feature reduces the logistical effort on test beds and it enables users to hook up optional devices on demand.

This requirement is fulfilled by introducing an appropriate Sequence in the CDH, which allows the automation system to detect known devices on arbitrary lines. Depending on the device identification, the automation system is able to link the appropriate Device Handler to the port where the device is connected.

6.3 Multi-Line Connection

At the moment, every CDH instance can be connected to the physical device only via one communication line. In order to support devices and subsystems, which communicate over more than one line (e.g. the communication via TCP/IP and UDP/IP port), it should be possible to define telegrams for different communication lines.

6.4 Binary Protocols

Currently, the CDH supports only ASCII protocol frames. Since there are a number of devices, which communicate with a binary protocol, this family of devices cannot be integrated using the CDH. Extending CDH with a capability to support binary protocol frames would significantly increase its versatility.

6.5 Capability Profiles

The CDH provides a minimum set of Sequences mapped to the transitions of the standard state machine of the Device Handler via the Device Framework interface, which is enough to synchronise devices (e.g. connect, start cyclic data acquisition, start/stop measurement). However, additional standard Sequences could be provided that would support standard profiles, such as device independent profiles (ASAM-DIP, 2002) defined by ASAM-GDI standard. This profile defines a general state model of a test run and the required transitions. Implementing this profile would imply that the Device Handler is interoperable on test bed systems of different vendors supporting this standard.

7 CONCLUSION

The integration of devices in automation systems is typically a complex procedure that requires not only a good knowledge of the device and the automation system, but also requires programming skills.

The concept of the CDH offers an alternative approach for the integration of less complex devices. The device integrator is able to focus on device functions and to integrate them into the automation system using a simple configuration procedure described in this document. Not only that the integration can be done at the customer site, but also the customer himself is in the position to integrate his or a third-party device and to maintain it.

The CDH was implemented in the automation system PUMA Open developed at AVL and has shown excellent results in the practice. The major number of in-house devices, specifically measurement devices, were integrated into the PUMA Open using the CDH and are productively in use by a number of OEM's and suppliers in the automotive industry. Moreover, customers have also integrated devices by means of the CDH by themselves and are using them in research and production.

The costs and the effort for the integration were significantly reduced and, at the same time, the quality of the integration has increased, since it was possible to focus on device capabilities and to work in the office with a device emulator.

Using this integration method, the device integration got very easy, if the device integrator understands the device well!

REFERENCES

Arbeitskreis der deutschen Automobilindustrie, 1991. UA Schnittstelle und Leitrechner, *V24/RS232 Schnittstelle – Grundsätzliches.*

ASAM e.v., 2001. *Introduction to ASAM-GDI*, rev. 10, www.asam.net/03_standards_05.php.

ASAM e.v., 2001. *ASAM-GDI Part A: Specification of the Device Capability Description of an ASAM-GDI Driver*, rev. 4.2.

ASAM e.v., 2001. *ASAM-GDI Part B: Specification of the Interface of an ASAM-GDI Driver*, rev. 4.2.

ASAM e.v., 2002. *Device Independent Profile (DIP) Specification*, rev. 4.2.

AVL List GmbH, 2004. *PUMAopen Test Bed Automation Software*, www.avl.com.

Capital Equipment Corporation, 2003. *Test Point*, version 5, www.cec488.com.

Clark, S., 1999. *VBScript : Programmer's Reference*, Wrox.

Friedl, J.E.F, 2002. *Mastering Regular Expressions*, 2nd ed., O'Reilly & Associates.

Hetland, M.L., 2002. *Practical Python*, 1st ed., APress

ISO, *1990. Overview and Architecture*, IEEE Std 802-1990.

IVI Foundation, 2003. Driver Architecture Specification, rev. 1.2, www.ivifoundation.org.

National Instruments, 2003. *LabVIEW*, version 7, www.labview.com/labview.

Schneider, F., 2001. T.A. Powell, *JavaScript: The Complete Reference*, McGraw-Hill Osborne Media.

Wall, L., Christiansen, T., Orwant, J., 2000. *Programming Perl*, 3rd ed., O'Reilly & Associates.

NON LINEAR SPECTRAL SDP METHOD FOR BMI-CONSTRAINED PROBLEMS : APPLICATIONS TO CONTROL DESIGN

Jean-Baptiste Thevenet
ONERA-CERT, 2 av. Edouard Belin, 31055 Toulouse, France
and UPS-MIP (Mathmatiques pour l'Industrie et la Physique), CNRS UMR 5640
118, route de Narbonne 31062 Toulouse, France, thevenet@cert.fr

Dominikus Noll
UPS-MIP, noll@mip.ups-tlse.fr

Pierre Apkarian
ONERA-CERT and UPS-MIP, apkarian@cert.fr

Keywords: Bilinear matrix inequality, spectral penalty function, trust-region, control synthesis.

Abstract: The purpose of this paper is to examine a nonlinear spectral semidefinite programming method to solve problems with bilinear matrix inequality (BMI) constraints. Such optimization programs arise frequently in automatic control and are difficult to solve due to the inherent non-convexity. The method we discuss here is of augmented Lagrangian type and uses a succession of unconstrained subproblems to approximate the BMI optimization program. These tangent programs are solved by a trust region strategy. The method is tested against several difficult examples in feedback control synthesis.

1 INTRODUCTION

Minimizing a linear objective function subject to bilinear matrix inequality (BMI) constraints is a useful way to describe many robust control synthesis problems. Many other problems in automatic control lead to BMI feasibility and optimization programs, such as filtering problems, synthesis of structured or reduced-order feedback controllers, simultaneous stabilization problems and many others. Formally, these problems may be described as

$$\begin{array}{ll} \text{minimize} & c^T x, \ x \in \mathbb{R}^n \\ \text{subject to} & \mathcal{B}(x) \preceq 0, \end{array} \qquad (1)$$

where $\preceq 0$ means negative semi-definite and

$$\mathcal{B}(x) = A_0 + \sum_{i=1}^{n} x_i A_i + \sum_{1 \leq i < j \leq n} x_i x_j B_{ij} \qquad (2)$$

is a bilinear matrix-valued operator with data $A_i, B_{ij} \in \mathbb{S}^m$.

The importance of BMIs was recognized over the past decade and different solution strategies have been proposed. The earliest ideas use interior point methods, imported from semidefinite programming (SDP), concave programming or coordinate descent methods employing successions of SDP subproblems. The success of these methods is up to now very limited, and even moderately sized programs with no more than 100 variables usually make such methods fail.

Significant progress was achieved by two recent approaches, both using successions of SDP tangent subproblems. In (Fares et al., 2002), a sequential semidefinite is used, programming (S-SDP) algorithm, expanding on the sequential quadratic programming (SQP) technique in traditional nonlinear programming. In (Noll et al., 2002; Apkarian et al., 2003; Fares et al., 2001) an augmented Lagrangian strategy is proposed, which was since then successfully tested against a considerable number of difficult synthesis problems (Fares et al., 2000; Apkarian et al., 2002; Apkarian et al., 2004; Noll et al., 2002). This approach is particularly suited when BMI problems are transformed to LMI-constrained programs with an additional nonlinear equality constraints:

$$\begin{array}{ll} \text{minimize} & c^T x, \ x \in \mathbb{R}^n \\ \text{subject to} & \mathcal{A}(x) \preceq 0 \\ & g(x) = 0. \end{array} \qquad (3)$$

Here $\mathcal{A}$ is an affine operator (i.e., (2) with $B_{ij} = 0$) with values in $\mathbb{S}^m$, and $g : \mathbb{R}^n \to \mathbb{R}^p$ incorporates a finite number of equality constraints. Notice that *every* BMI problem may in principle be transformed to the form (3), but the change will only be fruitful if it is motivated from the control point of view. The latter is for instance the case in robust or reduced order synthesis, where the projection lemma is successfully brought into play. For semidefinite programming, (Zibulevsky, 1996; Ben-Tal and Zibulevsky, 1997) propose a modified augmented Lagrangian method,

J. Braz et al. (eds.), Informatics in Control, Automation and Robotics I, 61–72.
© 2006 *Springer. Printed in the Netherlands.*

which in (Kocvara and Stingl, 2003) has been used to build a software tool for SDP called PENNON. The idea is further expanded in (Henrion et al., 2003) to include BMI-problems.

The present paper addresses cases where the passage from (1) to (3) seems less suited and a direct approach is required. Our algorithm solves (1) by a sequence of unconstrained subproblems via a trust region technique. Using trust regions is of the essence, since we have to bear in mind that (1) is usually highly non-convex. In consequence, local optimal methods risk failure by getting stalled at a local minimum of constraint violation. This refers to points x satisfying $\mathcal{B}(x)/\not\preceq 0$, where the algorithm makes no further progress. Such a case means failure to solve the underlying control problem, which requires at least x with $\mathcal{B}(x) \preceq 0$. This phenomenon may be avoided (or at least significantly reduced) with genuine second order strategies. In particular, trust region techniques do not require convexification of the tangent problem Hessian, a major advantage over line search methods. (Notice that in this situation, any approach to (1) using a succession of SDPs is bound to face similar difficulties due to the need to convexify the Hessian. On the other hand, if LMI-constrained subproblems with non-convex quadratic objectives $f(x)$ can be handled, such an approach is promising. This idea has been developed successfully in (Apkarian et al., 2004).

While our technique could be applied basically to *any* BMI-constrained program, we presently focus on automatic control issues. In particular, we give applications in static output feedback control under $\mathcal{H}_2$, $\mathcal{H}_\infty$ or mixed $\mathcal{H}_\infty/\mathcal{H}_2$ performance indicies. Two other difficult cases, multi-model and multi-objective synthesis, and synthesis of structured controllers, are presented.

The structure of the paper is as follows. In section 2, we present our spectral SDP method and discuss its principal features. Section 3 presents theory needed to compute first and second derivatives of spectral functions. Finally, in section 4, numerical experiments on automatic control issues are examined.

NOTATION

Our notation is standard. We let $\mathbb{S}^m$ denote the set of $m \times m$ symmetric matrices, M^T the transpose of the matrix M and $\mathrm{Tr}\, M$ its trace. We equip $\mathbb{S}^m$ with the euclidean scalar product $\langle X, Y \rangle = X \bullet Y = \mathrm{Tr}\,(XY)$. For symmetric matrices , $M \succ N$ means that $M - N$ is positive definite and $M \succeq N$ means that $M - N$ is positive semi-definite. The symbol $\otimes$ stands for the usual Kronecker product of matrices and vec stands for the columnwise vectorization on

matrices. We shall make use of the properties:

$$\mathrm{vec}\,(AXB) \;=\; (B^T \otimes A)\,\mathrm{vec}\,X, \qquad (4)$$

$$\mathrm{Tr}\,(AB) \;=\; \mathrm{vec}^T A^T \,\mathrm{vec}\,B, \qquad (5)$$

which hold for matrices A, X and B of compatible dimensions. The Hadamard or Schur product is defined as

$$A \circ B = ((A_{ij}B_{ij})). \qquad (6)$$

The following holds for matrices of the same dimension:

$$\mathrm{vec}\,A \;\circ\; \mathrm{vec}\,B = \mathrm{diag}(\mathrm{vec}\,A)\mathrm{vec}\,B, \qquad (7)$$

where the operator diag forms a diagonal matrix with $\mathrm{vec}\,A$ on the main diagonal.

2 NONLINEAR SPECTRAL SDP METHOD

In this chapter we present and discuss our method to solve BMI-constrained optimization problems. Section 2.1 gives the general out-set, while the subsequent sections discuss theoretical and implementational aspects of the method.

2.1 General Outline

The method we are about to present applies to more general classes of objective functions than (1). We shall consider matrix inequality constrained programs of the form

$$\begin{array}{ll} \text{minimize} & f(x),\, x \in \mathbb{R}^n \\ \text{subject to} & \mathcal{F}(x) \preceq 0, \end{array} \qquad (8)$$

where f is a class $\mathcal{C}^2$ function, $\mathcal{F} : \mathbb{R}^n \to \mathbb{S}^m$ a class $\mathcal{C}^2$ operator. We will use the symbol $\mathcal{B}$ whenever we wish to specialize to bilinear operators (2) or to the form (1), and we use $\mathcal{A}$ for affine operators.

Our method is a penalty/barrier multiplier algorithm as proposed in (Zibulevsky, 1996; Mosheyev and Zibulevsky, 2000). According to that terminology, a spectral penalty (SP) function for (1) is defined as

$$F(x,p) = f(x) + \mathrm{Tr}\,\Phi_p(\mathcal{F}(x)), \qquad (9)$$

where $\varphi_p : \mathbb{R} \to \mathbb{R}$ is a parametrized family of scalar functions, which generates a family of matrix-valued operators $\Phi_p : \mathbb{S}^m \to \mathbb{S}^m$ upon defining:

$$\Phi_p(X) := S\mathrm{diag}\left[\varphi_p\left(\lambda_i(X)\right)\right] S^T. \qquad (10)$$

Here $\lambda_i(X)$ stands for the ith eigenvalue of $X \in \mathbb{S}^m$ and S is an orthonormal matrix of associated eigenvectors. An alternative expression for (9) using (10) is:

$$F(x,p) = f(x) + \sum_{i=1}^{m} \varphi_p(\lambda_i(\mathcal{F}(x))). \qquad (11)$$

As φ_p is chosen strictly increasing and satisfies $\varphi_p(0) = 0$, each of the following programs (parametrized through $p > 0$)

$$\begin{array}{ll} \text{minimize} & f(x), \, x \in \mathbb{R}^n \\ \text{subject to} & \Phi_p(\mathcal{F}(x)) \preceq 0 \end{array} \qquad (12)$$

is equivalent to (8). Thus, $F(x,p)$ may be understood as a penalty function for (8). Forcing $p \to 0$, we expect the solutions to the unconstrained program $\min_x F(x,p)$ to converge to a solution of (8).

It is well-known that pure penalty methods run into numerical difficulties as soon as penalty constants get large. Similarly, using pure SP functions as in (9) would lead to ill-conditioning for small $p > 0$. The epoch-making idea of Hestenes (Hestenes, 1969) and Powell (Powell, 1969), known as the augmented Lagrangian approach, was to avoid this phenomenon by including a linear term carrying a Lagrange multiplier estimate into the objective. In the present context, we follow the same line, but incorporate Lagrange multiplier information by a nonlinear term. We define the augmented Lagrangian function associated with the matrix inequality constraints in (1) as

$$\begin{aligned} L(x,V,p) &= f(x) + \mathrm{Tr}\, \Phi_p(V^T \mathcal{F}(x)V), \quad (13) \\ &= f(x) + \sum_{i=1}^{m} \varphi_p(\lambda_i(V^T \mathcal{F}(x)V)). \end{aligned}$$

In this expression, the matrix variable V has the same dimension as $\mathcal{F}(x) \in \mathbb{S}^m$ and serves as a factor of the Lagrange multiplier variable $U \in \mathbb{S}^m$, $U = VV^T$ (for instance, a Cholesky factor). This has the immediate advantage of maintaining non-negativity of $U \succeq 0$. In contrast with classical augmented Lagrangians, however, the Lagrange multiplier U is not involved linearly in (13). We nevertheless reserve the name of an augmented Lagrangian for $L(x,V,p)$, as its properties resemble those of the classical augmented Lagrangian. Surprisingly, the non-linear dependence of $L(x,V,p)$ on V is not at all troublesome and a suitable first-order update formula $V \to V^+$, generalizing the classical one, will be readily derived in section 3.2.

Let us mention that the convergence theory for an augmented Lagrangian method like the present one splits into a local and a global branch. Global theory gives weak convergence of the method towards critical points from arbitrary starting points x, if necessary by driving $p \to 0$. An important complement is provided by local theory, which shows that as soon as the iterates reach a neighbourhood of attraction of one of those critical points predicted by global theory, and if this critical point happens to be a local minimum satisfying the sufficient second order optimality conditions, then the sequence will stay in this neighbourhood, converge to the minimum in question, and the user will not have to push the parameter p below

a certain threshold $\bar{p} > 0$. Proofs for global and local convergence of the augmented Lagrangian (AL) exist in traditional nonlinear programming (see for instance (Conn et al., 1991; Conn et al., 1996; Conn et al., 1993b; Conn et al., 1993a; Bertsekas, 1982)). Convergence theory for matrix inequality constraints is still a somewhat unexplored field. A global convergence result which could be adapted to the present context is given in (Noll et al., 2002). Local theory for the present approach covering a large class of penalty functions φ_p will be presented in a forthcoming article. Notice that in the convex case a convergence result has been published in (Zibulevsky, 1996). It is based on Rockafellar's idea relating the AL method to a proximal point algorithm. Our present testing of nonconvex programs (1) shows a similar picture. Even without convexity the method converges most of the time and the penalty parameter is in the end stably away from 0, $p \geq \bar{p}$.

Schematically, the augmented Lagrangian technique is as follows:

Spectral augmented Lagrangian algorithm

1. Initial phase. Set constants $\gamma > 0, \rho < 1$. Initialize the algorithm with x_0, V_0 and a penalty parameter $p_0 > 0$.

2. Optimization phase. For fixed V_j and p_j solve the unconstrained subproblem

$$\text{minimize}_{x \in \mathbb{R}^n} \quad L(x,V_j,p_j) \qquad (14)$$

Let x_{j+1} be the solution. Use the previous iterate x_j as a starting value for the inner optimization.

3. Update penalty and multiplier. Apply first-order rule to estimate Lagrange multiplier:

$$\begin{aligned} V_{j+1}V_{j+1}^T = V_j S \left[\mathrm{diag}\varphi_p' \left(\lambda_i \left(V_j^T \right.\right.\right. \\ \left.\left.\left. \cdot \mathcal{F}(x_{j+1})V_j)\right)\right] S^T V_j^T, \quad (15) \end{aligned}$$

where S diagonalizes $V_j^T \mathcal{F}(x_{j+1})V_j$. Update the penalty parameter using:

$$p_{j+1} = \left\{ \begin{array}{ll} \rho p_j, & \text{if } \lambda_{\max}(0, \mathcal{F}(x_{j+1})) \\ & > \gamma \lambda_{\max}(0, \mathcal{F}(x_j)) \\ p_j, & \text{else} \end{array} \right. \qquad (16)$$

Increase j and go back to step 2.

In our implementation, following the recommendation in (Mosheyev and Zibulevsky, 2000), we have used the log-quadratic penalty function $\varphi_p(t) = p\varphi_1(t/p)$ where

$$\varphi_1(t) = \left\{ \begin{array}{ll} t + \frac{1}{2}t^2 & \text{if } t \geq -\frac{1}{2} \\ -\frac{1}{4}\log(-2t) - \frac{3}{8} & \text{if } t < -\frac{1}{2}, \end{array} \right. \qquad (17)$$

but other choices could be used (see for instance (Zibulevsky, 1996) for an extended list).

The multiplier update formula (15) requires the full machinery of differentiability of the spectral function $\mathrm{Tr}\,\Phi_p$ and will be derived in section 3.2. Algorithmic aspects of the subproblem in step 2 will be discussed in section 3.3. We start by analyzing the idea of the spectral AL-method.

2.2 The mechanism of the Algorithm

In order to understand the rationale behind the AL-algorithm, it may be instructive to consider classical nonlinear programming, which is a special case of the scheme if the values of the operator $\mathcal{F}$ are diagonal matrices: $\mathcal{F}(x) = \mathrm{diag}\,[c_1(x), \ldots, c_m(x)]$. Then the Lagrange multiplier U and its factor V may also be restricted to the space of diagonal matrices, $U = \mathrm{diag}\,u_i$, $V = \mathrm{diag}\,v_i$, and we recover the situation of classical polyhedral constraints. Switching to a more convenient notation, the problem becomes

$$\begin{aligned}
\text{minimize} \quad & f(x) \\
\text{subject to} \quad & c_j(x) \leq 0, \ j = 1, \ldots, m
\end{aligned}$$

With $u_i = v_i^2$, we obtain the analogue of the augmented Lagrangian (13):

$$L(x, v, p) \;=\; f(x) + \sum_{i=1}^{m} p\varphi_1(v_i^2 c_i(x)/p).$$

Here we use (17) or another choice from the list in (Zibulevsky, 1996). Computing derivatives is easy here and we obtain

$$\nabla L(x, v, p) = \nabla f(x)$$
$$+ \sum_{i=1}^{m} \varphi_1'(v_i^2 c_i(x)/p) v_i^2 \nabla c_i(x).$$

If we compare with the Lagrangian:
$L(x, u) = f(x) + u^T c(x),$
and its gradient:
$\nabla L(x, u) = \nabla f(x) + \sum_{i=1}^{m} u_i \nabla c_i(x),$
the following update formula readily appears:

$$u_i^+ = v_i^{+2} = v_i^2 \varphi_1'(v_i^2 c_i(x^+)/p),$$

the scalar analogue of (15). Suppose now that φ_1 has the form (17). Then for sufficiently small $v_i^2 c_i(x)$ we expect $v_i^2 c_i(x)/p > -\frac{1}{2}$, so

$$\begin{aligned}
p\,\varphi_1(v_i^2 c_i(x)/p) \;=\; & p(v_i^2 c_i(x)/p \\
& + \frac{1}{2} v_i^4 c_i(x)^2/p^2) \\
=\; & u_i c_i(x) + \frac{1}{2}(v_i^4/p) c_i(x)^2.
\end{aligned}$$

If we recall the form of the classical augmented Lagrangian for inequality constraints (cf. (Bertsekas, 1982))

$$\begin{aligned}
\mathcal{L}(x, u, \mu) = {} & f(x) \\
& + \frac{\mu}{2} \sum_{i=1}^{m} \left(\max\left[0, u_i + c_i(x)/\mu\right]^2 - u_i^2 \right),
\end{aligned}$$

we can see that the term v_i^4/p takes the place of the penalty parameter $1/\mu$ as long as the $v_i's$ remain bounded. This suggests that the method should behave similarly to the classical augmented Lagrangian method in a neighbourhood of attraction of a local minimum. In consequence, close to a minimum the updating strategy for p should resemble that for the classical parameter μ. In contrast, the algorithm may perform very differently when iterates are far away from local minima. Here the smoothness of the functions φ_1 may play favorably. We propose the update (16) for the penalty parameter, but other possibilities exist, see for instance Conn et al., and need to be tested and compared in the case of matrix inequality constraints.

3 DERIVATIVES OF SP FUNCTIONS

In this section we obtain formulas for the first- and second-order derivatives of the augmented Lagrangian function (13). This part is based on the differentiability theory for spectral functions developed by Lewis et al. (Lewis, 1996; Lewis, 2001; Lewis and S.Sendov, 2002). To begin with, recall the following definition from (Lewis and S. Sendov, 2002).

Definition 3.1 *Let $\lambda : \mathbb{S}^m \to \mathbb{R}^m$ denote the eigenvalue map $\lambda(X) := (\lambda_1(X), \ldots, \lambda_m(X))$. For a symmetric function $\psi : \mathbb{R}^m \to \mathbb{R}$, the spectral function Ψ associated with ψ is defined as $\Psi : \mathbb{S}^m \to \mathbb{R}$, $\Psi = \psi \circ \lambda$.*

First order differentiability theory for spectral functions is covered by (Lewis, 1996). We will need the following result from (Lewis, 1996):

Lemma 3.2 *Let ψ be a symmetric function defined on an open subset D of $\mathbb{R}^n$. Let $\Psi = \psi \circ \lambda$ be the spectral function associated with ψ. Let $X \in \mathbb{S}^m$ and suppose $\lambda(X) \in D$. Then Ψ is differentiable at X if and only if ψ is differentiable at $\lambda(X)$. In that case we have the formula:*

$$\nabla \Psi(X) = \nabla(\psi \circ \lambda)(X) = S\,\mathrm{diag}\,\nabla\psi\,(\lambda(X))\,S^T$$

for every orthogonal matrix S satisfying $X = S\,\mathrm{diag}\,\lambda(X)\,S^T$. $\qquad\square$

The gradient $\nabla\Psi(X)$ is a (dual) element of $\mathbb{S}^m$, which we distinguish from the differential, $d\Psi(X)$. The latter acts as a linear form on tangent vectors $dX \in \mathbb{S}^m$ as

$$
\begin{aligned}
d\Psi(X)[dX] &= \nabla\Psi(X) \bullet dX \\
&= \mathrm{Tr}\left(S \operatorname{diag}\nabla\psi\left(\lambda(X)\right)S^T dX\right).
\end{aligned}
$$

It is now straightforward to compute the first derivative of a composite spectral functions $\Psi \circ \mathcal{F}$, where $\mathcal{F} : \mathbb{R}^n \to \mathbb{S}^m$ is sufficiently smooth. We obtain

$$
\nabla\left(\Psi \circ \mathcal{F}\right)(x) = d\mathcal{F}(x)^*\left[\nabla\Psi\left(\mathcal{F}(x)\right)\right] \in \mathbb{R}^n, \quad (18)
$$

where $d\mathcal{F}(x)$ is a linear operator mapping $\mathbb{R}^n \to \mathbb{S}^m$, and $d\mathcal{F}(x)^*$ its adjoint, mapping $\mathbb{S}^m \to \mathbb{R}^n$. With the standing notation

$$
F := \mathcal{F}(x), \quad F_i := \frac{\partial\mathcal{F}(x)}{\partial x_i} \in \mathbb{S}^m,
$$

we obtain the representations

$$
\begin{aligned}
d\mathcal{F}(x)[\delta x] &= \sum_{i=1}^{n} F_i \delta x_i, \\
d\mathcal{F}(x)^*[dX] &= \left(F_1 \bullet dX, \ldots, F_n \bullet dX\right).
\end{aligned}
$$

Then the ith element of the gradient is

$$
\begin{aligned}
&\left(\nabla\left(\Psi \circ \mathcal{F}\right)(x)\right)_i \\
&\quad = \mathrm{Tr}\left(F_i S \operatorname{diag}\nabla\psi\left(\lambda(F)\right)S^T\right). \quad (19)
\end{aligned}
$$

Observe that $F_i = A_i + \sum_{j<i} B_{ji}x_j + \sum_{i<j} B_{ij}x_j$ if $\mathcal{B}$ is of the form (2), $F_i = A_i$ for affine operators $\mathcal{A}$.

We are now ready for the first derivative of the augmented Lagrangian function $L(x, V, p)$. We consider a special class of symmetric functions $\psi : \mathbb{R}^n \to \mathbb{R}$,

$$
\psi(x) = \sum_{i=1}^{n} \varphi(x_i), \quad (20)
$$

generated by a single scalar function $\varphi : \mathbb{R} \to \mathbb{R}$. Here the spectral function $\Psi = \psi \circ \lambda$ is of the form $\Psi = \mathrm{Tr}\,\Phi$ with $\Phi : \mathbb{S}^m \to \mathbb{S}^m$ the matrix-valued operator

$$
\Phi(X) = S \operatorname{diag}\varphi\left(\lambda(X)\right)S^T,
$$

S being an orthogonal matrix diagonalizing X. Using φ_p, ψ_p and Ψ_p instead of φ, ψ and Ψ, the penalty term in (13) takes the form

$$
\begin{aligned}
\Psi_p(V^T\mathcal{F}(x)V)) &= \psi_p \circ \lambda(V^T\mathcal{F}(x)V)) \\
&= \sum_{i=1}^{m} \varphi_p(\lambda_i(V^T\mathcal{F}(x)V)) \\
&= \mathrm{Tr}\,\Phi_p(V^T\mathcal{F}(x)V),
\end{aligned}
$$

where Φ_p is given in (10). Now we apply Lemma 3.2, respectively formula (18) or even (19) to $\Psi_p =$

Tr Φ_p and the operator $\widetilde{\mathcal{F}}(x) = V^T\mathcal{F}(x)V$, where we observe *en passant* that

$$
\begin{aligned}
d\widetilde{\mathcal{F}}(x)^*[X] &= d\left(\left\{V^T\mathcal{F}V\right\}(x)\right)^*[X] \\
&= d\mathcal{F}(x)^*[VXV^T],
\end{aligned}
$$

or put differently, $\widetilde{F}_i = V^T F_i V$. This gives the following

Proposition 3.3 *With the above notations the gradient of $L(x, V, p)$ is given as*

$$
\begin{aligned}
\frac{\partial}{\partial x_i}L(x, V, p) &= \frac{\partial}{\partial x_i}f(x) \\
&+ \mathrm{Tr}\left(F_i\, V\, S \operatorname{diag}[\varphi_p'(\lambda_i(V^T FV))]\, S^T\, V^T\right).
\end{aligned}
$$

In vector notation

$$
\begin{aligned}
\nabla L(x, V, p) &= \nabla f(x) + \left[\operatorname{vec}\left(F_1\right), \ldots, \operatorname{vec}\left(F_n\right)\right]^T \\
&\cdot \operatorname{vec}\left(VS \operatorname{diag}[\varphi_p'(\lambda_i(V^T FV))]\, S^T V^T\right),
\end{aligned}
$$

where S is orthogonal and diagonalizes $V^T FV$. $\quad\square$

3.1 Second Derivatives

The second order theory of spectral function is more involved and presented in (Lewis, 2001; Lewis and S.Sendov, 2002). We require the following central observation.

Lemma 3.4 *With the same notations as above, $\Psi = \psi \circ \lambda$ is twice differentiable at $X \in \mathbb{S}^m$ if and only if ψ is twice differentiable at $\lambda(X) \in D$.* $\quad\square$

We specialize again to the class of spectral functions $\Psi = \mathrm{Tr}\,\Phi$ generated by a single scalar function φ, see (20). Then we have the following result imported from (Shapiro, 2002):

Lemma 3.5 *Let $\varphi : \mathbb{R} \to \mathbb{R}$ be a function on the real line, and let ψ be the symmetric function on $\mathbb{R}^n$ generated by (20) above. Let $\Psi = \mathrm{Tr}\,\Phi$ denote the spectral function associated with ψ. Suppose φ is twice differentiable on an open interval I. Then Ψ is twice differentiable at X whenever $\lambda_i = \lambda_i(X) \in I$ for every i. Moreover, if we define the matrix $\varphi^{'[1]} = \varphi^{'[1]}(X) \in \mathbb{S}^m$ as:*

$$
\varphi_{ij}^{'[1]} := \begin{cases} \frac{\varphi'(\lambda_i) - \varphi'(\lambda_j)}{\lambda_i - \lambda_j} & \text{if } \lambda_i \neq \lambda_j \\ \varphi''(\lambda_i) & \text{if } \lambda_i = \lambda_j \end{cases},
$$

then the following formula defines the second order derivative of Ψ at X

$$
d^2\Psi(X))[dX, dX] = \mathrm{Tr}\left(S^T dX S \varphi^{'[1]} \circ \{S^T dX S\}\right),
$$

where S is orthogonal and diagonalizes X. $\quad\square$

In order to obtain second derivatives of composite functions, $\Psi \circ \mathcal{F}$, we will require the chain rule, which for the second differential forms $d^2\Psi(X)[dX, dX]$ of Ψ and $d^2\mathcal{F}(x)[\delta x, \delta x]$ of $\mathcal{F}$ amounts to:

$$\begin{aligned}
d^2&(\Psi \circ \mathcal{F})(x)[\delta x, \delta x] \\
&= d\Psi(\mathcal{F}(x))\left[d^2\mathcal{F}(x)[\delta x, \delta x]\right] \\
&\quad + d^2\Psi(\mathcal{F}(x))\left[d\mathcal{F}(x)[\delta x], d\mathcal{F}(x)[\delta x]\right]. \quad (21)
\end{aligned}$$

For bilinear $\mathcal{B}$, the second order form $d^2\mathcal{B}(x)$ is of course independent of x and given as

$$d^2\mathcal{B}(x)[\delta x, \delta x] = \sum_{i<j} \delta x_i \delta x_j B_{ij}.$$

In particular, $d^2\mathcal{A} = 0$ for affine operators $\mathcal{A}$. During the following we shall use these results to obtain second derivatives of the augmented Lagrangian function $L(x, V, p)$.

As in the previous section, we apply the results to $\Psi_p = \operatorname{Tr} \Phi_p$ generated by φ_p and to the operator $\widetilde{\mathcal{F}}(x) = V^T \mathcal{F}(x) V$. As we have already seen, for $\widetilde{\mathcal{F}}$ we have the formula

$$\begin{aligned}
d\widetilde{\mathcal{F}}(x)[\delta x] &= \sum_{i=1}^n V^T F_i V \delta x_i \\
&= V^T \sum_{i=1}^n F_i \delta x_i V \\
&= V^T d\mathcal{F} V.
\end{aligned}$$

Therefore, using Lemma 3.5, the second term in (21) becomes

$$\begin{aligned}
d^2&\Psi_p(V^T FV)\left[V^T d\mathcal{F} V, V^T d\mathcal{F} V\right] \\
&= \operatorname{Tr}\left(S^T V^T d\mathcal{F} V S \, \varphi_p'^{[1]} \circ \{S^T V^T d\mathcal{F} V S\}\right),
\end{aligned}$$

where S is orthogonal and diagonalizes $V^T FV$ and $\varphi_p'^{[1]} = \varphi_p'^{[1]}(V^T FV)$. This may now be transformed to vector-matrix form using the operator vec and the relations (4) - (7).

$$\begin{aligned}
d^2&\Psi_p(V^T FV)[V^T d\mathcal{F} V, V^T d\mathcal{F} V] \\
&= \operatorname{vec}^T((VS)^T d\mathcal{F}(VS)) \\
&\quad \cdot \operatorname{vec}(\varphi_p'^{[1]} \circ \{(VS)^T d\mathcal{F}(VS)\}) \\
&= \operatorname{vec}^T(d\mathcal{F}) \, \mathcal{K}_p \, \operatorname{vec}(d\mathcal{F}), \quad (22)
\end{aligned}$$

where the matrix $\mathcal{K}_p$ is

$$\begin{aligned}
\mathcal{K}_p &= [(VS) \otimes (VS)]\operatorname{diag}\left\{\operatorname{vec} \varphi_p'^{[1]}\right\} \\
&\quad \cdot [(VS)^T \otimes (VS)^T].
\end{aligned}$$

Using $d\mathcal{F} = \sum_{i=1}^n F_i \delta x_i$, the last line in (22) gives

$$\begin{aligned}
d^2&\Psi_p(V^T FV)[V^T d\mathcal{F} V, V^T d\mathcal{F} V] \\
&= \delta x^T \left[\operatorname{vec}(F_1), \ldots, \operatorname{vec}(F_n)\right]^T \mathcal{K}_p \\
&\quad \cdot \left[\operatorname{vec}(F_1), \ldots, \operatorname{vec}(F_n)\right] \delta x,
\end{aligned}$$

so we are led to retain the symmetric $n \times n$ matrix

$$\begin{aligned}
H_1 &= \left[\operatorname{vec}(F_1), \ldots, \operatorname{vec}(F_n)\right]^T \mathcal{K}_p \\
&\quad \cdot \left[\operatorname{vec}(F_1), \ldots, \operatorname{vec}(F_n)\right]. \quad (23)
\end{aligned}$$

Let us now consider the second part of the Hessian $\nabla^2 L(x, V, p)$, coming from the term $d\Psi(\mathcal{F}(x))\left[d^2\mathcal{F}(x)[\delta x, \delta x]\right]$ in (21). First observe that trivially

$$\begin{aligned}
d^2\widetilde{\mathcal{F}}(x)[\delta x, \delta x] &= d^2\{V^T \mathcal{F} V\}(x)[\delta x, \delta x] \\
&= V^T d^2\mathcal{F}(x)[\delta x, \delta x] \, V.
\end{aligned}$$

Hence

$$\begin{aligned}
d&\Psi_p(V^T FV)\left[V^T d^2\mathcal{F}(x)[\delta x, \delta x]V\right] \\
&= \operatorname{Tr}\left(S \operatorname{diag}\varphi_p'\left(\lambda_i(V^T FV)\right) S^T \right. \\
&\qquad\qquad \left. \cdot V^T d^2\mathcal{F}(x)[\delta x, \delta x]V\right) \\
&= \operatorname{Tr}\left(V S \operatorname{diag}\varphi_p'\left(\lambda_i(V^T FV)\right) S^T V^T \right. \\
&\qquad\qquad\qquad \left. \cdot d^2\mathcal{F}(x)[\delta x, \delta x]\right). \quad (24)
\end{aligned}$$

Introducing matrices $F_{ij} = F_{ij}(x) \in \mathbb{S}^m$ to represent

$$d^2\mathcal{F}(x)[\delta x, \delta x] = \sum_{i,j=1}^n F_{ij}\delta x_i \delta x_j,$$

relation (24) may be written as

$$\begin{aligned}
d&\Psi_p(V^T FV)\left[V^T d^2\mathcal{F}(x)[\delta x, \delta x]V\right] \\
&= \sum_{i,j=1}^n \delta x_i \delta x_j \operatorname{Tr}\left(F_{ij} V S \operatorname{diag}\varphi_p'\left(\lambda_i(V^T \right.\right. \\
&\qquad\qquad\qquad\qquad \left.\left. \cdot FV)\right) S^T V^T\right).
\end{aligned}$$

Hence the interest to define a second symmetric $n \times n$ matrix H_2 by

$$(H_2)_{ij} = \operatorname{Tr}\left(F_{ij} V S \operatorname{diag}\varphi_p'\left(\lambda_i(V^T FV)\right) S^T V^T\right), \quad (25)$$

where as before S is orthogonal and diagonalizes $V^T FV$. We summarize by the following

Proposition 3.6 *The Hessian of the augmented Lagrangian function $L(x, V, p)$ is of the form*

$$\nabla^2 L(x, V, p) = \nabla^2 f(x) + H_1 + H_2,$$

where H_1, given by (23), is positive semidefinite. H_2 is given by (25).

Proof. The only element which remains to be proved is non-negativity of H_1, which hinges on the non-negativity of $\mathcal{K}_p$, and hence on that of $\varphi_p'^{[1]}$. The latter is a consequence of the convexity of φ_p, hence the result. $\qquad\square$

3.2 Multiplier Update Rule

The first-order multiplier update rule is derived similarly to the classical case of polyhedral constraints. For our analysis we will require the traditional Lagrangian of problem (8), which is

$$L(x, U) = f(x) + U \bullet \mathcal{F}(x). \qquad (26)$$

Suppose x^+ is the solution of problem (14) in step 2 of our algorithm. Let $U = VV^T$ the current Lagrange multiplier estimate. A new multiplier U^+ and therefore a new factor estimate V^+ with $U^+ = V^+V^{+T}$ is then obtained by equating the gradients of the augmented Lagrangian in (13) and the traditional Lagrangian of problem (1). Writing $L(x, U) = f(x) + \mathrm{Tr}\,(U\mathcal{F}(x)) = f(x) + \mathrm{Tr}\,(VV^T\mathcal{F}(x)) = f(x) + \mathrm{Tr}\,(V^T\mathcal{F}(x)V)$, and computing its gradient at x^+ and $U^+ = V^+V^{+T}$ implies

$$\nabla L(x^+, U^+) = \nabla f(x^+)$$
$$+ [\mathrm{vec}\,(F_1), \dots, \mathrm{vec}\,(F_n)]^T \mathrm{vec}\,(V^+V^{+T}).$$

Comparing with

$$\nabla L(x^+, V^+, p)$$
$$= \nabla f(x^+) + [\mathrm{vec}\,(F_1), \dots, \mathrm{vec}\,(F_n)]^T$$
$$\cdot \mathrm{vec}\,\left(V\,S\,\mathrm{diag}\,\varphi_p'(\lambda_i(V^T\mathcal{F}(x^+)V)S^TV^T\right)$$

suggests the update rule

$$V^+V^{+T} = VS\,[\mathrm{diag}\,\varphi_p'(\lambda_i(V^T\mathcal{F}(x^+)V))]\,(VS)^T, \qquad (27)$$

where S is orthogonal and diagonalizes $V^T\mathcal{F}(x^+)V$. This was already presented in our algorithm. The argument suggests the equality to hold only modulo the null space of $[\mathrm{vec}\,(F_1), \dots, \mathrm{vec}\,(F_n)]^T$, but the limiting case gives more information. Notice that equality

$$\nabla L(\bar{x}, \bar{U}) = \nabla L(\bar{x}, \bar{V}, p)$$

holds at a Karush-Kuhn-Tucker pair $(\bar{x}, \bar{U})$ and $\bar{U} = \bar{V}\bar{V}^T$ for every $p > 0$, a formula whose analogue in the classical case is well-known. That means we would expect (27) to hold at $x^+ = \bar{x}$, $V = V^+ = \bar{V}$. This is indeed the case since by complementarity, $\bar{V}^T\mathcal{F}(\bar{x})\bar{V} = 0$. With $\varphi_p'(0) = 1$ for every $p > 0$, the right hand side in (27) is $\bar{V}^TS^TS\bar{V} = \bar{V}^T\bar{V}$, hence equals the left hand side. We may therefore read (27) as some sort of fixed-point relation, which is satisfied in the limit $V = V^+$ if iterates converge. Notice, however, that convergence has to be proved by an extra argument, because (27) is not of contraction type.

Notice that by construction the right hand term in (27) is indeed positive definite, so V^+ could for instance be chosen as a Cholesky factor.

3.3 Solving the Subproblem - Implementational Issues

Efficient minimization of $L(x, V_j, p_j)$ for fixed V_j, p_j is at the core of our approach and our implementation is based on a Newton trust region method, following the lines of Lin and Moré (Lin and More, 1998). As compared to Newton line search algorithms or other descent direction methods, trust regions can take better advantage of second-order information. This is witnessed by the fact that negative curvature in the tangent problem Hessian, frequently arising in BMI-minimization when iterates are still far away from any neighbourhood of local convergence, may be taken into account. Furthermore, trust region methods often (miraculously) find good local minima, leaving the bad ones behind. This additional benefit is in contrast with what line search methods achieve, and is explained to some extent by the fact that, at least over the horizon specified by the current trust region radius, the minimization in the tangent problem is a global one.

As part of our testing, and at an early stage of the development, we studied the behavior of our code on a class of LMI problems, where we compared line search and trust region approaches. Even in that situation, where line search is supposedly at its best due to convexity, the trust region variant appeared to be more robust, though often slower. This is significant when problems get ill-conditioned, which is often the case in automatic control applications. Then the Newton direction, giving descent in theory, often fails to do so due to ill-conditioning. In that event recourse to first-order methods has to be taken. This is where a trust region strategy continues to perform reasonably well.

As currently implemented, both the trust region and line search variant of our method apply a Cholesky factorization to the tangent problem Hessian. This serves either to build an efficient preconditioner, or to compute the Newton direction in the second case. When the Hessian is close to becoming indefinite, a modified Cholesky factorization is used ($\nabla^2 L + E = JJ^T$, with J a lower triangular matrix and E a shift matrix of minimal norm). This is a critical point of the algorithm, particularly for our hard problems, because the condition number of the shifted Hessian has to be reasonable to avoid large computational round-off errors. However, trading a larger condition number may be often desirable, for example when the Hessian at the solution is ill-conditioned. We found that the revised modified Cholesky algorithm proposed by (Schnabel and Eskow, 1999) achieves this goal fairly well.

There are a few more features of our approach, which in our opinion merit special attention as they seem to be typical for automatic control applications.

A first issue is the magnitude of the decision variables, which may vary greatly within a given problem. In particular the range of the Lyapunov variables may show differences of several orders of magnitude, a phenomenon which is difficult to predict in advance, as the physical meaning of the Lyapunov variables is often lacking. Properly scaling these variables is therefore difficult in general.

A similar issue is to give prior bounds on the controller gains. This is mandatory because these values are to be useful in the hard-ware implementations of the theoretically computed controllers. Exceedingly large gains would bear the risk to amplify measurement noise and thwart the process. This may also be understood as a quest on the well-posedness of the initial control problem, which should include explicit gain bound constraints.

Incomplete knowledge about the numerical range of the different variables is in our opinion the source of most numerical difficulties and a method to balance data is extremely important.

4 NUMERICAL EXAMPLES

In this section we test our code on a variety of difficult applications in automatic control. The first part uses examples from static output feedback control. We then discuss applications like sparse linear constant output-feedback design, simultaneous state-feedback stabilization with limits on feedback gains and finally multi-objective $\mathcal{H}_2/\mathcal{H}_\infty$-controller design. Here we import examples from (Hassibi et al., 1999). In all these cases our method achieves closed-loop stability, while performance indicies like $\mathcal{H}_\infty$, $\mathcal{H}_2$ or mixed performances previously published in the literature are significantly improved. We always start our algorithm at $x_0 = 0$ in order to capture a realistic scenario, where no such prior information is available.

4.1 Static Output Feedback Controller Synthesis

We first recall a BMI characterization of the classical output feedback synthesis problem. To this end let $P(s)$ be an LTI (Linear Time Invariant) system with state-space equations:

$$P(s): \quad \begin{bmatrix} \dot{x} \\ z \\ y \end{bmatrix} = \begin{bmatrix} A & B_1 & B_2 \\ C_1 & D_{11} & D_{12} \\ C_2 & D_{21} & D_{22} \end{bmatrix} \begin{bmatrix} x \\ w \\ u \end{bmatrix}, \quad (28)$$

where

- $x \in \mathbb{R}^{n_1}$ is the state vector,
- $u \in \mathbb{R}^{m_2}$ is the vector of control inputs,
- $w \in \mathbb{R}^{m_1}$ is a vector of exogenous inputs,

- $y \in \mathbb{R}^{p_2}$ is the vector of measurements,
- $z \in \mathbb{R}^{p_1}$ is the controlled or performance vector.
- $D_{22} = 0$ is assumed without loss of generality.

Let $T_{w,z}(s)$ denote the closed-loop transfer functions from w to z for some static output-feedback control law:

$$u = Ky. \quad (29)$$

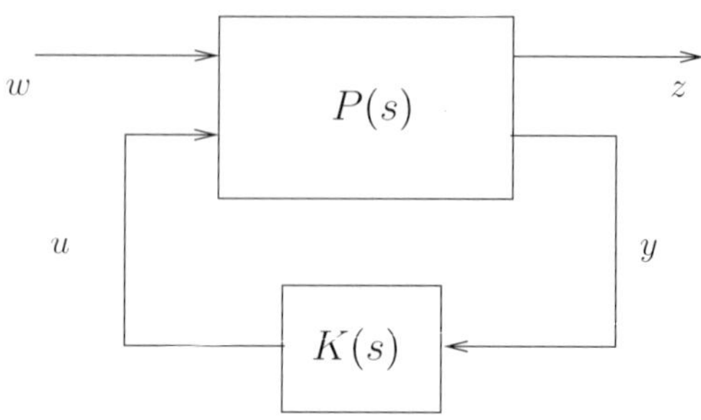

Figure 1: Output-feedback Linear Fractional Transform.

Our aim is to compute K subject to the following constraints:

- *internal stability:* for $w = 0$ the state vector of the closed-loop system (28) and (29) tends to zero as time goes to infinity.

- *performance:* the $\mathcal{H}_\infty$ norm $\|T_{w,z}(s)\|_\infty$, respectively the $\mathcal{H}_2$ norm $\|T_{w,z}(s)\|_2$, is minimized where the closed-loop transfer $T_{w,z}(s)$ is described as:

$$\begin{cases} \dot{x} = (A + B_2KC_2)x + (B_1 + B_2KD_{21})w \\ z = (C_1 + D_{12}KC_2)x + (D_{11} + D_{12}KD_{21})w. \end{cases}$$

Recall that in the $\mathcal{H}_2$ case, some feedthrough terms must be nonexistent in order for the $\mathcal{H}_2$ performance to be well defined. We have to assume $D_{11} = 0$, $D_{21} = 0$ for the plant in (28). This guaranteed, both static $\mathcal{H}_\infty$ and $\mathcal{H}_2$ indicies are then transformed into a matrix inequality condition using the bounded real lemma (Anderson and Vongpanitlerd, 1973). This leads to well-know characterizations:

Proposition 4.1 *A stabilizing static output feedback controller K with $\mathcal{H}_\infty$ gain $\|T_{w,z}(s)\|_\infty \leq \gamma$ exists provided there exists $X \in \mathbb{S}^{n_1}$ such that:*

$$\begin{bmatrix} (A+B_2KC_2)^T X + * & * & * \\ (B_1+B_2KD_{21})^T X & -\gamma I & * \\ (C_1+D_{12}KC_2) & (D_{11}+D_{12}KD_{21}) & -\gamma I \end{bmatrix} \prec 0 \quad (30)$$

$$X \succ 0. \quad (31)$$

Proposition 4.2 *A stabilizing static output feedback controller K with $\mathcal{H}_2$ performance $\|T_{w,z}(s)\|_2 \leq \sqrt{\gamma}$ exists provided there exist $X, M \in \mathbb{S}^{n_1}$ such that:*

$$\begin{bmatrix} (A+B_2KC_2)^T X+* & * \\ (C_1+D_{12}KC_2) & -I \end{bmatrix} \prec 0 \qquad (32)$$

$$\begin{bmatrix} M & * \\ X(B_1+B_2KD_{21}) & X \end{bmatrix} \succ 0 \qquad (33)$$

$$\mathrm{Tr}(M) \leq \gamma. \qquad (34)$$

In order to convert these programs to the form (1), we have to replace strict inequalities $\prec 0$ by $\preceq -\epsilon$ for a suitable threshold. The choice of $\epsilon > 0$ poses no problem in practice.

4.1.1 Transport Airplane (TA)

We consider the design of a static $\mathcal{H}_\infty$ controller for a transport airplane from (Gangsaas et al., 1986). Our algorithm computes a static controller K after solving 29 unconstrained minimization subproblems (14):

$$K_\infty = \begin{bmatrix} 0.69788 \\ -0.64050 \\ -0.83794 \\ 0.09769 \\ 1.57062 \end{bmatrix}^T.$$

The associated $\mathcal{H}_\infty$ performance is 2.22 and represents 30% improvement over the result in (Leibfritz, 1998).

4.1.2 VTOL Helicopter (VH)

The state-space data of the VTOL helicopter are borrowed from (Keel et al., 1988). They are obtained by linearizing the helicopter rigid body dynamics at given flight conditions. This leads to a fourth-order model. Both $\mathcal{H}_\infty$ and $\mathcal{H}_2$ static controllers are computed with the spectral SDP method. The algorithm converges after solving 29 tangent problems (14):

$$K_\infty = \begin{bmatrix} 0.50750 \\ 10.00000 \end{bmatrix}.$$

The final $\mathcal{H}_\infty$ performance is 1.59e-1, which is about 40% improvement over the performance obtained in (Leibfritz, 1998).

$$K_2 = \begin{bmatrix} 0.13105 \\ 5.95163 \end{bmatrix},$$

which gives a $\mathcal{H}_2$ performance of 9.54e-2 after solving 28 subproblems.

4.1.3 Chemical Reactor (CR)

A model description for the chemical reactor can be found in (Hung and MacFarlane, 1982). Both $\mathcal{H}_\infty$

Problem	n	m	iter	cpu	index
TA ($\mathcal{H}_\infty$)	51	30	29	39	2.22
VH ($\mathcal{H}_\infty$)	13	12	29	0.3	1.59e-1
VH ($\mathcal{H}_2$)	16	13	28	0.5	9.54e-2
CR ($\mathcal{H}_\infty$)	15	18	28	1.3	1.17
CR ($\mathcal{H}_2$)	25	19	41	24.6	1.94
PA ($\mathcal{H}_\infty$)	19	13	30	13.5	2.19e-3

Table 1: Static output feedback synthesis. n and m give size of (1), iter counts instances of (14), cpu in seconds, index gives the corresponding $\mathcal{H}_\infty$ or $\mathcal{H}_2$ performance.

and $\mathcal{H}_2$ static controllers are computed with the spectral SDP method, which requires solving respectively 28 and 41 subproblems. The resulting gains are:

$$K_\infty = \begin{bmatrix} -10.17039 & -31.54413 \\ -26.01021 & -100.00000 \end{bmatrix},$$

with $\mathcal{H}_\infty$-performance $\|T_{w,z}(s)\|_\infty = 1.17$.

$$K_2 = \begin{bmatrix} 0.35714 & -2.62418 \\ 2.58159 & 0.77642 \end{bmatrix},$$

with $\mathcal{H}_2$ performance $\|T_{w,z}(s)\|_2 = 1.94$.

4.1.4 Piezoelectric Actuator (PA)

In this section, we consider the design of a static controller for a piezoelectric actuator system. A comprehensive presentation of this system can be found in (Chen, 1998). Our algorithm computes a static $\mathcal{H}_\infty$ controller after solving 30 instances of (14):

$$K_\infty = \begin{bmatrix} 0.09659 & -1.45023 & -100.00 \end{bmatrix}.$$

The computed $\mathcal{H}_\infty$-performance 2.19e-3 improves significantly over the value 0.785 given in (Leibfritz, 1998).

4.2 Miscellaneous Examples

4.2.1 Sparse Linear Constant Output-Feedback Design

In many applications it is of importance to design sparse controllers with a small number of nonzero entries. This is for instance the case when it is mandatory to limit the number of arithmetic operations in the control law, or when reducing the number of sensor/actuator devices is critical. In such a case one may attempt to synthesize a controller with an imposed sparsity structure. A more flexible idea is to let the problem find its own sparsity structure by using a ℓ_1-norm cost function. We solve the following (BMI) problem:

$$\begin{aligned} \min \quad & \sum_{\substack{1 \le i \le n, \\ 1 \le j \le n}} |K_{ij}|, \\ \text{s. t.} \quad & P \succeq \epsilon I \\ & (A + BKC)^T P + P(A + BKC) \\ & \qquad \preceq -2\alpha P - \epsilon I \end{aligned}$$

where ϵ is the usual threshold to adapt strict inequalities to the form (1) or (8). This corresponds to designing a sparse linear constant output feedback control $u = Ky$ for the system $\dot{x} = Ax + Bu$, $y = Cx$ with an imposed decay rate of at least α in the closed-loop system.

Minimizing the ℓ_1-norm of the feedback gain is a good heuristic for obtaining sparse feedback gain matrices. This may be explained by the fact that the function $\phi(t) = |t|$ is the natural continuous interpolant of the counting function $c(t) = \lfloor t \rfloor$. Minimizing the function $\sum_{ij} \lfloor K_{ij} \rfloor$ or even counting the nonzero entries K_{ij} in K is what we would really like to do, but this is a discrete optimization problem subject to matrix inequality constraints.

There are many more applications to this heuristic. Finding sparse feedback gain matrices is also a good way of solving the actuator/sensor placement or controller topology design problems. In our example the method works nicely and a sparsity pattern is apparent. If small but nonzero values occur, one may set these values to zero if below a certain threshold, derive a sparsity pattern, and attempt a structured design with that pattern imposed. This leads to another class of BMI problems.

In our example, the system is initially unstable with rate of $\alpha_0 = -0.2832$. This value is obtained by computing the smallest negative real part of the eigenvalues of A (data available from (Hassibi et al., 1999)).

In a first test similar to the one in (Hassibi et al., 1999), we seek a sparse K so that the decay rate of the closed-loop system is not less than 0.35. We obtain the following gain matrix:

$$K_\infty = \begin{bmatrix} 0.2278 & 0.0000 & 0.0000 & 0.0000 \\ 0.0000 & 0.0000 & 0.0000 & 0.0000 \\ 0.0000 & 0.0000 & 0.0000 & 0.0000 \\ 0.0000 & 0.0000 & 0.0000 & 0.0000 \\ 0.0000 & 0.3269 & 0.0000 & 0.0000 \end{bmatrix},$$

with only two non-zeros elements. The path-following method presented in (Hassibi et al., 1999) gives 3 non-zeros elements. The spectral method gives once again a significant improvement, while maintaining the closed-loop stability.

If now in a second test a higher decay rate ($\alpha \ge 0.5$) (and so a better damped closed-loop system), is required, our method computes

$$K_\infty = \begin{bmatrix} 0.1072 & 0.0000 & 0.0000 & 0.0000 \\ -0.0224 & 0.0000 & 0.0000 & 0.0000 \\ 0.3547 & 0.0000 & 0.0000 & 0.0000 \\ 0.0000 & 0.0000 & 0.0000 & 0.0000 \\ 0.0186 & 0.1502 & 0.0000 & 0.0000 \end{bmatrix},$$

with 5 non-zeros entries. As expected, the sparsity decreases when α increases.

4.2.2 Simultaneous State-Feedback Stabilization with Limits on Feedback Gains

One attempts here to stabilize three different linear systems using a common linear constant state-feedback law with limits on the feedback gains. Specifically, suppose that:

$$\dot{x} = A_k x + B_k u, \quad u = K x, \quad k = 1, 2, 3.$$

The goal is to compute K such that $|K_{ij}| \le K_{ij,max}$ and the three closed-loop systems:

$$\dot{x} = (A_k + B_k K) x, \quad k = 1, 2, 3,$$

are stable. Under these constraints the worst decay-rate of the closed-loop systems is maximized.

Proposition 4.3 *The feedback gain K stabilizes the above systems simultaneously if and only if there exist $P_k \succ 0$, $\alpha_k > 0$, $k = 1, 2, 3$ such that*

$$(A_k + B_k K)^T P_k + P_k(A_k + B_k K) \prec -2\alpha_k P_k,$$
$$k = 1, 2, 3.$$

Introducing our usual threshold parameter $\epsilon > 0$, we have to check whether the value of the following BMI program is positive:

$$\begin{aligned} \text{maximize} \quad & \min_{k=1,2,3} \alpha_k \\ \text{subject to} \quad & |K_{ij}| \le K_{ij,max}, \\ & (A_k + B_k K)^T P_k + P_k(A_k + B_k K) \\ & \qquad \preceq -2\alpha_k P_k - \epsilon I \\ & P_k \succeq \epsilon I, \quad k = 1, 2, 3. \end{aligned}$$

By looking in (Hassibi et al., 1999) and computing the corresponding eigenvalues, the reader will notice that all three systems are unstable. We chose, as proposed by Hassibi, How and Boyd, $K_{ij,max} = 50$. We obtained a solution after 19 unconstrained minimization subproblems, in 1.8 seconds. This experiment gave the best improvement among our present tests. The worst decay-rate was increased to $\alpha = 6.55$, an improvement of more than 500% over the value 1.05 obtained in (Hassibi et al., 1999). The corresponding gain matrix is the following:

$$K_\infty = \begin{bmatrix} -17.1926 & -29.8319 & -50 \\ 50 & 18.7823 & -5.3255 \end{bmatrix}.$$

4.2.3 $\mathcal{H}_2/\mathcal{H}_\infty$ Controller Design

Mixed $\mathcal{H}_2/\mathcal{H}_\infty$-performance indicies represent one of those prominent cases where the projection Lemma fails and direct BMI-techniques are required. Here we consider the case where a static state-feedback controller $u = Kx$ for the open loop system

$$\dot{x} = Ax + B_1 w + Bu$$
$$z_1 = C_1 x + D_1 u$$
$$z_2 = C_2 x + D_2 u$$

is sought for. The purpose is to find K such that the $\mathcal{H}_2$ norm $w \to z_2$ is minimized, while the $\mathcal{H}_\infty$-norm $w \to z_1$ is fixed below a performance level γ. After introducing our usual threshold $\epsilon > 0$, the following matrix inequality optimization problem arises:

minimize η^2
subject to

$$\begin{bmatrix} \begin{pmatrix} (A + BK)^T P_1 + P_1(A + BK) \\ + (C_1 + D_1 K)^T (C_1 + D_1 K) \end{pmatrix} & * \\ B_1{}^T P_1 & -\gamma^2 I \end{bmatrix}$$

$$\preceq -\epsilon I$$

$$\begin{bmatrix} (A + BK)^T P_2 + P_2(A + BK) & * \\ B_1^T P_2 & -I \end{bmatrix} \preceq -\epsilon I$$

$$\begin{bmatrix} P_2 & * \\ C_2 + D_2 K & Z \end{bmatrix} \succeq \epsilon I, \qquad \mathrm{Tr}\,(Z) \leq \eta^2,$$

$$P_1 \succeq \epsilon I, \qquad P_2 \succeq \epsilon I.$$

The first block contains quadratic terms, which may be converted to BMI-from using a Schur complement. With $\gamma = 2$, the algorithm needs 21 instances of (14) to compute the compensator:

$$K_{2,\infty} = [1.9313 \quad 0.3791 \quad -0.2038],$$

with corresponding $\mathcal{H}_2$-norm 0.7489. While the authors mention a performance of 0.3286, the compensator provided in (Hassibi et al., 1999) gives to us a $\mathcal{H}_2$-norm of 0.8933. In this case the improvement over the result in (Hassibi et al., 1999) is negligible, which we interpret in the sense that this result was already close to the global optimum. (Nota bene, one cannot have any guarantee as to the truth of this statement. Indeed, in this example we found that many other local minima could be computed).

5 CONCLUSION

A spectral penalty augmented Lagrangian method for matrix inequality constrained nonlinear optimization programs was presented and applied to a number of small to medium-size test problems in automatic control.

Problem	n	m	iter	cpu
Sparse design 1	56	51	29	2.9
Sparse design 2	56	51	19	7.3
Simult. stab.	28	34	19	1.8
$\mathcal{H}_2/\mathcal{H}_\infty$	17	27	21	0.8

Table 2: Miscellaneaous examples.

The algorithm performs robustly if parameters are carefully tuned. This refers to the choice of the initial penalty value, p, the control of the condition number of the tangent problem Hessian $\nabla^2 L(x, V, p)$ in (14), and to the stopping tests (complementarity, stationarity), where we followed the lines of (Henrion et al., 2003). For the updates $p \to p^+$ and $V \to V^+$ we found that it is vital to avoid drastic changes, since the new subproblem (14) may be much more difficult to solve. This is particularly so for the nonconvex BMI case.

We remind the reader that similar to our previous approaches, the proposed algorithm is a local optimization method, which gives no certificate as to finding the global optimal solution of the control problem. If fact, such a method may in principle even fail completely by converging to a local minimum of constraint violation. For all that, our general experience, confirmed in the present testing, is that this type of failure rarely occurs. We consider the local approach largely superior for instance to a very typical class of methods in automatic control, where hard problems are solved by a tower of LMI problems with growing size, N, where a solution is guaranteed as $N \to \infty$. Such a situation is often useless, since the size of the LMI problem needed to solve the underlying hard problem is beyond reach, and since a prior estimate is not available or too pessimistic. The major advantage of local methods is that the subproblems, on whose succession the problem rests, all have the same dimension.

We finally point out that the local convergence of the algorithm will be proved in a forthcoming article.

REFERENCES

Anderson, B. and Vongpanitlerd, S. (1973). *Network analysis and synthesis: a modern systems theory approach.* Prentice-Hall.

Apkarian, P., Noll, D., Thevenet, J. B., and Tuan, H. D. (2004). A Spectral Quadratic-SDP Method with Applications to Fixed-Order $\mathcal{H}_2$ and $\mathcal{H}_\infty$ Synthesis. In *Asian Control Conference, to be published in European Journal of Control.*

Apkarian, P., Noll, D., and Tuan, H. D. (2002). Fixed-order $\mathcal{H}_\infty$ control design via an augmented Lagrangian method . *Rapport Interne 02-13, MIP,*

UMR 5640, Maths. Dept. - Paul Sabatier University . http://mip.ups-tlse.fr/publi/publi.html.

Apkarian, P., Noll, D., and Tuan, H. D. (2003). Fixed-order $\mathcal{H}_\infty$ control design via an augmented Lagrangian method . *Int. J. Robust and Nonlinear Control*, 13(12):1137–1148.

Ben-Tal, A. and Zibulevsky, M. (1997). Penalty/barrier multiplier methods for convex programming problems. *SIAM J. on Optimization*, 7:347–366.

Bertsekas, D. P. (1982.). *Constrained optimization and Lagrange multiplier methods*. Academic Press, London.

Chen, B. M. (1998). H_∞ *Control and Its Applications*, volume 235 of *Lectures Notes in Control and Information Sciences*. Springer Verlag, New York, Heidelberg, Berlin.

Conn, A. R., Gould, N., and Toint, P. L. (1991). A globally convergent augmented Lagrangian algorithm for optimization with general constraints and simple bounds. *SIAM J. Numer. Anal.*, 28(2):545 – 572.

Conn, A. R., Gould, N. I. M., Sartenaer, A., and Toint, P. L. (1993a). Global Convergence of two Augmented Lagrangian Algorithms for Optimization with a Combination of General Equality and Linear Constraints. Technical Report TR/PA/93/26, CERFACS, Toulouse, France.

Conn, A. R., Gould, N. I. M., Sartenaer, A., and Toint, P. L. (1993b). Local convergence properties of two augmented Lagrangian algorithms for optimization with a combination of general equality and linear constraints. Technical Report TR/PA/93/27, CERFACS, Toulouse, France.

Conn, A. R., Gould, N. I. M., Sartenaer, A., and Toint, P. L. (1996). Convergence properties of an augmented Lagrangian algorithm for optimization with a combination of general equality and linear constraints. *SIAM J. on Optimization*, 6(3):674 – 703.

Fares, B., Apkarian, P., and Noll, D. (2000). An Augmented Lagrangian Method for a Class of LMI-Constrained Problems in Robust Control Theory. In *Proc. American Control Conf.*, pages 3702–3705, Chicago, Illinois.

Fares, B., Apkarian, P., and Noll, D. (2001). An Augmented Lagrangian Method for a Class of LMI-Constrained Problems in Robust Control Theory. *Int. J. Control*, 74(4):348–360.

Fares, B., Noll, D., and Apkarian, P. (2002). Robust Control via Sequential Semidefinite Programming. *SIAM J. on Control and Optimization*, 40(6):1791–1820.

Gangsaas, D., Bruce, K., Blight, J., and Ly, U.-L. (1986). Application of modern synthesis to aircraft control: Three case studies. *IEEE Trans. Aut. Control*, AC-31(11):995–1014.

Hassibi, A., How, J., and Boyd, S. (1999). A path-following method for solving bmi problems in control. In *Proc. American Control Conf.*, pages 1385–1389.

Henrion, D., M.Kocvara, and Stingl, M. (2003). Solving simultaneous stabilization BMI problems with PEN-NON. In *IFIP Conference on System Modeling and Optimization*, volume 7, Sophia Antipolis, France.

Hestenes, M. R. (1969). Multiplier and gradient method. *J. Optim. Theory Appl.*, 4:303 – 320.

Hung, Y. S. and MacFarlane, A. G. J. (1982). *Multivariable feedback: A classical approach*. Lectures Notes in Control and Information Sciences. Springer Verlag, New York, Heidelberg, Berlin.

Keel, L. H., Bhattacharyya, S. P., and Howze, J. W. (1988). Robust control with structured perturbations. *IEEE Trans. Aut. Control*, 36:68–77.

Kocvara, M. and Stingl, M. (2003). A Code for Convex Nonlinear and Semidefinite Programming. *Optimization Methods and Software*, 18(3):317–333.

Leibfritz, F. (1998). Computational design of stabilizing static output feedback controller. Rapports 1 et 2 Mathematik/Informatik, Forschungsbericht 99-02, Universitt Trier.

Lewis, A. (1996). Derivatives of spectral functions. *Mathematics of Operations Research*, 21:576–588.

Lewis, A. (2001). Twice differentiable spectral functions. *SIAM J. on Matrix Analysis and Applications*, 23:368–386.

Lewis, A. and S.Sendov, H. (2002). Quadratic expansions of spectral functions. *Linear Algebra and Appl.*, 340:97–121.

Lin, C. and More, J. (1998). Newton's method for large bound–constrained optimization problems. Technical Report ANL/MCS-P724–0898, Mathematics and Computer Sciences Division, Argonne National Laboratory.

Mosheyev, L. and Zibulevsky, M. (2000). Penalty/barrier multiplier algorithm for semidefinite programming. *Optimization Methods and Software*, 13(4):235–261.

Noll, D., Torki, M., and Apkarian, P. (2002). Partially Augmented Lagrangian Method for Matrix Inequality Constraints. *submitted*. Rapport Interne , MIP, UMR 5640, Maths. Dept. - Paul Sabatier University.

Powell, M. J. D. (1969). A method for nonlinear constraints in minimization problem. In Fletcher, R., editor, *Optimization*. Academic Press, London, New York.

Schnabel, R. B. and Eskow, E. (1999). A revised modified cholesky factorization algorithm. *SIAM J. on Optimization*, 9(4):1135–1148.

Shapiro, A. (2002). On differentiability of symmetric matrix valued functions. *School of Industrial and Systems Engineering, Georgia Institute of Technology*. Preprint.

Zibulevsky, M. (1996). *Penalty/Barrier Multiplier Methods for Large-Scale Nonlinear and Semidefinite Programming*. Ph. D. Thesis, Technion Isral Institute of Technology.

A STOCHASTIC OFF LINE PLANNER OF OPTIMAL DYNAMIC MOTIONS FOR ROBOTIC MANIPULATORS

Taha Chettibi, Moussa Haddad, Samir Rebai
Mechanical Laboratory of Structures, EMP, B.E.B., BP17, 16111, Algiers, Algeria
Email: tahachettibi@yahoo.fr; moussa.haddad@caramail.com, srebai@yahoo.fr

Abd Elfath Hentout
Laboratory of applied mathematics, EMP, B.E.B., BP17, 16111, Algiers, Algeria
Email: Hentout@hotmail.com

Keywords: Robotic manipulator, Motion planning, Stochastic optimization, Obstacles avoidance.

Abstract: We propose a general and simple method that handles free (or point-to-point) motion planning problem for redundant and non-redundant serial robots. The problem consists of linking two points in the operational space, under constraints on joint torques, jerks, accelerations, velocities and positions while minimizing a cost function involving significant physical parameters such as transfer time and joint torque quadratic average. The basic idea is to dissociate the search of optimal transfer time T from that of optimal motion parameters. Inherent constraints are then easily translated to bounds on the value of T. Furthermore, a stochastic optimization method is used which not only may find a better approximation of the global optimal motion than is usually obtained via traditional techniques but that also handles more complicated problems such as those involving discontinuous friction efforts and obstacle avoidance.

1 INTRODUCTION

Motion planning constitutes a primordial phase in the process of robotic system exploitation. It is a challenging task because the robot behaviour is governed by highly non linear models and is subjected to numerous geometric, kinematic and dynamic constraints (Latombe, 1991) (Angeles, 1997) (Chettibi, 2001). Two categories of motions can be distinguished (Angeles, 1997) (Chettibi, 2000). The first covers motions along prescribed geometric path and correspond, for example, to continuous welding or glowing operations (Bobrow, 1985) (Kang, 1986) (Pfeiffer, 1987) (Chettibi, 2001b). The second, which is the focus of this paper, concerns point-to-point (or free) motions involved, for example, in discrete welding or pick-and-place operations (Bessonnet, 1992) (Mitsi, 1995) (Lazrak, 1996) (Danes, 1998) (Chettibi, 2001a). In general, many different ways are possible to perform the same task. This freedom of choice can be exploited judiciously to optimize a given performance criterion. Hence, motion generation becomes an optimization problem. It is

here referred to as the optimal free motion planning problem (OFMPP).

In the specialized literature, various resolutions methods have been proposed to handle the OFMPP. They can be grouped in two main families; namely: *direct* and *indirect* methods (Hull, 1997) (Betts, 1998). The indirect methods are, in general, applications of optimal control theory and in particular *Pontryagin Maximum Principle* (PMP) (Pontryagin, 1965). Optimality conditions are stated under the form of a boundary value problem that is generaly too difficult to solve (Bessonnet, 1992) (Lazrak, 1996) (Chettibi, 2000). Several techniques, such as the phase plane method (Bobrow, 1985) (Kang, 1986) (Jaques, 1986) (Pfeiffer, 1987), exploit the structure only of the optimal solution given by PMP and get numerical solutions via other means. In general, such techniques are applied to limited cases and have several drawbacks resumed below:

- They require the solution of a N.L multi-point shooting problem (David, 1997) (John, 1998),
- They require analytical computing of gradients (Lazrak, 1996) (Bessonnet, 1992),

J. Braz et al. (eds.), Informatics in Control, Automation and Robotics I, 73–80.
© 2006 *Springer. Printed in the Netherlands.*

- The region of convergence may be small (Chettibi, 2001) (Lazrak, 1996),
- Path inequality are difficult to handle (Danes, 1998),
- They introduce new variables known as co-state variables that are, in general, difficult to estimate (Lazrak, 1996) (Bessonnet, 1992) (Danes, 1998) (Pontryagin, 1965).
- In minimum time transfer problems, they lead to discontinuous controls (bang-bang) that may create many practical problems (Ola, 1994) (Chettibi, 2001*a*). In fact, the controller must work in saturation for long periods. The optimal control leaves no control authority to compensate for any tracking error caused by either unmodeled dynamics or delays introduced by the on-line feedback controller

To overcome these difficulties, direct methods have been proposed. They are based on discretisation of dynamic variables (states, controls). They seek to solve directly a parameter optimization problem. Then, N.L. programming (Tan, 1988) (Martin, 1997) (Martin, 1999) (Chettibi, 2001*a*) or stochastic optimization techniques (Chettibi, 2002*b*) are applied to compute optimal values of parameters. Other ways of discretisation can be found in (Richard, 1993) (Macfarlane, 2001). These techniques suffer, however, from numerical explosion when treating high dimension problems. Although they have been used successfully to solve a large variety of problems, techniques based on N.L. programming (Fletcher, 1987) (David, 1997) (Danes, 1998) (John, 1998) (Chettibi, 2000) have two essential drawbacks:

- They are easily attracted by local minima ;
- They generally require information on gradient and hessian that are difficult to get analytically. In addition, continuity of second order must be ensured, while realistic physical models may include some discontinuous terms (frictions).

In parallel to these methods, that take into account both kinematics and dynamics aspects of the problem, numerous pure geometric planners have been proposed to find solutions for the simplified problem that consists of finding only feasible geometric paths (Piano movers problem) (Latombe, 1991) (Overmars, 1992) (Barraquand, 1992) (Kavraki, 1994) (Barraquand, 1996) (Kavraki, 1996) (Latombe, 1999) (Garber, 2002). In spite of this simplification, the problem still remains quite complex with exponential computational time in the degree of freedom (d.o.f.). Of course, any extension (presence of obstacles, for example) adds in computational complexity. Even so, various practical planners have been proposed. Reference (Latombe, 1991) gives an excellent overview of early methods (before 1991) such as: potential field, cell decomposition and roadmap methods, some of which have shown their limits. For instance, a potential field based planner is quickly attracted by local minima (Khatib, 1986) (Latombe, 1991) (Barraquand, 1992). Cell decomposition methods often require difficult and quite expensive geometric computations and data structures tend to be very large (Latombe, 1991) (Overmars, 1992). The key issue for roadmap methods is the construction of the roadmap. Various techniques have been proposed that produce different sorts of roadmaps based on visibility and Voronoi graphs (Latombe, 1991).

During the last decade, interest was given to stochastic techniques to solve various forms of optimal motion planning problems. In particular, powerful algorithms were proposed to solve the basic geometric problem. Probabilistic roadmaps (PRM) or Probabilistic Path Planners (PPP) were introduced in (Overmars, 1992) (Barraquand, 1996) (Kavraki, 1994) (Kavraki, 1996) and applied successfully to complex situations. They are generally executed in two steps: first a roadmap is constructed, according to a stochastic process, then the motion planning query is treated. Due to the power of this kind of schemes, many perspectives are expected as shown in (Latombe, 1999). However, there are few attempts to apply them to solve the complete OFMPP. References (LaValle, 1998) (LaValle, 1999) propose the method of Rapidly exploring Random Trees (RRTs) as an extention of PPP to optimize feasible trajectories for NL systems. Dynamic model and inherent constraints are taken into account.

In (Chettibi, 2002*a*), we introduced a different scheme using a sequential stochastic technique to solve the OFMPP. We present here this simple and versatile method and how it can be used to handle complex situations involving both friction efforts and obstacle avoidance.

2 PROBLEM STATEMENT

Let us consider a serial redundant or non-redundant manipulator with n d.o.f.. Motion equations can be derived using Lagrange's formalism or Newton-Euler formalism (Dombre, 1988) (Angeles, 1997):

$$M\left(q\right)\ddot{q} + Q\left(q,\dot{q}\right) + G\left(q\right) = \tau \qquad (1a),$$

q, $\dot{q}$ and $\ddot{q}$ are respectively joints position, velocity, acceleration vectors. $\mathbf{M}(\boldsymbol{q})$ is the inertia matrix. $\boldsymbol{Q}(q,\dot{q})$ is the vector of centrifugal and *Coriolis* forces in which joints velocities appear under a quadratic form. $\mathbf{G}(\mathbf{q})$ is the vector of potential forces and $\boldsymbol{\tau}$ is the vector of actuator efforts.

In order to make the dynamic model more realistic, we may introduce, for the ith joint, friction efforts as follows:

$$\sum_{j=1}^{n} M_{ij}(q(t))\ddot{q}_j(t) + Q_i(q(t),\dot{q}(t)) \qquad (1b),$$

$$+ G_i(q(t)) + F_i^{\,v}\dot{q}(t) + F_i^{\,s} sign\,(\dot{q}(t)) = \tau_i(t)$$

$F_i^{\,V}$ and $F_i^{\,s}$ are, respectively, sec and viscous friction coefficients of the ith joint.

The robot is required to move freely from an initial state $\mathbf{P}_i$ to a final state $\mathbf{P}_f$, both of which are specified in the operational space. In addition to solving for $\tau(t)$ and transfer time T, we must find the trajectory defined by $\mathbf{q}(t)$ such as the initial and the final state are matched, constraints are respected and a cost function is minimized.

The cost function adopted here is a balance between transfer time T and the quadratic average of actuator efforts:

$$F_{obj} = \mu\,T + \frac{1-\mu}{2}\int_0^T\sum_{i=1}^{n}\left(\frac{\tau_i(t)}{\tau_i^{max}}\right)^2 dt \qquad (2).$$

μ is a weighting coefficient chosen from [0,1] and according to the relative importance we would like to give to the minimization of T or to the quadratic average of actuator efforts. The case $\mu=1$ corresponds to the optimal time free motion planning problem.

Constraints that must be satisfied during the entire transfer ($0 \leq t \leq T$) are summarized bellow: for $i = 1, \ldots, n$ we have bounds on:

- *Joint torques:*
$$|\tau_i(t)| \leq \tau_i^{max} \qquad (3a);$$
- *Joint jerks :*
$$|\dddot{q}_i(t)| \leq \dddot{q}_i^{\,max} \qquad (3b);$$
- *Joint accelerations:*
$$|\ddot{q}_i(t)| \leq \ddot{q}_i^{\,max} \qquad (3c);$$
- *Joint velocities:*
$$|\dot{q}_i(t)| \leq \dot{q}_i^{\,max} \qquad (3d);$$
- *Joint positions:*
$$|q_i(t)| \leq q_i^{\,max} \qquad (3e).$$

Of course, non-symmetrical bounds on the above physical quantities can also be handled without any new difficulty.

Relations (3*a, b, c, d* and *e*) traduce the fact that not all motions are tolerable and that power resources are limited and must be used rationally in order to control correctly the robot dynamic behavior. Also, since joint position tracking errors increase with jerk, constraints (3*b*) are introduced to limit excessive wear and hence to extend the robot life-span (Latombe, 1991) (Piazzi, 1998) (Macfarlane, 2001).

In the case where obstacles are present in the robot workspace, motion must be planned in such a way collision is avoided between links and obstacles. Therefore, the following constraint has to be satisfied :

$$C(\mathbf{q})=False \qquad (3f).$$

The Boolean function C indicates whether or not the robot at configuration $\mathbf{q}$ is in collision with an obstacle. This function uses a distance function $D(\mathbf{q})$ that supplies for any robotic configuration the minimal distance to obstacles.

3 REFORMULATION OF THE PROBLEM

The normalization of the time scale, initially used to make the problem with fixed final time, is exploited to reformulate the problem and to make it propitious for a stochastic optimization strategy. Details are shown bellow.

3.1 Scaling

We introduce a normalized time scale as follows:
$$t = x.T \quad \text{with} \quad x\in[0,1] \qquad (4).$$
Hereafter, we will use the prime symbol to indicate derivations with respect to x :

$$q'=\frac{dq(x)}{dx}, \qquad q''=\frac{d^2q(x)}{dx^2} \quad , \qquad q'''=\frac{d^3q(x)}{dx^3} \qquad (5).$$

Relations (1*a*) and (1*b*) can be written as follows:

$$(1a) \Rightarrow \qquad \psi_i(x)=\frac{1}{T^2}H_i + \overline{G}_i \qquad (6a)$$

$$(1b) \Rightarrow \psi_i = \frac{1}{T^2}H_i + \overline{G}_i + \frac{1}{T}\overline{F}_i^{\,v}q_i' + \overline{F}_i^{\,s} sign(q') \qquad (6b)$$

where:

$$\psi_i(x)=\frac{\tau_i(x)}{\tau_i^{max}}, \quad \overline{M}_{ij}=\frac{M_{ij}}{\tau_i^{max}}, \quad \overline{Q}_i=\frac{Q_i}{\tau_i^{max}},$$

$$\overline{G}_i=\frac{G_i}{\tau_i^{max}}, \quad H_i=\sum_{j=1}^{n}\overline{M}_{ij}q''_j + \overline{Q}_i \qquad (7)$$

and:
$$\overline{F}_i^{\,v}=\frac{F_i^{\,v}}{\tau_i^{max}}, \quad \overline{F}_i^{\,s}=\frac{F_i^{\,s}}{\tau_i^{max}} \qquad (8)$$

3.2 Cost Function

With the previous notations, the cost function (2) becomes without friction efforts:

$$F_{obj} = T\left(S_0 + \frac{S_2}{T^2} + \frac{S_4}{T^4}\right) \quad (9),$$

Where S_0, S_2 and S_4 are given by:

$$S_0 = \mu + \frac{1-\mu}{2}\int_0^1 \sum_{i=1}^n \overline{G}_i^2\, dx$$

$$S_2 = (1-\mu)\int_0^1 \sum_{i=1}^n H_i \overline{G}_i\, dx \quad (10).$$

$$S_4 = \frac{1-\mu}{2}\int_0^1 \sum_{i=1}^n H_i^2\, dx$$

It must be noted that S_0, S_2 and S_4 are real coefficients that do depend on the joint evolution profile $q(x)$ but that do not depend on T. Also, S_0 and S_4 are always positive. Expression (9) represents a family of curves whose general shape, for any feasible motion, is shown in figure 1*a*. The minimum of each of these curves is reached when T takes on the value $T = T_m$ given by :

$$T_m = \left(\frac{S_2 + \sqrt{S_2^2 + 12 S_0 S_4}}{2 S_0}\right)^{1/2} \quad (11).$$

If friction efforts are taken into account, we introduce the following quantities:

$$K_i = \overline{F}_i^v q', \qquad \overline{\overline{G}}_i = \overline{G}_i + \overline{F}_i^s sign(q') \quad (12)$$

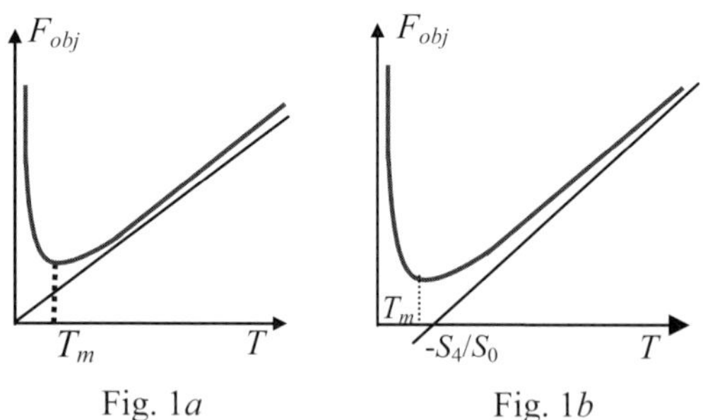

Figure 1: General shape of the cost function; (a) without friction efforts , (b) with friction efforts.

The expression of (2) becomes then :

$$F_{obj} = T\left(S_0 + \frac{S_1}{T} + \frac{S_2}{T^2} + \frac{S_3}{T^3} + \frac{S_4}{T^4}\right) \quad (13)$$

where:

$$S_0 = \mu + \frac{1-\mu}{2}\int_0^1 \sum_{i=1}^n \overline{\overline{G}}_i^2\, d\lambda$$

$$S_1 = (1-\mu)\int_0^1 \sum_{i=1}^n K_i \overline{\overline{G}}_i\, d\lambda \quad (14)$$

$$S_2 = \frac{1-\mu}{2}\int_0^1 \sum_{i=1}^n \left(K_i^2 + 2 H_i \overline{\overline{G}}_i\right) d\lambda$$

$$S_3 = (1-\mu)\int_0^1 \sum_{i=1}^n H_i K_i\, d\lambda$$

$$S_4 = \frac{1-\mu}{2}\int_0^1 \sum_{i=1}^n H_i^2\, d\lambda$$

For a given profile $q(x)$, (13) represents a family of curves whose general shape is shown in figure 1*b*, but now the asymptotic line intersects the time axis at $T = -S_1/S_0$. Furthermore, T_m has to be computed numerically since (11) is no longer applicable.

3.3 Effects of Constraints

Constraints imposed on the robot motion will be handled sequentially within the iterative process of minimization described in the next section. Already, we can group constraints into several categories according to the stage of the iterative process at which they will be handled.

3.3.a Constraints of the First Category

In the first category, we have constraints that will not add any restriction on the value of T. For example, joint position constraints (3*e*) become:

$$|q_i(x)| \le q_i^{max} \quad \forall x \in [0,1] \quad i = 1,\dots,n \quad (15),$$

and those due to obstacles presence (3*f*) become :

$$C(q(x)) = False \quad \forall x \in [0,1] \quad (16)$$

In both cases, only the joint position profiles $q(x)$ are determinant.

3.3.b Constraints of Second Category

In the second category, we have constraints that can be transformed into explicit *lower* bounds on T. For example joint velocity constraints lead to:

$$\frac{1}{T}|q'_i(x)| \le \dot{q}_i^{max} \Rightarrow T \ge \frac{|q'_i(x)|}{\dot{q}_i^{max}} \quad i = 1,\dots,n$$

$$\text{so:} \quad T \ge T_v \,, \quad T_v = \max_{i=1,\dots,n}\left[\max_{[0,1]} \frac{|q'_i(x)|}{\dot{q}_i^{max}}\right] \quad (17).$$

Joint acceleration and jerk constraints are transformed in the same way to give:

For accelerations:

$$T \geq T_a \;, \quad T_a = \max_{i=1,\ldots,n} \left[\max_{[0,1]} \left(\frac{\left|q''_i(x)\right|}{\ddot{q}_i^{\,max}} \right)^{1/2} \right] \quad (18),$$

and for jerks:

$$T \geq T_J \;, \quad T_J = \max_{i=1,\ldots,n} \left[\max_{[0,1]} \left(\frac{\left|q'''_i(x)\right|}{\dddot{q}_i^{\,max}} \right)^{1/3} \right] \quad (19).$$

Thus, (17), (18) and (19) define three lower bounds on transfer period. In consequence; T must satisfy the following condition:

$$T \geq T^* \;, \qquad T^* = max(T_v, T_a, T_J) \quad (20),$$

This type of constraints defines a forbidden region as shown in figure 2. Note that two cases are possible.

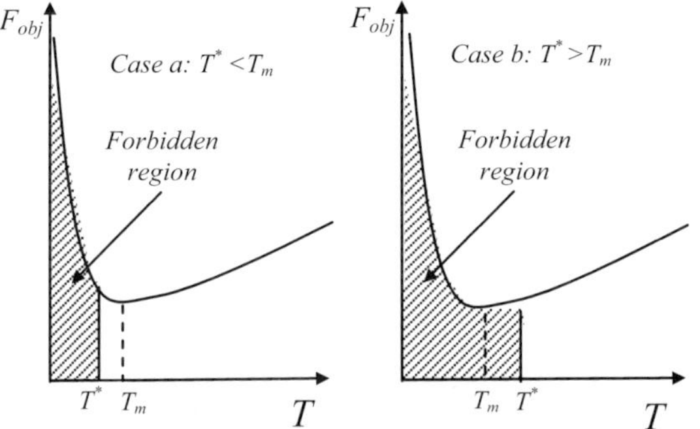

Figure 2: Bounds on transfer time value due to constraints of second category.

3.3.c Constraints of Third Category

In the third category, we have constraints that can be transformed into explicit bilateral bounds on T. For example those imposed on the value of joint torques (3a) define, in general, bracketing bounds on T, namely: T_L and T_R. In consequence,

$$T \in [T_L, T_R] \quad (21).$$

A fourth category might be included and would concern any other constraint that *does* add restrictions on T but that cannot be easily translated into simple bounds on T.

4 STRATEGY OF RESOLUTION

The iterative process of minimization proposed here includes the following steps:

Step 1: Generate a random (or guessed) temporal evolution shape $q_i(x)$ for each of the joint variables, taking into account any constraints of the first category (15), (16) as well as any conditions imposed on the initial and the final state.

Step 2: Get the S coefficients from (10) or (14) and T_m from (11) or by numerical means. If $F(T_m)$ is greater than F_{best} obtained so far, then there is no need to continue and hence, return to *Step* 1. Otherwise, a first bracketing interval $[T_1, T_2]$ is deduced (Fig. 3) in which F is decreasing from T_1 to T_m and increasing from T_m to T_2.

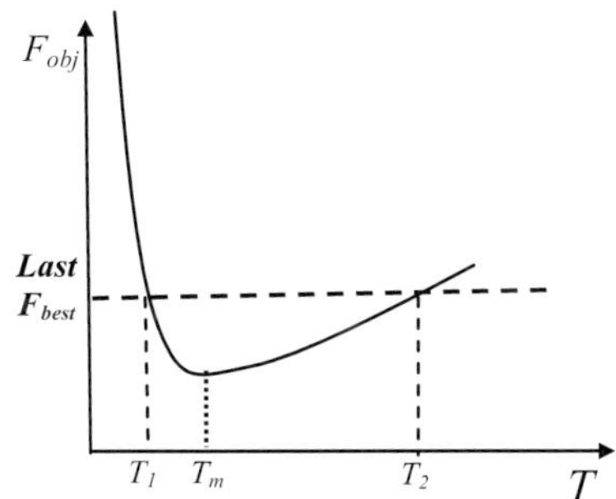

Figure 3: New exploration region defined by a new lower value of F_{best}.

The remaining steps will simply consist of changing T_1, T_m or T_2 while keeping this bracketing.

Step 3: Get T_a, T_v, T_j from (17, 18, 19) and T^* from (20). If $T^* > T_2$ then return to *Step* 1 else modify T_1 and/or T_m according to Fig. 2. That is: in case (*a*) $T_1 \leftarrow T^*$ while in (*b*) $T_1 \leftarrow T^*$ and $T_m \leftarrow T^*$.

Step 4: Get $[T_L, T_R]$ from (21). If $T_L > T_2$ or $T_R < T_1$ then return to *Step* 1. Otherwise, we have a new improved F_{best} :

> If $T_m \in [T_L, T_R]$ then
> $\quad F_{best} \leftarrow F(T_m)$
> Else if $T_m < T_L$ then
> $\quad F_{best} \leftarrow F(T_L)$
> Else
> $\quad F_{best} \leftarrow F(T_R)$
> End if

The above steps can be imbedded in a stochastic optimization strategy to determine better profiles $q_i(x)$, $i = 1,\ldots n$, leading to lower values of the objective function.

One way to get a guessed temporal evolution shape $q_i(x)$ for the joint variables, at any stage of optimization process, is to use randomly generated clamped cubic spline functions with nodes distributed for $x \in [0,1]$ (Fig. 4).

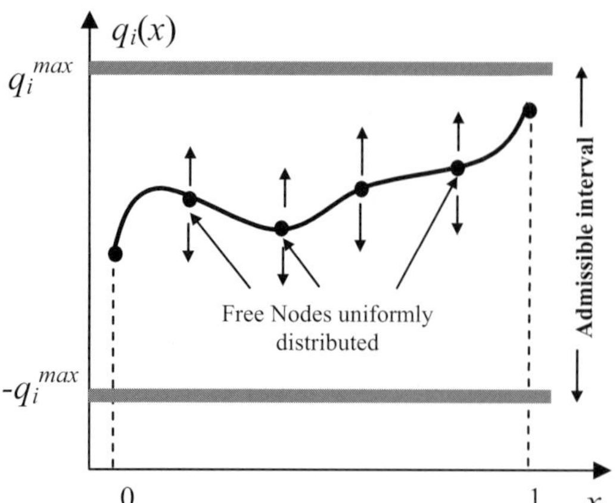

Figure 4: Approximation of joint position temporal evolution.

5 NUMERICAL RESULTS

We consider here a redundant planar robot constituted of four links connected by revolute joints. The corresponding geometric and inertial characteristics are listed in Appendix A. It is asked to move among two static obstacles disposed in it's work space at respectively (2, 1.5) and (-1.5, 1.5) with both unity radius. The robot begin at ($\pi/4$, -$\pi/2$, $\pi/4$, 0) and stops at ($\pi/2$, 0, 0, 0). Boundary velocities are null. The numerical results are obtained with μ=0.5 for both cases: with and without friction efforts. The corresponding optimal motions are depicted in Figures 5a, b, c, d, e and f.

In fact, without introducing friction effort we get : F_{obj} = 2.7745(s) and T_{opt} = 4.9693 (s).In the presence of friction efforts we get a different result: F_{obj} = 3.1119 (s) and T_{opt} = 5.3745 (s). Hence, to achieve the same task, we need more time and more effort in the presence of friction efforts.

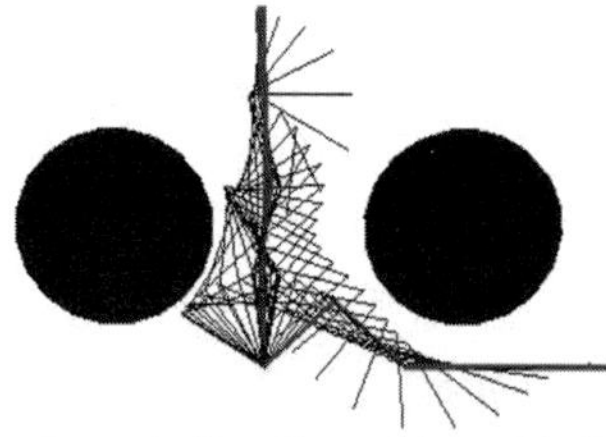

Figure 5a: Aspect of motion without friction effect.

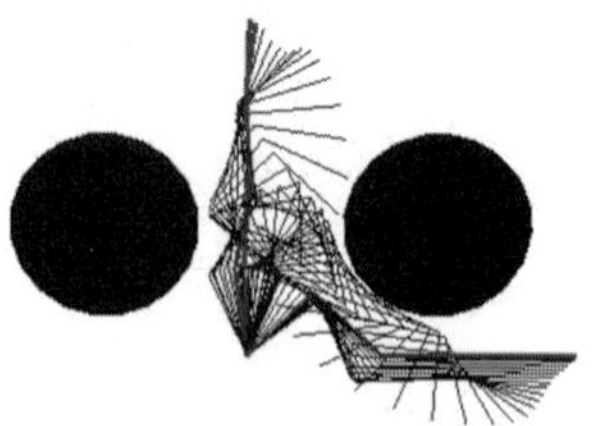

Figure 5b: Aspect of motion with friction effect.

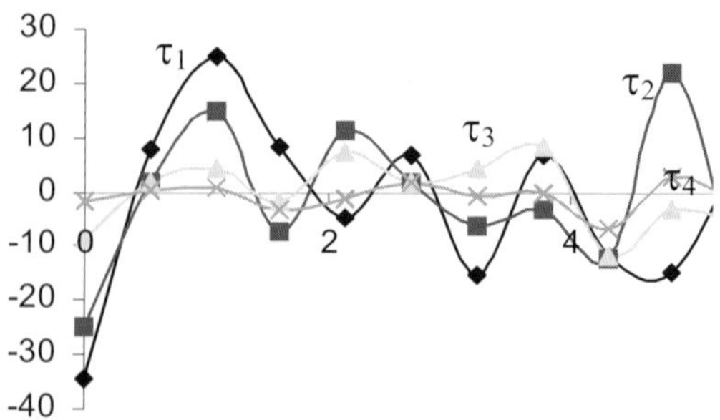

Figure 5c: Evolution of joint torques with friction effect.

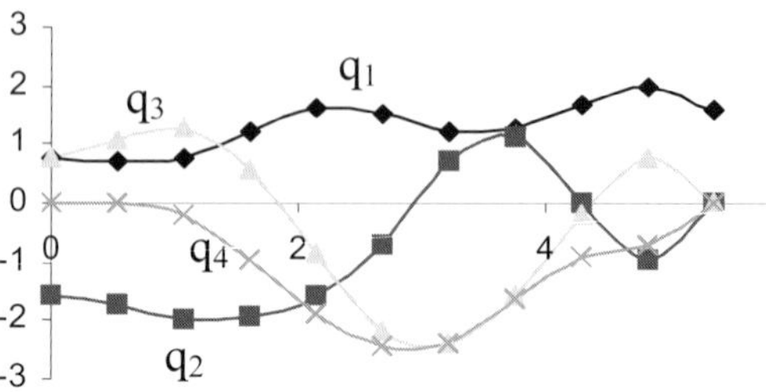

Figure 5d. Evolution of joint positions with friction effect.

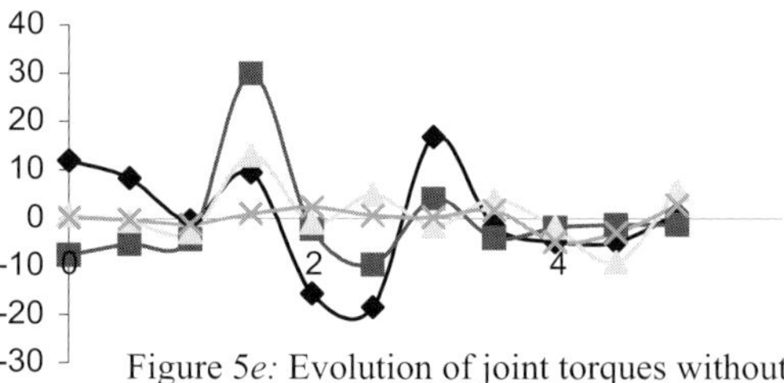

Figure 5e: Evolution of joint torques without friction effect.

Figure 5f: Evolution of joint positions without friction effect.

6 CONCLUSION

In this paper we have presented a simple trajectory planner of point-to-point motions for robotic arms. The problem is highly non-linear due first to the complex robot dynamic model that must be verified during the entire transfer, then to the non-linearity of the cost function to be minimized and finally to numerous constraints to be simultaneously respected. The OFMPP is originally an optimal control one and has been transformed into a parametric optimization problem. The optimization parameters are time transfer T and the position of nodes defining the shape of joint variables. The research of T has been separated from that of the others parameters in order to make the computing process efficient and to handle constraints easily by transforming them into explicit bounds on T possible values. In fact, the various possible constraints have been regrouped in four families according to their possible effects on T values and then have been handled sequentially during each optimization step. Nodes, defining $q(x)$ shape, are connected by cubic spline functions and their positions are perturbed inside a stochastic process until the objective function value is sufficiently reduced while all constraints are all satisfied. This ensured smoothness of resulted profiles. The objective function has been written under a weighting form permitting to make balance between reducing T and magnitude of implied torques.

Numerical examples, where a stochastic optimization process, implementing the proposed approach, has been used along with cubic spline approximations, and dealing with complex problems, such as those involving discontinuous friction efforts and obstacle avoidance, have been presented to show the efficiency of this technique. Others successful tests have been made in parallel for complex robotic architectures, like biped robots, will be presented in a future paper.

ACKNOWLEDGEMENTS

We thank Prof. H. E. Lehtihet for his suggestions and helpful discussions.

REFERENCES

Angeles J., 1997, *Fundamentals of robotic mechanical systems. Theory, methods, and algorithms,"* Springer Edition.

Barraquand J., Langlois B., Latomb J. C., 1992, Numerical Potential Field Techniques for robot path planning, *IEEE Tr. On Sys., Man, and Cyb.*, 22(2):224-241.

Barraquand J., Kavraki L., Latombe J. C., Li T. Y., Motwani R., Raghavan P., 1996, A random Sampling Scheme for path planning, 7th *Int. conf. on Rob. Research ISRR.*

Bobrow J.E., Dubowsky S., Gibson J.S., 1985, Time-Optimal Control of robotic manipulators along specified paths, *The Int. Jour. of Rob. Res.*, 4 (3), pp. 3-16.

Bessonnet G., 1992, *Optimisation dynamique des mouvements point à point de robots manipulateurs*, Thèse d'état, Université de Poitiers, France.

Betts J. T., 1998, Survey of numerical methods for trajectory optimization, *J. Of Guidance, Cont. & Dyn.*, 21(2), 193-207.

Chen Y., Desrochers A., 1990, A proof of the structure of the minimum time control of robotic manipulators using Hamiltonian formulation, *IEEE Trans. On Rob. & Aut.* 6(3), pp. 388-393.

Chettibi T., 2000, *Contribution à l'exploitation optimale des bras manipulateurs*, Magister thesis, EMP, Algiers, Algeria.

Chettibi T., 2001*a*, Optimal motion planning of robotic manipulators, *Maghrebine Conference of Electric Engineering*, Constantine, Algeria.

Chettibi T., Yousnadj A., 2001*b*, Optimal motion planning of robotic manipulators along specified geometric path, *International Conference on productic*, Algiers.

Chettibi T., Lehtihet H. E., 2002*a*, A new approach for point to point optimal motion planning problems of robotic manipulators, *6th Biennial Conf. on Engineering Syst. Design and Analysis (ASME)*, Turkey, APM10.

Chettibi T., 2002*b*, Research of optimal free motions of manipulator robots by non-linear optimization, *Séminaire international sur le génie Mécanique*, Oran, Algeria.

Danes F., 1998, *Critères et contraintes pour la synthèse optimale des mouvements de robots manipulateurs. Application à l'évitement d'obstacles*, Thèse d'état, Université de Poitiers.

Dombre E. & Khalil W., 1988, *Modélisation et commande des robots*, First Edition, Hermes.

Fletcher R., 1987, *Practical methods of optimization*, Second Edition, Wiley Interscience Publication.

Garber M., Lin. M.C., 2002, Constrained based motion planning for virtual prototyping, *SM'02*, Germany.

Glass K., Colbaugh R., Lim D., Seradji H., 1995, Real time collision avoidance for redundant manipulators, *IEEE Trans. on Rob. & Aut.*, 11(3), pp. 448-457.

Hull D. G., 1997, Conversion of optimal control problems into parameter optimization problems, *J. Of Guidance, Cont. & Dyn.*, 20(1), 57-62.

Jaques J., Soltine E., Yang H. S., 1989, Improving the efficiency of time-optimal path following algorithms, *IEEE Trans. on Rob. & Aut.*, 5 (1).

Kang G. S., McKay D. N., 1986, Selection of near minimum time geometric paths for robotic manipulators, *IEEE Trans. on Aut. & Contr.*, AC31(6), pp. 501-512.

Kavraki L., Latombe J. C., 1994, Randomized Preprocessing of Configuration Space for Fast Path Planning, *Proc. Of IEEE Int. Conf. on Rob. & Aut.*, pp. 2138-2139, San Diego.

Kavraki L., Svesta P., Latombe J. C., Overmars M., 1996, Probabilistic Roadmaps for Path Planning in High Dimensional Configuration space, *IEEE trans. Robot. Aut.*, 12:566-580.

Khatib O., 1986, Real-time Obstacle Avoidance for Manipulators and Mobile Robots, *Int. Jour. of Rob. Research*, vol. 5(1).

Latombe J. C., 1991, *Robot Motion Planning*, Kluwer Academic Publishers.

Latombe J. C., 1999, Motion planning: A journey of, molecules, digital actors and other artifacts, *Int. Jour. Of Rob. Research*, 18(11), pp. 1119-1128.

LaValle S. M., 1998, Rapidly exploring random trees: A new tool for path planning, *TR98-11*, Computer Science Dept., Iowa State University. http://janowiec.cs.iastate.edu/papers/rrt.ps.

LaValle S. M., Kuffner J. J., 1999, Randomized kinodynamic planning, *Proc. IEEE Int. Conf. On Rob. & Aut.*, pp. 473-479.

Lazrak M., 1996, *Nouvelle approche de commande optimale en temps final libre et construction d'algorithmes de commande de systèmes articulés*, Thèse d'état, Université de Poitiers.

Macfarlane S., Croft E. A., 2001, Design of jerk bounded trajectories for on-line industrial robot applications, *Proceeding of IEEE Int. Conf. on rob. & Aut.* pp. 979-984.

Martin B. J., Bobrow J. E., 1997, Minimum effort motions for open chain manipulators with task dependent end-effector constraints, *Proc. IEEE Int. Conf. Rob. & Aut.* Albuquerque, New Mexico, pp. 2044-2049.

Martin B. J., Bobrow J. E., 1999, Minimum effort motions for open chain manipulators with task dependent end-effector constraints, Int. Jour. Rob. of Research, 18(2), pp. 213-224.

Mao Z. , Hsia T. C., 1997, Obstacle avoidance inverse kinematics solution of redundant robots by neural networks, *Robotica*, 15, 3-10.

Mayorga R. V., 1995, A framework for the path planning of robot manipulators, *IASTED third Int. Conf. on Rob. and Manufacturing*, pp. 61-66.

Mitsi S., Bouzakis K. D., Mansour G., 1995, Optimization of robot links motion in inverse kinematics solution considering collision avoidance and joints limits, *Mach. & Mec. Theory,*30 (5), pp. 653-663.

Ola D., 1994, Path-Constrained robot control with limited torques. Experimental evaluation, *IEEE Trans. On Rob. & Aut.*, 10(5), pp. 658-668.

Overmars M. H., 1992, A random Approach to motion planning, *Technical report RUU-CS-92-32*, Utrecht University.

Piazzi A., Visioli A., 1998, Global minimum time trajectory planning of mechanical manipulators using internal analysis, *Int. J. Cont.*, 71(4), 631-652.

Pfeiffer F., Rainer J., 1987, A concept for manipulator trajectory planning, *IEEE J. of Rob. & Aut.*, Ra-3 (3): 115-123.

Powell M. J., 1984, Algorithm for non-linear constraints that use Lagrangian functions, *Mathematical programming*, 14, 224-248.

Pontryagin L., Boltianski V., Gamkrelidze R., Michtchenko E., 1965, *Théorie mathématique des processus optimaux*, Edition Mir.

Rana A.S., Zalazala A. M. S., 1996, Near time optimal collision free motion planning of robotic manipulators using an evolutionary algorithm, *Robotica*, 14, pp. 621-632.

Richard M. J., Dufourn F., Tarasiewicz S., 1993, Commande des robots manipulateurs par la programmation dynamique, *Mech. Mach. Theory*, 28(3), 301-316.

Rebai S., Bouabdallah A., Chettibi T., 2002, Recherche stochastique des mouvements libres optimaux des bras manipulateurs, *Premier congrès international de génie mécanique*, Constantine, Algérie.

Tan H. H., Potts R. B., 1988, Minimum time trajectory planner for the discrete dynamic robot model with dynamic constraints, *IEEE Jour. Of Rob. & Aut.* 4(2), 174-185.

Appendix A: *Characteristics of the 4R robot*

Joint N°	1	2	3	4
α (rad)	0	0	0	0
d(m)	0	1	1	1
r(m)	0	0	0	0
a(m)	0	0	0	0
M(kg)	5	4	3	2
Izz(kg.m^2)	1	0.85	0	0
τ(N.m)	25	20	15	5
F_s (N.m)	0.7	0.2	0.5	0.2
F_v (N.m.s)	1	0.2	0.5	0.2

FUZZY MODEL BASED CONTROL APPLIED TO IMAGE-BASED VISUAL SERVOING

Paulo J. Sequeira Gonçalves
Escola Superior de Tecnologia de Castelo Branco
Av. Empresário, 6000-767 Castelo Branco, Portugal
Email: pgoncalves@est.ipcb.pt

L.F. Mendonça, J.M. Sousa, J.R. Caldas Pinto
Technical University of Lisbon, IDMEC / IST
Av. Rovisco Pais, 1049-001 Lisboa, Portugal
{mendonca,j.sousa,jcpinto}@dem.ist.utl.pt

Keywords: visual servoing, robotic manipulators, inverse fuzzy control, fuzzy compensation, fuzzy modeling.

Abstract: A new approach to eye-in-hand image-based visual servoing based on fuzzy modeling and control is proposed in this paper. Fuzzy modeling is applied to obtain an inverse model of the mapping between image features velocities and joints velocities, avoiding the necessity of inverting the Jacobian. An inverse model is identified for each trajectory using measurements data of a robotic manipulator, and it is directly used as a controller. As the inversion is not exact, steady-state errors must be compensated. This paper proposes the use of a fuzzy compensator to deal with this problem. The control scheme contains an inverse fuzzy model and a fuzzy compensator, which are applied to a robotic manipulator performing visual servoing, for a given profile of image features velocities. The obtained results show the effectiveness of the proposed control scheme: the fuzzy controller can follow a point-to-point pre-defined trajectory faster (or smoother) than the classic approach.

1 INTRODUCTION

In eye-in-hand image-based visual servoing, the Jacobian plays a decisive role in the convergence of the control, due to its analytical model dependency on the selected image features. Moreover, the Jacobian must be inverted on-line, at each iteration of the control scheme. Nowadays, the research community tries to find the right image features to obtain a diagonal Jacobian (Tahri and Chaumette, 2003). The obtained results only guarantee the decoupling from the position and the orientation of the velocity screw. This is still a hot research topic, as stated very recently in (Tahri and Chaumette, 2003).

In this paper, the previous related problems in the Jacobian are addressed using fuzzy techniques, to obtain a controller capable to control the system. First, a fuzzy model to derive the inverse model of the robot is used to compute the joints and end-effector velocities in a straightforward manner. Second, the control action obtained from the inverse model is compensated to nullify a possible steady-state error by using a *fuzzy compensation*.

A two degrees of freedom planar robotic manipulator is controlled, based on eye-in-hand image-based visual servoing using fuzzy control systems.

The paper is organized as follows. Section 2 describes briefly the concept of image-based visual servoing. The problem statement for the control problem tackled in this paper is presented in Section 3. Section 4 presents fuzzy modeling and identification. Fuzzy compensation of steady-state errors is described in Section 5. Section 6 presents the experimental setup. The obtained results are presented in Section 7. Finally, Section 8 present the conclusions and the possible future research.

2 IMAGE-BASED VISUAL SERVOING

In image-based visual servoing (Hutchinson et al., 1996), the controlled variables are the image features, extracted from the image containing the object. The choice of different image features induces different control laws, and its number depends also on the number of degrees of freedom (DOF) of the robotic manipulator under control. The robotic manipulator used as test-bed in this paper is depicted in Fig. 1, and it has 2 DOF. Thus, only two features are needed to perform the control. The image features, s, used are the coordinates x and y of one image point.

81

J. Braz et al. (eds.), Informatics in Control, Automation and Robotics I, 81–88.
© 2006 *Springer. Printed in the Netherlands.*

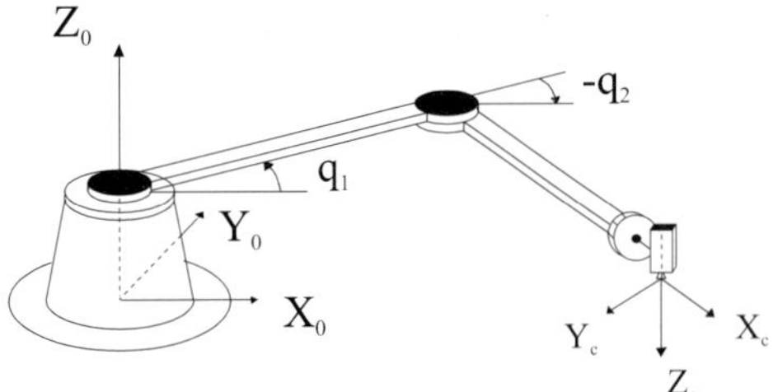

Figure 1: Planar robotic manipulator with eye-in-hand, camera looking down, with joint coordinates, and world and camera frames.

2.1 Modeling the Image-Based Visual Servoing System

In this paper, the image-based visual servoing system used is the eye-in-hand system, where the camera is fixed at the robotic manipulator end-effector. The kinematic modeling of the transformation between the image features velocities and the joints velocities is defined as follows. Let ^eP be the pose of the end-effector (translation and rotation), and ^cP be the pose of the camera. Both depend on the robot joint variables q. Thus, the transformation from the camera velocities and the end-effector velocities (Gonçalves and Pinto, 2003a) is given by:

$$^c\dot{P} = {}^cW_e \cdot {}^e\dot{P},\qquad(1)$$

where

$$^cW_e = \begin{bmatrix} {}^cR_e & S\left({}^ct_e\right)\cdot{}^cR_e \\ 0_{3\times3} & {}^cR_e \end{bmatrix}\qquad(2)$$

and $S(^ct_e)$, is the skew-symmetric matrix associated with the translation vector ^ct_e, and ^cR_e is the rotation matrix between the camera and end-effector frames needed to be measured.

The joint and end-effector velocities are related in the end-effector frame by:

$$^e\dot{P} = {}^eJ_R(q)\cdot\dot{q}\qquad(3)$$

where ^eJ_R is the robot Jacobian for the planar robotic manipulator (Gonçalves and Pinto, 2003a), and is given by:

$$^eJ_R(q) = \begin{bmatrix} l_1\cdot\sin(q_2) & l_2 + l_1\cdot\cos(q_2) & 0 & 0 & 0 & 1 \\ 0 & l_2 & 0 & 0 & 0 & 1 \end{bmatrix}^T\qquad(4)$$

and l_i, with $i = 1,2$ are the lengths of the robot links. The image features velocity, $\dot{s}$ and the camera velocity, $^c\dot{P}$ are related by:

$$\dot{s} = J_i(x,y,Z)\cdot{}^c\dot{P}\qquad(5)$$

where $J_i(x,y,Z)$ is the image Jacobian, which is derived using the pin-hole camera model and a single image point as the image feature (Gonçalves and

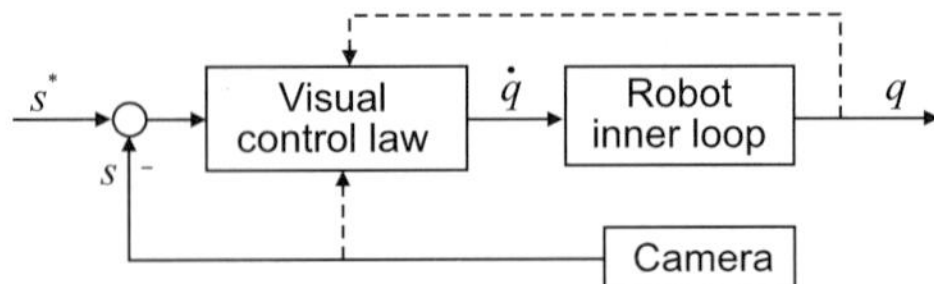

Figure 2: Control loop of image-based visual servoing, using a PD control law.

Pinto, 2003a), s, and is defined as

$$J_i(x,y,Z) = \begin{bmatrix} -\frac{1}{Z} & 0 & \frac{x}{Z} & x\cdot y & -\left(1+x^2\right) & y \\ 0 & -\frac{1}{Z} & \frac{y}{Z} & 1+y^2 & -x\cdot y & -x \end{bmatrix}\qquad(6)$$

where Z is the distance between the camera and the object frames.

The relation between the image features velocity and the joints velocities can now be easily derived from (1), (3) and (5):

$$\dot{s} = J(x,y,Z,q)\cdot\dot{q},\qquad(7)$$

where J is the total Jacobian, defined as:

$$J(x,y,Z,q) = J_i(x,y,Z)\cdot{}^cW_e\cdot{}^eJ_R(q)\qquad(8)$$

2.2 Controlling the Image-Based Visual Servoing System

One of the classic control scheme of robotic manipulators using information from the vision system, is presented in (Espiau et al., 1992). The global control architecture is shown in Fig. 2, where the block *Robot inner loop* law is a PD control law.

The robot joint velocities $\dot{q}$ to move the robot to a predefined point in the image, s^* are derived using the *Visual control law*, (Gonçalves and Pinto, 2003a), where an exponential decayment of the image features error is specified:

$$\dot{q} = -K_p\cdot\hat{J}^{-1}(x,y,Z,q)\cdot(s - s^*).\qquad(9)$$

K_p is a positive gain, that is used to increase or decrease the decayment of the error velocity.

3 PROBLEM STATEMENT

To derive an accurate global Jacobian, J, a perfect modeling of the camera, the image features, the position of the camera related to the end-effector, and the depth of the target related to the camera frame must be accurately determined.

Even when a perfect model of the Jacobian is available, it can contain singularities, which hampers the application of a control law. To overcome these difficulties, a new type of differential relationship between the features and camera velocities was proposed in

(Suh and Kim, 1994). This approach estimates the variation of the image features, when an increment in the camera position is given, by using a relation G. This relation is divided into G_1 which relates the position of the camera and the image features, and F_1 which relates their respective variation:

$$s + \delta s = G(^cP + \delta^c P) = G_1(^cP) + F_1(^cP, \delta^c P) \tag{10}$$

The image features velocity, $\dot{s}$, depends on the camera velocity screw, $^c\dot{P}$, and the previous position of the image features, s, as shown in Section 2. Considering only the variations in (10):

$$\delta s = F_1(^cP, \delta^c P), \tag{11}$$

let the relation between the camera position variation $\delta^c P$, the joint position variation, δq and the previous position of the robot q be given by:

$$\delta^c P = F_2(\delta q, q). \tag{12}$$

The equations (11) and (12) can be inverted if a one-to-one mapping can be guaranteed. Assuming that this inversion is possible, the inverted models are given by:

$$\delta^c P = F_1^{-1}(\delta s, {}^cP) \tag{13}$$

and

$$\delta q = F_2^{-1}(\delta^c P, q) \tag{14}$$

The two previous equations can be composed because the camera is rigidly attached to the robot end-effector, i.e., knowing q, ^cP can easily be obtained from the robot direct kinematics, ^cR_e and ^ct_e. Thus, an inverse function F^{-1} is given by:

$$\delta q = F^{-1}(\delta s, q), \tag{15}$$

and it states that the joint velocities depends on the image features velocities and the previous position of the robot manipulator. Equation (15) can be discretized as

$$\delta q(k) = F_k^{-1}(\delta s(k+1), q(k)). \tag{16}$$

In image-based visual servoing, the goal is to obtain a joint velocity, $\delta q(k)$, capable of driving the robot according to a desired image feature position, $s(k+1)$, with an also desired image feature velocity, $\delta s(k+1)$, from any position in the joint spaces. This goal can be accomplished by modeling the inverse function F_k^{-1}, using inverse fuzzy modeling as presented in section 4. This new approach to image-based visual servoing allows to overcome the problems stated previously regarding the Jacobian inverse, the Jacobian singularities and the depth estimation, Z.

4 INVERSE FUZZY MODELING

4.1 Fuzzy Modeling

Fuzzy modeling often follows the approach of encoding expert knowledge expressed in a verbal form in a collection of if–then rules, creating an initial model structure. Parameters in this structure can be adapted using input-output data. When no prior knowledge about the system is available, a fuzzy model can be constructed entirely on the basis of systems measurements. In the following, we consider data-driven modeling based on fuzzy clustering (Sousa et al., 2003).

We consider rule-based models of the Takagi-Sugeno (TS) type. It consists of fuzzy rules which each describe a local input-output relation, typically in a linear form:

$$R_i : \textbf{If } x_1 \text{ is } A_{i1} \textbf{ and } \ldots \textbf{ and } x_n \text{ is } A_{in}$$
$$\textbf{then } y_i = \mathbf{a}_i \mathbf{x} + b_i, \quad i = 1, 2, \ldots, K. \tag{17}$$

Here R_i is the ith rule, $\mathbf{x} = [x_1, \ldots, x_n]^T$ are the input (antecedent) variable, $A_{i1}, \ldots, A_{in}$ are fuzzy sets defined in the antecedent space, and y_i is the rule output variable. K denotes the number of rules in the rule base, and the aggregated output of the model, $\hat{y}$, is calculated by taking the weighted average of the rule consequents:

$$\hat{y} = \frac{\sum_{i=1}^{K} \beta_i y_i}{\sum_{i=1}^{K} w_i \beta_i}, \tag{18}$$

where β_i is the degree of activation of the ith rule: $\beta_i = \Pi_{j=1}^n \mu_{A_{ij}}(x_j)$, $i = 1, 2, \ldots, K$, and $\mu_{A_{ij}}(x_j) : \mathbb{R} \rightarrow [0,1]$ is the membership function of the fuzzy set A_{ij} in the antecedent of R_i.

To identify the model in (17), the regression matrix X and an output vector $\mathbf{y}$ are constructed from the available data: $X^T = [\mathbf{x}_1, \ldots, \mathbf{x}_N]$, $\mathbf{y}^T = [y_1, \ldots, y_N]$, where $N \gg n$ is the number of samples used for identification. The number of rules, K, the antecedent fuzzy sets, A_{ij}, and the consequent parameters, $\mathbf{a}_i, b_i$ are determined by means of fuzzy clustering in the product space of the inputs and the outputs (Babuška, 1998). Hence, the data set Z to be clustered is composed from X and $\mathbf{y}$: $Z^T = [X, \mathbf{y}]$. Given Z and an estimated number of clusters K, the Gustafson-Kessel fuzzy clustering algorithm (Gustafson and Kessel, 1979) is applied to compute the fuzzy partition matrix U.

The fuzzy sets in the antecedent of the rules are obtained from the partition matrix U, whose ikth element $\mu_{ik} \in [0,1]$ is the membership degree of the data object $\mathbf{z}_k$ in cluster i. One-dimensional fuzzy sets A_{ij} are obtained from the multidimensional fuzzy sets defined point-wise in the ith row of the partition matrix by projections onto the space of the input variables x_j. The point-wise defined fuzzy sets A_{ij} are approximated by suitable parametric functions in order to compute $\mu_{A_{ij}}(x_j)$ for any value of x_j.

The consequent parameters for each rule are obtained as a weighted ordinary least-square estimate. Let $\theta_i^T = [\mathbf{a}_i^T; b_i]$, let X_e denote the matrix $[X; \mathbf{1}]$

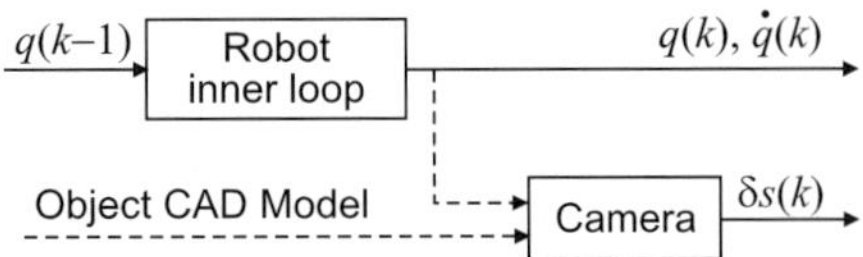

Figure 3: Robot-camera configuration for model identification.

and let W_i denote a diagonal matrix in $\mathbb{R}^{N \times N}$ having the degree of activation, $\beta_i(\mathbf{x}_k)$, as its kth diagonal element. Assuming that the columns of X_e are linearly independent and $\beta_i(\mathbf{x}_k) > 0$ for $1 \leq k \leq N$, the weighted least-squares solution of $\mathbf{y} = X_e\theta + \epsilon$ becomes

$$\theta_i = \left[X_e^T W_i X_e\right]^{-1} X_e^T W_i \mathbf{y}. \tag{19}$$

4.2 Inverse Modeling

For the robotic application in this paper, the inverse model is identified using input-output data from the inputs $\dot{q}(k)$, outputs $\delta s(k+1)$ and the state of the system $q(k)$. A commonly used procedure in robotics is to learn the trajectory that must be followed by the robot. From an initial position, defined by the joint positions, the robotic manipulator moves to the predefined end position, following an also predefined trajectory, by means of a PID joint position controller. This specialized procedure has the drawback of requiring the identification of a new model for each new trajectory. However, this procedure revealed to be quite simple and fast. Moreover, this specialized identification procedure is able to alleviate in a large scale the problems derived from the close-loop identification procedure. The identification data is obtained using the robot-camera configuration shown in Fig. 3.

Note that we are interested in the identification of the inverse model in (16). Fuzzy modeling is used to identify an inverse model, as e.g. in (Sousa et al., 2003). In this technique, only one of the states of the original model, $\dot{q}(k)$, becomes an output of the inverted model and the other state, $q(k)$, together with the original output, $\delta s(k + 1)$, are the inputs of the inverted model. This model is then used as the main controller in the visual servoing control scheme. Therefore, the inverse model must be able to find a joint velocity, $\dot{q}(k)$, capable to drive the robot following a desired image feature velocity in the image space, $\delta s(k + 1)$, departing from previous joint positions, $q(k)$.

5 FUZZY COMPENSATION OF STEADY-STATE ERRORS

Control techniques based on a nonlinear process model such as Model Based Predictive Control or control based on the inversion of a model, (Sousa et al., 2003), can effectively cope with the dynamics of nonlinear systems. However, steady-state errors may occur at the output of the system as a result of disturbances or a model-plant mismatch. A scheme is needed to compensate for these steady-state errors. A classical approach is the use of an integral control action. However, this type of action is not suitable for highly nonlinear systems because it needs different parameters for different operating regions of the system.

Therefore, this paper develops a new solution, called *fuzzy compensation*. The fuzzy compensator intends to compensate steady-state errors based on the information contained in the model of the system and allows the compensation action to change smoothly from an active to an inactive state. Taking the local derivative of the model with respect to the control action, it is possible to achieve compensation with only one parameter to be tuned (similar to the integral gain of a PID controller). For the sake of simplicity, the method is presented for nonlinear discrete-time SISO systems, but it is easily extended to MIMO systems. Note that this is the case of the 2-DOF robotic system in this paper.

5.1 Derivation of Fuzzy Compensation

In this section it is convenient to delay one step the model for notation simplicity. The discrete-time SISO regression model of the system under control is then given by:

$$y(k) = f(\mathbf{x}(k-1)), \tag{20}$$

where $\mathbf{x}(k - 1)$ is the state containing the lagged model outputs, inputs and states of the system. Fuzzy compensation uses a correction action called $u_c(k)$, which is added to the action derived from an inverse model-based controller, $u_m(k)$, as shown in Fig. 4. The total control signal applied to the process is thus given by,

$$u(k) = u_m(k) + u_c(k). \tag{21}$$

Note that the controller in Fig. 4 is an inverse model-based controller for the robotic application in this paper, but it could be any controller able to control the system, such as a predictive controller.

Taking into account the noise and a (small) offset error, a fuzzy set SS defines the region where the compensation is active, see Fig. 5. The error

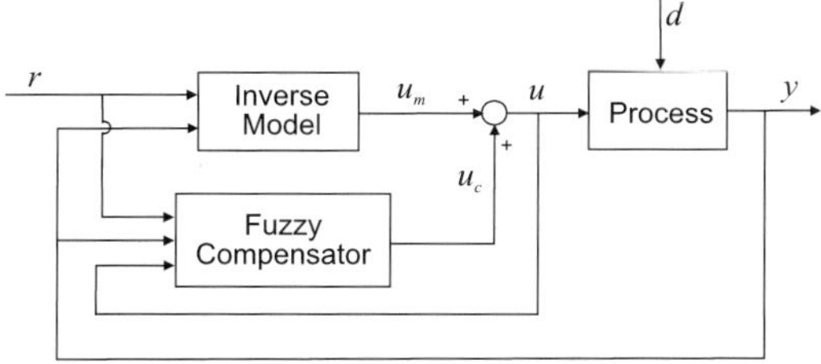

Figure 4: Fuzzy model-based compensation scheme.

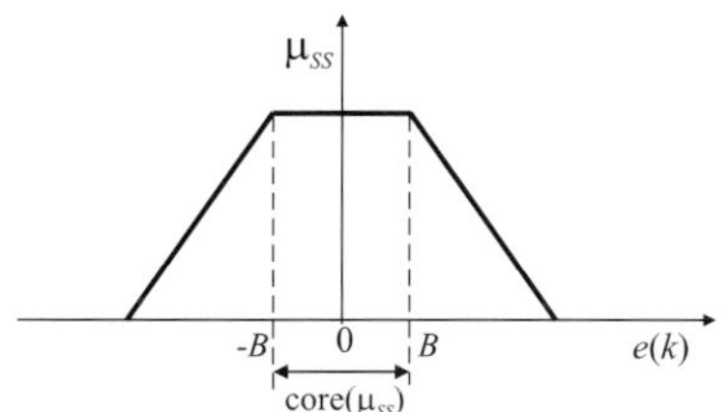

Figure 5: Definition of the fuzzy boundary SS where fuzzy compensation is active.

is defined as $e(k) = r(k) - y(k)$, and the membership function $\mu_{SS}(e(k))$ is designed to allow for steady-state error compensation whenever the support of $\mu_{SS}(e(k))$ is not zero. The value of B that determines the width of the $\mathrm{core}(\mu_{SS})$ should be an upper limit of the absolute value of the possible steady-state errors. Fuzzy compensation is fully active in the interval $[-B, B]$. The support of $\mu_{SS}(e(k))$ should be chosen such that it allows for a smooth transition from enabled to disabled compensation. This smoothness of μ_{SS} induces smoothness on the fuzzy compensation action $u_c(k)$, and avoids abrupt changes in the control action $u(k)$.

The compensation action $u_c(k)$ at time k is given by

$$u_c(k) = \mu_{SS}(e(k)) \left(\sum_{i=0}^{k-1} u_c(i) + K\, e(k)\, f_u^{-1} \right), \tag{22}$$

where $\mu_{SS}(e(k))$ is the error membership degree at time k, K_c is a constant parameter to be determined and

$$f_u = \left[\frac{\partial f}{\partial u(k-1)} \right]_{\mathbf{x}(k-1)} \tag{23}$$

is the partial derivative of the function f in (20) with respect to the control action $u(k-1)$, for the present state of the system $\mathbf{x}(k-1)$.

6 EXPERIMENTAL SETUP

The experimental setup can be divided in two subsystems, the vision system and the robotic manipulator system. The vision system acquires and process images at 50Hz, and sends image features, in pixels, to the robotic manipulator system via a RS-232 serial port, also at 50 Hz. The robotic manipulator system controls the 2 dof planar robotic manipulator, Fig. 6 and Fig. 1 using the image data sent from the vision system.

6.1 Vision System

The vision system performs image acquisition and processing under software developed in Visual C++^{TM} and running in an Intel Pentium IV^{TM}, at 1.7 GHz, using a Matrox Meteor^{TM} frame-grabber. The CCD video camera, Costar, is fixed in the robot end-effector looking up and with its optical axis perpendicular to the plane of the robotic manipulator. The planar target is parallel to the plane of the robot, and is above it. This configuration allows defining the Z variable as pre-measured constant in the image Jacobian (6) calculation, $Z = 1$. The target consists of one light emitting diode (LED), in order to ease the image processing and consequently diminish its processing time, because we are mainly concerned in control algorithms.

Following the work in (Reis et al., 2000) the image acquisition is performed at 50 Hz. It is well known that the PAL video signal has a frame rate of 25 frames/second. However, the signal is interlaced, i.e., odd lines and even lines of the image are codified in two separate blocks, known as fields. These fields are sampled immediately before each one is sent to the video signal stream. This means that each field is an image with half the vertical resolution, and the application can work with a rate of 50 frames/second. The two image fields were considered as separate images thus providing a visual sampling rate of 50 Hz.

When performing the centroid evaluation at 50 Hz, an error on its vertical coordinate will arise, due to the use of only half of the lines at each sample time. Multiplying the vertical coordinate by two, we obtain a simple approximation of this coordinate.

The implemented image processing routines extract the centroid of the led image. Heuristics to track this centroid, can be applied very easily (Gonçalves and Pinto, 2003b). These techniques allow us to calculate the image features vector s, i.e., the two image coordinates of the centroid. The image processing routines and the sending commands via the RS-232 requires less than 20ms to perform in our system. The RS-232 serial port is set to transmit at 115200 bits/second. When a new image is acquired, the pre-

Figure 6: : Experimental Setup, Planar Robotic Manipulator with eye-in-hand, camera looking up.

vious one was already processed and the image data sent via RS-232 to the robotic manipulator system.

6.2   Robotic Manipulator System

The robot system consists of a 2 dof planar robotic manipulator (Baptista et al., 2001) moving in a horizontal plane, the power amplifiers and an Intel Pentium^{TM} 200MHz, with a ServoToGo^{TM} I/O card. The planar robotic manipulator has two revolute joints driven by *Harmonic Drive Actuators - HDSH-14*. Each actuator has an electric d.c. motor, a harmonic drive, an encoder and a tachogenerator. The power amplifiers are configured to operate as current amplifiers. In this functioning mode, the input control signal is a voltage in the range $\pm 10V$ with current ratings in the interval $[-2; 2]V$. The signals are processed through a low cost ISA-bus I/O card from *ServoToGo, INC*. The I/O board device drivers were developed at the Mechanical Engineering Department from Instituto Superior Técnico. The referred PC is called in our overall system as the Target PC. It receives the image data from the vision system via RS-232, each 20 ms, and performs, during the 20 ms of the visual sample time, the visual servoing algorithms developed in the Host-PC.

It is worth to be noted that the robot manipulator joint limits are: $q \in \left[-\frac{\pi}{2}; \frac{\pi}{2}\right]$.

6.3   Systems Integration

The control algorithms for the robotic manipulator are developed in a third PC, called Host-PC. An executable file containing the control algorithms is then created for running in the Target-PC. The executable file created, containing the control algorithm, is then downloaded, via TCP/IP, to the Target-PC, where all the control is performed. After a pre-defined time for

execution, all the results can be uploaded to the Host-PC for analysis. During execution the vision system sends to the Target PC the actual image features as a visual feedback for the visual servoing control algorithm, using the RS-232 serial port.

7   RESULTS

This section presents the simulation results obtained for the robotic manipulator, presented in Section 6. First, the identification of the inverse fuzzy model of the robot is described. Then, the control results using the fuzzy model based controller introduced in this paper, i.e. the combination of inverse fuzzy control and fuzzy compensation, are presented.

7.1   Inverse Fuzzy Modeling

In order to apply the controller described in this paper, first an inverse fuzzy controller must be identified. Recall that a model must be identified for each trajectory. This trajectory is presented in Fig. 10. The profile chosen for the image features velocity moves the robot from the initial joints position $q = [-1.5; 0.3]$ to the final position $q = [-1.51; 1.52]$, in one second, starting and ending with zero velocity. An inverse fuzzy model (16) for this trajectory is identified using the fuzzy modeling procedure described in Section 4.1. The measurements data is obtained from a simulation of the planar robotic manipulator eye-in-hand system.

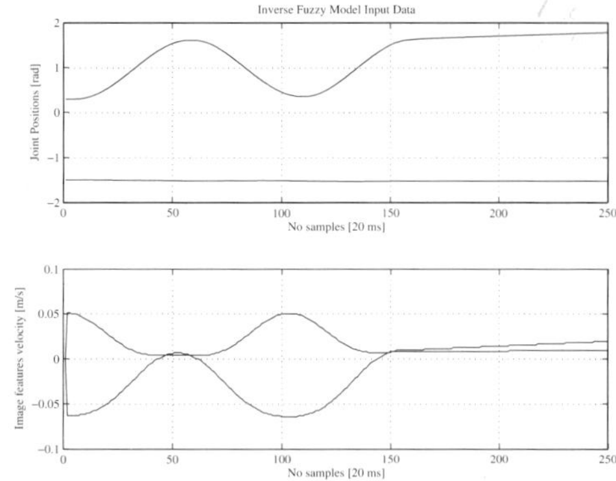

Figure 7: Input data for fuzzy identification. Top: joint positions, q. Bottom: image feature velocity, δs.

The set of identification data used to build the inverse fuzzy model contains 250 samples, with a sample time of 20ms. Figure 7 presents the input data, which are the joint positions $q_1(k)$ and $q_2(k)$, and the image features velocities $\delta s_x(k)$ and $\delta s_y(k)$, used for identification.

Note that to identify the inverse model, one cannot simply feed the inputs as outputs and the outputs as inputs. Since the inverse model (16) is a non-causal model, the output of the inverse model must be shifted one step, see (Sousa et al., 2003). The validation of the inverse fuzzy model is shown in Fig. 8, where the joint velocities δq are depicted. Note that two fuzzy models are identified, one for each velocity. It is clear that the model is quite good. Considering, e.g. the performance criteria *variance accounted for* (VAF), the models have the VAFs of 70.2% and 99.7%. When a perfect match occur, this measure has the value of 100%. Then, the inverse model for the joint velocity δq_2 is very accurate, but the inverse model for δq_1 is not so good. This was expectable as the joint velocity δq_1 varies much more than δq_2. However, this model will be sufficient to obtain an accurate controller, as is shown in Section 7.2.

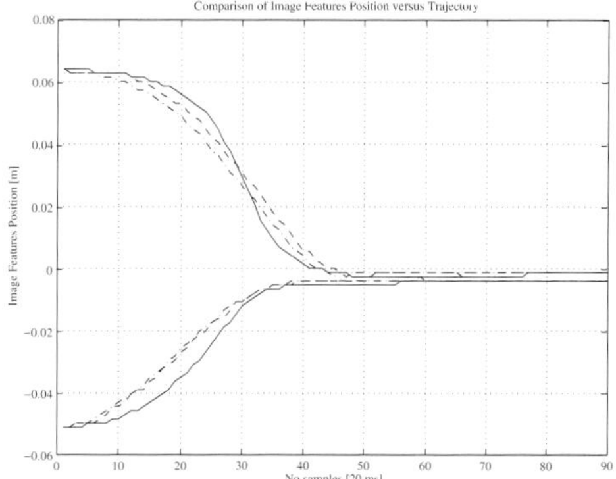

Figure 9: Image features trajectory, s. Solid – fuzzy visual servo control; dashed – classical visual servo control, and dash-dotted – desired trajectory.

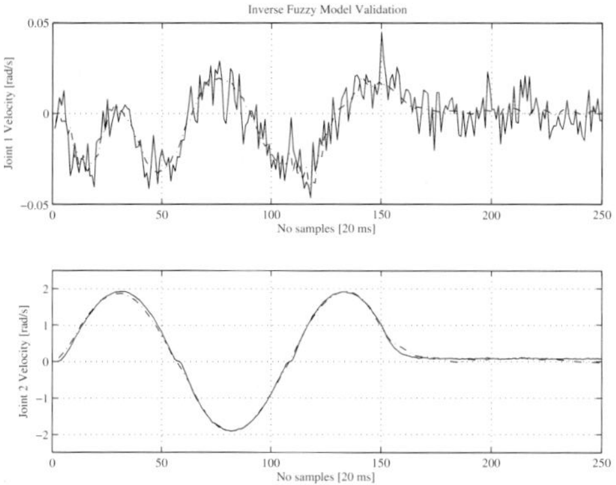

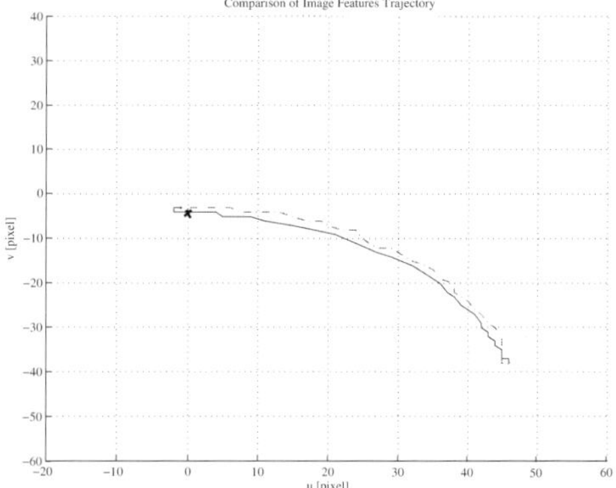

Figure 8: Validation of the inverse fuzzy model (joint velocities δq). Solid – real output data, and dash-dotted – output of the inverse fuzzy model.

Figure 10: Image features trajectory s, in the image plane. Solid – inverse fuzzy model control, and dash-dotted – classical visual servo control.

In terms of parameters, four rules (clusters) revealed to be sufficient for each output, and thus the inverse fuzzy model has 8 rules, 4 for each output, δq_1 and δq_2. The clusters are projected into the product-space of the space variables and the fuzzy sets A_{ij} are determined.

7.2 Control Results

This section presents the obtained control results, using the classical image-based visual servoing presented in Section 2, and the fuzzy model-based control scheme combining inverse model control presented in Section 4, and fuzzy compensation described in Sec ion 5. The implementation was developed in a simulation of the planar robotic manipulator eye-in-hand system. This simulation was developed and validated in real experiments, using classic visual servoing techniques, by the authors. Re-

call that the chosen profile for the image features velocity moves the robot from the initial joints position $q = [-1.5; 0.3]$ to the final position $q = [-1.51; 1.52]$ in one second, starting and ending with zero velocity.

The comparison of the image features trajectory for both the classic and the fuzzy visual servoing controllers is presented in Fig. 9. In this figure, it is shown that the classical controller follows the trajectory with a better accuracy than the fuzzy controller. However, the fuzzy controller is slightly faster, and reaches the vicinity of the target position before the classical controller. The image features trajectory in the image plane is presented in Fig. 10, which shows that both controllers can achieve the goal position (the times, $\times$, sign in the image) with a very small error. This figure shows also that the trajectory obtained with the inverse fuzzy model controller is smoother. Therefore, it is necessary to check the joint velocities in order to check their smoothness. Thus the joint veloc-

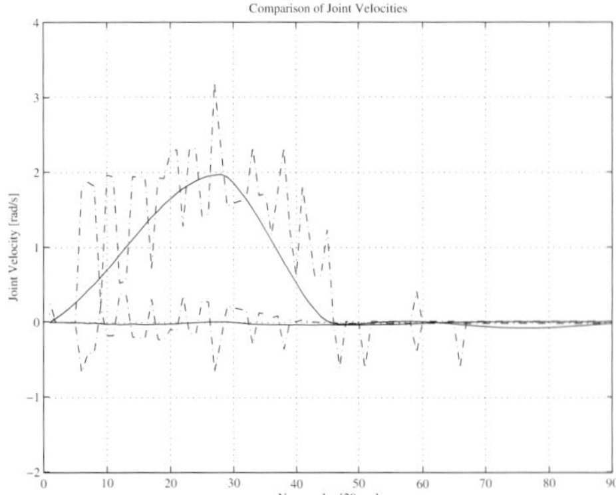

Figure 11: Joint velocities, δq. Solid – fuzzy visual servo, and dashed – classical visual servo.

ities are depicted in Fig. 11, where it is clear that the classical controller presents undesirable and dangerous oscillations in the joint velocities. This fact is due to the high proportional gain that the classical controller must have to follow the trajectory, which demands high velocity. This effect is easily removed by slowing down the classical controller. But in this case, the fuzzy controller clearly outperforms the classical controller. The classical controller can only follow the trajectory without oscillations in the joint velocities if the robot takes 1.5s to move from one point to the other. In this case, the classical controller is about 50% slower than the fuzzy model-based controller proposed in this paper.

8 CONCLUSIONS

This paper introduces an eye-in-hand image-based visual servoing scheme based on fuzzy modeling and control. The fuzzy modeling approach was applied to obtain an inverse model of the mapping between image features velocities and joints velocities. This inverse model is added to a fuzzy compensator to be directly used as controller of a robotic manipulator performing visual servoing, for a given image features velocity profile.

The obtained results showed that both the classical and the fuzzy controllers can follow the image features velocity profile. However, the proportional gain of the classic visual servoing must be very high. This fact justifies the oscillations verified in the joint velocities, which are completely undesirable in robot control. For that reason, the inverse fuzzy control proposed in this paper performs better.

As future work, the proposed fuzzy model based control scheme will be implemented in the experimental test-bed. Note that an off-line identification of the inverse fuzzy model must first be performed. The complete automation of this identification step is also under study.

ACKNOWLEDGEMENTS

This work is supported by the "Programa de Financiamento Plurianual de Unidades de I&D (POCTI) do Quadro Comunitário de Apoio III", by program FEDER, by the FCT project POCTI/EME/39946/2001, and by the "Programa do FSE-UE, PRODEP III, acção 5.3, no âmbito do III Quadro Comunitário de apoio".

REFERENCES

Babuška, R. (1998). *Fuzzy Modeling for Control*. Kluwer Academic Publishers, Boston.

Baptista, L., Sousa, J. M., and da Costa, J. S. (2001). Fuzzy predictive algorithms applied to real-time force control. *Control Engineering Practice*, pages 411–423.

Espiau, B., Chaumette, F., and Rives, P. (1992). A new approach to visual servoing in robotics. *IEEE Transactions on Robotics and Automation*, 8(3):313–326.

Gonçalves, P. S. and Pinto, J. C. (2003a). Camera configurations of a visual servoing setup, for a 2 dof planar robot. In *Proceedings of the 7th International IFAC Symposium on Robot Control, Wroclaw, Poland.*, pages 181–187, Wroclaw, Poland.

Gonçalves, P. S. and Pinto, J. C. (2003b). An experimental testbed for visual servo control of robotic manipulators. In *Proceedings of the IEEE Conference on Emerging Technologies and Factory Automation*, pages 377–382, Lisbon, Portugal.

Gustafson, D. E. and Kessel, W. C. (1979). Fuzzy clustering with a fuzzy covariance matrix. In *Proceedings IEEE CDC*, pages 761–766, San Diego, USA.

Hutchinson, S., Hager, G., and Corke, P. (1996). A tutorial on visual servo control. *IEEE Transactions on Robotics and Automation*, 12(5):651–670.

Reis, J., Ramalho, M., Pinto, J. C., and Sá da Costa, J. (2000). Dynamical characterization of flexible robot links using machine vision. In *Proceedings of the 5th Ibero American Symposium on Pattern Recognition*, pages 487–492, Lisbon, Portugal.

Sousa, J. M., Silva, C., and Sá da Costa, J. (2003). Fuzzy active noise modeling and control. *International Journal of Approximate Reasoning*, 33:51–70.

Suh, I. and Kim, T. (1994). Fuzzy membership function based neural networks with applications to the visual servoing of robot manipulators. *IEEE Transactions on Fuzzy Systems*, 2(3):203–220.

Tahri, O. and Chaumette, F. (2003). Application of moment invariants to visual servoing. In *Proceedings of the IEEE Int. Conf. on Robotics and Automation*, pages 4276–4281, Taipeh, Taiwan.

AN EVOLUTIONARY APPROACH TO NONLINEAR DISCRETE-TIME OPTIMAL CONTROL WITH TERMINAL CONSTRAINTS

Yechiel Crispin

Department of Aerospace Engineering
Embry-Riddle University
Daytona Beach, FL 32114
crispinj@erau.edu

Keywords: Optimal Control, Genetic Algorithms

Abstract: The nonlinear discrete-time optimal control problem with terminal constraints is treated using a new evolutionary approach which combines a genetic search for finding the control sequence with a solution of the initial value problem for the state variables. The main advantage of the method is that it does not require to obtain the solution of the adjoint problem which usually leads to a two-point boundary value problem combined with an optimality condition for finding the control sequence. The method is verified by solving the problem of discrete velocity direction programming with the effects of gravity and thrust and a terminal constraint on the final vertical position. The solution compared favorably with the results of gradient methods.

1 INTRODUCTION

A continuous-time optimal control problem consists of finding the time histories of the controls and the state variables such as to maximize (or minimize) an integral performance index over a finite period of time, subject to dynamical constraints in the form of a system of ordinary differential equations (Bryson, 1975). In a discrete-time optimal control problem, the time period is divided into a finite number of time intervals of equal time duration ΔT. The controls are kept constant over each time interval. This results in a considerable simplification of the continuous time problem, since the ordinary differential equations can be reduced to difference equations and the integral performance index can be reduced to a finite sum over the discrete time counter (Bryson, 1999). In some problems, additional constraints may be prescribed on the final states of the system. In this paper, we concentrate on the discrete-time optimal control problem with terminal constraints.Modern methods for solving the optimal control problem are extensions of the classical methods of the calculus of variations (Fox, 1950).

These methods are known as indirect methods and are based on the maximum principle of Pontryagin, which is astatement of the first order necessary conditions for optimality, and results in a two-point boundary value problem (TPBVP) for the state and adjoint variables (Pontryagin, 1962). It has been known, however, that the TPBVP is much more difficult to solve than the initial value problem (IVP). As a consequence, a second class of solutions, known as the direct method has evolved.

For example, attempts have been made to recast the original dynamic problem as a static optimization problem, also known as a nonlinear programming (NLP) problem.

This can be achieved by parameterisation of the state variables or the controls, or both. In this way, the original dynamical differential equations or difference equations are reduced to algebraic equality constraints. The problems with this approach is that it might result in a large scale NLP problem which has stability and convergence problems and might require excessive computing time. Also, the parameterisation might introduce spurious local minima which are not present in the original problem.

J. Braz et al. (eds.), Informatics in Control, Automation and Robotics I, 89–97.
© 2006 *Springer. Printed in the Netherlands.*

Several gradient based methods have been proposed for solving the discrete-time optimal control problem (Mayne, 1966). For example, Murray and Yakowitz (Murray, 1984) and (Yakowitz, 1984) developed a differential dynamic programming and Newton's method for the solution of discrete optimal control problems, see also the book of Jacobson and Mayne (Jacobson, 1970), (Ohno, 1978), (Pantoja, 1988) and (Dunn, 1989). Similar methods have been further developed by Liao and Shoemaker (Liao, 1991). Another method, the trust region method, was proposed by Coleman and Liao (Coleman, 1995) for the solution of unconstrained discrete-time optimal control problems. Although confined to the unconstrained problem, this method works for large scale minimization and is capable of handling the so called hard case problem.

Each method, whether direct or indirect, gradient-based or direct search based, has its own advantages and disadvantages. However, with the advent of computing power and the progress made in methods that are based on optimization analogies from nature, it became possible to achieve a remedy to some of the above mentioned disadvantages through the use of global methods of optimization. These include stochastic methods, such as simulated annealing (Laarhoven, 1989), (Kirkpatrick, 1983) and evolutionary computation methods (Fogel, 1998), (Schwefel, 1995) such as genetic algorithms (GA) (Michalewicz, 1992a), see also (Michalewicz, 1992b) for an interesting treatment of the linear discrete-time problem.

Genetic algorithms provide a powerful mechanism towards a global search for the optimum, but in many cases, the convergence is very slow. However, as will be shown in this paper, if the GA is supplemented by problem specific heuristics, the convergence can be accelerated significantly. It is well known that GAs are based on a guided random search through the genetic operators and evolution by artificial selection. This process is inherently very slow, because the search space is very large and evolution progresses step by step, exploring many regions with solutions of low fitness. What is proposed here, is to guide the search further, by incorporating qualitative knowledge about potential good solutions. In many problems, this might involve simple heuristics, which when combined with the genetic search, provide a powerful tool for finding the optimum very quickly.

The purpose of the present work, then, is to incorporate problem specific heuristic arguments, which when combined with a modified hybrid GA, can solve the discrete-time optimal control problem very easily. There are significant advantages to this approach. First, the need to solve the two-point boundary value problem (TPBVP) is completely avoided. Instead, only initial value problems (IVP) are solved. Second, after finding an optimal solution, we verify that it approximately satisfies the first-order necessary conditions for a stationary solution, so the mathematical soundness of the traditional necessary conditions is retained. Furthermore, after obtaining a solution by direct genetic search, the static and dynamic Lagrange multipliers (the adjoint variables) can be computed and compared with the results from a gradient method. All this is achieved without directly solving the TPBVP. There is a price to be paid, however, since, in the process, we are solving many initial value problems (IVPs). This might present a challenge in advanced difficult problems, where the dynamics are described by a higher order system of ordinary differential equations, or when the equations are difficult to integrate over the required time interval and special methods are required. On the other hand, if the system is described by discrete-time difference equations that are relatively well behaved and easy to iterate, the need to solve the initial value problem many times does not represent a serious problem. For instance, the example problem presented here , the discrete velocity programming problem (DVDP) with the combined effects of gravity, thrust and drag, together with a terminal constraint (Bryson, 1999), runs on a 1.6 GHz pentium 4 processor in less than a minute of CPU time.

In the next section, a mathematical formulation of the optimal control problem is given. The evolutionary approach to the solution is then described. In order to illustrate the method, a specific example, the discrete velocity direction programming (DVDP) is solved and results are presented and compared with the results of an indirect gradient method developed by Bryson (Bryson, 1999).

2 OPTIMAL CONTROL OF NONLINEAR DISCRETE TIME DYNAMICAL SYSTEMS

In this section, we describe a formulation of the nonlinear discrete-time optimal control problem subject to terminal constraints. Consider the nonlinear discrete-time dynamical system described by a system of difference equations with initial conditions

$$(1) \quad \boldsymbol{x}(i+1) = \boldsymbol{f}[\boldsymbol{x}(i), \boldsymbol{u}(i), i]$$

$$(2) \quad \boldsymbol{x}(0) = \boldsymbol{x_0}$$

where $\boldsymbol{x} \in \mathbb{R}^n$ is the vector of state variables, $\boldsymbol{u} \in \mathbb{R}^p$, $p < n$ is the vector of control variables and $i \in [0, N-1]$ is a discrete time counter. The function $\boldsymbol{f}$ is a nonlinear function of the state vector, the control vector and the discrete time i, i.e., $\boldsymbol{f} : \mathbb{R}^n \times \mathbb{R}^p \times \mathbb{R} \mapsto \mathbb{R}^n$. Define a performance index

$$(3)$$
$$J[\boldsymbol{x}(i), \boldsymbol{u}(i), i, N] = \phi[\boldsymbol{x}(N)]$$
$$+ \sum_{i=0}^{N-1} L[\boldsymbol{x}(i), \boldsymbol{u}(i), i]$$

where
$$\phi : \mathbb{R}^n \mapsto \mathbb{R}, \qquad L : \mathbb{R}^n \times \mathbb{R}^p \times \mathbb{R} \mapsto \mathbb{R}$$

Here L is the Lagrangian function and $\phi[\boldsymbol{x}(N)]$ is a function of the terminal value of the state vector $\boldsymbol{x}(N)$. Terminal constraints in the form of additional functions $\boldsymbol{\psi}$ of the state variables are prescribed as

$$(4) \quad \boldsymbol{\psi}[\boldsymbol{x}(N)] = 0 \qquad \boldsymbol{\psi} : \mathbb{R}^n \mapsto \mathbb{R}^k \quad k \leq n$$

The optimal control problem consists of finding the control sequence $\boldsymbol{u}(i)$ such as to maximize (or minimize) the performance index defined by (3), subject to the dynamical equations (1) with initial conditions (2) and terminal constraints (4). This is known as the Bolza problem in the calculus of variations (Bolza, 1904).

In an alternative formulation, due to Mayer, the state vector x_j, $j \in (1, n)$ is augmented by an additional variable x_{n+1} which satisfies the following initial value problem:

$$(5) \quad x_{n+1}(i+1) = x_{n+1}(i) + L[\boldsymbol{x}(i), \boldsymbol{u}(i), i]$$

$$(6) \quad x_{n+1}(0) = 0$$

The performance index can then be written as

$$(7)$$
$$J(N) = \phi[\boldsymbol{x}(N)] + x_{n+1}(N) \equiv \phi_a[\boldsymbol{x_a}(N)]$$

where $\boldsymbol{x_a} = [\boldsymbol{x} \ x_{n+1}]^T$ is the augmented state vector and ϕ_a the augmented performance function. In this paper, the Meyer formulation is used.

Define an augmented performance index with adjoint constraints $\boldsymbol{\psi}$ and adjoint dynamical constraints $\boldsymbol{f}(\boldsymbol{x}(i), \boldsymbol{u}(i), i) - \boldsymbol{x}(i+1) = 0$, with static and dynamical Lagrange multipliers $\boldsymbol{\nu}$ and $\boldsymbol{\lambda}$, respectively.

$$(8) \quad J_a = \phi + \boldsymbol{\nu}^T \boldsymbol{\psi} + \boldsymbol{\lambda}^T(0)[\boldsymbol{x_0} - \boldsymbol{x}(0)]$$
$$+ \sum_{i=0}^{N-1} \boldsymbol{\lambda}^T(i+1)\{\boldsymbol{f}[\boldsymbol{x}(i), \boldsymbol{u}(i), i] - \boldsymbol{x}(i+1)\}$$

Define the Hamiltonian function as

$$(9) \quad H(i) = \boldsymbol{\lambda}^T(i+1)\boldsymbol{f}[\boldsymbol{x}(i), \boldsymbol{u}(i), i]$$

Rewriting the augmented performance index in terms of the Hamiltonian, we get

$$(10)$$
$$J_a = \phi + \boldsymbol{\nu}^T\boldsymbol{\psi} - \boldsymbol{\lambda}^T(N)\boldsymbol{x}(N) + \boldsymbol{\lambda}^T(0)\boldsymbol{x_0}$$
$$+ \sum_{i=0}^{N-1} [H(i) - \boldsymbol{\lambda}^T(i)\boldsymbol{x}(i)]$$

A first order necessary condition for J_a to reach a stationary solution is given by the discrete version of the Euler-Lagrange equations

$$(11)$$
$$\boldsymbol{\lambda}^T(i) = H_{\boldsymbol{x}}(i) = \boldsymbol{\lambda}^T(i+1)\boldsymbol{f_x}[\boldsymbol{x}(i), \boldsymbol{u}(i), i]$$

with final conditions

$$(12) \quad \boldsymbol{\lambda}^T(N) = \phi_x + \boldsymbol{\nu}^T \boldsymbol{\psi_x}$$

and the control sequence $\boldsymbol{u}(i)$ satisfies the optimality condition:

$$(13) \quad H_{\boldsymbol{u}}(i) = \boldsymbol{\lambda}^T(i+1)\boldsymbol{f_u}[\boldsymbol{x}(i), \boldsymbol{u}(i), i] = 0$$

Define an augmented function Φ as

$$(14) \quad \Phi = \phi + \boldsymbol{\nu}^T \boldsymbol{\psi}$$

Then, the adjoint equations for the dynamical multipliers are given by

(15)
$$\boldsymbol{\lambda}^T(i) = H_{\boldsymbol{x}}(i) = \boldsymbol{\lambda}^T(i+1)\boldsymbol{f}_{\boldsymbol{x}}[\boldsymbol{x}(i), \boldsymbol{u}(i), i]$$

and the final conditions can be written in terms of the augmented function Φ

(16) $\boldsymbol{\lambda}^T(N) = \Phi_{\boldsymbol{x}} = \phi_{\boldsymbol{x}} + \boldsymbol{\nu}^T \boldsymbol{\psi}_{\boldsymbol{x}}$

If the indirect approach to optimal control is used, the state equations (1) with initial conditions (2) need to be solved together with the adjoint equations (15) and the final conditions (16), where the control sequence $u(i)$ is to be determined from the optimality condition (13). This represents a coupled system of nonlinear difference equations with part of the boundary conditions specified at the initial time $i = 0$ and the rest of the boundary conditions specified at the final time $i = N$. This is a nonlinear two-point boundary value problem (TPBVP) in difference equations. Except for special simplified cases, it is usually very difficult to obtain solutions for such nonlinear TPBVPs. Therefore, many numerical methods have been developed to tackle this problem, see the introduction for several references.

In the proposed approach, the optimality condition (13) and the adjoint equations (15) together with their final conditions (16) are not used in order to obtain the optimum solution. Instead, the optimal values of the control sequence $u(i)$ are found by genetic search starting with an initial population of solutions with values of $u(i)$ randomly distributed within a given domain. During the search, approximate, less than optimal values of the solutions $u(i)$ are available for each generation. With these approximate values known, the state equations (1) together with their initial conditions (2) are very easy to solve, by a straightforward iteration of the difference equations from $i = 0$ to $i = N - 1$. At the end of this iterative process, the final values $\boldsymbol{x}(N)$ are obtained, and the fitness function can be determined. The search than seeks to maximize the fitness function F such as to fulfill the goal of the evolution, which is to maximize $J(N)$, as given by the following Eq.(17), subject to the terminal constraints $\boldsymbol{\psi}[\boldsymbol{x}(N)] = 0$, as defined by Eq.(18).

(17)   maximize $J(N) = \phi[\boldsymbol{x}(N)]$

subject to the dynamical equality constraints, Eqs. (1-2) and to the terminal constraints (4), which are repeated here for convenience as Eq.(18)

(18) $\boldsymbol{\psi}[\boldsymbol{x}(N)] = 0 \qquad \boldsymbol{\psi} : \mathbb{R}^n \mapsto \mathbb{R}^k \quad k \leq n$

Since we are using a direct search method, condition (18) can also be stated as a search for a maximum, namely we can set a goal which is equivalent to (18) in the form

(19)   maximize $J_1(N) = -\boldsymbol{\psi}^{\mathbf{T}}[\boldsymbol{x}(N)]\boldsymbol{\psi}[\boldsymbol{x}(N)]$

The fitness function F can now be defined by

(20) $F(N) = \alpha J(N) + (1 - \alpha)J_1(N)$

$$= \alpha\phi[\boldsymbol{x}(N)] - (1 - \alpha)\boldsymbol{\psi}^{\mathbf{T}}[\boldsymbol{x}(N)]\boldsymbol{\psi}[\boldsymbol{x}(N)],$$

with $\alpha \in [0, 1]$ and $\boldsymbol{x}(N)$ determined from the following dynamical equality constraints:

(21) $\boldsymbol{x}(i+1) = \boldsymbol{f}[\boldsymbol{x}(i), \boldsymbol{u}(i), i]$, $i \in [0, N - 1]$

(22) $\boldsymbol{x}(0) = \boldsymbol{x_0}$

3   DISCRETE VELOCITY DIRECTION PROGRAMMING FOR MAXIMUM RANGE WITH GRAVITY AND THRUST

In this section, the above formulation is applied to a specific example, namely, the problem of finding the trajectory of a point mass subjected to gravity and thrust and a terminal constraint such as to achieve maximum horizontal distance, with the direction of the velocity used to control the motion. This problem has its roots in the calculus of variations and is related to the classical Brachistochrone problem, in which the shape of a wire is sought along which a bead slides without friction, under the action of gravity alone, from an initial point (x_0, y_0) to a final point (x_f, y_f) in minimum time t_f. The dual problem to the Brachistochrone problem consists of finding the shape of the wire such as the bead will reach a maximum horizontal distance x_f in a prescribed time t_f. Here, we treat the dual problem with the added effects of thrust and a terminal constraint where the final vertical position y_f is prescribed. The more difficult problem, where the body is moving in a viscous fluid and the effect of drag is taken into account was also solved, but due to lack of space, the results will be presented elsewhere. The reader interested in these problems can find an extensive discussion in Bryson's book (Bryson,1999).

Let point O be the origin of a cartesian system of coordinates in which x is pointing to the right and y is pointing down. A constant thrust force $\boldsymbol{F}$ is acting along the path on a particle of mass m which moves in a medium without friction. A constant gravity field with acceleration g is acting downward in the positive y direction. The thrust acts in the direction of the velocity vector $\boldsymbol{V}$ and its magnitude is $F = amg$, i.e. a times the weight mg. The velocity vector $\boldsymbol{V}$ acts at an angle γ with respect to the positive direction of x. The angle γ, which serves as the control variable, is positive when $\boldsymbol{V}$ points downward from the horizontal. The problem is to find the control sequence $\gamma(t)$ to maximize the horizontal range x_f in a given time t_f, provided the particle ends at the vertical location y_f. In other words, the velocity direction $\gamma(t)$ is to be programmed such as to achieve maximum range and fulfill the terminal constraint y_f. The equations of motion are

$$(23) \quad dV/dt = g(a + sin\gamma(t))$$

$$(24) \quad dx/dt = V cos\gamma(t)$$

$$(25) \quad dy/dt = V sin\gamma(t)$$

with initial conditions

$$(26) \quad V(0) = 0, \ x(0) = 0, \ y(0) = 0$$

and final constraint

$$(27) \quad y(t_f) = y_f.$$

We would like to formulate a discrete time version of this problem. The trajectory is divided into a finite number N of straight line segments of fixed time duration $\Delta T = t_f/N$, along which the direction γ is constant. We can increase N such as to approach the solution of the continuous trajectory. The velocity $\boldsymbol{V}$ is increasing under the influence of

a constant thrust ag and gravity $gsin\gamma$. The problem is to determine the control sequence $\gamma(i)$ at points i along the trajectory, where $i \in [0, N-1]$, such as to maximize x at time t_f, arriving at the same time at a prescribed elevation y_f. The time at the end of each segment is given by $t(i) = i\Delta T$, so i can be viewed as a time step counter at point i. Integrating the first equation of motion, Eq.(17) from a time $t(i) = i\Delta T$ to $t(i+1) = (i+1)\Delta T$, we get

$$(28) \quad V(i+1) = V(i) + g[a + sin\gamma(i)]\Delta T$$

Integrating the velocity V over a time interval ΔT, we obtain the length of the segment $\Delta d(i)$ connecting the points i and $i+1$.

$$(29)$$
$$\Delta d(i) = V(i)\Delta T + \tfrac{1}{2}g[a + sin\gamma(i)](\Delta T)^2$$

Once $\Delta d(i)$ is determined, it is easy to obtain the coordinates x and y as

$$(30) \quad x(i+1) = x(i) + \Delta d(i)cos\gamma(i)$$

$$= x(i) + V(i)cos\gamma(i)\Delta T$$

$$+ \tfrac{1}{2}g[a + sin\gamma(i)]cos\gamma(i)(\Delta T)^2$$

$$(31) \quad y(i+1) = y(i) + \Delta d(i)sin\gamma(i)$$

$$= y(i) + V(i)sin\gamma(i)\Delta T$$

$$+ \tfrac{1}{2}g[a + sin\gamma(i)]sin\gamma(i)(\Delta T)^2$$

We now develop the equations in nondimensional form. Introduce the following nondimensional variables denoted by primes:

$$(32) \quad t = t_f t', \ V = gt_f V', \ x = gt_f^2 x',$$
$$y = gt_f^2 y'$$

Since $t(i) = it_f/N$, the nondimensional time is $t'(i) = i/N$. The time interval was defined as $\Delta T = t_f/N = t_f(\Delta T)'$, so the nondimensional time interval becomes $(\Delta T)' = 1/N$. Substituting the nondimensional variables in the discrete equations of motion and omitting the prime notation, we obtain the nondimensional state equations:

(33) $V(i+1) = V(i) + [a + sin\gamma(i)]/N$

(34) $x(i+1) = x(i) + V(i)cos\gamma(i)/N$

$+ \frac{1}{2}[a + sin\gamma(i)]cos\gamma(i)/N^2$

(35) $y(i+1) = y(i) + V(i)sin\gamma(i)/N$

$+ \frac{1}{2}[a + sin\gamma(i)]sin\gamma(i)/N^2$

with initial conditions

(36) $V(0) = 0,\ x(0) = 0,\ y(0) = 0$

and terminal constraint

(37) $y(N) = y_f$

The optimal control problem now consists of finding the sequence $\gamma(i)$ such as to maximize the range $x(N)$, subject to the dynamical constraints (33-35), the initial conditions (36) and the terminal constraint (37), where y_f is in units of $g\ t_f^2$ and the final time t_f is given.

4 NECESSARY CONDITIONS FOR AN OPTIMUM

In this section we present the traditional indirect approach to the solution of the optimal control problem, which is based on the first order necessary conditions for an optimum. First, we derive the Hamiltonian function for the above DVDP problem. We then derive the adjoint dynamical equations for the adjoint variables (the Lagrange multipliers) and the optimality condition that needs to be satisfied by the control sequence $\gamma(i)$. Since we have used the symbol x for the horizontal coordinate, we denote the state variables by $\boldsymbol{\xi}$. So the state vector for this problem is

$$\boldsymbol{\xi} = [V\ x\ y]^T$$

The performance index and the terminal constraint are given by

(38) $J(N) = \phi[\boldsymbol{\xi}(N)] = x(N)$

(39) $\psi(N) = \psi[\boldsymbol{\xi}(N)] = y(N) - y_f = 0$

The Hamiltonian $H(i)$ is defined by

(40)

$H(i) = \lambda_V(i+1)\{V(i) + [a + sin\gamma(i)]/N\}$

$+ \lambda_x(i+1)\{x(i) + V(i)cos\gamma(i)/N$

$+ \frac{1}{2}[a + sin\gamma(i)]cos\gamma(i)/N^2\}$

$+ \lambda_y(i+1)\{y(i) + V(i)sin\gamma(i)/N$

$+ \frac{1}{2}[a + sin\gamma(i)]sin\gamma(i)/N^2\}$

The augmented performance index is given by

(41) $\Phi = \phi + \boldsymbol{\nu}^T \boldsymbol{\psi} = x(N) + \nu[y(N) - y_f]$

The discrete Euler-Lagrange equations are derived from the Hamiltonian function:

(42) $\lambda_V(i) = H_V(i) = \lambda_V(i+1)$

$+ cos\gamma(i)\lambda_x(i+1)/N + sin\gamma(i)\lambda_y(i+1)/N$

(43) $\lambda_x(i) = H_x(i) = \lambda_x(i+1)$

(44) $\lambda_y(i) = H_y(i) = \lambda_y(i+1)$

It follows from the last two equations that the multipliers $\lambda_x(i)$ and $\lambda_y(i)$ are constant. The final conditions for the multipliers are obtained from the augmented function Φ.

(45)

$\lambda_V(N) = \Phi_V(N) = \phi_V(N) + \nu\,\psi_V(N) = 0$

$\lambda_x(N) = \Phi_x(N) = \phi_x(N) + \nu\,\psi_x(N) = 1$

$\lambda_y(N) = \Phi_y(N) = \phi_y(N) + \nu\,\psi_y(N) = \nu$

Since λ_x and λ_y are constant, they can be set equal to their final values:

(46) $\lambda_x(i) = \lambda_x(N) = 1$,
$\lambda_y(i) = \lambda_y(N) = \nu$

With the values given in (46), the equation for $\lambda_V(i)$ becomes

(47)
$$\lambda_V(i) = \lambda_V(i+1) + cos\gamma(i)/N + \nu sin\gamma(i)/N$$

with final condition

(48) $\lambda_V(N) = 0$

The required control sequence $\gamma(i)$ is determined from the optimality condition

(49)
$$H_\gamma(i) = \lambda_V(i+1)cos\gamma(i)/N - V(i)sin\,\gamma(i)/N - asin\gamma(i)/(2N^2) + cos(2\gamma(i))/(2N^2) + \nu V(i)cos\gamma(i)/N + \nu acos\gamma(i)/(2N^2) + \nu sin(2\gamma(i))/(2N^2) = 0$$

$\lambda_V(i+1)$ is determined by the adjoint equation (47) and the Lagrange multiplier ν is determined from the terminal equality constraint $y(N) = y_f$.

5 AN EVOLUTIONARY APPROACH TO OPTIMAL CONTROL

We now describe the direct approach using genetic search. As was mentioned in Sec. , there is no need to solve the two-point boundary value problem described by the state equations (33-35) and the adjoint equation (47), together with the initial conditions (36), the final condition (48), the terminal constraint (37) together with the optimality condition (49) for the optimal control $\gamma(i)$. Instead, the direct evolutionary method allows us to evolve a population of solutions such as to maximize the objective function or fitness function $F(N)$. The initial population is built by generating a random population of solutions $\gamma(i)$, $i \in [0, N-1]$, uniformly distributed within a domain $\gamma \in [\gamma_{min}, \gamma_{max}]$. Typical values are $\gamma_{max} = \pi/2$ and either $\gamma_{min} = -\pi/2$ or $\gamma_{min} = 0$ depending on the problem. The genetic algorithm evolves this initial population using the operations of selection,

mutation and crossover over many generations such as to maximize the fitness function:

(50) $F(N) = \alpha J(N) + (1-\alpha)J_1(N)$

$$= \alpha\phi[\boldsymbol{\xi}(N)] - (1-\alpha)\boldsymbol{\psi}^{\mathbf{T}}[\boldsymbol{\xi}(N)]\boldsymbol{\psi}[\boldsymbol{\xi}(N)],$$

with $\alpha \in [0, 1]$ and $J(N)$ and $J_1(N)$ given by:

(51) $J(N) = \phi[\boldsymbol{\xi}(N)] = x(N)$

(52) $J_1(N) = \psi^2[\boldsymbol{\xi}(N)] = (y(N) - y_f)^2$

For each member in the population of solutions, the fitness function depends on the final values $x(N)$ and $y(N)$, which are determined by solving the initial value problem defined by the state equations (33-35) together with the initial conditions (36). This process is repeated over many generations. Here, we run the genetic algorithm for a predetermined number of generations and then we check if the terminal constraint (52) is fulfilled. If the constraint is not fulfilled, we can either increase the number of generations or readjust the weight $\alpha \in [0, 1]$. After obtaining an optimal solution , we can check the first order necessary conditions by first solving the adjoint equation (47) with its final condition (48). Once the control sequence is known, the solution of (47-48) is obtained by direct iteration backwards in time. We then check to what extent the optimality condition (49) is fulfilled by determining $H_\gamma(i) = e(i)$ for $i \in [0, N-1]$ and plotting the result as an error $e(i)$ measuring the deviation from zero.

The results for the DVDP problem with gravity and thrust, with $a = 0.5$ and the terminal constraint $y_f = -0.1$ are shown in Figs.(1-3). A value of $\alpha = 0.01$ was used. Fig.1 shows the evolution of the solution over 50 generations. The best fitness and the average fitness of the population are given. In all calculations the size of the population was 50 members.

The control sequence $\gamma(i)$, the velocity $V(i)$ and the optimal trajectory are given in Fig.2 where the sign of y is reversed for plotting. The trajectory obtained here was compared to that obtained by Bryson (Bryson, 1999) using a gradient method and the results are similar. In Fig. 3 we plot the expression for $dH/d\gamma(i)$ as given by the right-hand side of Eq. (49). Ideally, this should be equal to zero at every point i. However, since we are not using (49) to determine the control sequence, we obtain a small deviation from zero in our calculation. Finally, after determining the optimal solution, i.e. after the control and the trajectory are known, the adjoint variable $\lambda_V(i)$ can be estimated by using Eqs.(47-48). The result is shown in Fig.3.

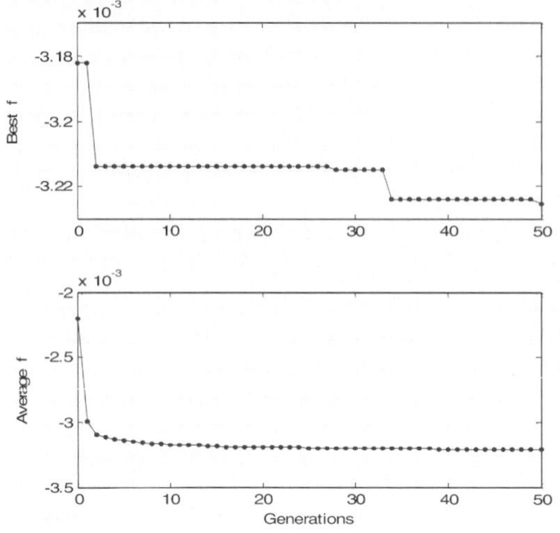

Figure 1: Convergence of the DVDP solution with gravity and thrust, $a = 0.5$ and terminal constraint $y_f = -0.1$.

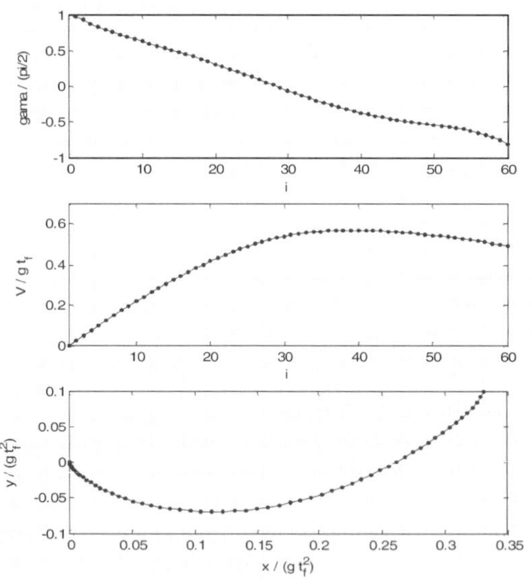

Figure 2: The control sequence $\gamma(i)$, the velocity $V(i)$ and the optimal trajectory for the DVDP problem with gravity g, thrust $a = 0.5$ and terminal constraint $y_f = -0.1$. The sign of y is reversed for plotting.

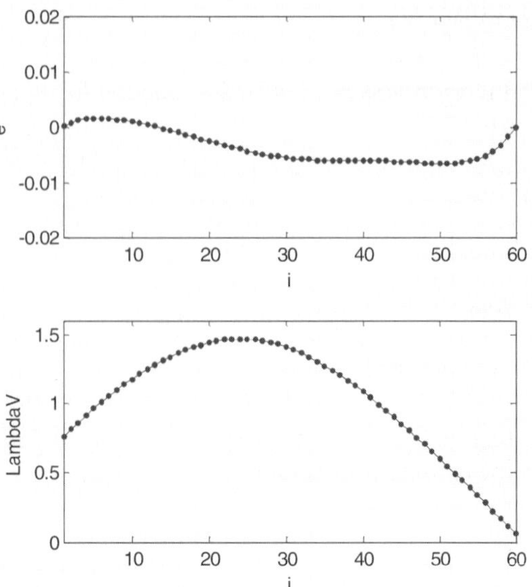

Figure 3: The error $e(i)$ measuring the deviation of the optimality condition $H_\gamma(i) = e(i)$ from zero, and the adjoint variable $\lambda_V(i)$ as a function of the discrete-time counter i. DVDP with gravity and thrust, $a = 0.5$. Here $N = 60$ time steps.

6 CONCLUSIONS

A new method for solving the discrete-time optimal control problem with terminal constraints was developed. The method seeks the best control sequence diretly by genetic search and does not make use of the first-order necessary conditions to find the optimum. As a consequence, the need to develop a Hamiltonian formulation and the need to solve a difficult two-point boundary value problem for finding the adjoint variables is completely avoided. This has a significant advantage in more advanced and higher order problems where it is difficult to solve the TPBVP with large systems of differential equations, but when it is still easy to solve the initial value problem (IVP) for the state variables. The method was demonstrated by solving a discrete-time optimal control problem, namely, the DVDP or the discrete velocity direction programming problem that was pioneered by Bryson using both analytical and gradient methods. This problem includes the effects of gravity and thrust and was solved easily using the proposed approach. The results compared favorably with those of

Bryson, who used analytical and gradient techniques.

REFERENCES

Bryson, A. E. and Ho, Y. C., Applied Optimal Control, Hemisphere, Washington, D.C., 1975.

Bryson, A.E., Dynamic Optimization, Addison-Wesley Longman, Menlo Park, CA, 1999.

Coleman, T.F. and Liao, A., An efficient trust region method for unconstrained discrete-time optimal control problems, Computational Optimization and Applications, 4, pp. 47-66, 1995.

Dunn, J., and Bertsekas, D.P., Efficient dynamic programming implementations of Newton's method for unconstrained optimal control problems, J. of Optimization Theory and Applications, 63 (1989), pp. 23-38.

Fogel, D.B., Evolutionary Computation, The Fossil Record, IEEE Press, New York, 1998.

Fox, C., An Introduction to the Calculus of Variations, Oxford University Press, London, 1950.

Jacobson, D. and Mayne, D. , Differential Dynamic Programming, Elsevier Science Publishers, 1970.

Kirkpatrick, and Gelatt, C.D. and Vecchi, Optimization by Simulated Annealing, Science, 220, 671-680, 1983.

Laarhoven, P.J.M. and Aarts, E.H.L., Simulated Annealing: Theory and Applications, Kluwer Academic, 1989.

Liao, L.Z. and Shoemaker, C.A., Convergence in unconstrained discrete-time differential dynamic programming, IEEE Trans. Automat. Contr., 36, pp. 692-706, 1991.

Mayne, D., A second-order gradient method for determining optimal trajectories of non-linear discrete time systems, Intnl. J. Control, 3 (1966), pp. 85-95.

Michalewicz, Z., Genetic Algorithms + Data Structures = Evolution Programs, Springer-Verlag, Berlin, 1992a.

Michalewicz, Z., Janikow, C.Z. and Krawczyk, J.B., A Modified Genetic Algorithm for Optimal Control Problems, Computers Math. Applications, 23(12), pp. 83-94, 1992b.

Murray, D.M. and Yakowitz, S.J., Differential dynamic programming and Newton's method for discrete optimal control problems, J. of Optimization Theory and Applications, 43, pp. 395-414, 1984.

Ohno, K., A new approach of differential dynamic programming for discrete time systems, IEEE Trans. Automat. Contr., 23 (1978), pp. 37-47.

Pantoja, J.F.A. de O. , Differential Dynamic Programming and Newton's Method, International Journal of Control, 53: 1539-1553, 1988.

Pontryagin, L. S. Boltyanskii, V. G., Gamkrelidze, R. V. and Mishchenko, E. F., The Mathematical Theory of Optimal Processes, Moscow, 1961, translated by Trirogoff, K. N. Neustadt, L. W. (Ed.), Interscience, New York, 1962.

Schwefel H.P., Evolution and Optimum Seeking, Wiley, New York, 1995.

Yakowitz, S.J., and Rutherford, B., Computational aspects of discrete-time optimal control, Appl. Math. Comput., 15 (1984), pp. 29-45.

A DISTURBANCE COMPENSATION CONTROL FOR AN ACTIVE MAGNETIC BEARING SYSTEM BY A MULTIPLE FXLMS ALGORITHM

Min Sig Kang

Department of mechanical engineering, Kyungwon University, Sungnam, Kyunggido, KOREA
Email: mskang@kyungwon.ac.kr

Joon Lyou

Dept. of Electronics Eng., Chungnam National Univ., Daejeon 305-764, KOREA
Email: jlyou@cnu.ac.kr

Keywords: Active magnetic bearing, Multiple filtered-x least mean square algorithm, Acceleration feedforward compensation.

Abstract: In this paper, a design technique is proposed for a disturbance feedforward compensation control to attenuate disturbance responses in an active magnetic bearing system, which is subject to base motion. To eliminate the sensitivity of model accuracy to disturbance responses, the proposed design technique is an experimental feedforward compensator, developed from an adaptive estimation, by means of the Multiple Filtered-x least mean square (MFXLMS) algorithm. The compensation control is applied to a 2-DOF active magnetic bearing system subject to base motion. The feasibility of the proposed technique is illustrated, and the results of an experimental demonstration are shown.

1 INTRODUCTION

Active magnetic bearing (AMB) systems are increasingly used in industrial applications. Unlike conventional bearings, AMB systems utilize magnetic fields to levitate and support a shaft in an air-gap within the bearing stator. When compared to conventional mechanical bearings, AMB offers the following unique advantages: non-contact, elimination of lubrication, low power loss, and controllability of bearing dynamic characteristics.

Recently, interest has increased regarding the application of AMB systems to the sight stabilization systems mounted on moving vehicles. When a vehicle is undergoing angular motion, the mirror axis of sight rotates relative to the vehicle, to stabilize the line of sight. In such systems, the friction of mechanical bearings that support the mirror axis may cause tracking errors and, hence, may deteriorate the quality of an image obtained through electro-optical equipment. To eliminate the undesirable effects of friction, an AMB system is used instead of mechanical bearings.

The main problem of a sight system levitated and stabilized by an AMB is the image scattering caused by base motion. One solution for reducing the effects of base motion is to expand the bandwidth of the control system by using feedback controls (Cole, 1998) such as PID control, state feedback control, H^{∞} control, and so on. A controller with a wider bandwidth, however, requires a higher sampling frequency, which often induces a mechanical resonance.

An alternative approach for disturbance attenuation is a feedforward compensation of the base acceleration. The effectiveness of this approach has been demonstrated in the field of hard disk drives, which are also subject to base motion (Jinzenji, 2001). Suzuki (1998) developed feedforward compensation based on a dynamic model of the AMB system and showed that increases in the vibration rejection can be achieved. In practice, however, a dynamic model is not reliably accurate, because of many problems associated with it, such as the non-linearity of AMB, approximation errors of the discrete equivalent to a continuous transfer function, and sensor dynamics.

Motivated to overcome these problems, in this work an alternative technique is proposed: a non-model based acceleration feedforward compensation control developed from an adaptive estimation, by means of the multiple filtered-x least mean square

99

(MFXLMS) algorithm (Kuo, 1996; White, 1997). The performance and the effectiveness of the proposed technique are demonstrated on a 2-DOF AMB system subject to base motion.

2 SYSTEM MODEL

The test rig used in this paper is an AMB system of 2-DOF shown in Fig. 1. Figure 2 is the photograph of the test rig. The test rig consists of two sets of AMB(left AMB: AMB-1, right AMB:AMB-2) and a circular shaft. Each end of the shaft is tied up by string wire such that the shaft moves only in the vertical plane. Each electromagnet is attached rigidly to each shaker(B&K-4808), which generates base motion resembling the vehicle motion. Two non-contacting proximity displacement sensors(AEC-5505) measure each air gap between the probe tip and the shaft surface, and the vertical acceleration of each electromagnet is measured by each accelerometer(Crossbow, CX04LP1Z).

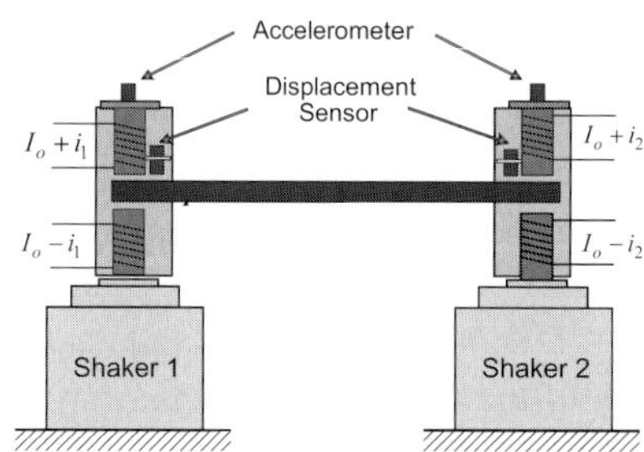

Figure 1: Schematic diagram of test rig.

From the free-body diagram of the system in Fig. 3, the equation of motion is given by.

$$\frac{m}{4}\begin{Bmatrix} \ddot{y}_1 + \ddot{y}_2 + \ddot{z}_1 + \ddot{z}_2 \\ \ddot{y}_1 + \ddot{y}_2 + \ddot{z}_1 + \ddot{z}_2 \end{Bmatrix} - \frac{J}{4a^2}\begin{Bmatrix} -\ddot{y}_1 + \ddot{y}_2 - \ddot{z}_1 + \ddot{z}_2 \\ \ddot{y}_1 - \ddot{y}_2 + \ddot{z}_1 - \ddot{z}_2 \end{Bmatrix} \\ + \frac{mg}{2}\begin{Bmatrix} 1 \\ 1 \end{Bmatrix} = \begin{Bmatrix} f_1 \\ f_2 \end{Bmatrix} \tag{1}$$

where m and J are the mass and the mass moment of inertia about the mass center of the shaft. y, $\ddot{z}$ and f mean the air gap, the vertical acceleration and the actuating force, respectively. The subscripts 1 and 2 denote the positions of the AMB-1 and the AMB-2, respectively. This definition is consistent hereafter.

Figure 2: Photograph of test rig.

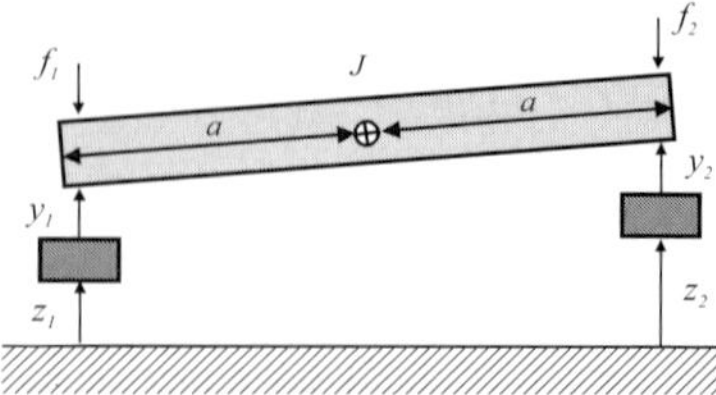

Figure 3: Free-body diagram of the levitated axis.

The magnetic attractive force is approximately proportional to the square of the coil current and inversely proportional to the square of gap. However the nonlinearity of the magnetic attractive force against the coil current is decreased with the bias current added to the coil current. Consequently the linearized model is given by

$$f_c = K_d y + K_i i_c \tag{2}$$

where K_y is the displacement stiffness and K_i is the current stiffness.

Since the time constant of the power amplifier-magnet coil can be designed to be small enough by current feedback control, the control current i_c can be assumed to be proportional to the applied voltage, u_c, to the amplifier, i.e.

$$i_c = K_a u_c \tag{3}$$

where K_a is the gain of the amplifier.

Substituting eqs. (2) and (3) into eq. (1) gives the linearized AMB system model as follows:

$$\begin{Bmatrix} \ddot{y}_1 \\ \ddot{y}_2 \end{Bmatrix} - \frac{1}{mJ}\begin{bmatrix} K_{d1}(J + ma^2) & K_{d2}(J - ma^2) \\ K_{d1}(J - ma^2) & K_{d2}(J + ma^2) \end{bmatrix}\begin{Bmatrix} y_1 \\ y_2 \end{Bmatrix} \\ = \frac{K_a}{mJ}\begin{bmatrix} K_{i1}(J + ma^2) & K_{i2}(J - ma^2) \\ K_{i1}(J - ma^2) & K_{i2}(J + ma^2) \end{bmatrix}\begin{Bmatrix} u_1 \\ u_2 \end{Bmatrix} - \begin{Bmatrix} \ddot{z}_1 + g \\ \ddot{z}_2 + g \end{Bmatrix} \tag{4}$$

It is clear from eq.(4) that the system is open-loop unstable, and the base acceleration and the gravitational force disturb the system.

3 CONTROLLER DESIGN

The system model in eq. (4) can be represented by the state space equation as

$$\dot{q} = Aq + Bu - d_a - f_g \qquad (5)$$
$$q = \{y_1 \ y_2 \ \dot{y}_1 \ \dot{y}_2\}^T, \ u = \{u_1 \ u_2\}^T$$

$$d_a = [0 \ 0 \ \ddot{z}_1 \ \ddot{z}_2]^T, \ f_g = [0 \ 0 \ g \ g]^T \qquad (6)$$

Since this system has no integrator, the state feedback control with integral is applied to eliminate the steady state error due to the gravity force.

$$u = -Kq - k_i \eta \qquad (7)$$

where K and k_i are the state feedback gain vectors, and η is the integration of y_1 and y_2, i.e., $\dot{\eta} = \{y_1 \ y_2\}^T$.

The feedback gains in eq. (7) can be design from various kinds of schemes. The closed-loop system stabilized by eq. (7) can be represented in discrete time domain as

$$A_1(q^{-1})y_1(k) = B_{11}(q^{-1})u_1(k) + B_{12}(q^{-1})u_2(k)$$
$$+ C_{11}(q^{-1})d_1(k) + C_{12}(q^{-1})d_2(k)$$
$$A_2(q^{-1})y_2(k) = B_{21}(q^{-1})u_1(k) + B_{22}(q^{-1})u_2(k) \qquad (8)$$
$$+ C_{21}(q^{-1})d_1(k) + C_{22}(q^{-1})d_2(k)$$

where variables with the index k mean the sampled variables. $A_i(q^{-1})$, $B_{ij}(q^{-1})$ and $C_{ij}(q^{-1})$ are the system polynomials. q^{-1} is the one step delay operator.

A general compensator for the system in eq.(8) is defined by

$$u_1(k) = W_{11}(q^{-1})d_1(k) + W_{12}(q^{-1})d_2(k)$$
$$u_2(k) = W_{21}(q^{-1})d_1(k) + W_{22}(q^{-1})d_2(k) \qquad (9)$$

Applying the compensator, eq. (9), to the system, eq. (8), yields the compensated system of the form

$$A_1 y_1(k) = [B_{11}W_{11} + B_{12}W_{21} + C_{11}]d_1(k)$$
$$+ [B_{11}W_{12} + B_{12}W_{22} + C_{12}]d_2(k)$$

$$A_2 y_2(k) = [B_{21}W_{11} + B_{22}W_{21} + C_{21}]d_1(k)$$
$$+ [B_{21}W_{12} + B_{22}W_{22} + C_{22}]d_2(k) \qquad (10)$$

Obviously, the perfect disturbance cancelling compensators W_{11}^*, W_{21}^*, W_{12}^*, W_{22}^* are derived from

$$\begin{Bmatrix} W_{11}^* \\ W_{21}^* \\ W_{12}^* \\ W_{22}^* \end{Bmatrix} = \frac{1}{B_{11}B_{22} - B_{12}B_{21}} \begin{Bmatrix} B_{22}C_{11} - B_{12}C_{21} \\ -B_{21}C_{11} + B_{11}C_{21} \\ B_{22}C_{12} - B_{12}C_{22} \\ -B_{21}C_{12} + B_{11}C_{22} \end{Bmatrix} \qquad (11)$$

Since the compensators given in eq. (11) are designed from the system model, the compensation performance should be sensitive to the accuracy of the model. In practice, however, this kind of perfect cancelling based on the model is not expected because of the problems such as inaccuracy of dynamic model, approximation error of a discrete equivalent to a continuous transfer function, and sensor dynamics. Motivated by these problems, an explicit optimal feedforward compensator design technique is proposed in this paper. By this technique, the feedforward compensator design can be separated into two parts.

1) Disturbance cancelling control for single harmonic base motion

It is clear from eq. (11) that the response to a harmonic base motion of the frequency ω_r can be exactly nullified by choosing the polynomial $W_{ij}(q^{-1})$ as the first order polynomial satisfying the relation

$$\left[W_{ij} = w_{ij}^1 + w_{ij}^2 q^{-1}\right]_{q=e^{jw_r T}} = W_{ij}^*\Big|_{q=e^{jw_r T}}, \ i,j = 1,2 \quad (12)$$

where T is the sampling interval.

Nullifying disturbance response by using feedforward compensator means physically matching the impedance from the base motion to the air-gap with the impedances from the base motions to the air-gaps through the AMB dynamics and of the feedforward compensators, so that the disturbance can be perfectly cancelled. However compensator design from the model is not suitable for practical applications. To get rid of the problems associated with the inaccurate model, adaptation of the feedforward compensator is proposed. This technique is an explicit design through experiments by using a multiple-FXLMS algorithm. The FXLMS algorithm has been extensively used in the field of active noise control(Kuo, 1996; Widrow and Stearns, 1985)). Figure 4 shows an example of the multiple-FXLMS algorithm to estimate the compensator polynomial W_{11}^* . The parameters of the compensators are estimated from the following update equation.

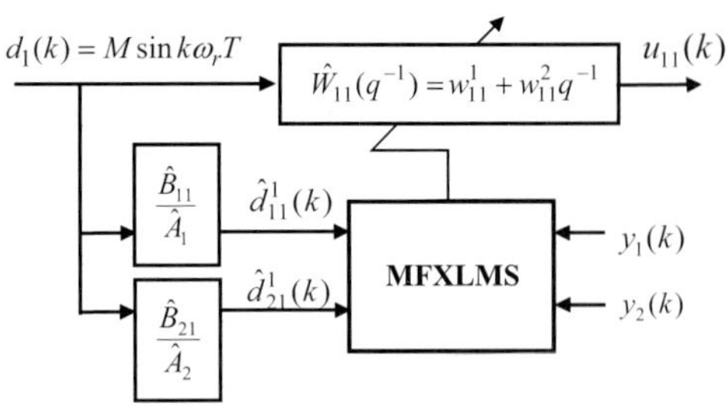

Figure 4: MFLMS algorithm for estimating $W_{11}(q^{-1})$.

$$\begin{Bmatrix} w_{ij}^1(k+1) \\ w_{ij}^2(k+1) \end{Bmatrix} = \begin{Bmatrix} w_{ij}^1(k) \\ w_{ij}^2(k) \end{Bmatrix} + \mu_{ij} \frac{\begin{Bmatrix} \hat{d}_{1i}^j(k) \\ \hat{d}_{1i}^j(k-1) \end{Bmatrix}}{D_{1i}^j(k)} y_1(k)$$

$$+ \eta_{ij} \frac{\begin{Bmatrix} \hat{d}_{2i}^j(k) \\ \hat{d}_{2i}^j(k-1) \end{Bmatrix}}{D_{2i}^j(k)} y_2(k) \quad (13)$$

$$D_{ij}^m(k) = \hat{d}_{ij}^m(k)^2 + \hat{d}_{ij}^m(k-1)^2, \hat{d}_{ij}^m(k) = \frac{\hat{B}_{ij}}{\hat{A}_i} d_m(k), \ i,j,m=1,2$$

where μ_{ij} and η_{ij} are the update gains, $\hat{A}_i(q^{-1})$ and $\hat{B}_{ij}(q^{-1})$ are the estimated system polynomials. All the compensator polynomials are estimated simultaneously from eq. (13).

By applying the MFXLMS algorithm meanwhile the exciters generate a stationary single harmonic base motion, the control parameters w_{ij}^1 and w_{ij}^2 in eq. (13) are estimated.

Since the base motion is single harmonic of frequency ω_r, the Fourier transforms $D_1(j\omega_r)$ and $D_2(j\omega_r)$ of d_1 and d_2 , respectively, would yield the relation $D_1(j\omega_r) = \alpha D_2(j\omega_r)$, where α is a complex number which represents the magnitude and phase relations between d_1 and d_2. The estimated polynomials are not unique but satisfy two independent relations in the following

$$\alpha \begin{Bmatrix} \hat{W}_{11} \\ \hat{W}_{21} \end{Bmatrix} + \begin{Bmatrix} \hat{W}_{12} \\ \hat{W}_{22} \end{Bmatrix} = \alpha \begin{Bmatrix} W_{11}^* \\ W_{21}^* \end{Bmatrix} + \begin{Bmatrix} W_{12}^* \\ W_{22}^* \end{Bmatrix} = \begin{Bmatrix} \beta_1 \\ \beta_2 \end{Bmatrix} \quad (14)$$

Thus it is necessary to have at least two sets of polynomials estimated from the experiments where

d_1 and d_2 have the same frequency but have different relations. For example, if a set of polynomials is estimated from the experiment where $\alpha = \alpha_1$ then one can determine β_1 and β_2 from eq. (14) as

$$\alpha_1 \begin{Bmatrix} \hat{W}_{11} \\ \hat{W}_{21} \end{Bmatrix} + \begin{Bmatrix} \hat{W}_{12} \\ \hat{W}_{22} \end{Bmatrix} = \begin{Bmatrix} \beta_1^1 \\ \beta_2^1 \end{Bmatrix} \quad (15)$$

Similarly, from another set of estimated polynomials obtained from another experiment where $\alpha = \alpha_2 \neq \alpha_1$, β_1 and β_2 are obtained as

$$\alpha_2 \begin{Bmatrix} \hat{W}_{11} \\ \hat{W}_{21} \end{Bmatrix} + \begin{Bmatrix} \hat{W}_{12} \\ \hat{W}_{22} \end{Bmatrix} = \begin{Bmatrix} \beta_1^2 \\ \beta_2^2 \end{Bmatrix} \quad (16)$$

From eqns. (14)-(16), the compensator polynomials that perfectly cancel any stationary harmonic base disturbance of the specified frequency ω_r can be determined as

$$\begin{Bmatrix} W_{11}^* \\ W_{21}^* \\ W_{12}^* \\ W_{22}^* \end{Bmatrix} = \begin{bmatrix} \alpha_1 & 0 & 1 & 0 \\ 0 & \alpha_1 & 0 & 1 \\ \alpha_2 & 0 & 1 & 0 \\ 0 & \alpha_2 & 0 & 1 \end{bmatrix}^{-1} \begin{Bmatrix} \beta_1^1 \\ \beta_2^1 \\ \beta_1^2 \\ \beta_2^2 \end{Bmatrix} \quad (17)$$

Repeating the above experimental procedures by changing the base motion frequency, sets of perfect cancelling compensator polynomials for each frequency are obtained.

2) Model fitting in frequency domain

From the sets of compensator parameters for each specified frequency, the FRF(frequency response function) of the disturbance cancelling feedforward compensators can be calculated. Based on this FRF, the compensators in eq. (9) are determined so as to minimize the cost function J

$$J_{ij} = \sum_{k=1}^{n} \lambda_{ij}(\omega_k) \left[\hat{W}_{ij}^*(q^{-1}) - W_{ij}(q^{-1}) \right]^2 \Big|_{q=e^{j\omega_k T}}, \ i,j=1,2 \quad (18)$$

where $\lambda_{lm}(\omega_k)$ is the frequency weighting and $\hat{W}_{lm}^*(q^{-1})$ is the estimated compensator obtained in the first step and $W_{lm}(q^{-1})$ is the compensator to be determined. To avoid unstable compensator, $W_{lm}(q^{-1})$ can have the form of FIR(finite impulse response) filter.

4 EXPERIMENTS

To verify the effectiveness of the proposed control scheme, experiments were conducted using the test apparatus shown in Fig. 2. All control algorithms were implemented on a digital computer equipped with a DSP(TI-DS-1104)) board. Throughout the experiments, the sampling frequency was kept at 2000Hz.

A pole placement feedback (FB) control was designed to have a closed-loop system with a damping ratio of $\varsigma = 0.8$ and natural frequency of $\omega_n = 80Hz$ in consideration of the spectral characteristics of the base motion. The vehicle motion is characterized by a band-limited random process of bandwidth 15Hz-60Hz.

To evaluate the convergence of the estimated compensator parameters and the corresponding disturbance rejection performance, a sequence of simple harmonic of frequency 30Hz was delivered to the shakers. The resultant base motion kept the relation $D_1(j\omega) = 1.023D_2(j\omega)$.

Figs. 5 and 6 show the estimated compensator parameters of $\hat{W}_{11}$ and the corresponding air-gap responses, respectively. We confirmed that all estimated parameters converged to their final values after 50 s. These figures reveal that the air-gap responses were consequently reduced, as the estimated parameters converged to their final values. The aforementioned convergence property and the disturbance rejection performance exhibit the feasibility of the proposed compensation control by means of the MFXLMS algorithm.

As explained in the above, at least, one more set of compensator parameters is necessary to determine the unique compensator polynomials which cancel the disturbance responses perfectly at $f = 30Hz$. The MFXLMS algorithm was applied to obtain another set of compensator parameters under the different base motion profile kept the relation $D_1(j\omega) = 1.465e^{j\pi/2}D_2(j\omega)$, $f = 30Hz$. Similar convergence and disturbance rejection properties to Figs. 6 and 7 were confirmed.

From the two sets of the parameters obtained, the FRF of the disturbance neutralizing compensator at $f = 30Hz$ was determined. The disturbance rejection performance of this compensator was evaluated under the base motion yielding the relation $D_1(j\omega) = 0.69e^{-j\pi/4}D_2(j\omega)$, $f = 30Hz$.

Fig. 7 shows the air-gap responses of the FB-control by itself and the FB with the compensation control. Fig. 7 reveals that the compensation control can almost neutralize any base motion responses of frequency 30Hz. Surprisingly, it was found that the control effort is reduced when the compensation was employed. The air-gap responses that remained after

employing the compensation came mainly from the inability of the shakers to produce a pure sinusoidal tone of motion.

Repeating the experiment, while changing the harmonic base motion frequency, sets of disturbance neutralizing compensator parameters for each frequency were obtained. The FRF $\hat{W}_{11}$ calculated from the estimated parameters is shown as an example in Fig. 8. Based on the FRF in Fig. 8, the best-fit compensator was determined to be the third-order polynomials.

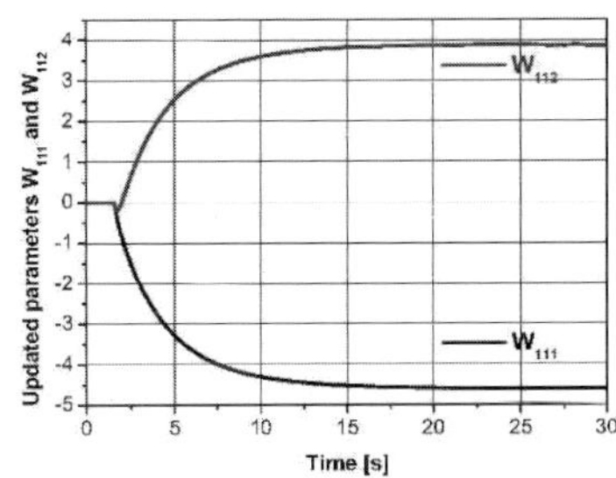

Figure 5: Estimated coefficients of $W_{11}(q^{-1})$.

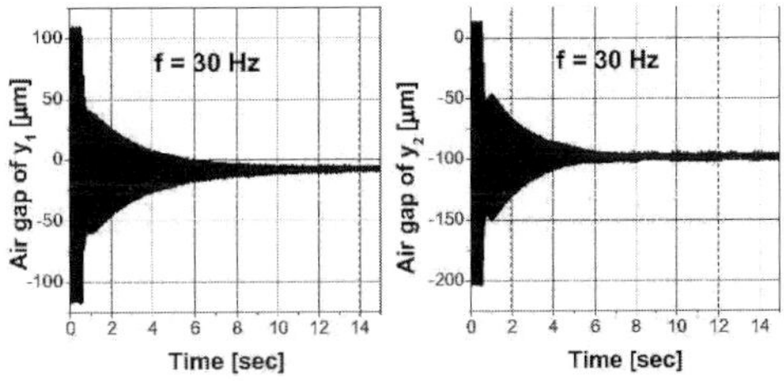

Figure 6: Air-gap responses during estimation by MFXLMS algorithm.

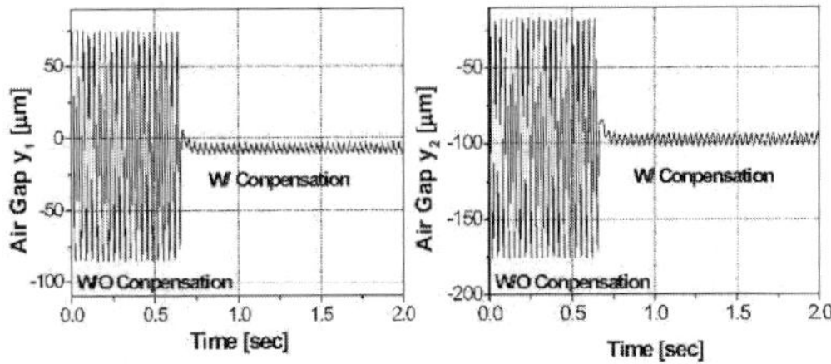

Figure 7: Compensated air gap responses.

To investigate the efficiency of the designed compensator, a comparison was made between the air-gap response with the compensation and without the compensation. During the control experiments, a sequence of band-limited random signals of bandwidth 15-60Hz was delivered to the shaker and the resultant base motion resembled that of the real vehicle.

As shown in Fig. 9, the air-gap responses were greatly reduced by applying the feedforward compensation. For y_1, the standard deviations of the air-gap with compensation and without compensation were calculated to be $\sigma = 14.53\,\mu m$ and $1.43\,\mu m$, respectively. For y_2, the standard deviations of the air-gap with compensation and without compensation were calculated to be $\sigma = 13.13\,\mu m$ and $1.08\,\mu m$, respectively. The control voltages were slightly reduced after employing compensation. Figure 10 shows the spectra of the air-gap responses in Fig. 9. The disturbance attenuation ratio is approximately –20db within the frequency band of the base motion.

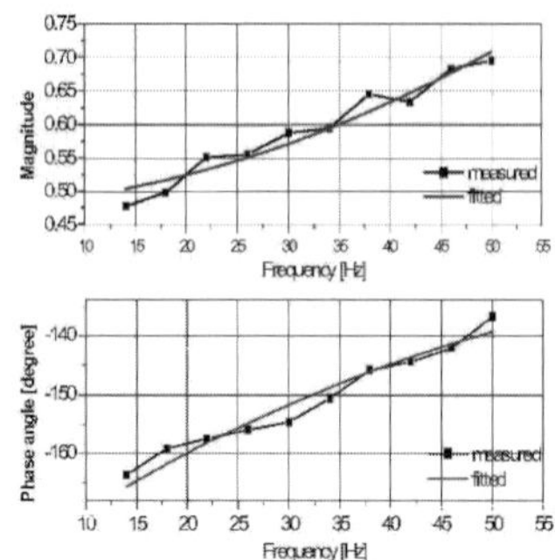

Figure 8: Measured and fitted FRF of $\hat{W}_{11}$.

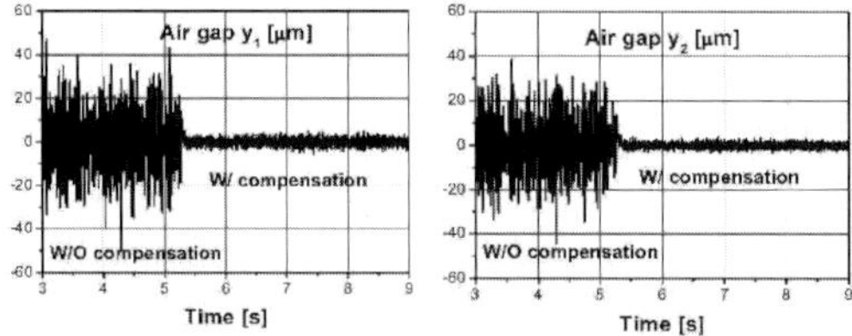

Figure 9: Air-gap responses w/ and w/o compensation.

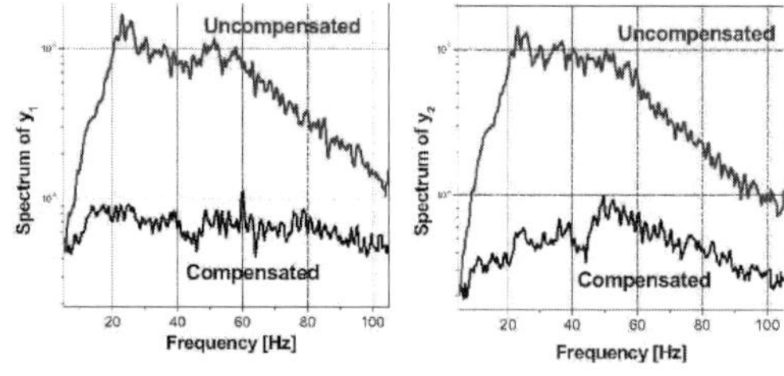

Figure 10: Spectra of air-gap with and without compensation.

5 CONCLUSION

In this work, an experimental feedforward compensator design technique, developed from an adaptive estimation by means of the Multiple Filtered-x least mean square (MFXLMS) algorithm has been proposed. The feasibility of the proposed technique has been verified by an experimental study, by using a 2-DOF active magnetic bearing system subject to base motion. The experimental results showed that the standard deviation of the compensated response was reduced to less than 10% of that by feedback control alone.

ACKNOWLEDGEMENTS

This work was supported by grant no.(R01-2003-000-10857-0) from the Basic Research Program of the Korea Science & Engineering Foundation.

REFERENCES

Brunet, M., 1998. Practical Applications of Active Magnetic Bearing to the Industrial World. *1'st International Symposium on Magnetic Bearing*. Zurich, pp. 225-244.

Cole, M. O. T., Keogh, P. S. and Burrows, C. R. , 1998. Control and Non-linear Compensation of a Rotor/Magnetic Bearing System Subject to base Motion. *6th Int. Symposium on Magnetic Bearings*. Cambridge, MA, pp. 618-627.

Jinzenji, A., Sasamoto, T., Aikawa, K., Yoshida, S. and Aruga, K., 2001. Acceleration feedforward control Against Rotational Disturbance in hard Disk Drives. *IEEE Trans. On Magnetics*. Vol. 37, No. 2, pp. 888-893.

Kasada, M.E., Clements, J., Wicks, A. L., Hall, C. D., and Kirk, R. G., 2000, Effect of sinusoidal base motion on a magnetic bearing, *Proc. IEEE International Conference on Control Applications*, pp. 144-149.

Kuo, S. M. and Morgan, D. R., 1996. *Active Noise Control Systems*. A Wiley-Interscience Publication, John Wiley Sons, Inc.

Suzuki, Y., 1998. Acceleration Feedforward Control for Active Magnetic Bearing Excited by Ground Motion. *IEEE Proc. Control Theory Appl.* Vol. 145, pp. 113-118.

Wang, A.K. and Ren, W., 1999, Convergence analysis of the multiple-variable filtered-x LMS algorithm with application to active noise control, *IEEE Trans. On Signal Processing*, Vol. 47, No. 4, pp. 1166-1169.

White, M. T. and Tomizuka, M., 1997. Increased Disturbance Rejection in Magnetic Disk Drives by Acceleration Feedforward Control and Parameter Adaptation. *Control Engineering Practice*. vol. 5, no. 6. pp. 741-751.

Widrow, B. and Stearns, S. D., 1985. *Adaptive Signal Processing*. Prentice Hall. Englewood Cliffs, NJ.

AN INTELLIGENT RECOMMENDATION SYSTEM BASED ON FUZZY LOGIC

Shi Xiaowei

Philips (China) Investment Co.,Ltd. Shanghai R&D Centre
38F,Tower1 Kerry Everbright City   218 Tian Mu Xi Road   Shanghai, P.R.C.200070
Email: nancy.shi@philips.com   Tel.: 86-21-63541088-5917   Fax: 86-21-63544954

Keywords: Recommendation, User profile, Fuzzy-logic, Multi-agent, Metadata

Abstract: An intelligent recommendation system for a plurality of users based on fuzzy logic is presented. The architecture of a multi-agent recommendation system is described. How the system simulates human intelligence to provide recommendation to users is explained. The recommendation system is based on the fuzzy user profile, fuzzy filtering and recommendation agents. The user profile is updated dynamically based on the feedback information. Fuzzy logic is used in fuzzy filtering to integrate different types of features together for a better simulation of human intelligence. Ambiguity problems can be solved successfully in this system, e.g., deducing whether a programme with both interesting features and uninteresting features is worth recommending or not. The application scenario shows that it is more convenient for users to find programmes of their interest with the proposed recommendation system.

1 INTRODUCTION

Due to the digitalisation of television, in the near future we will be able to receive hundreds of channels via satellites, terrestrial antenna, cable and even phone lines. At the same time, it is becoming increasingly challenging for television viewers to identify television programmes of their interest. Recommendation systems, such as TV-Advisor, Personal TV, etc, have been studied to help viewers to find, personalize, and organize the contents [1,2,3,4,5].

The following recommendation process is considered: matching Electronic Programme Guide (EPG) metadata to the user preference knowledge, filtering out the tedious programmes and recommending interesting programmes to users, and updating the user profiles based on feedback information.

The traditional methods to evaluate if a programme is good enough to be recommended are based on the explicit (i.e. non-fuzzy) inference. In other words, the programme evaluation result can either be "interesting" or "non-interesting". As we know, explicit mathematics cannot intelligently simulate human's flexible inference. Especially, it is difficult to decide whether a programme with both interesting features and uninteresting features should be recommended by the existing filtering and recommendation methods.

This paper presents an algorithm about the integration of fuzzy logic into a multi-agent recommendation system to better recommend interesting programmes to users. The architecture of the fuzzy recommendation system is described. Moreover, the reason why fuzzy logic is adopted in the recommendation system is explained. The fuzzy recommendation algorithm is also developed. Finally, the recommendation process is illustrated with an example.

2 RECOMMENDATION SYSTEM

2.1 System Architecture

The recommendation system is provided to generate programme recommendations for multiple users based on programme metadata and user profiles. The architecture of this system is illustrated in Figure 1. The system uses a central recommendation server unit that includes a Fuzzy User Profile Database, Fuzzy Filtering Agent, Fuzzy Recommendation Agent, Profiling Agent, and Interface Agent.

J. Braz et al. (eds.), Informatics in Control, Automation and Robotics I, 105–109.
© 2006 *Springer. Printed in the Netherlands.*

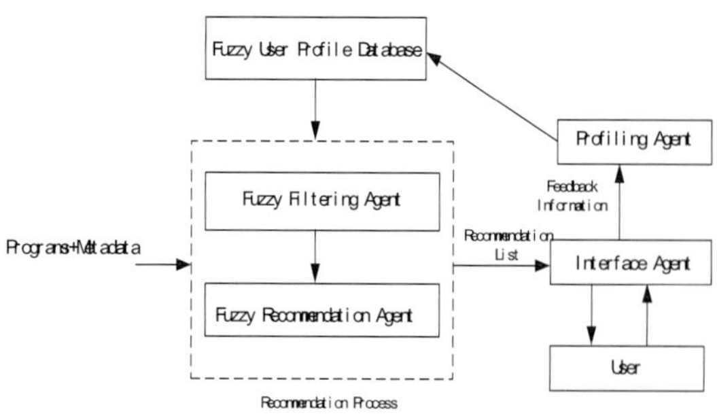

Figure 1: System architecture.

- Fuzzy Filtering means to gather and make the incoming live programme broadcasted based on programme metadata match to fuzzy user profiles, which is based on fuzzy logic, and then filter the programmes based on fuzzy threshold;
- Fuzzy Recommendation generates an optimal recommendation list to every user according to the learned user preference knowledge, and transmits it to user terminals;
- Profiling updates the user profile based on both the explicit and implicit feedback information from the Interface Agent;
- The Interface Agent handles the interactions with the user.

2.2 Information Description

In the context of TV-Anytime [7], metadata consists of two kinds of information: (a) content description metadata; (b) consumer metadata. Programme content description metadata includes attributes of a television programme, such as the title, genre, list of actors/actresses, language, etc. These data are used to make a search.

The user profile metadata in this system defines the details of a user's preferences and aversion. These descriptions are closely correlated with media descriptions, and thus enable the user to search, filter, select and consume the desired content efficiently.

2.3 Application of Fuzzy Logic Control Theory in a Recommendation System

There are many factors that influence a user if he/she wants to view a programme or not. The user's attitude to a programme is the result of some complicated reaction. In other words, it is difficult for a user to describe their emotion about a programme in quantity. Fuzzy theory can simulate human intelligence. It owns the advantage of describing this kind of indefinite object, providing a possibleway to solve theproblem. Hence, fuzzy theory is used in programme recommendation in this paper.

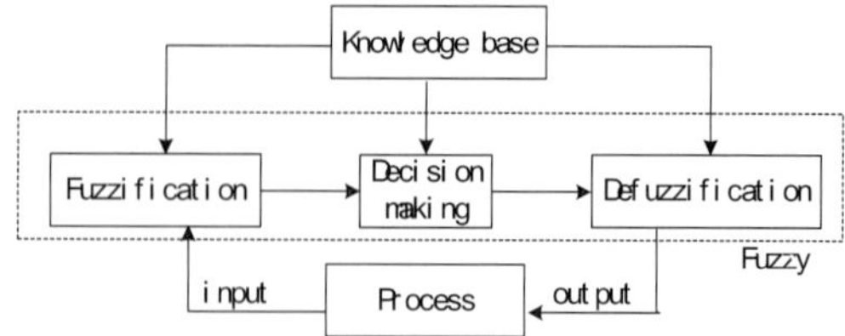

Figure 2: General structure of fuzzy inference system.

Fuzzy theory includes a series of procedures for representing set membership, attributes, and relationships that cannot be described by single point numeric estimates. The structure of a fuzzy inference system is shown in Figure 2.

Where,

Knowledge base: parameters of membership functions and definitions of rules;

Fuzzification: transformation of crisp inputs into membership values;

Decision-making: fuzzy inference operations on the rules;

Defuzzification: transformation of the fuzzy result of the inference into a crisp output

3 FUZZY RECOMMENDATION

3.1 Fuzzy Information Database

3.1.1 The Fuzzy User Profile

The user profile, UP, can be represented by a vector of these 3-tuples. If there are m distinct terms in the profile, it will be represented by:

$$UP = ((t_i, ld_i, w_i),....(t_i, ld_i, w_i)....., (t_m, ld_m, w_m)) \quad (3\text{-}1)$$

Where: t_i is a term; ld_i is the "Like_degree" of term t_i; w_i is the Weight of term t_i; i is the order of t_i in the profile.

The fuzzy user profile transforms the crisp parameters ("Like_degree", Weight) into membership values. "Like_degree" means the degree the user likes a feature. The shape and location may be different for different problems. If e_1 and e_2 represent "Like_degree" and "Weight" respectively, the fuzzy memberships can be described as Figure 3. It is known from Figure 3 that

Figure 3: The fuzzy membership function of user profile.

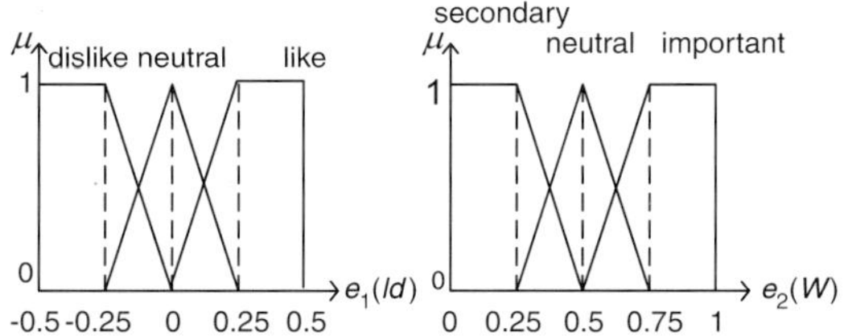

e_2 is always greater than 0. A larger e_2 indicates that this feature is more important. When $e_1 \geq 0$, it means "like" and a larger e_1 means the user likes the feature more; if $e_1 \leq 0$, it indicates that the user "dislike" this feature and the smaller e_1, the more the user dislikes it.

3.1.2 EPG Metadata

A programme can be represented by a vector of size n

$$C = (t_1, t_2, ..., t_i, ...t_n) \tag{3-2}$$

Where t_i is the ith feature of the programme.

3.2 Similarity Matching

The Fuzzy Filtering Agent calculates the similarity between the programme and the user profile. If the calculated similarity is above the preset threshold, the user is considered to be interested in the programme; then the programme metadata is transferred to the Recommendation Agent.

In this system, the fuzzy similarity matching process can be divided into two steps:
- Feature matching;
- Programme matching.

3.2.1 Feature Matching

Feature matching decides how a feature of a programme is related to a user's preference. In order to get the feature "interest_degree" , the procedures is executed as shown in Figure 2.

Here, Like_degree and Weight are set as input and the feature of "interest_degree", which means how much a user likes the programme, is set as output.

The step of decision making is based on the fuzzy inference rules such as following,

I.If Like_degree is "dislike" And Weight is "secondary" Then f_i is "disgusted";

II.If Like_degree is "dislike" And Weight is "important" Then f_i is "very disgusted";

III.If Like_degree is "neutral" And Weight is "important" Then f_i is "neutral";

...

The method of "centre of gravity" takes more useful factors into consideration. It is adopted in this system to defuzzicate the feature's interest_degree.

3.2.2 Programme Matching

Programme matching is to evaluate the programme interest_degree. It can be calculated by the average interest_degree of the features related with the programme.

3.3 Filtering & Ranking

Next step is to set a threshold to filter the coming programme metadata, select the interesting programme, and then rank and recommend them based on the programme interest_degree. In this system, ranking and recommendation processes are performed by the Fuzzy Recommendation Agent.

A threshold is set in the Fuzzy Filtering Agent. The threshold can be a crisp value, or a fuzzy value such as "how much does the user like". If the programme interest_degree is greater than the threshold, which means the programme is what the user wants to watch, then the Filtering Agent will transfer the programme metadata to the Fuzzy Recommendation Agent.

Based on the learned user preference knowledge, the Fuzzy Recommendation Agent generates an optimal recommendation list and sends it to the user according to interest_degree.

3.4 Profiling Agent

In the system, the feedback can be explicitly given by the user or implicitly derived from observations of the users' reaction to a recommended programme. So, the Profiling Agent revises the user profile based on both the explicit and implicit feedback information. In this section, how to update the user profile by the implicit feedback information is mainly explained.

For a recommended programme, the user always has two attitudes to it: skipping over, or watching. In other words, the user will skip (delete) the programme he/she dislikes, watch the programme he/she likes or he/she is not sure. If the user has watched the programme for a period of time, the user's profile will be refined and revised based on the viewing behaviour.

In this system, for programme i, the algorithm for updating user profile is depicted as follows:

$$Weight'_i = Weight_i + \alpha \bullet \frac{(WD_i - \theta)}{RD_i} \tag{3-3}$$

$$Like\_degree'_i = Like\_degree_i + \beta \bullet \frac{(WD_i - \theta)}{RD_i} \quad\quad (3\text{-}4)$$

Where, WD_i : The time duration watched;

RD_i : The real time duration of the programme;

θ : The threshold of the time duration. If WD_i is less than θ, that means the user is not interested in that programme;

α and β : are less than 1. They are used to slow down the change of Weight and Like_degree. Because Weight is more stable than Like_degree, $\alpha \leq \beta$.

If $Weight'_i$ is larger than its higher-boundary, let $Weight'_i = higher\_boundary$;

If $Weight'_i$ is less than its lower-boundary, let $Weight'_i = lower\_boundary$;

If $Like\_degree'_i$ is larger than its higher-boundary, let $Like\_degree'_i = higher\_boundary$;

If $Like\_degree'_i$ is less than its lower-boundary, let $Like\_degree'_i = lower\_boundary$.

4 EXAMPLE

4.1 Similarity Matching and Preference Learning Example

Table 1 shows an assumed user profile A and upcoming programme. The problem to be solved is to determine whether the user likes the programme and how much he/she does.

Based on the above described procedures, the programme fuzzy filtering inference can be illustrated as Figure 4.

For the user profile, the actor LiQinqin's Like_degree is -0.125, which indicates the user's emotion about him is between "dislike" and "neutral". In this case, both values of $\mu_{ld=neutral}$ and $\mu_{ld=dislike}$ are 0.5. In addition, the feature of "Actor" 's Weight is 0.8, so this feature is "important" and $\mu_{important} = 1$.

Matching by fuzzy logic rules, the actor LiQinqin meets both rule II and III. For rule II , μ_{fi} is 0.5, which means the user is "very disgust" at this feature; for rule III, μ_{fi} is 0.5 and the user feels "neutral" about this feature.

Through defuzzification, interest_degree f_2 for the actor LiQinqin is about -0.4. It shows that the user's emotion about this feature is mainly "much disgusted";

Considering other features, the calculated value of the programme interest_degree P is 0.45. From Figure 4, when P (0.45) is mapped into its fuzzy

Table 1: Initial conditions.

User profile A	Programme"Cala is a dog"
Genre Weight=0.9 (Movie Like_degree=0.5 ...); Actor Weight=0.8 (Ge you Like_degree=0.5 ..... LiQinqin Like_degree=-0.125);	Genre is movie Actors are GeYou, LiQinqin Duration=2hour

membership , the emotion of the user can be obtained. In the case discussed, it is between "much interested" and "interested" ($\mu_{interested} \approx 0.2$, $\mu_{much\ interested} \approx 0.8$).

According to the user's behaviour, the user profile is updated. For the discussed programme, duration RD_i is 2 hours. Assumed, the threshold of the watching time duration $\theta = 20$ minutes, the time duration watched $WD_i =2$ hours, $\alpha =0.01$, $\beta =0.1$, then:

$$\frac{(WD_i - \theta)}{RD_i} = 0.83$$

For the fuzzy model of user profile A assumed, the user profile is updated as following:

Genre   Weight=0.9083
(Movie   Like_degree=0.5083
Comedy   Like_degree=0.3
News   Like_degree=-0.2);
Actor   Weight=0.8083
(XuJing   Like_degree=0.1
GeYou   Like_degree=0.583=0.5
LiQinqin   Like_degree=-0.125+0.083=-0.042);

4.2 Application Scenario

An application scenario is provided. Figure 5 shows the main user interface of the fuzzy recommendation system. Functions are listed in the left column. A programme recommendation list is on the topside of middle column. The description of a selected programme is presented in the bottom of middle column. For a selected programme, three choices are provided to the user, which are "Display", "Delete", and "Skip".

5 CONCLUSION

An intelligent Multi-agent recommendation system is developed to provide programme recommendations for multiple users based on programme metadata and fuzzy user profiles.

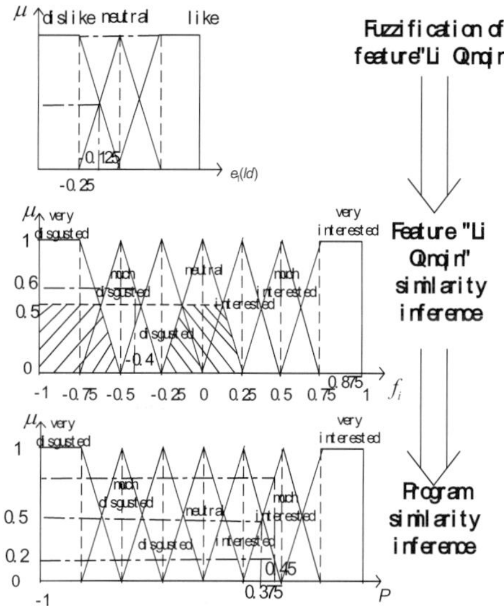

Figure 4: The programme filtering inference.

Different from traditional methods, the user profile is not based on the viewing history but on a compact and low-cost structure, including all terms that user "likes" and "dislikes". The user profile is updated dynamically, e.g. "increase" or "decrease" the corresponding preference parameters according to the feedback information. The filtering agent uses fuzzy logic to integrate different types of features together for a better simulation of human intelligence. This system shows a better capability in solving problems of ambiguities in programme recommendation, e.g., deducing whether a programme with both the interesting features and uninteresting features is worth recommending or not. The application scenario shows that this recommendation system can help users enjoy life more freely.

Figure 5: Main interface.

REFERENCES

Kaushal Kurapati, Srinivas Gutta, 2001. A Multi-Agent TV Recommender. *http://www.di.unito.it/~liliana/UM01/kurapati.pdf.*

Belkin, N.J. & Croft, W.B., 1994. Information filtering and information retrieval: two sides of the same coin. *Communications of the ACM,1994.*

Foltz, P. W. and Dumais, S. T., 1992. Personalized information delivery: An analysis of information filtering methods. *Communications of the ACM, 1992.*

Ardissono, L. and Buczak, 2002. A. Personalisation in future TV". *Proceedings of TV'02, the 2nd workshop on personalization in future TV.*

Masthoff, J., 2002. Modeling a group of television viewers, *Proceedings of the workshop Future TV, 2002.*

Babuska, R., 1998. Fuzzy Modeling for Control". *International Series in Intelligent Technologies,* Kluwer Academic Publishers.

TV-Anytime Forum, 2003. Metadata Specification S-3 Part A: Metadata Schemas, Version 1.3, *http://www,tv-anytime.org*

MODEL REFERENCE CONTROL IN INVENTORY AND SUPPLY CHAIN MANAGEMENT
The implementation of a more suitable cost function

Heikki Rasku, Juuso Rantala, Hannu Koivisto

Institute of Automation and Control, Tampere University of Technology, P.O. Box 692, Tampere, Finland
Email: heikki.rasku@tut.fi, juuso.rantala@tut.fi, hannu.koivisto@tut.fi

Keywords: Model reference control, model predictive control, inventory management.

Abstract: A method of model reference control is investigated in this study in order to present a more suitable method of controlling an inventory or a supply chain. The problem of difficult determining of the cost of change made in the control in supply chain related systems is studied and a solution presented. Both model predictive controller and a model reference controller are implemented in order to simulate results. Advantages of model reference control in supply chain related control are presented. Also a new way of implementing supply chain simulators is presented and used in the simulations.

1 INTRODUCTION

In recent years model predictive control (MPC) has gained a lot of attention in supply chain management and in inventory control. It has been found to be a suitable method to control business related systems and very promising results has been shown in many studies. The main idea in MPC has remained the same in most studies but many variations of the cost function can be found. Basically these cost functions, used in studies concerning MPC in supply chain management, can be separated in two different categories: quadratic and linear cost functions. The use of a linear cost function can be seen appropriate as it can take advantage of actual unit costs determined in the case. On the other hand these costs need to be fairly accurate to result as an effective control. Examples of studies using linear cost functions in supply chain control can be found in (Ydstie, Grossmann et al., 2003) and (Hennet, 2003). In this study we will no longer study the linear form of the cost function but concentrate on the quadratic form. The quadratic form of the cost function is used in, for example, (Tzafestas et al., 1997) and (Rivera et al., 2003). In supply chain management the question is not only about how to control the chain but also about what is being controlled. The traditional quadratic form of the cost function used in MPC has one difficulty when it comes to controlling an inventory or a supply chain. The quadratic form involves penalizing of changes

in the controlled variable. Whether this variable is the order rate or the inventory level or some other actual variable in the business, it is always very difficult to determine the actual cost of making a change in this variable. In this study we present an effective way of controlling an inventory with MPC without the problem of determining the cost of changing the controlled variable. The method of model reference control will be demonstrated in inventory control and results presented. The structure of this paper is as follows. In Chapter 2 we will take a closer view on model predictive control and on the theory behind model reference control. In Chapter 3 we present simulations with both model predictive control and model reference control and do some comparisons between those two. Finally we conclude the results from our study in the last chapter, Chapter 4.

2 MODEL PREDICTIVE CONTROL

Model predictive control originated in the late seventies and has become more and more popular ever since. MPC itself is not an actual control strategy, but a very wide range of control methods which make use of a model of the process. MPC was originally developed for the use of process control but has diversified to a number of other areas of control, including supply chain management and

J. Braz et al. (eds.), Informatics in Control, Automation and Robotics I, 111–116.
© 2006 *Springer. Printed in the Netherlands.*

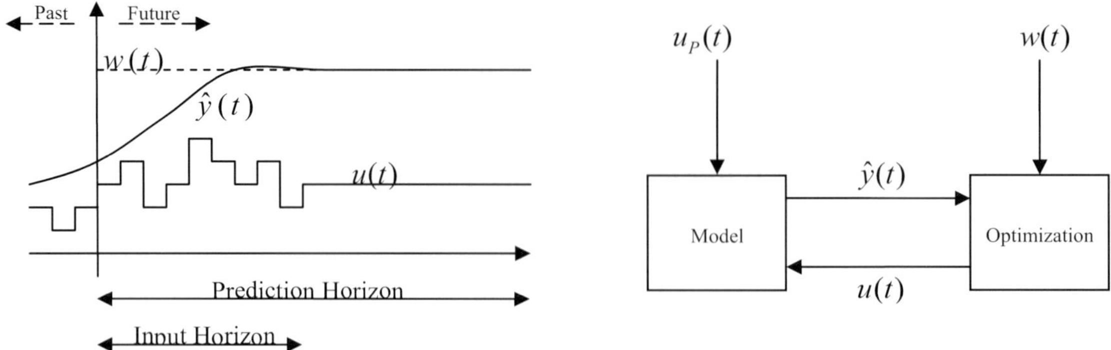

Figure 1: The main idea and the implementation of MPC.

inventory control in which it has gained a lot of attention. Today MPC is the only modern control method to have gained success in real world applications. (Camacho and Bordons 2002), (Maciejowski 2002).

As stated earlier, Model Predictive Control is a set of control algorithms that use optimization to design an optimal control law by minimizing an objective function. The basic form of the objective function can be written as

$$
\begin{aligned}
J&\left(N_1, N_2, N_u\right) \\
&= \sum_{j=N_1}^{N_2} \delta(j)\left[\hat{y}(t+j \mid t) - w(t-j)\right]^2 \\
&\quad + \sum_{j=1}^{N_u} \lambda(j)\left[\Delta u(t+j-1)\right]^2,
\end{aligned}
\tag{1}
$$

where N_1 and N_2 are the minimum and maximum cost horizons and N_u is the control horizon. $\delta(j)$ and $\lambda(j)$ can be seen as the unit costs of the control. $w(t)$, $\hat{y}(t)$ and $\Delta u(t)$ are the reference trajectory, the predicted outputs and the change between current predicted control signal and previous predicted control signal, respectively. (Camacho and Bordons 2002).

The algorithm consists of two main elements, an optimization tool and a model of the system. The optimizer contains a cost function with constraints and receives a reference trajectory $w(t)$ to which it tries to lead the outputs as presented in Figure 1. The actual forecasting in MPC is done with the model which is used to predict future outputs $\hat{y}(t)$ on the basis of the previous inputs $u_P(t)$ and future inputs $u(t)$ the optimizerhas solved as presented in Figure 1. Theseforecasts are then used to evaluate the

control and a next optimization on the horizon is made. After all the control signals on the horizon are evaluated, only the first control signal is used in the process and the rest of the future control signals are rejected. This is done because on the next optimizing instant, the previous output from the process is already known and therefore a new, more accurate forecast can be made due to new information being available. This is the key point in the receding horizon technique as the prediction gets more accurate on every step of the horizon but also is the source of heavy computing in MPC. The receding horizon technique also allows the algorithm to handle long time delays. (Camacho and Bordons 2002).

2.1 Implementing the Cost Function

As presented in equation 1, the basic form of a MPC cost function penalizes changes made in control weighted with a certain parameter λ. This kind of damping is not very suitable for controlling an inventory or a supply chain due to the difficulty of determining the parameter λ as it usually is either the cost of change in inventory level or the cost of change in ordering. On the other hand the parameter λ cannot be disregarded as it results as minimum-variance control which most definitely is not the control desired. Another problem with the basic form of MPC used in inventory control is the fact that it penalizes the changes made in ordering and not in inventory levels, which can cause unnecessary variations in the inventory level as will be shown later in this study.

In this studywe present a more suitableway to form the cost function used in a model predictive controller. The problematic penalizing of changes in the control is replaced with a similar way to the one presented in, for example, (Lambert, 1987) and used, for example, in (Koivisto *et al.*, 1991). An inverted discrete filter is implemented in the cost function so that the resulting cost function can be written as

$$J = \sum_{i=N_1}^{N_2} \left(y^*(i) - P(q^{-1}) \cdot \hat{y}(i) \right)^2, \qquad (2)$$

where $y^*(i)$ = Target output,
   $\hat{y}(i)$ = Predicted output
   $P(q^{-1})$ = Inverted discrete filter

The filter $P(q^{-1})$ used can be written as

$$P(q^{-1}) = \frac{1 - p_1 q^{-1} - p_2 q^{-2} - \dots}{1 - p_1 - p_2 - \dots} \qquad (3)$$

As can be seen, the number of tuneable parameters can be reduced as the simplest form of the cost function consists of only one tuneable parameter, p_1 which is used in the filter. Naturally the reduction of tuneable parameters is a definite improvement it self.

Thedampeningperformed by the model reference control is also an advantage concerning bullwhip effect as over ordering has been found one of the major causes of this problem. (Towill, 1996) When the model reference control is applied to an inventory level controller the most basic form of the cost function results as

$$J = \sum_{i=N_1}^{N_2} \left(I^*(i) - P(q^{-1}) \cdot \hat{I}(i) \right)^2, \qquad (4)$$

where $I^*(i)$ = Desired inventory level
   $\hat{I}(i)$ = Predicted inventory level
   $P(q^{-1})$ = Inverted discrete filter as in
   equation (3)

3 SIMULATIONS

The simulations in this study were made using MATLAB® and Simulink®. The goal in the simulations was to show the advantages of a model reference controller in inventory control compared to a traditional model predictive controller. To construct the simulators a set of universal supply chain blocks was used. The main idea in these blocks is the ability to construct any supply chain desired without programming the whole chain from scratch. The basic structure of a desired chain can be implemented with basic drag and drop operations and actual dynamics can be programmed afterwards. The set of blocks consists of three different elements which are inventory block, production block and a so called dummy supplier block. These blocks are the actual interface for programming each individual element. With these blocks the whole supply chain can be constructed and simulated with a high level of visibility and clarity.

3.1 Simulator Implementation Tool

The main idea in the universal production block can be seen in Figure 2. The submodules Stock, Control and Demand forecast can all be implemented

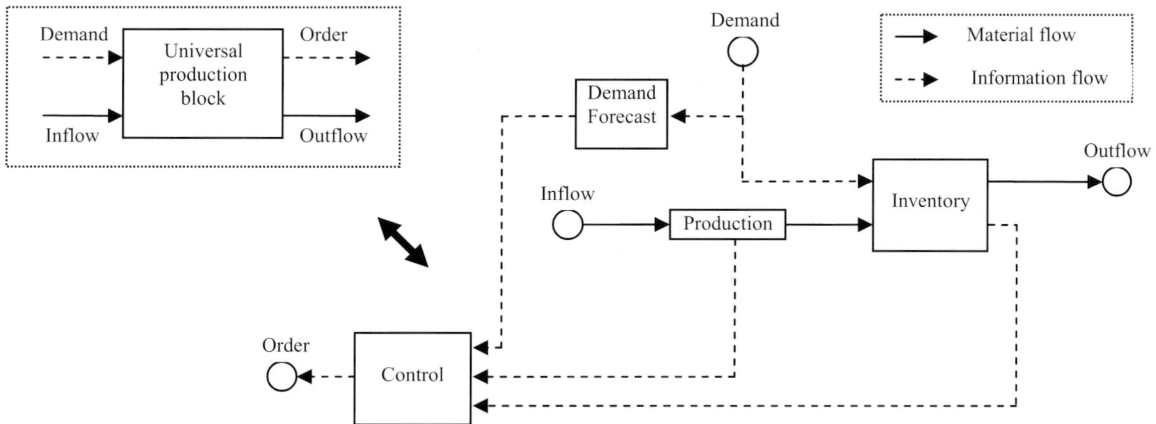

Figure 2: Structure of the universal production block.

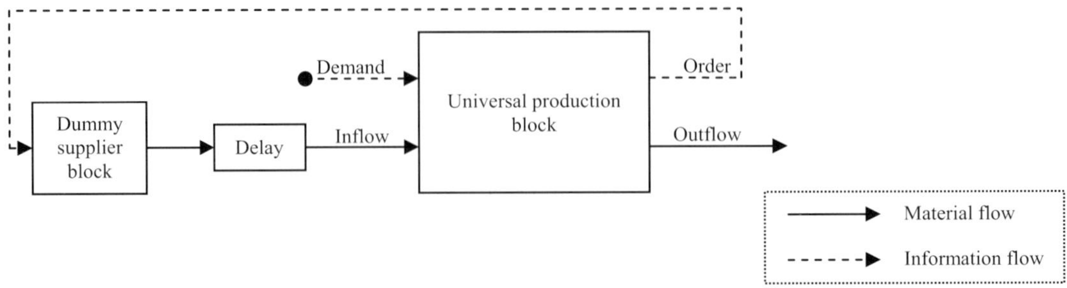

Figure 3: The structure of the simulator used in this study.

uniquely. For example the Inventory block can be constructed to operate linearly in order to test more theoretical control methods or it can even be implemented as realistic as possible to study the performance of a real world supply chain. The inventory element represents an end product inventory of a production plant. Also different control and demand forecasting methods can be tested and tuned via the Control and Demand forecast elements in the block, respectively. The universal production block has naturally a submodule called Production which consists of production dynamics in the simulated factory.

The universal inventoryblock isbasically the universal production block but without the Production submodule. The universal inventory block can be used as a traditional warehouse or as a whole saler or even as a material inventory for a production plant. As it consists of the same control related elements as the universal production block, it can have a control method and a inventory policy of its own independently from the production plant.

The dummysupplier blockis very differentfromthe rest of the set. It is used to solve the problem of long supply chains. Usually one does not want to model the whole supply chain as it can consist of tens of companies. Most of the upstream companies are also irrelevant in the simulations from the end products point of view. Therefore it is necessary to replace the companies in the upstream of the chain with a dummy supplier block. This block takes the order from its customer as an input and supplies this order with certain alterations as an output. These alterations can be anything from basic delay to consideration of decay. Once again, this block can be seen as an interface for the programmer who can decide the actual operations within the block.

3.2 Inventory Control Simulations

To present the advantages in the model reference control used in inventory control, a very simple model was constructed using the universal block set presented earlier. The structure of the simulator can be seen in Figure 3. Both the universal production block and the dummy supplier block are as presented earlier. To keep the model as simple as possible, all delays in the model are constant. Each block in the model consists of a unit delay so that total delay in the model is 3 units. Also no major plant-model mismatch is involved in the controller and no constraints are set. Both controllers also receive identical accurate demand forecasts. With this model we present two simulations with different demand patterns.

For the traditional model predictive controlled inventory the following cost function was implemented

$$ J = \sum_{i=N_1}^{N_2} \left(\delta \left(I^*(i) - \hat{I}(i) \right)^2 + \lambda \left(\Delta O(i) \right)^2 \right) \cdot \text{(5)} $$

where $I^*(i)$ = Desired inventory level
$\hat{I}(i)$ = Predicted inventory level
$\Delta O(i)$ = Change in order rate
δ, λ = Weight parameters

For model reference controlled inventory we used the cost function presented in equation 4 with the most basic form of the filter so that the only parameter to tune is p_1. As mentioned earlier, the number of tunable parameters in model reference control is reduced by one when compared to the

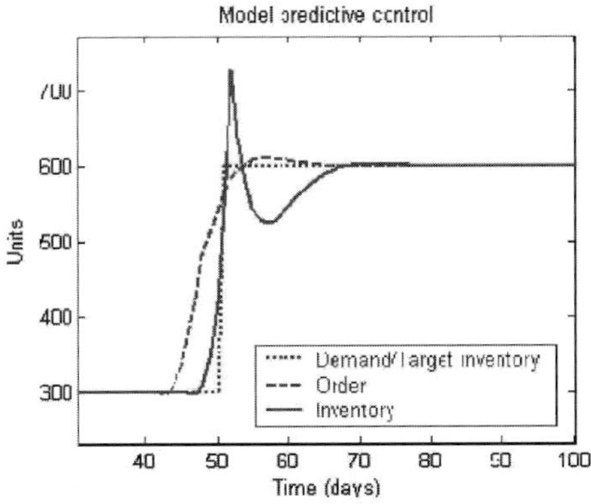
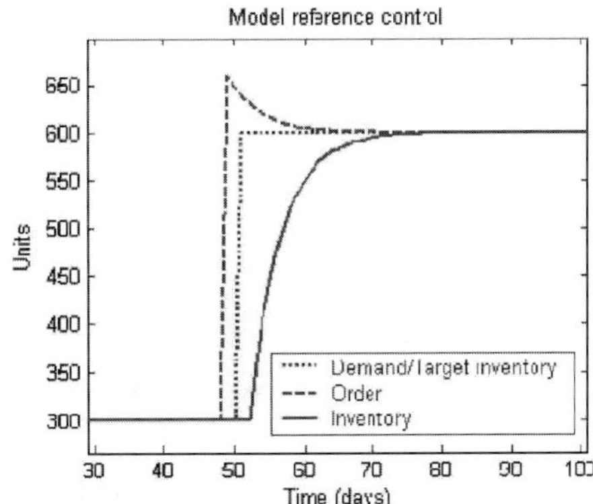

Figure 4: Step responses of an MPC-controlled inventory. In the left-hand figure the cost function is in the form presented in equation 5, and in the right-hand figure the cost function is in the form presented in equation 4.

traditional model predictive control. This is obvious when we look at the cost functions presented in this simulation, as the cost function used in model reference controller has only one parameter p_1 instead of the two parameters, δ and λ included in the model predictive controller.

3.3 Step Response Simulations

The first simulation was a basic step response test which demonstrated the difference between a model refence controlled inventory and a traditional model predictive controlled inventory. The step was implemented in both demand and target inventory levels at the same time to cause a major change in the system. In other words, target inventory level was set to follow the demand so that every day the company had products in stock worth one day's sale. In this simulation the controller parameters were chosen to be as follows: $\delta = 0.1$, $\lambda = 0.9$ and $p_1 = 0.8$, with the control horizon of 10 days. The results can be seen in Figure 4 where a step in demand has occured at the moment of 50 days. As can be seen, the response of the model reference controller is much more smoother than the step response of a traditional form of the model predictive controller. This is due to the fact that the model reference controller dampens directly the changes made in the inventory level instead of dampening the changes made in ordering. Model predictive controller is forced to start ordering excessively in advance to the step due to the limitations in changing the control action. The more reasonably implemented model reference controller orders exactly the amount needed every instant. It becomes obvious that the penalizing of control actions is not a suitable way to

control supply chain related tasks as it causes additional variations in the system. Model reference control is much more efficient in achieving what was beeing pursued in this study – smooth control method which is simple to tune and implement.

3.4 Simulations with a More Realistic Demand Pattern

A more practical simulation was also completed with more realistic demand pattern and forecasting error. The demand involved also noise which made the control task even more realistic. The control horizon was kept at 10 days in both controllers but the parameters δ and λ needed to be retuned as the parameters δ and λ used in the previous simulation resulted as very poor responses. New parameters were chosen as follows: $\delta = 0.3$ and $\lambda = 0.7$. The model reference controller did not need any retuning as it survived both simulations very satisfyingly with same parameter $p_1 = 0.8$. The target inventory level was kept constant in the level of 100 units. This is propably not the most cost efficient way of managing an inventory but was used nevertheless to keep the results easy to understand.

The demand curve and inventory response can be seen in Figure 5. The demand curve is identical in both pictures but as can be seen there were major differences in inventory levels. Inventory levels in the model predictive controlled case showed major variations at the same time as demand rapidly increased. No such variations were found in the model reference case. Once again the penaling of control actions forced the controller to order excessive amounts in order to avoid stock-out.

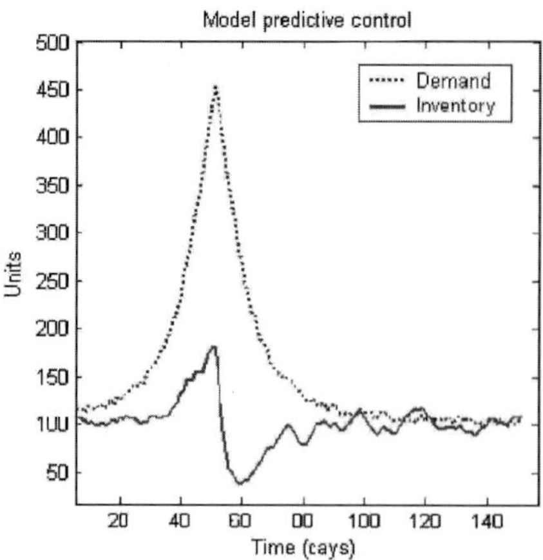

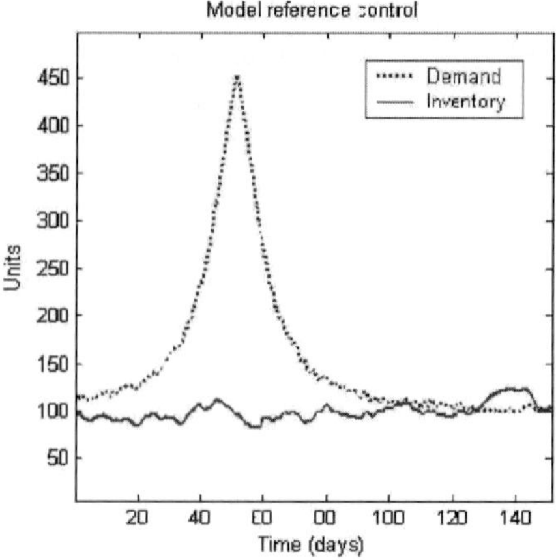

Figure 5: The inventory responses of both controllers to a gaussian-shaped demand pattern.

4 CONCLUSIONS

In this study we presented a solution to the problematic determining of the cost parameter penalizing changes made in the control in model predictive controller used in business related systems. The model reference control was studied and simulations performed to demonstrate the abilities and advanteges of this control method. It has been shown, that the model reference control method is an effective way to control an inventory and most of all that the method allows us to avoid the problematic parameter λ in the equation 1. This reduction of a very problematic parameter is most definitely inevitable if any kind of practical solutions are ever desired. Therefore all future research concerning business related control should concider this. It should also be kept in mind that any reduction of tuneable parameters can be seen as an advantage.

Also we showed in thisstudy that the model reference control is at least as applicable in inventory control as model predictive control if not even better. The simple, yet effective and smooth response model reference control provides suits perfectly to the unstable and variating environment of business related systems. It can also be said that the filter-like behaviour is desireable in order to reduce the bullwhip behaviour, but further research is needed on this field.

A new supply chain simulator interfacewas also presented and used in the simulations of this study. The set of universal supply chain blocks gives an opportunity of testing and simulating the perforance of different control methods or even different forms of supply chains without reprogramming the basic elements of inventory and production.

REFERENCES

Camacho, E. F. and Bordons, C., 2002, *Model Predictive Control*. Springer-Verlag London Limited

Hennet, J.-C., 2003. A bimodal scheme for multi-stage production and inventory control. In *Automatica, Vol 39, pp. 793-805.*

Koivisto, H., Kimpimäki, P., Koivo, H., 1991. Neural predictive control – a case study. In *1991 IEEE International Symposium on Intelligent Control.* IEEE Press.

Lambert, E. P., 1987. *Process Control Applications of Long-Range Prediction*. PhD Thesis, University of Oxford. Trinity.

Maciejowski, J. M., 2002, *Predictive Control with Constraints*. Pearson Education Limited

Rivera, D. E., Braun, M. W., Flores, M. E., Carlyle, W. M., Kempf, K. G., 2003. A Model predictive control framework for robust management of multi-product, multi-echelon demand networks. In *Annual Reviews in Control, Vol. 27, pp. 229-245.*

Towill, D. R., 1996. Industrial dynamics modeling of supply chains. In *International Journal of Physical Distribution & Logistics Management, Vol. 26, No. 2, pp.23-42.*

Tzafestas, S., Kapsiotis, G., Kyriannakis, E., 1997. Model-based predictive control for generalized production problems. In *Computers in Industry, Vol. 34, pp. 201-210.*

Ydstie, B. E., Grossmann, I. E., Perea-López, E., 2003, A Model predictive control strategy for supply chain optimization. In *Computers and Chemical Engineering, Vol. 27, pp. 1202-1218.*

AN LMI OPTIMIZATION APPROACH FOR GUARANTEED COST CONTROL OF SYSTEMS WITH STATE AND INPUT DELAYS

O.I. Kosmidou, Y.S. Boutalis and Ch. Hatzis

Department of Electrical and Computer Engineering Democritus University of Thrace
67100 Xanthi, Greece
kosmidou@ee.duth.gr

Keywords: Uncertain systems, time-delays, optimal control, guaranteed cost control, LMI.

Abstract: The robust control problem for linear systems with parameter uncertainties and time-varying delays is examined. By using an appropriate uncertainty description, a linear state feedback control law is found ensuring the closed-loop system's stability and a performance measure, in terms of the *guaranteed cost*. An LMI objective minimization approach allows to determine the "optimal" choice of free parameters in the uncertainty description, leading to the minimal guaranteed cost.

1 INTRODUCTION

Model uncertainties and time-delays are frequently encountered in physical processes and may cause performance degradation and even instability of control systems (Malek-Zavarei and Jamshidi, 1987), (Mahmoud, 2000), (Niculescu, 2001). Hence, stability analysis and robust control problems of uncertain dynamic systems with delayed states have been studied in recent control systems literature; for details and references see e.g. (Niculescu et al., 1998), (Kolmanovskii et al., 1999), (Richard, 2003). Since control input delays are often imposed by process design demands as in the case of transmission lines in hydraulic or electric networks, it is necessary to consider uncertain systems with both, state and input time-delays. Moreover, when the delays are imperfectly known, one has to consider uncertainty on the delay terms, as well. In recent years, LMIs are used to solve complex analysis and control problems for uncertain delay systems (e.g. (Li and de Souza, 1997), (Li et al., 1998), (Tarbouriech and da Silva Jr., 2000), (Bliman, 2001), (Kim, 2001), and related references).

The purpose of the present paper is to design control laws for systems with uncertain parameters and uncertain time-delays affecting the state and the control input. Although the uncertainties are assumed to enter linearly in the system description, it is well known that they may be time-varying and nonlinear in nature, in most physical systems. Consequently, the closed-loop system's stability has to be studied

in the Lyapunov-Krasovskii framework; the notion of quadratic stability is then extended to the class of time-delay systems (Barmish, 1985),(Malek-Zavarei and Jamshidi, 1987). On the other hand, it is desirable to ensure some performance measure despite uncertainty and time-delay, in terms of guaranteed upper bounds of the performance index associated with the dynamic system. The latter specification leads to the guaranteed cost control (Chang and Peng, 1972), (Kosmidou and Bertrand, 1987).

In the proposed approach the uncertain parameters affecting the state, input, and delay matrices are allowed to vary into a pre-specified range. They enter into the system description in terms of the so-called *uncertainty matrices* which have a given structure. Different unity rank decompositions of the uncertainty matrices are possible, by means of appropriate scaling. This description is convenient for many physical system representations (Barmish, 1994). An LMI optimization solution (Boyd et al., 1994)is then sought in order to determine the appropriate uncertainty decomposition; the resulting guaranteed cost control law ensures the minimal upper bound. The closed-loop system's quadratic stability follows as a direct consequence.

The paper is organized as follows: The problem formulation and basic notions are given in Section 2. Computation of the solution in the LMI framework is presented in Section 3. Section 4 presents a numerical example. Finally, conclusions are given in Section 5.

117

J. Braz et al. (eds.), Informatics in Control, Automation and Robotics I, 117–123.

2   PROBLEM STATEMENT AND DEFINITIONS

Consider the uncertain time-delay system described in state-space form,

$$
\dot{x}(t) = [A_1 + \Delta A_1(t)]x(t)
$$
$$
+[A_2 + \Delta A_2(t)]x(t - d_1(t)) + [B_1 + \Delta B_1(t)]u(t)
$$
$$
+[B_2 + \Delta B_2(t)]u(t - d_2(t)) \tag{1}
$$

for $t \in [0, \infty)$ and with $x(t) = \phi(t)$ for $t < 0$.

In the above description, $x(t) \in \mathbf{R}^n$ is the state vector, $u(t) \in \mathbf{R}^m$ is the control vector and A_1, $A_2 \in \mathbf{R}^{n \times n}$, B_1, $B_2 \in \mathbf{R}^{n \times m}$ are constant matrices.

The model uncertainty is introduced in terms of

$$
\Delta A_1(t) = \sum_{i=1}^{k_1} A_{1i} r_{1i}(t), \qquad | r_{1i}(t) | \leq \bar{r}_1
$$

$$
\Delta A_2(t) = \sum_{i=1}^{k_2} A_{2i} r_{2i}(t), \qquad | r_{2i}(t) | \leq \bar{r}_2
$$

$$
\Delta B_1(t) = \sum_{i=1}^{l_1} B_{1i} p_{1i}(t), \qquad | p_{1i}(t) | \leq \bar{p}_1
$$

$$
\Delta B_2(t) = \sum_{i=1}^{l_2} B_{2i} p_{2i}(t), \qquad | p_{2i}(t) | \leq \bar{p}_2 \tag{2}
$$

where A_{1i}, A_{2i}, B_{1i}, B_{2i} are given matrices with constant elements determining the uncertainty structure in the state, input, and delay terms.

The uncertain parameters r_{1i}, r_{2i}, p_{1i}, p_{2i} are Lebesgue measurable functions, possibly time-varying, that belong into pre-specified bounded ranges (2), where $\bar{r}_1$, $\bar{r}_2$, $\bar{p}_1$, $\bar{p}_2$ are positive scalars; since their values can be taken into account by the respective uncertainty matrices, it is assumed that $\bar{r}_1 = \bar{r}_2 = \bar{p}_1 = \bar{p}_2 = 1$, without loss of generality. Moreover, the advantage of the affine type uncertainty description (2) is that it allows the uncertainty matrices to have unity rank and thus to be written in form of vector products of appropriate dimensions,

$$
\begin{aligned}
A_{1i} &= d_{1i}e_{1i}^T, & i &= 1, ..., k_1 \\
A_{2i} &= d_{2i}e_{2i}^T, & i &= 1, ..., k_2 \\
B_{1i} &= f_{1i}g_{1i}^T, & i &= 1, ..., l_1 \\
B_{2i} &= f_{2i}g_{2i}^T, & i &= 1, ..., l_2
\end{aligned} \tag{3}
$$

Obviously, the above decomposition is not unique; hence, the designer has several degrees of freedom in choosing the vector products, in order to achieve the design objectives. By using the vectors in (3), define

the matrices,

$$
\begin{aligned}
D_1 &:= [d_{11}...d_{1k_1}], & E_1 &:= [e_{11}...e_{1k_1}] \\
D_2 &:= [d_{21}...d_{2k_2}], & E_2 &:= [e_{21}...e_{2k_2}] \\
F_1 &:= [f_{11}...f_{1l_1}], & G_1 &:= [g_{11}...g_{1l_1}] \\
F_2 &:= [f_{21}...f_{2l_2}], & G_2 &:= [g_{21}...g_{2l_2}]
\end{aligned} \tag{4}
$$

which will be useful in the proposed guaranteed cost approach.

The time delays in (1) are such that,

$$
\begin{aligned}
0 \leq d_1(t) \leq \bar{d}_1 < \infty, & \qquad \dot{d}_1(t) \leq \beta_1 < 1 \\
0 \leq d_2(t) \leq \bar{d}_2 < \infty, & \qquad \dot{d}_2(t) \leq \beta_2 < 1
\end{aligned} \tag{5}
$$

$\forall t \geq 0$.

Associated with system (1) is the quadratic cost function

$$
J(x(t), t) = \int_0^\infty [x^T(t)Qx(t) + u^T(t)Ru(t)]dt \tag{6}
$$

with $Q > 0$, $R > 0$, which is to be minimized for a linear constant gain feedback control law of the form

$$
u(t) = Kx(t) \tag{7}
$$

by assuming (A_1, B_1) stabilizable, $(Q^{1/2}, A_1)$ detectable and the full state vector $x(t)$ available for feedback.

In the absence of uncertainty and time-delay, the above formulation reduces to the optimal quadratic regulator problem (Anderson and Moore, 1990). Since uncertainties and time-delays are to be taken into account, the notions of *quadratic stability* and *guaranteed cost control* have to be considered. The following definitions are given.

Definition 2.1

The uncertain time-delay system (1)-(5) is *quadratically stabilizable independent of delay*, if there exists a static linear feedback control of the form of (7), a constant $\theta > 0$ and positive definite matrices $P \in \mathbf{R}^{n \times n}$, $R_1 \in \mathbf{R}^{n \times n}$ and $R_2 \in \mathbf{R}^{m \times m}$, such that the time derivative of the Lyapunov-Krasovskii functional

$$
\mathcal{L}(x(t), t) = x^T(t)Px(t)
$$
$$
+ \int_{t-d_1(t)}^t x^T(\tau)R_1 x(\tau)d\tau
$$
$$
+ \int_{t-d_2(t)}^t u^T(\tau)R_2 u(\tau)d\tau \tag{8}
$$

satisfies the condition

$$
\dot{\mathcal{L}}(x(t), t) = \frac{d\mathcal{L}(x(t), t)}{dt} \leq -\theta \parallel x(t) \parallel^2 \tag{9}
$$

along solutions $x(t)$ of (1) with $u(t) = Kx(t)$, for all $x(t)$ and for all admissible uncertainties and time-delays, i.e. consistent with (2), (3) and (5), respectively.

The resulting closed-loop system is called *quadratically stable* and $u(t) = Kx(t)$ is a *quadratically stabilizing control law*.

Definition 2.2

Given the uncertain time-delay system (1)-(5) with quadratic cost (6), a control law of the form of (7) is called a *guaranteed cost control*, if there exists a positive number $\mathcal{V}(x(0), \phi(-\bar{d}_1), \phi(-\bar{d}_2))$, such that

$$J(x(t), t) \leq \mathcal{V}(x(0), \phi(-\bar{d}_1), \phi(-\bar{d}_2)) \qquad (10)$$

for all $x(t)$ and for all admissible uncertainties and time-delays. The upper bound $\mathcal{V}(.)$ is then called a *guaranteed cost*.

The following Proposition provides a sufficient condition for quadratic stability and guaranteed cost control.

Proposition 2.3

Consider the uncertain time-delay system (1)-(5) with quadratic cost (6). Let a control law of the form of (7) be such that the derivative of the Lyapunov-Krasovskii functional (8) satisfies the condition

$$\dot{\mathcal{L}}(x(t), t) \leq -x^T(t)[Q + K^T RK]x(t) \qquad (11)$$

for all $x(t)$ and for all admissible uncertainties and time-delays. Then, (7) is a guaranteed cost control law and

$$\mathcal{V}(x(0), \phi(-\bar{d}_1), \phi(-\bar{d}_2)) = x^T(0)Px(0)$$
$$+ \int_{-\bar{d}_1}^{0} \phi^T(\tau)R_1\phi(\tau)d\tau$$
$$+ \int_{-\bar{d}_2}^{0} \phi^T(\tau)K^T R_2 K\phi(\tau)d\tau \qquad (12)$$

is a guaranteed cost for (6). Moreover, the closed-loop system is quadratically stable.

Proof

By integrating both sides of (11) one obtains

$$\int_0^T \dot{\mathcal{L}}(x(t), t)dt \leq - \int_0^T [x^T(t)Qx(t) + u^T(t)Ru(t)]dt \qquad (13)$$

and thus

$$\mathcal{L}(x(T), T) - \mathcal{L}(x(0), 0) = x^T(T)Px(T)$$
$$+ \int_{T-d_1(t)}^{T} x^T(\tau)R_1 x(\tau)d\tau$$
$$+ \int_{T-d_2(t)}^{T} x^T(\tau)K^T R_2 Kx(\tau)d\tau$$
$$- x^T(0)Px(0) - \int_{-d_1(t)}^{0} x^T(\tau)R_1 x(\tau)d\tau$$
$$- \int_{-d_2(t)}^{0} x^T(\tau)K^T R_2 Kx(\tau)d\tau \leq$$
$$- \int_0^T [x^T(t)Qx(t) + u^T(t)Ru(t)]dt \qquad (14)$$

Since (11) is satisfied for $\mathcal{L} > 0$, it follows from Definition 2.1 that (9) is also satisfied for $\theta = \lambda_{min}(Q + K^T RK)$. Consequently $u(t) = Kx(t)$ is a quadratically stabilizing control law and thus $\mathcal{L}(x(t), t) \to 0$ as $T \to \infty$ along solutions $x(.)$ of system (1). Consequently, the above inequality yields

$$\int_0^\infty [x^T(t)Qx(t) + u^T(t)Ru(t)]dt$$
$$\leq x^T(0)Px(0) + \int_{-\bar{d}_1}^{0} \phi^T(\tau)R_1\phi(\tau)d\tau$$
$$+ \int_{-\bar{d}_2}^{0} \phi^T(\tau)K^T R_2 K\phi(\tau)d\tau \qquad (15)$$

and thus

$$J(x(t), t) \leq \mathcal{V}(x(0), \phi(-\bar{d}_1), \phi(-\bar{d}_2)) \qquad (16)$$

for $\mathcal{V}(x(0), \phi(-\bar{d}_1), \phi(-\bar{d}_2))$ given by (12).

Remark 2.4

The condition resulting from Proposition 2.3 is a *delay-independent* condition. It is well known (Li and de Souza, 1997), (Kolmanovskii et al., 1999)that control laws arising from delay-independent conditions are likely to be conservative. Besides, quadratic stability and guaranteed cost approaches provide conservative solutions, as well. A means to reduce conservatism consists in minimizing the upper bound of the quadratic performance index by finding the appropriate guaranteed cost control law. Such a solution will be sought in the next Section.

3 SOLUTION IN THE LMI FRAMEWORK

In order to determine the unity rank uncertainty decomposition in an "optimal" way such that the guaranteed cost control law minimizes the corresponding guaranteed cost bound, an LMI objective minimization problem will be solved. The uncertainty decomposition (3) can be written as (Fishman et al., 1996),

$$
\begin{aligned}
A_{1i} &= (s_{1i}^{1/2} d_{1i})(s_{1i}^{-1/2} e_{1i})^T & i &= 1, ..., k_1 \\
A_{2i} &= (s_{2i}^{1/2} d_{2i})(s_{2i}^{-1/2} e_{2i})^T & i &= 1, ..., k_2 \\
B_{1i} &= (t_{1i}^{1/2} f_{1i})(t_{1i}^{-1/2} g_{1i})^T & i &= 1, ..., l_1 \\
B_{2i} &= (t_{2i}^{1/2} f_{2i})(t_{2i}^{-1/2} g_{2i})^T & i &= 1, ..., l_2
\end{aligned} \tag{17}
$$

where s_{ij}, t_{ij} are positive scalars to be determined during the design procedure. For this purpose, diagonal positive definite matrices are defined,

$$
\begin{aligned}
S_1 &:= diag(s_{11}, ..., s_{1k_1}) \\
S_2 &:= diag(s_{21}, ..., s_{2k_2}) \\
T_1 &:= diag(t_{11}, ..., t_{1k_1}) \\
T_2 &:= diag(t_{21}, ..., t_{2k_2})
\end{aligned} \tag{18}
$$

The existence of a solution to the guaranteed cost control problem is obtained by solving an LMI feasibility problem. The following Theorem is presented:

Theorem 3.1

Consider the uncertain time-delay system (1)-(5) with quadratic cost (6). Suppose there exist positive definite matrices S_1, S_2, T_1, T_2, W, $\bar{R}_1$, $\bar{R}_2$, such that the LMI

$$
\begin{bmatrix}
\Lambda(.) & WE_1 & WE_2 & B_1 R^{-1} G_1 \\
E_1^T W & -S_1 & 0 & 0 \\
E_2^T W & 0 & -S_2 & 0 \\
G_1^T R^{-1} B_1^T & 0 & 0 & -T_1 \\
G_2^T R^{-1} B_1^T & 0 & 0 & 0 \\
R^{-1} B_1^T & 0 & 0 & 0 \\
W & 0 & 0 & 0 \\
W & 0 & 0 & 0
\end{bmatrix}
$$

$$
\begin{bmatrix}
B_1 R^{-1} G_2 & B_1 R^{-1} & W & W \\
0 & 0 & 0 & 0 \\
0 & 0 & 0 & 0 \\
0 & 0 & 0 & 0 \\
-T_2 & 0 & 0 & 0 \\
0 & -\frac{1}{\beta_2}\bar{R}_2 & 0 & 0 \\
0 & 0 & -\frac{1}{\beta_1}\bar{R}_1 & 0 \\
0 & 0 & 0 & -\hat{Q}
\end{bmatrix} < 0 \tag{19}
$$

where

$$
\begin{aligned}
\Lambda(.) &= A_1 W + W A_1^T - B_1 R^{-1} B_1^T + A_2 A_2^T \\
&+ B_2 B_2^T + D_1 S_1 D_1^T + D_2 S_2 D_2^T + F_1 T_1 F_1^T \\
&+ F_2 T_2 F_2^T + B_1 R^{-1} R^{-1} B_1^T
\end{aligned} \tag{20}
$$

and

$$
\hat{Q} = (I_n + Q)^{-1} \tag{21}
$$

$$
\bar{R}_1 = R_1^{-1} \tag{22}
$$

$$
\bar{R}_2 = R_2^{-1} \tag{23}
$$

admits a feasible solution $(S_1, S_2, T_1, T_2, W, \bar{R}_1, \bar{R}_2)$. Then, the control law

$$
u(t) = -R^{-1} B_1^T P x(t) \tag{24}
$$

with

$$
P := W^{-1} \tag{25}
$$

is a guaranteed cost control law. The corresponding guaranteed cost is

$$
\begin{aligned}
\mathcal{V}(x(0), \phi(-\bar{d}_1), \phi(-\bar{d}_2)) &= x^T(0) P x(0) \\
&+ \int_{-\bar{d}_1}^{0} \phi^T(\tau) R_1 \phi(\tau) d\tau \\
&+ \int_{-\bar{d}_2}^{0} \phi^T(\tau) P B_1 R^{-1} R_2 R^{-1} B_1^T P \phi(\tau) d\tau
\end{aligned} \tag{26}
$$

Proof

The time derivative of the Lyapunov-Krasovskii functional is

$$
\begin{aligned}
\dot{\mathcal{L}}(x(t), t) &= 2x^T(t) P \dot{x}(t) + x^T(t) R_1 x(t) \\
&- (1 - \dot{d}_1(t)) x^T(t - d_1(t)) R_1 x(t - d_1(t)) \\
&+ x^T(t) K^T R_2 K x(t) - (1 - \dot{d}_2(t)) \\
&\quad x^T(t - d_2(t)) K^T R_2 K x(t - d_2(t))
\end{aligned} \tag{27}
$$

By using (1), (5) and (17), one obtains the inequality

$$
\begin{aligned}
\dot{\mathcal{L}}(x(t), t) &\leq 2x^T(t) P\{[A_1 + \Delta A_1(t)] x(t) \\
&+ [A_2 + \Delta A_2(t)] x(t - d_1(t)) \\
&- [B_1 + \Delta B_1(t)] R^{-1} B_1^T P x(t) \\
&- [B_2 + \Delta B_2(t)] R^{-1} B_1^T P x(t - d_2(t))\} \\
&+ x^T(t) R_1 x(t) - (1 - \beta_1) x^T(t - d_1(t)) R_1 x(t - d_1(t)) \\
&+ x^T(t) P B_1 R^{-1} R_2 R^{-1} B_1^T P x(t) - (1 - \beta_2) \\
&\quad x^T(t - d_2(t)) P B_1 R^{-1} R_2 R^{-1} B_1^T P x(t - d_2(t))
\end{aligned} \tag{28}
$$

which is to be verified for all $x(.) \in \mathbf{R}^n$. Furthermore, by using the identity $2 \mid ab \mid \leq a^2 + b^2$, for any a, b, $\in \mathbf{R}^n$, as well as (2)-(4), (17) and (18), the following quadratic upper bounding functions are derived,

$$2x^T(t)P\Delta A_1(t)x(t) \leq\mid 2x^T(t)P\Delta A_1(t)x(t)\mid$$

$$\leq \sum_{i=1}^{k_1}\mid 2x^T(t)PA_{1i}r_{1i}(t)x(t)\mid$$

$$\leq \sum_{i=1}^{k_1}\mid 2x^T(t)P(d_{1i}s_{1i}^{1/2})(e_{1i}s_{1i}^{-1/2})^T x(t)\mid$$

$$\leq x^T(t)P\sum_{i=1}^{k_1} s_{1i}d_{1i}d_{1i}^T Px(t)$$

$$+ x^T(t)\sum_{i=1}^{k_1} s_{1i}^{-1}e_{1i}e_{1i}^T x(t)$$

$$= x^T(t)PD_1S_1D_1^T Px(t)$$
$$+ x^T(t)E_1S_1^{-1}E_1^T x(t) \tag{29}$$

$\forall x(.) \in \mathbf{R}^n$. In a similar way one obtains

$$2x^T(t)PA_2x(t-d_1(t)) \leq x^T(t)PA_2A_2^T Px(t)$$
$$+x^T(t-d_1(t))x(t-d_1(t)) \tag{30}$$

$$2x^T(t)P\Delta A_2(t)x(t-d_1(t))$$
$$\leq x^T(t)PD_2S_2D_2^T Px(t)$$
$$+ x^T(t-d_1(t))E_2S_2^{-1}E_2^T x(t-d_1(t)) \tag{31}$$

$$-2x^T(t)P\Delta B_1(t)R^{-1}B_1^T Px(t)$$
$$\leq x^T(t)PF_1T_1F_1^T Px(t)$$
$$+ x^T(t)PB_1R^{-1}G_1T_1^{-1}G_1^T R^{-1}B_1^T Px(t) \tag{32}$$

$$-2x^T(t)PB_2R^{-1}B_1^T Px(t-d_2(t))$$
$$\leq x^T(t)PB_2R^{-1}B_2^T Px(t)$$
$$+ x^T(t-d_2(t))PB_1R^{-1}B_1^T Px(t-d_2(t)) \tag{33}$$

$$-2x^T(t)P\Delta B_2(t)R^{-1}B_1^T Px(t-d_2(t))$$
$$\leq x^T(t)PF_2T_2F_2^T Px(t) +$$
$$x^T(t-d_2(t))PB_1R^{-1}G_2T_2^{-1}G_2^T R^{-1}B_1^T P$$
$$x(t-d_2(t)) \tag{34}$$

The above inequalities are true $\forall x(.) \in \mathbf{R}^n$. By using these quadratic upper bounding functions, it is straightforward to show that the guaranteed cost condition (11) is satisfied if

$$A_1 + A_1^T P - PB_1R^{-1}B_1^T P + Q$$
$$+P[D_1S_1D_1^T + F_1T_1F_1^T + D_2S_2D_2^T$$
$$+F_2T_2F_2^T + A_2A_2^T + B_2B_2^T]P$$
$$+PB_1R^{-1}[G_1T_1^{-1}G_1^T + G_2T_2^{-1}G_2^T$$
$$+I_m + \beta_2R_2]R^{-1}B_1^T P + I_n$$
$$+E_1S_1^{-1}E_1^T + E_2S_2^{-1}E_2^T + \beta_1R_1 < 0 \tag{35}$$

By multiplying every term of the above matrix inequality on the left and on the right by $P^{-1} := W$ and by applying Shur complements, the LMI form (19) is obtained.

In the sequel, a minimum value of the guaranteed cost is sought, for an "optimal" choice of the rank-1 uncertainty decomposition. For this purpose, the minimization of the guaranteed cost is obtained by solving an objective minimization LMI problem.

Theorem 3.2

Consider the uncertain time-delay system (1)-(5) with quadratic cost (6). Suppose there exist positive definite matrices S_1, S_2, T_1, T_2, W, $\bar{R}_1$, $\bar{R}_2$, M_1, M_2, M_3, such that the following LMI objective minimization problem

$$min_{\mathcal{X}}\bar{J} = min_{\mathcal{X}}(Tr(M_1) + Tr(M_2) + Tr(M_3)) \tag{36}$$
$$\mathcal{X} = (S_1, S_2, T_1, T_2, W, \bar{R}_1, \bar{R}_2, M_1, M_2, M_3), \text{ with}$$

LMI constraints (19) and

$$\begin{bmatrix} -M_1 & I_n \\ I_n & -W \end{bmatrix} < 0 \tag{37}$$

$$\begin{bmatrix} -M_2 & I_n \\ I_n & -\bar{R}_1 \end{bmatrix} < 0 \tag{38}$$

$$\begin{bmatrix} -M_3 & I_m \\ I_m & -\bar{R}_2 \end{bmatrix} < 0 \tag{39}$$

has a solution $\mathcal{X}(.)$. Then, the control law

$$u(t) = -R^{-1}B_1^T Px(t) \tag{40}$$

with

$$P := W^{-1} \tag{41}$$

is a guaranteed cost control law. The corresponding guaranteed cost

$$\mathcal{V}(x(0), \phi(-\bar{d}_1), \phi(-\bar{d}_2)) = x^T(0)Px(0)$$

$$+\int_{-\bar{d}_1}^0 \phi^T(\tau)R_1\phi(\tau)d\tau$$

$$+\int_{-\bar{d}_2}^0 \phi^T(\tau)PB_1R^{-1}R_2R^{-1}B_1^T P\phi(\tau)d\tau \tag{42}$$

is minimized over all possible solutions.

Proof

According to Theorem (3.1), the guaranteed cost (42) for the uncertain time-delay system (1)-(5) is ensured by any feasible solution $(S_1, S_2, T_1, T_2, W, \bar{R}_1, \bar{R}_2)$ of the convex set defined by (19). Furthermore, by taking the Shur complement of (37) one has $-M_1 + W^{-1} < 0 \implies M_1 > W^{-1} = P \implies Trace(M_1) >$

$Trace(P)$. Consequently, minimization of the trace of M_1 implies minimization of the trace of P. In a similar way it can be shown that minimization of the traces of M_2 and M_3 implies minimization of the traces of R_1 and R_2, respectively. Thus, minimization of $\bar{J}$ in (36) implies minimization of the guaranteed cost of the uncertain time-delay system (1)-(5) with performance index (6). The optimality of the solution of the optimization problem (36) follows from the convexity of the objective function and of the constraints.

4 EXAMPLE

Consider the second order system of the form of equation (1)with nominal matrices,

$$A_1 = \begin{bmatrix} -2 & 1 \\ 0 & 1 \end{bmatrix}, \ A_2 = \begin{bmatrix} -0.2 & 0.1 \\ 0 & 0.1 \end{bmatrix},$$
$$B_1 = \begin{bmatrix} 0 \\ 1 \end{bmatrix}, B_2 = \begin{bmatrix} 0 \\ 0.1 \end{bmatrix}$$

The uncertainty matrices on the state and control as well as on the delay matrices are according to (2),

$$\Delta A_1 = \begin{bmatrix} 0 & 0 \\ 0.1 & 0.1 \end{bmatrix}, \ \Delta A_2 = \begin{bmatrix} 0 & 0 \\ 0.03 & 0.03 \end{bmatrix},$$
$$\Delta B_1 = \begin{bmatrix} 0 \\ 0.1 \end{bmatrix}, \Delta B_2 = \begin{bmatrix} 0 \\ 0.03 \end{bmatrix}$$

and hence, the rank-1 decomposition may be chosen such that,

$$D_1 = \begin{bmatrix} 0 \\ 0.1 \end{bmatrix}, D_2 = \begin{bmatrix} 0 \\ 0.03 \end{bmatrix}, E_1 = E_2 = \begin{bmatrix} 1 \\ 1 \end{bmatrix},$$
$$F_1 = \begin{bmatrix} 0 \\ 0.1 \end{bmatrix}, F_2 = \begin{bmatrix} 0 \\ 0.03 \end{bmatrix}, G_1 = G_2 = 1$$

The state and control weighting matrices are, respectively, $Q = I_2$, $R = 4$. It is $\beta_1 = 0.3$, $\beta_2 = 0.2$.

By solving the LMI objective minimization problem one obtains the guaranteed cost $\bar{J} = 25.1417$. The optimal rank-1 decomposition is obtained with $s_1 = 0.6664$, $s_2 = 2.2212$, $t_1 = 2.5000$, $t_2 = 8.3333$. Finally, the corresponding control gain is $K = [-0.2470 \ -6.0400]$.

5 CONCLUSIONS

In the previous sections, the problem of guaranteed cost control has been studied for the class of uncertain linear systems with state and input time varying delays. A constant gain linear state feedback control law has been obtained by solving an LMI feasibility problem. The closed-loop system is then quadratically stable and preserves acceptable performance for all parameter uncertainties and time varying delays of a given class. The system performance deviates from the optimal one, in the sense of LQR design of the nominal system, due to uncertainties and time delays. However, the performance deterioration is limited and this is expressed in terms of a performance upper bound, namely the guaranteed cost. In order to make the GC as small as possible, one has to solve an LMI minimization problem. The minimal upper bound corresponds to the "optimal" rank-1 decomposition of the uncertainty matrices.

REFERENCES

Anderson, B. and Moore, J. (1990). *Optimal Control:Linear Quadratic Methods*. Prentice Hall, Englewood Cliffs, N.J., 2nd edition.

Barmish, B. (1985). Necessary and sufficient conditions for quadratic stabilizability of uncertain linear systems. *Journal of Optimization Theory and Application*, 46: 399–408.

Barmish, B. (1994). *New tools for robustness of linear systems*. Macmillan, NY.

Bliman, P.-A. (2001). Solvability of a lyapunov equation for characterization of asymptotic stability of linear delay systems. In *Proceedings of the ECC '01 European Control Conference,Porto, Portugal*, pages 3006–3011. ICEIS Press.

Boyd, S., Ghaoui, L. E., Feron, E., and Balakrishnan, V. (1994). *Linear matrix inequalities in system and control theory*. SIAM Studies in Applied Mathematics, Philadelphia.

Chang, S. and Peng, T. (1972). Adaptive guaranteed cost control of systems with uncertain parameters. *I.E.E.E. Trans.on Automatic Control*.

Fishman, A., Dion, J.-M., Dugard, L., and Trofino-Neto, A. (1996). A linear matrix inequality approach for guaranteed cost control. In *Proceedings of the 13th IFAC World Congress, San Francisco, USA*.

Kim, J.-H. (2001). Delay and its time-derivative dependent robust stability of time-delayed linear systems with uncertainty. *I.E.E.E. Trans.on Automatic Control*, 46: 789–792.

Kolmanovskii, V., Nikulescu, S., and Richard, J.-P. (1999). On the liapunov-krasovskii functionals for stability analysis oflinear delay systems. *International Journal of Control*, 72: 374–384.

Kosmidou, O. and Bertrand, P. (1987). Robust controller design for systems with large parameter variations. *International Journal of Control*, 45: 927–938.

Li, H., Niculescu, S.-I., Dugard, L., and Dion, J.-M., editors (1998). *Robust guaranteed cost control for uncertain linear time-delay systems*, pages 283–301. In

. Stability and control of time-delay systems. Springer-Verlag, London.

Li, X. and de Souza, C. (1997). Delay-dependent robust stability and stabilization of uncertain linear delay systems: a linear matrix inequality approach. *IEEE Trans. on Automatic Control*, 42: 1144–1148.

Mahmoud, M. (2000). *Robust control and filtering for time-delay systems*. Marcel Dekker, NY.

Malek-Zavarei, M. and Jamshidi, M. (1987). *Time-delay systems:analysis, optimization and applications*. North Holland, Amsterdam.

Niculescu, S. (2001). *Delay effects on stability*. Springer-Verlag, London.

Niculescu, S., Verriest, E., Dugard, L., and Dion, J.-M., editors (1998). *Stability and robust stability of time-delay systems: A guided tour*, pages 1–71. In Stability and control of time-delay systems. Springer-Verlag, London.

Richard, J.-P. (2003). Time-delay systems: an overview of some recent advances and open problems. *Automatica*, 39: 1667–1694.

Tarbouriech, S. and da Silva Jr., J. G. (2000). Synthesis of controllers for continuous-time delay systems with saturating controls via lmi's. *IEEE Trans. on Automatic Control*, 45: 105–111.

USING A DISCRETE-EVENT SYSTEM FORMALISM FOR THE MULTI-AGENT CONTROL OF MANUFACTURING SYSTEMS

Guido Maione

DIASS, Politecnico di Bari, Viale del Turismo, 8, 74100, Taranto, Italy
Email: gmaione@poliba.it

David Naso

DEE, Politecnico di Bari, Via Re David, 200, 70125, Bari, Italy
Email: naso@deemail.poliba.it

Keywords: Multi-Agent Systems, Discrete Event Dynamic Systems, Distributed Manufacturing Control, Heterarchical Manufacturing Systems.

Abstract: In the area of Heterarchical Manufacturing Systems modelling and control, a relatively new paradigm is that of Multi-Agent Systems. Many efforts have been made to define the autonomous agents concurrently operating in the system and the relations between them. But few results in the current literature define a formal and unambiguous way to model a Multi-Agent System, which can be used for the real-time simulation and control of flexible manufacturing environments. To this aim, this paper extends and develops some results previously obtained by the same authors, to define a discrete event system model of the main distributed agents controlling a manufacturing system. The main mechanism of interaction between three classes of agents is presented.

1 INTRODUCTION

Nowadays, the study of appropriate tools for modelling and designing Multi-Agent Systems (MAS) technologies is a key-issue involving all their application areas, including telecommunication and computer networks, communities of intelligent robots, web-based agents for information retrieval, to mention a few. Moreover, considerable research efforts have been devoted to the definition of standards and to the development of platforms for unambiguous agent specification, especially in the context of software engineering.

Focusing on the specific context of industrial manufacturing, this paper proposes an approach based on the Discrete EVent System (DEVS) specification (Zeigler et al., 2000) to obtain a complete and unambiguous characterization of a multi-agent control system. By using the DEVS formalism, we describe agents as atomic dynamic systems, subject to external inputs from (and generating outputs to) other agents. Furthermore, we directly obtain the model of the entire network of agents by specifying the relationships between the atomic agents. The DEVS technique is fully compatible with the heterarchical design principles, and leads to MAS where all information and control functions are distributed across autonomous entities. In particular, the DEVS formalism is an interesting alternative to other recently proposed tools for MAS specification, e.g. the UML (Huhns and Stephens, 2001), Petri Nets (Lin and Norrie, 2001). The success of this formalism is due to its suitability for developing useful models both for discrete event simulation, and for implementation of the software controllers on plant's operating system. Namely, the DEVS formalism can effectively model many recent MAS architectures, such as part-driven heterarchical manufacturing systems (Duffie and Prabhu, 1996, Prabhu and Duffie, 1999) and schemes inspired by the Contract Net protocol (Smith, 1980, Parunak, 1994, Sousa and Ramos, 1999).

As in most MAS for manufacturing control (Heragu, 2002, Shen and Norrie, 1999), in our model all the agents use decision algorithms emulating micro-economic environments. Each agent uses a fictitious currency to buy services from other seller agents which, on their turn, use pricing strategies. Sellers and buyers have to reach an equilibrium between conflicting objectives, i.e. to maximize profit and to minimize costs, respectively.

J. Braz et al. (eds.), Informatics in Control, Automation and Robotics I, 125–132.
© 2006 *Springer. Printed in the Netherlands.*

Recently, there have been efforts to develop analytical models of negotiation processes using, for instance, Petri nets (Hsieh, 2004), underlining the need of a systematical analysis and validation method for distributed networks of autonomous control entities. Many other researches have focused on the experimental validation of MAS on distributed simulation platforms (Shattenberg and Uhrmacher, 2001, Logan and Theodoropoulos, 2001), which allow to perform detailed investigations on the interaction schemes. Sharing the same motivations with the mentioned researches, our work focuses on the development of DEVS models of MAS, which combines the rigor of a tool suitable for performing the theoretical analysis of structural properties of the MAS, with its efficiency in directly translating the model in a detailed simulation environment, and its flexibility in testing both the individual and collective dynamics of the autonomous entities. Namely, our main goal is to find a multi-agent design platform that allows users to assess the relative effectiveness of agents' decision and interaction strategies, with special interest to adaptive learning mechanisms that allow agents to improve their performance during their operation (Maione and Naso, 2003a).

In this paper, we develop a detailed model of the interactions between the main agents in a manufacturing system. This contribution extends previous researches by the authors, in which, for sake of simplicity, the interactions with transport units were not considered in detail, and illustrates the basic mechanisms of the modelling procedure. The paper also outlines other main directions of the research in progress. Section 2 introduces the basic components of the proposed MAS, and specifies their roles and relations. Section 3 specifies how to model agents as atomic DEVS. Section 4 focuses on the main interactions between agents, describing the negotiation for a manufacturing process on a part. Sections 5 and 6 give some experimental results, overview the advantages of the approach, and enlighten the issues open for further research.

2 THE MULTI-AGENT SYSTEMS CONTROL APPROACH

We consider each *Part Agent* (PA) as a control unit connected to the corresponding physical part (piece) flowing through the system. In accordance with the related literature (Duffie and Prabhu, 1996, Heragu, 2002, Prabhu and Duffie, 1999), we associate each part into a batch of identical items in process with a PA that identifies in real-time (i.e. shortly before a part is ready for a new process) the most suitable

workstation for the imminent operation on that part and, consequently, the most suitable vehicle to transfer it to the station. The selection is based on real-time updated information directly received from the alternative stations and transport vehicles, through an exchange of messages with other agents. Namely, a *Workstation Agent* (WA) is a software entity controlling a workstation or a cell of machines performing the same operations, and a *Transport Agent* (TA) is associated with the transport system or with a single or group of transport units.

At each operation step, by interacting with WAs and TAs, a PA chooses the machine servicing the associated part and the transport device moving the piece from its current position (the output buffer of the occupied machine) to the chosen workstation.

In this framework, one can consider also specific agents designed to execute other tasks necessary for the global operation of the manufacturing plant (Maione and Naso, 2003b). In particular, one can associate an *Order Agent* (OA) with each different order issued by the management level. The OA retains information on quantity and type of products to be processed. Similarly, one can define a *Loading Agent* (LA) to manage the loading/unloading station where the raw/completed parts enter/exit the system.

The global control of the manufacturing floor emerges from the concurrent actions of the various agents in the system. The careful analysis of their interactions is fundamental to understand how to obtain the desired global behaviour. For instance, the OA interacts with PAs to control the release of the quantity and type of raw parts necessary to realize that order. The PAs interact with the LA to negotiate the loading/unloading of raw/completed parts. Here, we concentrate on the interactions between a PA and WAs and TAs when a part is ready for a new process and its PA has to decide the resources (workstation and transport device) necessary to fulfil the operations, among a set of available alternatives.

The high-level agents' decisions are executed by low-level plant controllers that are not modelled here. One can also view the network of interacting agents as a distributed controller supervising and synchronizing the underlying physical hardware.

3 THE DISCRETE-EVENT MODELLING FRAMEWORK

The agents operating in our MAS model interact one with another by exchanging event messages. Outputs from an agent become inputs for other agents. The agent state is updated by external input events (inputs) and internal events. Each event in the life of an agent is considered an instantaneous or "atomic" action without duration. Time-consuming actions are represented by a pair of events, the first denoting its start and the second denoting its finish.

So, unambiguous models for the agents in the system are identified by all the classified events which affect the dynamics of each type of agent. The *internal events* are generated by the internal dynamics of the agent, and the *exogenous events* are *inputs* which are not determined by the agent. Finally, the external output events (or *outputs*) represent the reaction of the agents. Then, it is important to define the *sequential state* of each agent. Namely, events change the state. An agent stays in a state until either it receives an input, or an amount of time determined by a time advance function elapses. In the latter case, an internal event occurs to change state according to a specified internal transition function. Otherwise, if an agent receives an input, an external transition function specifies the state transition according to the current *total state* of the agent, defined by the sequential state, the time elapsed since the last transition and some additional information. Finally, agents generate outputs according to an output function. Delays and faults in the communication process are also considered in our model. Although the effects of these phenomena are often neglected in technical literature, we evaluate their effects both on overall production performance and on the efficiency of the MAS, expressed by ad-hoc performance measures. This allows us to track, monitor and optimize the interaction among agents.

To conclude, each agent may be modeled as an atomic DEVS as follows:

$$A = < \mathbf{X}, \mathbf{Y}, \mathbf{S}, \delta_{int}, \delta_{ext}, \lambda, ta > \qquad (1)$$

where $\mathbf{X}$ is the set of inputs, $\mathbf{Y}$ is the set of outputs, $\mathbf{S}$ is the set of sequential states, $\delta_{int}: \mathbf{S} \rightarrow \mathbf{S}$ is the internal transition function, $\delta_{ext}: \mathbf{Q} \times \mathbf{X} \rightarrow \mathbf{S}$ is the external transition function, $\lambda: \mathbf{S} \rightarrow \mathbf{Y}$ is the output function, $ta: \mathbf{S} \rightarrow \mathfrak{R}_0^+$ is the time advance function ($\mathfrak{R}_0^+$ set of positive real numbers with 0 included), $\mathbf{Q} = \{(\mathbf{s}, e, \mathbf{DL}) \mid \mathbf{s} \in \mathbf{S}, 0 \leq e \leq ta(\mathbf{s})\}$ is the total state set.

The sequential state $\mathbf{s}$ contains the main information on the *status*, specifying the condition of an agent between two successive events. Other data strictly depend on the type of the agent. E.g., for a PA, one can consider the current residual sequence of operation steps necessary to complete the procedure, the set of available machines for the negotiated operation, and prospected time in current state before the next internal event. For a WA, $\mathbf{s}$ includes the queues of the requests received from PAs for availability, information and confirmation of negotiated operations (see below), and the time before the next internal event. For a TA, $\mathbf{s}$ may contain similar queues of requests received from PAs, and the time before the next internal event.

The total state $\mathbf{q}$ depends on $\mathbf{s}$, the time e elapsed since the last transition and the decision law $\mathbf{DL}$ used by the agent to select and rank the offers received from other agents and to decide its action.

Usually, to build the models, one observes that each agent may require or offer a service to other agents. A precise mechanism usually defines the negotiation protocols working according to a cycle *"announce-offer-reward-confirm"*: an agent starts a bid by requiring availability for a service to one or more agents, requests the available agents information representing an offer for the negotiated service, collects the replies from the queried agents, selects the best offer, sends a rewarding message, waits for a confirmation and finally acquires the service from the rewarded agent.

In this paper, we focus on the interactions of a PA with WAs and TAs when contracting for an operation in the procedure to be accomplished for a part in process. We describe the main part of the PA DEVS model, by concentrating on the mechanism ruling the *status-transitions* of a PA, which are triggered by inputs or internal events, and the outputs generated for each *status*. We don't go into the details of the **DL** used by each agent. To this aim, we exploit the models already defined and developed in precedent papers (Maione and Naso, 2003a,b) for PAs, WAs and TAs, but we expand and better clarify them to put them together.

4 THE INTERACTIONS OF A PA WITH WAS AND TAS

To accomplish the manufacturing tasks, each PA interacts with WAs to choose the workstation for the next operation and with TAs to select the vehicle moving the part from the station currently occupied to the next one. We assume that the PA firstly communicates exclusively with WAs, then with TAs only.

For $t < t_{P0}$ let the PA associated with a generic part, say P, be in a quiescent *status* (QUIESC) and let it begin its activity at t_{P0} (event X_{P0}). Then P spends the interval $[t_{P0}, t_{P1}]$ to send outputs Y_{P01}, $Y_{P02},..., Y_{P0w}$ at instants $t_{01} > t_{P0}, t_{02},..., t_{0w} = t_{P1}$. These messages request the availability to all the WAs of the alternative stations (w in number) that can serve the part. The sequence of requests cannot be interrupted by any internal or external occurrence. For sake of simplicity, instead of modelling a sequence of w *status*-values, we refer to REQWAV for the whole duration of the activity and assume that P makes transition at t_{P1} (event I_{P1}).

In $[t_{P1}, t_{P2}]$ P waits for answers (WAIWAV). Namely, the request P transmits to each WA may queue up with similar ones sent by other PAs. Next transition occurs at t_{P2} when either P receives all the answers from the queried WAs (X_{P1}), or a specified time-out of WAIWAV expires before P receives all the answers. In case it receives no reply within the timeout (I_{P2}), P returns to REQWAV and repeats the request procedure. In case of time-out expiration and some replies received (I_{P3}), P considers only the received answers to proceed. The repeated lack of valid replies may occur for system congestion, for machine failures or communication faults, or for other unpredictable circumstances. In all cases permanent waits or deadlocks may occur. To avoid further congestion and improve system fault-tolerance, we use time-outs and let P repeat the cycle REQWAV-WAIWAV only a finite number of times, after which P is unloaded from the system.

If all or some replies are received before the time-out expiration, P starts requesting service to the $m \leq w$ available WAs at t_{P2}. In [t_{P2}, t_{P3}] P requests information to these WAs by sending them Y_{P11}, Y_{P12}, ..., Y_{P1m} at instants $t_{11} > t_{P2}$, t_{12},..., $t_{1m} = t_{P3}$. If the sequence of requests cannot be interrupted, we refer to REQWSE for the whole activity. We assume that at t_{P3} P makes transition (I_{P4}).

Then, P spends [t_{P3}, t_{P4}] waiting for offers from the available WAs (WAIWOF), as the request P transmits to each WA may queue up with those sent by other PAs. Next transition occurs at t_{P4} when either P receives all the answers from the queried WAs (X_{P2}) or a time-out of WAIWOF expires. In case it receives no reply within the timeout (I_{P5}), P returns to REQWSE and repeats the procedure. In case of time-out expiration and some replies are received (I_{P6}), P considers only the received offers to select the next server. Again, to avoid congestion, P repeats the cycle REQWSE-WAIWOF a finite number of times, then it is discharged.

Once received the offers from WAs, P utilizes [t_{P4}, t_{P5}] to take a decision for selecting the workstation (TAKWDE). At t_{P5} the decision algorithm ends (I_{P7}), after selecting a WA and building a queue to rank all the offers of other WAs.

Subsequently, P reserves the chosen machine by transmitting a booking message (Y_{P2}) to the corresponding WA. So P takes [t_{P5}, t_{P6}] for communicating the choice to the WA (COMCHW). At t_{P6} the communication ends (I_{P8}). Now, the WA has to send a rejection if there is a conflict with another PA or a booking confirmation (X_{P5}). Hence, P uses [t_{P6}, t_{P7}] to wait for a confirmation from the selected WA (WAIWCO). The confirmation is necessary because during the decision interval the condition of the plant can be modified by actions of other PAs, and the selected server can be no longer available. If P receives a rejection (X_{P3}), or does not receive any reply within a time-out (I_{P9}), it returns to

COMCHW, sends a new request of confirmation to the second WA in the decision rank. If P has no other alternative destinations and the rejection (X_{P4}) or the time-out (I_{P10}) occurs, it returns to REQWAV and repeats the negotiation. Also WAIWCO, WAIWAV and WAIWOF cannot lead to deadlocks, thanks to the time-out.

At t_{P7}, after receiving a confirmation from the selected WA, P starts the negotiation with TAs for a device to carry the part from the current position to the input buffer of the selected workstation, where the last negotiated operation is to be made. Then P opens the bid and spends [t_{P7}, t_{P8}] to send Y_{P31}, Y_{P32}, ..., Y_{P3v} at instants $t_{31} > t_{P7}$, t_{32},..., $t_{3v} = t_{P8}$ to all the v possible TAs to request their availability (REQTAV). In [t_{P8}, t_{P9}] after the end of transmission (I_{P11}), P waits for availability-answers (WAITAV) until a time-out expires: if no reply is received, P gets back to REQTAV (I_{P12}) to repeat the request. Otherwise, if all replies are received before the time-out expiration (X_{P6}) or $u \leq v$ replies are received and the time-out expires (I_{P13}), at t_{P9} P starts requesting service to the u available TAs (REQTSE).

Then P uses [t_{P9}, t_{P10}] to send outputs Y_{P41}, Y_{P42}, ..., Y_{P4u} at instants $t_{41} > t_{P9}$, t_{42},..., $t_{4u} = t_{P10}$ to all the available TAs and, after the transmission is completed (I_{P14}), P waits for offers from TAs (WAITOF) in [t_{P10}, t_{P11}] until a time-out expires. If no offer is received (I_{P15}), the PA repeats the request. If only some offers arrive and the time-out expires (I_{P16}) or all offers arrive before the time-out (X_{P7}), P can take a transport-decision (TAKTDE) for selecting the best offering TA in [t_{P11}, t_{P12}]. After selection (I_{P17}), in [t_{P12}, t_{P13}] P communicates its choice by sending Y_{P5} to this TA (COMCHT). After this communication (I_{P18}), P waits for a rejection or a confirmation from the selected TA (WAITCO) until a time-out expires. If no reply is received in the waiting period [t_{P13}, t_{P14}] and a time-out expires (I_{P19}) or a rejection is received (X_{P8}), in case other offers from TAs are available P gets back to COMCHT and selects a new TA; in case no other TA is available and there is a time-out expiration (I_{P20}) or a rejection (X_{P9}), the availability request is repeated and P gets back to REQTAV.

If a confirmation is received (X_{P10}), P makes a transition to issue a transport command (TRANSP). It takes the interval [t_{P14}, t_{P15}] to issue the command Y_{P6} to load the part on the vehicle associated with the selected TA and to start the transport process. When, at time t_{P15}, the command is complete (I_{P21}), P gets back to QUIESC. In case of the last operation, Y_{P6} also signals the completion of the working procedure to a controller, which influences and adapts the **DL** the PA uses for ranking the offers (Maione and Naso, 2003a). In this case, P leaves the system.

In general, from t_{P15} to the beginning of the next operation (if any), P stops making decisions, receiving and sending messages and remains quiescent. The associated part is loaded on the transport vehicle and transferred to the next workstation where it is downloaded in the input buffer. Here, it waits in queue until receiving service, after a proper set-up. After the operation, the part reaches the output buffer and is ready for the next destination. All the above processes are driven by low-level controllers and do not involve agent activities. So, only when the processes are over, P is again ready to start a new negotiation phase. If for $t>t_{P15}$ faults occur to the selected machine or vehicle, P remains in QUIESC and there is no need to restart negotiations with WAs or TAs. In fact, the plant controllers manage the repair process: when the normal operating conditions are restored, the part is transported to the selected machine.

Note that, after the negotiation cycle is complete, when the chosen and confirmed WA (or TA) signals to the PA the end of the operation (or transport)

process, the PA can take into account its new availability. If, at this time, the PA is requesting or waiting for availability or information from other WAs (or TAs), or is taking a decision for operation (transport) on other parts, the received messages from the past-selected WA (or TA) wait in a queue until the PA gets to REQWSE (or REQTSE). In this case, the PA will send an output Y_{P1m+1} (or Y_{P4u+1}) also to this new available WA (TA).

Figure 1 depicts all this complex interaction dynamics. Circles represent the PA *status*-values, arrows represent the events, and the outputs, directly associated with *status*-values, are encapsulated into the circles. As the figure shows, the PA may receive confirmation from a WA (or a TA) after several successive couples COMCHW-WAIWCO (or COMCHT-WAITCO). Also, time-outs can bring the PA back to REQWAV (from WAIWAV when no answer is received from WAs or from WAIWCO after a WA-rejection) or to REQTAV (from WAITAV when no answer is received from TAs or from WAITCO after a TA-rejection).

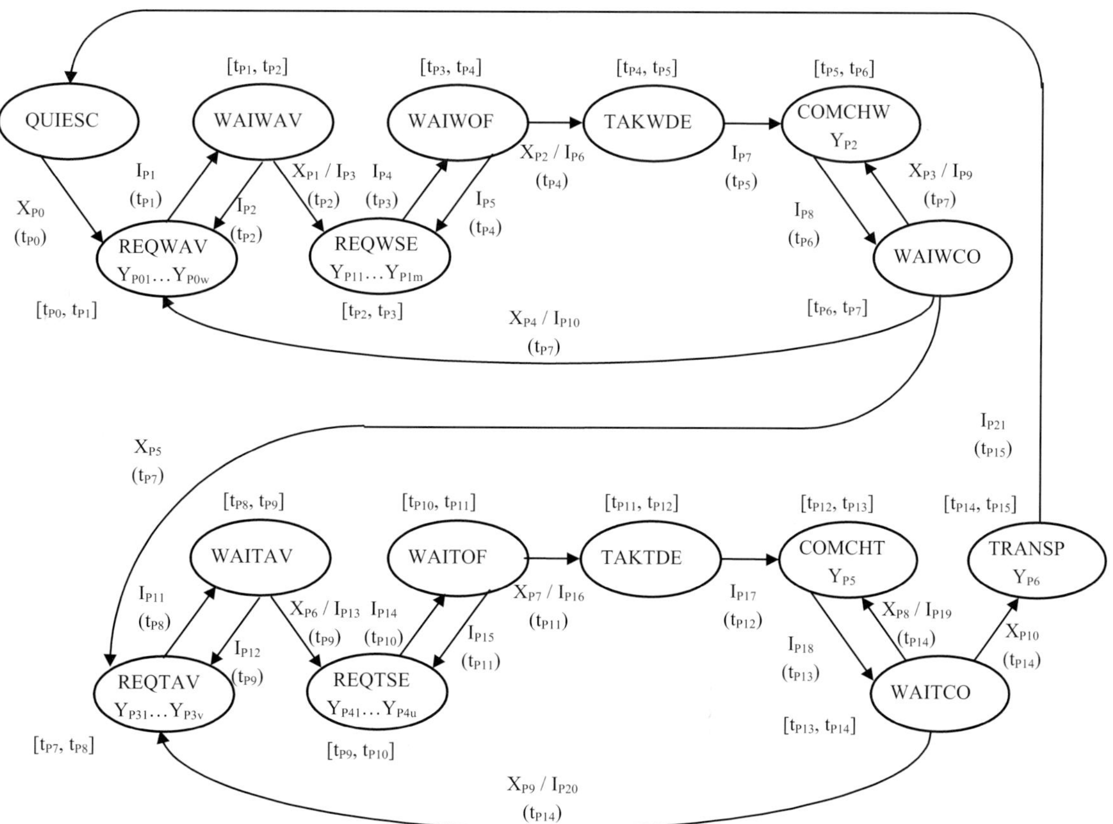

Figure 1: dynamics of a PA when negotiating with WAs and TAs.

On one hand, one could simplify the model by merging REQWAV, WAIWAV, REQWSE and WAIWOF, i.e. by considering the PA sending requests for availability and information all together. So, each WA would offer availability and the information necessary to the PA decision at the same time. Only two *status*-values would be necessary, the first for the request, the second for the wait. The same reduction is possible for REQTAV, WAITAV, REQTSE and WAITOF.

On the other hand, the more detailed model of the PA activities, which considers two different time-outs for the two different waiting conditions previously defined, can be more effective in reducing the PA waiting times and in avoiding deadlocks. In fact, the effect of delays and losses of messages due to workstation or transport unavailability (faults, malfunctions, overloaded workstations, etc.) and to communication faults are reduced. Also, the cyclic repetition of requests and waits and the consequent delays in the decision processes are limited. As a consequence, the risk of discharging PAs from the system is reduced.

To better enlighten the negotiation mechanism, we summarize in Tables I-III the *status*-values, inputs, internal events and outputs.

Figure 1 shows that the negotiation mechanism maintains a well defined structure with other agents participating to a negotiation process.

In a similar way, a DEVS model can be defined for other interactions between classes of agents (Maione and Naso, 2003b). The common structure of the negotiation mechanism is advantageous for building up complex models in a modular way.

The DEVS model of the agents' interactions is particularly suitable for developing a complete simulation platform for the analysis of the dynamics of the complex MAS controlling a manufacturing plant. In particular, our model allows the simulation of both the plant processes and their macroscopic hardware components (machines, AGVs, parts, etc.), and the details of the control activities performed by agents (inputs, outputs, states, time-outs). So, we can evaluate the classical indices of a manufacturing system performance (throughput, number of completed items, lateness, etc.), but also the effects of agents and their decision policies and the MAS efficiency (number of negotiation cycles, number of requests). Also, we can measure the agents' behavior in steady-state operating conditions and their adaptation to abrupt disturbances (shortages of materials, workload changes, hardware faults, etc.).

In this sense, we made all these measures when agents were using different decision policies, to see how they dynamically react to disturbances (Maione and Naso, 2003a,c). We compared other MAS that use conventional decision heuristics (based on the delivery time of parts to machines, the distance to the next workstation, the required set-up time) with our MAS, both with and without adaptation. We let agents use a set of decision rules for a limited amount of time (the agent life-cycle) and then we replace the rules by using the most successful ones. The replacement at the end of life-cycle was guided by a mechanism emulating the 'survival of the fittest' natural selection process and propagating the fittest rules to the next population of agents. The fitness of each decision rule was the average lateness of the parts controlled (Maione and Naso, 2003a).

Table I: *status*-values for a Part Agent.

Status	Agent's Activity Description
QUIESC	Agent inactive
REQWAV	Request availability to all possible WAs
WAIWAV	Wait for availability signal from WAs
REQWSE	Request service to available WAs
WAIWOF	Wait for offers from available WAs
TAKWDE	Take decision for the best WA offer
COMCHW	Communicate choice to selected WA
WAIWCO	Wait confirm./reject. from selected WA
REQTAV	Request availability to all possible TAs
WAITAV	Wait for availability signal from TAs
REQTSE	Request service to available TAs
WAITOF	Wait for offers from available TAs
TAKTDE	Take decision for the best TA offer
COMCHT	Communicate choice to selected TA
WAITCO	Wait confirm./reject. from selected TA
TRANSP	Command selected TA to move the part

Table II: inputs received and outputs sent by a PA when negotiating with WAs and TAs.

Inputs	Time	Description
X_{P0}	t_{P0}	Start negotiation for a new operation
X_{P1}	t_{P2}	Last reply for WA availability received
X_{P2}	t_{P4}	Last reply for WA offer received
X_{P3}	t_{P7}	Rejection & alternative WAs in the PA rank
X_{P4}	t_{P7}	Rejection & no alternative WA in the PA rank
X_{P5}	t_{P7}	Confirmation from a WA
X_{P6}	t_{P9}	Last reply for TA availability
X_{P7}	t_{P11}	Last reply for TA offer
X_{P8}	t_{P14}	Rejection & alternative TAs in the PA rank
X_{P9}	t_{P14}	Rejection & no alternative TA in the PA rank
X_{P10}	t_{P14}	Confirmation from a TA

Outputs	Description
$Y_{P01}\ Y_{P02}\ \dots\ Y_{P0w}$	Requests of availability
$Y_{P11}\ Y_{P12}\ \dots\ Y_{P1m}$	Requests of service to available WAs
Y_{P2}	Choice communication to the selected WA
$Y_{P31}\ Y_{P32}\ \dots\ Y_{P3v}$	Requests of availability TAs
$Y_{P41}\ Y_{P42}\ \dots\ Y_{P4u}$	Requests of service to available TAs
Y_{P5}	Choice communication to the selected TA
Y_{P6}	Transport command

Table III: internal events of a PA when negotiating with WAs and TAs.

Internal Event	Time	Description
I_{P1}	t_{P1}	End of WA availability request process
I_{P2}	t_{P2}	Time-out and no availability signal received from WAs
I_{P3}	t_{P2}	Time-out and some (m) availability signals received from WAs
I_{P4}	t_{P3}	End of WA service request process
I_{P5}	t_{P4}	Time-out and no offer received from the m available WAs
I_{P6}	t_{P4}	Time-out and some offers ($o<m$) received from the available WAs
I_{P7}	t_{P5}	End of workstation-decision process
I_{P8}	t_{P6}	End of choice communication to the selected WA
I_{P9}	t_{P7}	Time-out and no confirmation received from the selected WA when other ranked WA offers are available
I_{P10}	t_{P7}	Time-out and no confirmation received from the selected WA when no other ranked WA offers are available
I_{P11}	t_{P8}	End of TA availability request process
I_{P12}	t_{P9}	Time-out and no availability signal received from TAs
I_{P13}	t_{P9}	Time-out and some (u) availability signals received from TAs
I_{P14}	t_{P10}	End of TA service request process
I_{P15}	t_{P11}	Time-out and no offer received from the u available TAs
I_{P16}	t_{P11}	Time-out and some offers ($o<u$) received from the available TAs
I_{P17}	t_{P12}	End of transport-decision process
I_{P18}	t_{P13}	End of choice communication to the selected TA
I_{P19}	t_{P14}	Time-out and no confirmation received from the selected TA when other ranked TA offers are available
I_{P20}	t_{P14}	Time-out and no confirmation received from the selected TA when no other ranked TA offers are available
I_{P21}	t_{P15}	End of transport command

5 SOME EXPERIMENTAL RESULTS AND FUTURE PLANS

The DEVS model is particularly suitable for developing a simulation platform for the analysis of the complex dynamics of distributed multi-agent control systems. Differently from traditional discrete event models of manufacturing plants, mainly devoted to simulate the macroscopic components, our model considers also the detailed dynamics of the software agents (exchanged event messages, internal events and outputs). In this way, we may study also the effects of hardware faults, congestion of the communication network, message losses or similar circumstances.

For instance, it is possible to compare various decision policies used by the various agents to negotiate operations in a detailed simulation model, as done in (Maione and Naso, 2003a).

The simulation model also allows us to perform comparative analysis in dynamic conditions, and define reactive policies that minimize the effects of disturbances, e.g. a workstation fault. For example Figure 2 compares four different agents' decision policies by throughput: minimizing the distance

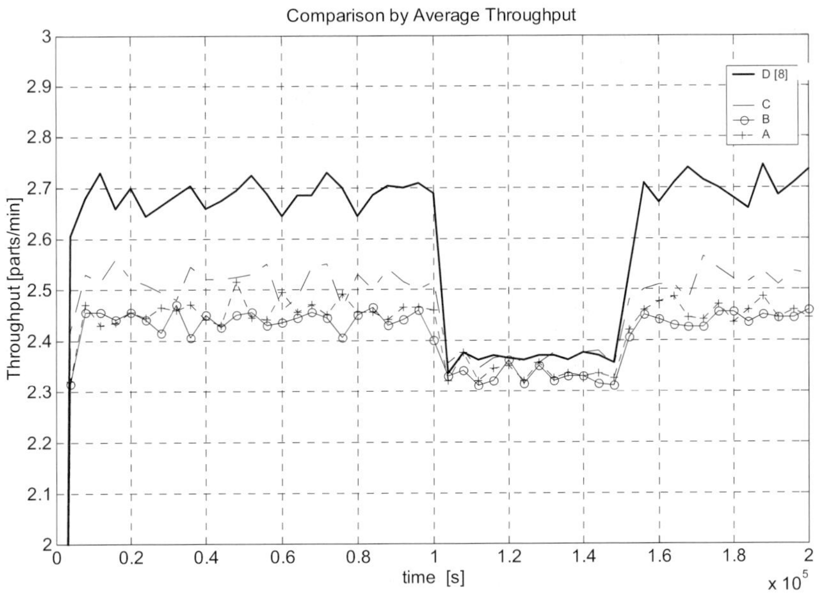

Figure 2: comparison of performance in dynamic conditions (workstation fault at t = 1 0^5 s).

between consecutive workstations (A), minimizing the wait for set-up of the workstation (B), minimizing the wait in queue at the workstation (C), and using a learning strategy (D). It can be noted how the reactive policy (D), designed with the aid of the DEVS model, outperforms the common (static) decision rules using MAS.

Future research aims at applying the proposed approach to other complex distributed control and optimization problems, such as those involved in large-scale logistic or supply chains.

6 CONCLUDING REMARKS

In this paper, we used the DEVS formalism to specify the model of the main agents operating in a MAS controlling part-driven heterarchical manufacturing systems. In this context, we detailed the interactions guiding the negotiations related to a generic step in a working procedure associated with a part process. The model respects heterarchical principles and can be used in a simulation platform which allows us to analyze both the classical performance indices of a manufacturing system and the effectiveness of the MAS, using decision policies which implement adaptation strategies.

The proposed method leaves many interesting issues open for further research. The next step toward the experimentation of the multi-agent control system on an automated manufacturing plant, is to test the DEVS model on a distributed network of computers, each hosting one or more of the agents in the heterarchical network. This aims at investigating and properly addressing the critical issues related to distributed autonomous controllers that cannot be examined when simulating the whole set of agents on a centralized platform.

REFERENCES

Duffie, N.A., Prabhu, V.V., 1996. Heterarchical control of highly distributed manufacturing systems. International Journal of Computer Integrated Manufacturing, Vol. 9, No. 4, pp. 270-281.

Han, W., Jafari, M.A., 2003. Component and Agent-Based FMS Modeling and Controller Synthesis. IEEE Trans. Sys., Man, and Cybernetics – Part C, Vol. 33, No. 2, pp. 193-206.

Heragu, S.S., Graves, R.J., Kim, B.-I., Onge, A.St., 2002. Intelligent Agent Based Framework for Manufacturing Systems Control. IEEE Trans. Sys., Man, and Cyber. - Part A, Vol. 32, No. 5.

Hsieh, F.-S., 2004. Model and control holonic manufacturing systems based on fusion of contract nets and Petri nets. Automatica, 40, pp. 51-57.

Huhns, M.N., Stephens, L.M., 2001. Automating supply chains. IEEE Int. Comput., Vol. 5, No. 4, pp. 90-93.

Kotak, D., Wu, S., Fleetwood, M., Tamoto, H., 2003. Agent-based holonic design and operations environment for distributed manufacturing. Computers in Industry, 52, pp. 95-108.

Lee, J.S., Hsu, P.L., 2004. Design and Implementation of the SNMP Agents for Remote Monitoring and Control via UML and Petri Nets. IEEE Trans. Control Sys. Techn., Vol. 12, No. 2, pp. 293-302.

Lin, F., Norrie, D.H., 2001. Schema-based conversation modeling for agent-oriented manufacturing systems. Computers in Industry, Vol. 46, pp. 259-274.

Logan, B., Theodoropoulos, G., 2001. The distributed Simulation of Multiagent systems. Proceedings of the IEEE, Vol. 89, No.2, pp. 174-185.

Maione, G., Naso, D., 2003a. A Genetic Approach for Adaptive Multi-Agent Control in Heterachical Manufacturing Systems. IEEE Trans. Sys., Man, and Cyb. – Part A: Spec. Issue Collective Intelligence in Multi-Agent Systems, Vol. 33, No. 5, pp. 573-588.

Maione, G., Naso, D., 2003b. A Discrete-Event System Model for Multi-Agent Control of Automated Manufacturing Systems. In IEEE SMC'03, Int. Conf. on Sys., Man and Cyb., Washington D.C., USA.

Maione, G., Naso, D., 2003c. A Soft Computing Approach for Task Contracting in Multi-Agent Manufacturing Control. Comp. Ind., Vol. 52, No. 3, pp. 199-219.

Parunak, H.V.D., 1994. Applications of Distributed Artificial Intelligence in Industry. In O'Hare, Jennings, (Eds.), Foundations of Distributed Artificial Intelligence, Wiley-Inter-Science.

Prabhu, V., Duffie, N., 1999. Nonlinear dynamics in distributed arrival time control of heterarchical manufacturing systems, IEEE Trans. Control Systems Technology, Vol. 7, No. 6, pp. 724-730.

Shattenberg, B., Uhrmacher, A.M., 2001. Planning Agents in James. Proc. of the IEEE, Vol. 89, No. 2, pp. 158-173.

Shen, W., Norrie, D.H., 1999. Agent-Based Systems for Intelligent Manufacturing: A State-of-the-Art Survey, Knowledge and Information Systems, an International Journal, Vol. 1, No. 2, pp. 129-156.

Smith, R.G., 1980. The Contract Net Protocol: High Level Communication and Control in a Distributed Problem Solver. IEEE Trans. Computers, Vol. 29, No. 12, pp. 1104-1113.

Sousa, P., Ramos, C., 1999. A distributed architecture and negotiation protocol for scheduling in manufacturing systems. Computers in Industry, Vol. 38, pp. 103-113.

Zeigler, B.P., Praehofer, H., Kim, T.G., 2000. Theory of Modelling and Simulation, Academic Press, 2nd edit.

PART 2

Robotics and Automation

FORCE RIPPLE COMPENSATOR FOR A VECTOR CONTROLLED PM LINEAR SYNCHRONOUS MOTOR

Markus Hirvonen and Heikki Handroos

Institute of Mechatronics and Virtual Engineering, Lappeenranta University of Technology, Lappeenranta, Finland
E-mail: markus.hirvonen@lut.fi, heikki.handroos@lut.fi

Olli Pyrhönen

Department of Electrical Engineering, Lappeenranta University of Technology, Lappeenranta, Finland
E-mail: olli.pyrhonen@lut.fi

Keywords: Cogging, Disturbance Observation, Linear Motor, Velocity Control, Vibration Suppression.

Abstract: A dynamic model including non-idealities for a permanent magnet linear synchronous motor (PMLSM) is postulated and verified. The non-idealities acting on the physical linear motor are measured and analyzed. These experimental results are utilized in the model. The verified simulation model is used in developing a force disturbance compensator for the velocity controller of the motor. The force non-idealities, such as the cogging force, friction and load force variation, are estimated using a disturbance observer. The acceleration signal in the observer is derived through the use of a low-acceleration estimator. The significant effects of the disturbance compensator on the simulated and measured dynamics of the motor are shown.

1 INTRODUCTION

The linear motor is an old invention but it is only recently that, as a result of the development of permanent magnets and their decreased costs, permanent magnet linear motors have become a viable alternative to rotating motors fitted with linear transmissions. In machine automation, linear movement has traditionally been transmitted from a rotary actuator by means of a ball screw, rack and pinion or belt. The linear motor simplifies the mechanical structure, eliminating the contact-type nonlinearities caused by backlash, friction, and compliance. In addition, the main benefits of a linear motor include its high-power density, reliability and efficiency.

Nowadays, the controllers commercially available, mainly PID algorithms with fixed gains, are unable to compensate for the undesirable phenomena that reduce the precision of motion such as backlash, static friction, load variations etc. Large controller gains are needed in order to maintain the stiff control required when suppressing load disturbances that tend to reduce the stability of a system. Therefore, extended methods for the compensation of disturbance have become an important topic of research. By compensating for an unknown time-varying force based on the estimation of such a force, faster speed responses and smaller speed ripples can be achieved.

In disturbance compensation, the compensation technique itself is a very simple feed-forward control, but the difference arises from the different disturbance estimation algorithms. In (Castillo-Castaneda et al., 2001), the friction compensation has been studied using model-based estimation. One disadvantage of this technique is that it is suitable for tracking only, since the desired velocity must be known in advance. Kim et al. (2002) and Tan et al. (2002) have studied sliding mode estimators in compensation feedback, while Godler et al. (1999), Deur et al. (2000), Bassi et al. (1999) and Hong et al. (1998) have studied the disturbance observer of a more general algorithm. Godler et al. (1999) compared load disturbance compensation with an acceleration control loop inside a speed loop. They have found that control implemented using an acceleration control loop can better tolerate parameter variation as well as disturbance in comparison to robust control with a disturbance observer. On the other hand, Deur et al. (2000) suggested the use of a disturbance observer in industrial applications due to its simple

135

J. Braz et al. (eds.), Informatics in Control, Automation and Robotics I, 135–142.
© 2006 *Springer. Printed in the Netherlands.*

implementation as well as simple design which has no robustness constraints.

The motor model discussed in the above-mentioned papers is a simplified model. The development of computers and software has made it feasible to simulate the more detailed dynamical behavior of machine systems. This paper discusses the use of a more detailed non-linear dynamic model for the analysis of a linear transmission system. The motor studied in this paper is a commercial three-phase linear synchronous motor application. The moving part (the mover) consists of a slotted armature and a three phase windings, while the surface permanent magnets (the SPMs) are mounted along the whole length of the path (the stator).

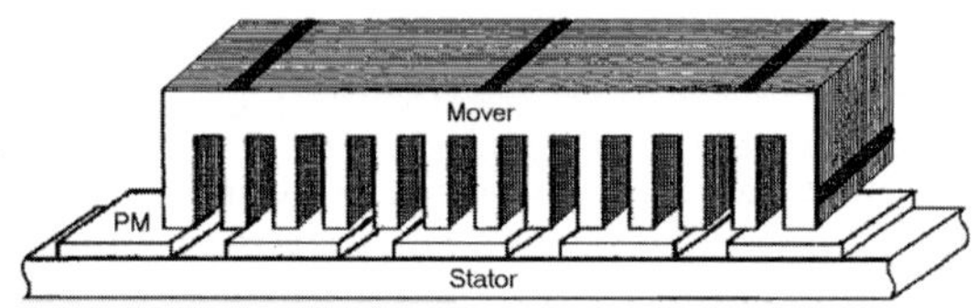

Figure 1: The structure of the studied linear motor.

First of all, the equations for modelling the vector-controlled motor drive and non-idealities are derived. Then, the simulation model is verified with the measurements from physical linear motor applications, and the comparison of the responses is shown in the study. Finally, a disturbance observer based on (Hong et al., 1998), (Godler et al., 1999), (Deur et al., 2000), and (Bassi et al., 1999) is implemented in the physical motor system after being tested in the simulation model, and the results and conclusions are presented.

2 SIMULATION MODEL

2.1 Model of LSM

The modeling of the dynamics of the linear synchronous motor examined in this paper is based on the space-vector theory. The time-varying parameters are eliminated and all the variables expressed on orthogonal or mutually decoupled direct and quadrature axes, which move at a synchronous speed of ω_s. The d- and q-axes equivalent to the circuit of the PMLSM are shown in figure 2, and the corresponding equations are (1) and (2), respectively.

The voltage equations for the synchronous

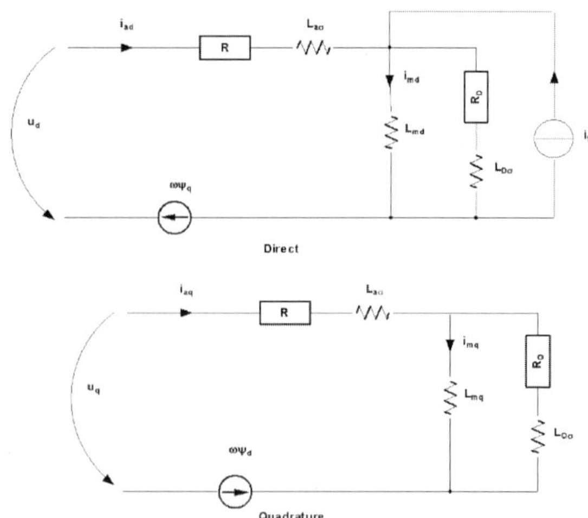

Figure 2: Two axial models of the linear synchronous motor.

machines are

$$u_d = Ri_{ad} + \frac{d\psi_d}{dt} - \omega_s\psi_q,\tag{1}$$

$$u_q = Ri_{aq} + \frac{d\psi_q}{dt} + \omega_s\psi_d,\tag{2}$$

where u_d and u_q are the d- and q-axis components of the terminal voltage, i_{ad} and i_{aq} the d- and q-axis components of the armature current, R is the armature winding resistance and ψ_d, ψ_q are the d- and q-axis flux linkage components of the armature windings. The synchronous speed can be expressed as $\omega_s=\pi v_s/\tau$, where v_s is the linear synchronous velocity and τ the pole pitch. Although the physical system does not contain a damper, which in PMLSM usually takes the form of an aluminum cover on the PMs, virtual damping must be included in the model, due to eddy currents. The voltage equations of the short-circuited damper winding are

$$0 = R_D i_D + \frac{d\psi_D}{dt},\tag{3}$$

$$0 = R_Q i_Q + \frac{d\psi_Q}{dt},\tag{4}$$

where R_D and R_Q are the d- and q-axis components of the damper winding resistance and i_D and i_Q the d- and q-axis components of the damper winding current. The armature and damper winding flux linkages in the above equations are

$$\psi_d = L_{ad}i_{ad} + L_{md}i_D + \psi_{pm}, \tag{5}$$

$$\psi_q = L_{aq}i_{aq} + L_{mq}i_Q, \tag{6}$$

$$\psi_D = L_{md}i_{ad} + L_D i_D + \psi_{pm}, \tag{7}$$

$$\psi_Q = L_{mq}i_{aq} + L_Q i_Q, \tag{8}$$

where L_{ad} and L_{aq} are the d- and q-axis components of the armature self-inductance, L_D and L_Q the d- and q-axis components of the damper winding inductance, L_{md} and L_{mq} the d- and q-axis components of the magnetizing inductance and ψ_{pm} is the flux linkage per phase of the permanent magnet. By solving the flux linkage differential equations from (1) to (4) and substituting the current equations from (5) to (8) into these equations, the equations for the simulation model of the linear motor can be derived. The electromagnetic thrust of a PMLSM is (Morimoto et al., 1997)

$$F_{dx} = \frac{p_e}{v_s} = \frac{\pi}{\tau}(\psi_d i_{aq} - \psi_q i_{ad}), \tag{9}$$

where p_e is the electrical power.

2.2 Non-idealities of PMLSM

The force ripple of the PMLSM is larger than that of rotary motors because of the finite length of the stator or mover and the wide slot opening. In the PMLSM, the thrust ripple is caused mainly by the detent force generated between the PMs and the armature. This type of force can be divided into two components, for tooth- and core-type detent force. In the tooth detent force, the force component is generated between the PMs and the primary teeth, while the core-type detent force component is generated between PMs and the primary core. The wavelength of the core component is usually one pole pitch, while that of the teeth component is one slot pitch. The core-type detent force can be efficiently reduced by optimizing the length of the moving part or smooth-forming the edges of the mover, and the tooth-type detent force can be reduced by skewing the magnets and chamfering the edges of the teeth (Jung et al., 2002), (Hyun et al., 1999), (Inoue et al., 2000), (Zhu et al., 1997), (Hor et al., 1998).

The detent force effect tends to move the mover to a position in which the energy of the magnetic circuit is at its minimum. This phenomenon attempts to stall the mover at the stator pole positions and is always present, even when no current is flowing through the motor coils (Otten et al., 1997). The ripple of the detent force produces both vibrations and noise and reduces controllability (Chun et al., 2000). The force ripple is dominant at low velocities and accelerations. At higher velocities, the cogging force is relatively small and the influence of dynamic effects (acceleration and deceleration) is more dominant (Otten et al., 2000). In this study, the detent force was measured in the reference system. The force ripple can be described by sinusoidal functions of the load position, x, with a period of φ and an amplitude of A_r, i.e.,

$$F_{ripple} = A_{r1}\sin(\varphi_1 x)\left[A_{r1} + A_{r2}\sin(\varphi_2 x)\right]. \tag{10}$$

In figure 3, the results of the simulation are compared with those measured in the reference system.

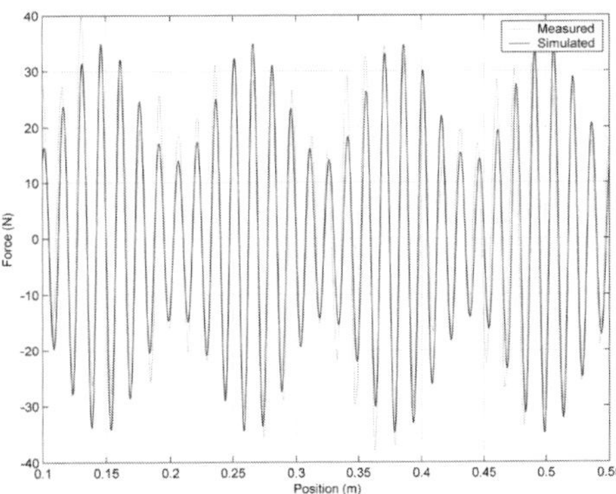

Figure 3: Comparison of measured and simulated detent forces.

The reluctance force is another phenomenon that occurs in linear synchronous motors. A lot of research has been carried out into the reluctance in linear induction machines, in which the phenomenon depends on velocity. The reluctance force in PMLSMs has been studied to a lesser extent. The reluctance force is due to the variations in the self-inductance of the windings with respect to the relative position between the mover and the magnets (Tan et al., 2002). The reluctance force was observed to be relatively small in the reference system and, therefore, has not been included in the model.

The model also takes into account the effect of friction. Friction is very important for control engineering, for example, in the design of drive systems, high-precision servomechanisms, robots, pneumatic and hydraulic systems and anti-lock brakes for cars. Friction is highly nonlinear and may result in steady-state errors, limit cycles and poor performance (Olsson et al., 1998). Friction was modeled using the simple gradient method in which the linear motor system was set in a tilted position and the moving part allowed to slide down freely. Friction was measured at several tilt angles and the results obtained were used to plot the friction function as a function of speed. The friction model took into account the Coulomb (static) and viscous (dynamic) components

$$F_\mu = sign(v)\left[F_{coulomb} + abs(v)F_{viscous}\right], \quad (11)$$

where v is the velocity of the motor. The friction function was incorporated into the simulation model in such a way that it acts between the stator and the mover. In the simulation, the smoothing of the friction function was used to obtain a numerically efficient model in order to improve simulation rate.

The effect of load variation was also taken into consideration. This was carried out using a forced vibrating non-homogenous two-mass model, where the load force variation ΔF_l is calculated using a spring-mass equation, which is the sum of the spring force, F_s, and the damping force, F_d. The equation of motion for such a system is

$$\Delta F_l = F_s + F_d = k(x_1 - x_2) + c(\dot{x}_1 - \dot{x}_2) = m_2\ddot{x}_2 , (12)$$

where k is the spring constant, c the damping coefficient, and $x_i, \dot{x}_i$ and $\ddot{x}_i$ are the displacement, velocity and acceleration of masses m_i, respectively.

The total disturbance force equation can be described using the equations of detent force, friction force and load force variation; i.e., the disturbance force, F_{dist}, is

$$F_{dist} = F_{ripple} + F_\mu + \Delta F_l \qquad (13)$$

This resultant disturbance force component is added to the electromotive force to influence the dynamical behavior of the linear motor system.

2.3 Current Controller of the Linear Motor

The current control of the system is implemented in the form of vector control. Vector control is based on the space vector theory of electrical machines and, therefore, can be easily implemented in the motor model that is also based on the space vector theory. Vector control is suitable for the force (torque) control of both induction and synchronous motors. Generally in the vector control theory, the force and flux components are analyzed separately from the motor currents using the mathematical model of the machine, and control algorithms control these components separately. In the vector control used in this study, the direct axis current, i_{ad}, is set to zero (i_{ad}=0) assuming that it does not influence the generation of force; i.e. equation (9) transforms to

$$F_{dx} = \frac{\pi}{\tau}(\psi_d i_{aq}) . \qquad (14)$$

This means that angle ψ, between the armature current and q-axis, always remains at 0° and that the thrust is proportional to the armature current, i_a=i_{aq}. The drawback of vector control is its low robustness for changes in the machine parameters. The resistance values change considerably due to temperature variations, and the inductances rapidly reach their saturation levels. However, a vector controller is appropriate for applications in which good dynamics and/or accurate velocity control is needed.

In literature, vector control is presented in many ways. In this study, we have used a simulation principle in which the incoming thrust command, F^*, is converted to the i_q current component by dividing the force value by the force constant of the motor, K_m. The current control algorithms are executed in the rotor flux coordinates and the outputs of the controllers are transformed back into the stator reference frame, and these values, u_{sa}, u_{sb}, and u_{sc}, are the inputs of the control inverter. In the simulation model, the modulation technique used is sinusoidal pulse width modulation (SPWM) with ideal switches.

2.4 Verification of the Simulation Model

The simulation model was implemented and analyzed in the MatLab/Simulink® software using the previously mentioned equations. The PWM

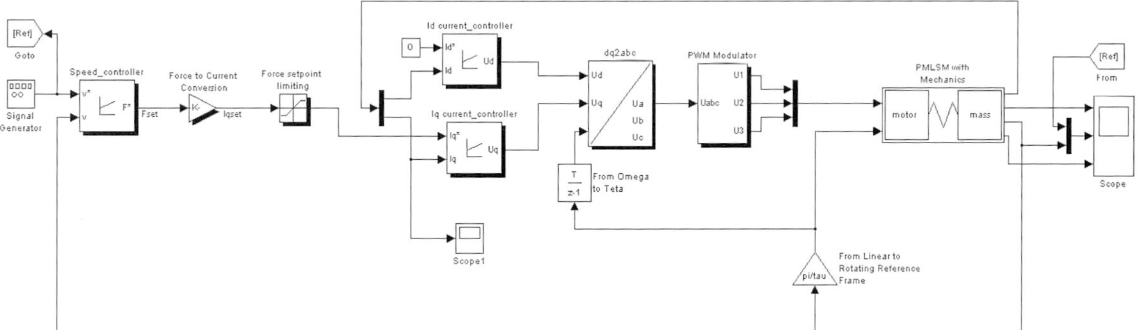

Figure 4: The Simulink model of the motor system.

inverter is modeled as an ideal voltage source and common Simulink blocks are used for the model. The time step of the integrator in the analysis was 10 µs, except for the velocity controller, which had a time step of 1 ms. The parameters used in the simulation are introduced in table I in the appendix. Figure 4 shows the Simulink model of the system.

The simulation results were compared with those measured in the reference system. The motor studied in this paper is a commercial three-phase linear synchronous motor application with a rated force of 675 N. The moving part is set up on an aluminum base with four recirculating roller bearing blocks on steel rails. The position of the linear motor was measured using an optical linear encoder with a resolution of approximately one micrometer. The parameters of the linear motor are given in table II in the appendix.

The spring-mass mechanism was built on a tool base in order to act as a flexible tool (for example, a picker that increases the level of excitation). The mechanism consists of a moving mass, which can be altered in order to change the natural frequency of the mechanism and a break spring, which is connected to the moving mass on the guide. The purpose of the mechanism is to increase the level of excitation when the motor's vibrational frequency is equal to the mechanism's natural frequency, which was calculated at being 9.1 Hz for a mass of 4 kg. The motion of the moving mass was measured using an accelerometer.

The physical linear motor application was driven in such a way that the PI velocity controller was implemented in Simulink to gain the desired velocity reference. The derived algorithm was transferred to C code for dSpace's digital signal processor (DSP) to use in real-time. The force command, F*, was fed into the drive of the linear motor using a DS1103 I/O card. The computational time step for the velocity controller was 1 ms, while the current controller cycle was 31.25 µs. The measured and simulated velocity responses and force generating quadrature currents are compared in figure 5. Sine and step functions were used as the reference velocity.

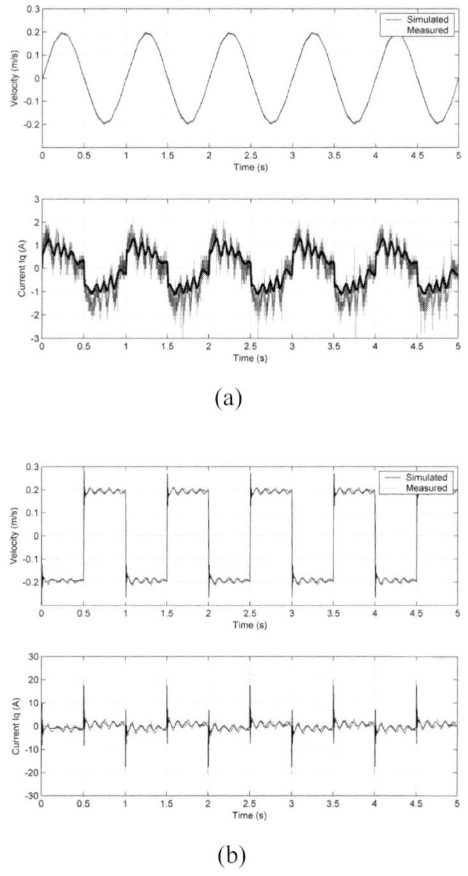

Figure 5: A comparison of the measured and simulated quadrature currents and velocity responses in the case of (a) a sine velocity reference and (b) a step velocity reference.

3 DISTURBANCE COMPENSATION

Disturbance compensation is applied to the motor model to reduce a detrimental force ripple. Force ripple compensation improves the speed response and robustness of the system. The force ripple, i.e. the detent force, friction, and load variation, is estimated using a disturbance observer, or in other words, a load force observer. Figure 6 shows the construction of the compensator.

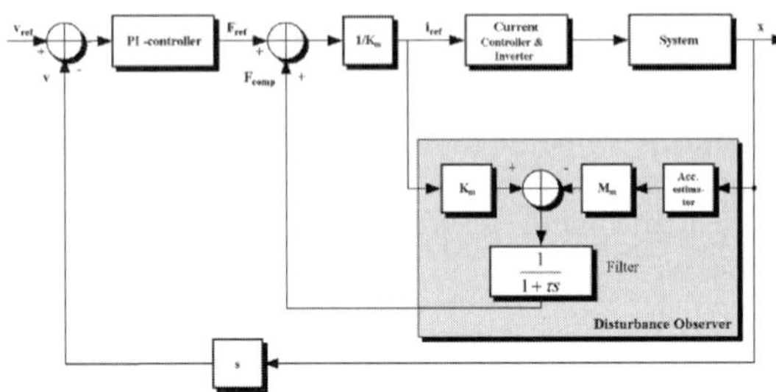

Figure 6: Disturbance compensation scheme utilizing load force estimation.

The concept of the observer is based on the comparison of the actual input force with the ideal one. This gives rise to an error, which after proper filtering, is used to produce the compensation current, i_{comp}. The filter implemented in this study is a second-order Butterworth digital low-pass filter with a cut off frequency of 50 Hz. The main function of the filter is to reduce high-frequency noise due to input signal derivation but also break the algebraic loop in the simulation model between the currents i_{ref} and i_{comp}. Unfortunately, the time delay of the filter also limits the performance of robust control, since it delays the estimated force disturbance (Godler et al., 1999); therefore, the cut-off frequency should be as high as possible. Hong et al. (1998) suggested that an artificial delay be used in the filtering bath of the observer in order to improve dynamic behavior.

The limitations of this method are highlighted by the fact that acceleration, which is needed in the disturbance observer, is generally not available. Usually, acceleration is calculated as the time derivative of the output of the pulse encoder, although the signal becomes easily erroneous due to the high noise ratio in the encoder signal, and the filtering of this kind of signal increases the undesirable time delay, which leads to an unstable response. In this study, acceleration is estimated using an acceleration estimator, which is based on the construction introduced in (Lee et al., 2001). This so-called low-acceleration estimator is based on

the fact that the displacement signal from the encoder is accurate and numerical integration provides more stable and accurate results than does numerical differentiation. Figure 7 shows the structure of the accelerated estimator.

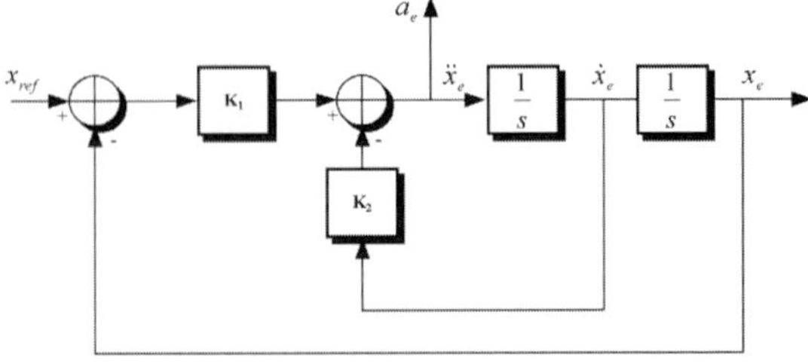

Figure 7: Structure of low-acceleration estimator.

The estimation of acceleration, a_e, is calculated from the displacement signal, x, of the encoder using a double integrator. The estimator takes the form of a PD controller, in which the estimated displacement, x_e, is set to follow the actual displacement, x; i.e. the acceleration estimate is

$$a_e = K_1\left(x - x_e\right) - K_2\frac{dx_e}{dt} \qquad (15)$$

and the transfer function from x to x_e is

$$\frac{x_e}{x} = \frac{K_1}{s^2 + K_2 s + K_1} = \frac{\omega_b^2}{s^2 + 2\zeta\omega_b s + \omega_b^2}, \qquad (16)$$

where ω_b represents the bandwidth of the acceleration estimator and ζ is the damping ratio. Gains K_1 and K_2 from the PD controller can be determined from the required bandwidth of the estimator. Lee et al. (2001) propose that a good guideline for the damping ratio is 0.707, which corresponds to critical damping.

The proposed disturbance regulator has been tested in the simulation model introduced earlier and implemented in the physical application. The algorithm of the controller was run in real-time with a frequency of 1 kHz. The damping ratio of the acceleration estimator, ζ, was 0.707 and the bandwidth ω_b 1000 Hz, i.e. the gains are $K_1{=}10e^5$ and $K_2{=}1414$. Figure 8 shows a comparison of the velocity errors between the reference and actual velocities in compensated and non-compensated systems, when the amplitude of the reference velocity sine signal was 0.1 m/s.

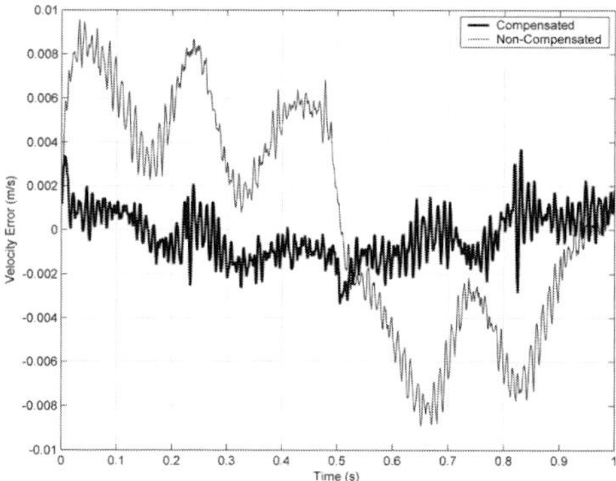

Figure 8: The error between the actual and reference velocities in the compensated and non-compensated systems, when the reference velocity signal is sine with an amplitude of 0.1 m/s.

4 CONCLUSION

This paper presented and verified a dynamic model for a PMLSM including non-idealities. The model appeared to be an effective tool for designing the controller of such a system. A disturbance estimator, which included a low-acceleration estimator, was proposed and successfully implemented in the control of the motor. By means of an accurate simulation model, it was possible to design and test the controller without fear of physical damage. The implementation of the proposed controller was easily carried out by using a DSP system that supported the used simulation software. Preliminary parameter tuning was performed by using the simulation model and final tuning was carried out in the physical linear motor application.

It was observed that mechanical non-idealities have important effect on the dynamics of the motor system. This effect can be reduced by constructional modifications and/or a suitable control algorithm. As mentioned before, the acceleration signal for disturbance estimation or another control algorithm is not usually available. The double derivation of the encoder signal produces a very noisy signal, and filtering this kind of signal leads to undesirable stability problems. With the proposed method, acceleration can easily be estimated from the position signal.

APPENDIX

TABLE I
SIMULATION PARAMETERS

Symbol	Parameter	Value
f_N	Nominal motor frequency	50 Hz
ψ_{PM}	Magnetic flux	0.8 Wb
L_{ad}	D-axis armature inductance	30 mH
L_{ad}	Q-axis armature inductance	40 mH
L_{aq}	D-axis magnetizing inductance	20 mH
L_{md}	Q-axis magnetizing inductance	20 mH
L_D	D-axis damper winding inductance	30 mH
L_Q	Q-axis damper winding inductance	40 mH
R	Winding resistance	4.8 Ω
R_D	D-axis damper winding resistance	2.4 Ω
R_Q	Q-axis damper winding resistance	2.4 Ω

TABLE II
LINEAR MOTOR DATA BY THE MANUFACTURER

Symbol	Parameter	Value
F_N	Nominal force	675 N
A_N	Nominal current	7.2 A
K_m	Motor constant	94 N/A
R_P	Electric resistance	4.8 Ω
L_P	Winding inductance	20 mH
V_n	Max. speed at nominal thrust	2.1 m/s
τ_M	Pole pitch (180° electrical)	15 mm
P_N	Nominal power	3910 W
L	Stroke	1000 mm
Kp	P gain of vel. controller	10000 Ns/m
Ki	I gain of vel. controller	0.01 s
Kp	P gain of current controller	50.3 V/A
Ki	I gain of current controller	0.002 s
f_{mod}	Modulation frequency (PWM)	4000 Hz
V	Inverter bus voltage	720 V

REFERENCES

Bassi, E., Benzi, F., Buja, G. and Moro, F., 1999. "Force disturbance compensation for an A.C. brushless linear motor," in Conf. Rec. ISIE'99, Bled, Slovenia, 1999, pp.1350-1354.

Castillo-Castaneda, E. and Okazaki, Y., 2001. "Static friction compensation approach to improve motion accuracy of machine tool linear stages," Instrumentation and development, vol. 5, no. 1, March 2001.

Chun, J.-S., Jung, H.-K. and Jung, J.-S., 2000. "Analysis of the end-effect of permanent magnet linear synchronous motor energized by partially exited primary," ICEM, International Conference on Electrical Machines, Espoo, Finland, vol. 1, pp. 333-337, 2000.

Deur, J. and Peric, N., 2000. "A Comparative study of servosystems with acceleration feedback," IEEE Ind. Applicat. Conf., Italy, 2000, pp. 1533-1540.

Godler, I., Inoue, M. and Ninomiya, T., 1999. "Robustness comparison of control schemes with disturbance observer and with acceleration control loop," in Conf. Rec. ISIE'99, Slovenia, 1999, pp. 1035-1040.

Gieras, J. F. and Piech, Z. J., 2001. Linear Synchronous Motors: Transportation and Automation Systems. Boca Raton, USA: CRC, 2001.

Hong, K. and Nam, K., 1998. "A Load torque compensation scheme under the speed measurement delay," IEEE Trans. Ind. Electron., vol. 4, no. 2, April 1998.

Hor, P. J., Zhu., Z. Q., Howe, D. and Rees-Jones, J., 1998. "Minimization of cogging force in a linear permanent magnet motor", IEEE Trans. Magn., vol. 34, no. 5, pp. 3544-3547, Sept. 1998.

Hyun, D.-S., Jung, I.-S., Shim, J.-H. and Yoon, S.-B., 1999. "Analysis of forces in a short primary type permanent magnet linear synchronous motor," IEEE Trans. Energy Conversion, vol. 14, no. 4, pp. 1265-1270, 1999.

Inoue, M. and Sato, K., 2000. "An approach to a suitable stator length for minimizing the detent force of permanent magnet linear synchronous motors," IEEE Trans. Magn., vol. 36, no. 4, pp. 1890-1893, July 2000.

Jung, S.-Y. and Jung, H.-K., 2002. "Reduction of force ripples in permanent magnet linear synchronous motor," Proceeding of International Conference on Electrical Machines, Brugge, Belgium pp. 452, Aug 2002.

Kim, B. K., Chung, W. K. and Oh, S. R., 2002. "Disturbance Observer Based Approach to the Design of Sliding Mode Controller for High Performance Positioning Systems," 15th IFAC World Congress on Automatic Control, 2002.

Lee, S.-H. and Song, J.-B., 2001. "Acceleration estimator for low-velocity and low-acceleration regions based on encoder position data," IEEE/ASME Trans. Mechatron., vol. 6, no. 1, March 2001.

Morimoto, S., Sanada, M. and Takeda, Y., 1997. "Interior Permanent Magnet Synchronous Motor for High Performance Drives," IEEE Trans. Ind. Applicat., vol. 33, Issue 4, July-Aug. 1997.

Olsson, H., Åström, K. J., de Wit, C. C., Gäfvert, M. and Olsson, P. L., 1998. "Friction Models and Friction Compensation," European Journal of Control, 1998, Vol. 4, No. 3.

Otten, G., de Vries, T.J.A., van Amorengen, J., Rankers, A.M. and Gaal, E.W., 1997. "Linear motor motion control using a learning feedforward controller," IEEE/ASME Trans. Mechatron., vol. 2, no. 3, pp. 179-187, 1997.

Tan, K. K., Huang, S. N. and Lee, T. H., 2002. "Robust adaptive numerical compensation for friction and force ripple in permanent-magnet linear motors," IEEE Trans. Magn., vol. 38, no. 1, January 2002.

Zhu, Z. Q., Xia, Z. P., Howe, D. and Mellor, P. H., 1997. "Reduction of cogging force in slotless linear permanent magnet motors", Proc.IEEE Electr. Power Appl., vol. 144, no. 4, pp. 277-282, July 1997.

HYBRID CONTROL DESIGN FOR A ROBOT MANIPULATOR IN A SHIELD TUNNELING MACHINE

Jelmer Braaksma, Ben Klaassens, Robert Babuška
Delft Center for Systems and Control, Delft University of Technology
Mekelweg 2, 2628 CD Delft, the Netherlands, e-mail: r.babuska@dcsc.tudelft.nl

Cees de Keizer
IHC Systems B.V., P.O. Box 41, 3360 AA Sliedrecht, the Netherlands

Keywords: Robotic manipulator, shield tunneling, hybrid control, collision, environment model, feedback linearization.

Abstract: In a novel shield tunneling method, the tunnel lining is produced by an extrusion process with continuous shield advance. A robotic manipulator is used to build the tunnel supporting formwork, consisting of rings of steel segments. As the shield of the tunnel-boring machine advances, concrete is continuously injected in the space between the formwork and the surrounding soil. This technology requires a fully automated manipulator, since potential human errors may jeopardize the continuity of the tunneling process. Moreover, the environment is hazardous for human operators. The manipulator has a linked structure with seven degrees of freedom and it is equipped with a head that can clamp the formwork segment. To design and test the controller, a dynamic model of the manipulator and its environment was first developed. Then, a feedback-linearizing hybrid position-force controller was designed. The design was validated through simulation in Matlab/Simulink. Virtual reality visualization was used to demonstrate the compliance properties of the controlled system.

1   INTRODUCTION

With conventional shield tunnel boring machines (TBM), the tunnel lining is built of pre-fabricated concrete segments (Tanaka, 1995; K. Kosuge et al., 1996). These segments must be transported from the factory to the construction site, which requires considerable logistics effort. Delays in delivery and damage of the segments during their transport are not uncommon.

To improve the process, IHC Tunneling Systems are developing a novel shield tunneling method in which the tunnel lining is produced by extruding concrete in the space between the drilled hole and the supporting steel formwork. This formwork is built of rings, each consisting of eight segments.

The main functions of the formwork are: i) support the extruded concrete tunnel lining until it is hardened and becomes self-supporting, ii) provide the thrust force to the TBM shield which actually excavates the tunnel. The thrust is supplied by 16 individually controlled hydraulic thrust cylinders, evenly spread around the shield's circumference.

A robotic manipulator (erector) is used to place the segments at the desired position. The erector is a linked robot mounted such that it can rotate around the TBM spine, see Fig. 1. It is equipped with a head that clamps the segment and picks it up from the conveyer. At the back end of the TBM, the concrete is already hardened and the formwork can be dismantled by a similar manipulator (derector). The segments are cleaned and inspected for possible damage before they can be transported to the front end of the TBM and re-used.

As in the proposed method the tunnel-boring process is continuous (maximum speed of 2 m/h), the segments have to be placed within a limited time interval, and simultaneously with the TBM forward movement. To comply with the stringent timing requirements and to ensure the continuity of the boring process, the erector must be controlled automatically (as opposed to the current practice of using operator-controlled manipulators). By automating the process, the probability of a human error is limited and also the hazard for the operator is considerably reduced.

The controller of the erector must guarantee stable and safe operation in three different modes (Fig. 2):

1. The segment is in free space and the main task is to bring it close to the desired position.

2. The segment is in contact with the hydraulic thrust cylinders which push it toward the already placed formwork ring.

143

J. Braz et al. (eds.), Informatics in Control, Automation and Robotics I, 143–150.
© 2006 *Springer. Printed in the Netherlands.*

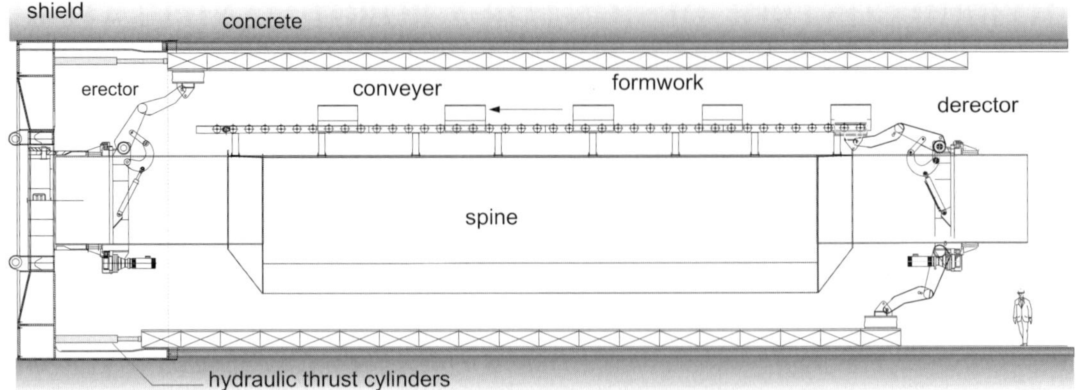

Figure 1: Cross-section of the shield tunneling machine. Two working configurations of the derector are shown.

3. The segment is in contact with the hydraulic thrust cylinders and the formwork ring. The force delivered by the thrust cylinders will align the segment in the desired position.

While the first mode is characterized by traditional trajectory-following control, the latter two modes require compliant control, capable of limiting the contact forces and correcting a possible mis-alignment due to a different orientation of the segment and the already placed ring. Note that no absolute measurement of the segment position is available, it can only be indirectly derived from the measurements of the individual angles of the manipulator links.

To analyze the process and to design the control system, a detailed dynamic model of the erector has been developed. This model includes collision models for operation modes 2 and 3. A hybrid control scheme is then applied to control certain DOFs by a position control law and the remaining ones by a force control law. Input-output feedback linearization is applied in the joint space coordinates. Simulation results are reported.

2   MODELING

The mathematical model consists of three parts: the manipulator, the environment and the hydraulic thrust cylinders. This section describes these elements one by one.

2.1   Manipulator Model

The erector was designed as a linked structure with three arms and a fine-positioning head, see Fig. 3. The first arm is attached to the base ring (body0) that can rotate around the spine. This ring is actuated by an electric motor. The arms are actuated by hydraulic

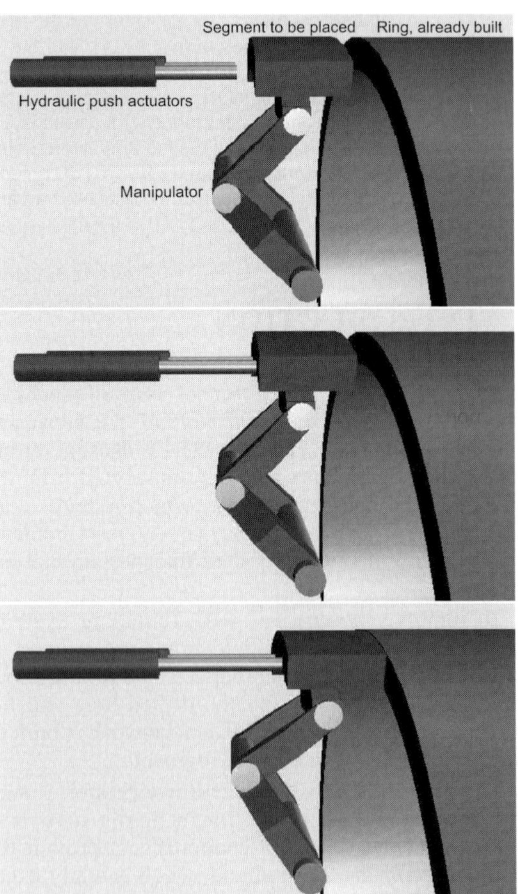

Figure 2: The erector manipulates segments in three different modes: in free space (top), in contact with the hydraulic thrust cylinders (middle), in contact with the trust cylinders and the segment ring (bottom).

actuators (not modeled here). At the top of the third arm (body3), the head is attached, which has three degrees of freedom (DOF). The entire manipulator has seven DOF (q_0 to q_6) in total.

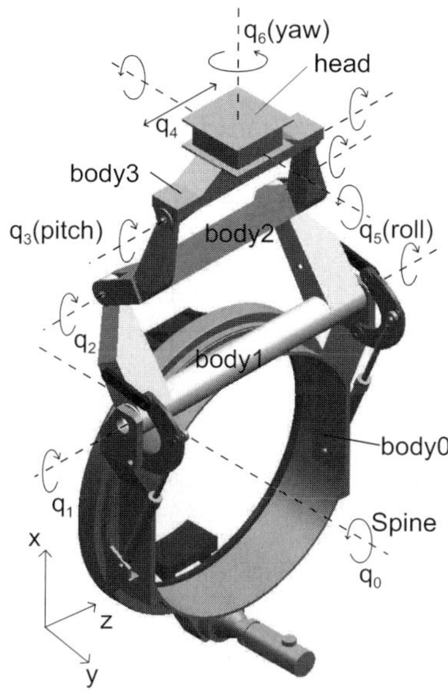

Figure 3: The structure and the corresponding degrees of freedom of the erector.

The bodies are assumed rigid bodies and the DOF in body0 is not taken into account. This is justified, as the rotation around the spine is fixed during the positioning of the head. Angle q_0 is still an input to the model, fixed to a value between 0 and 2π rad. The control inputs of the robot are assumed to be perfect torque/force generators in the joints.

The dynamic equation for the manipulator is:

$$D(\mathbf{q})\ddot{\mathbf{q}} + C(\mathbf{q}, \dot{\mathbf{q}})\dot{\mathbf{q}} + F_v\dot{\mathbf{q}} + \phi(\mathbf{q}) = \boldsymbol{\tau} - J_0^{b4}(\mathbf{q})^T\mathbf{f} \tag{1}$$

where $\mathbf{f}$ is the vector of contact forces between the hydraulic thrust cylinders and the segment plus the contact forces between the segment and the ring. The matrix $D(\mathbf{q})$ is the inertia matrix, $C(\mathbf{q}, \dot{\mathbf{q}})$ is the Coriolis and centrifugal matrix, F_v is the friction matrix and $\phi(\mathbf{q})$ is the gravity vector. J_0^{b4} is the Jacobian from frame $b4$ to frame 0, and q is the vector of the joint angles. The manipulator is actuated with the torque/force vector $\boldsymbol{\tau}$.

2.2 Environment Model

The environment consists of the segment ring (Fig. 4). An accurate environment model, very important to the contact tasks in robotics, is usually difficult to obtain in an analytic form. In the literature, usually a simplified linear model is used (Vukobratović,

Figure 4: Two consecutive rings of the formwork. Each ring consists of eight segments.

1994). The environment is modeled as a mass-spring-damper-system. The surface of the segment's side is considered flat and only the normal force acting on the side of the segment is taken into account. Tangential friction forces are neglected. The collision detection is implemented with the following smooth switching function

$$g(x) = \frac{1}{\pi}\operatorname{atan}(\alpha(x - x_e)) + \frac{1}{2} \tag{2}$$

where α is the steepness of the function, x is the position of the segment and x_e is the position of the environment. The environment force is described by the following equation:

$$f_e = g(x)(K_e(x - x_e) + D_e(\dot{x} - \dot{x}_e)) \tag{3}$$

For stationary environments, equation (3) can be simplified by taking $\dot{x}_e = 0$. K_e is the stiffness and D_e is the damping.

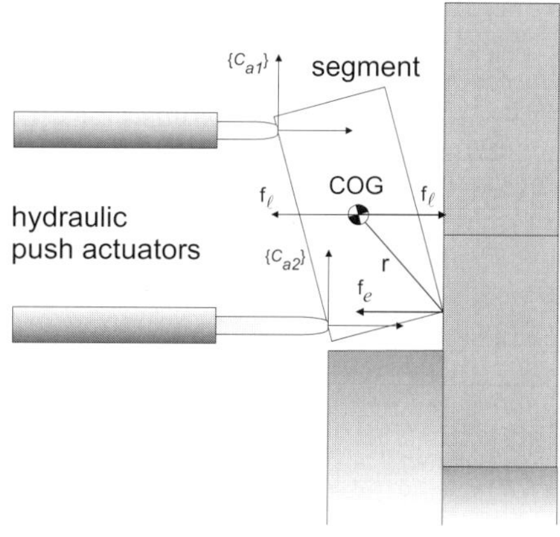

Figure 5: Decomposition of contact force $\mathbf{f}_e$.

In the Cartesian space, the dynamic behavior depends on the direction vector of the force and the location of the collision. Fig. 5 shows an unaligned collision of the segment with the formwork ring. In this figure, a transformation is applied, which decomposes a force $\mathbf{f}_e$ into a force $\mathbf{f}_l$ and torque $\boldsymbol{\tau}$ acting on the center of gravity (COG). The following formulas can be applied.

$$\mathbf{f}_l = \mathbf{f}_e \tag{4}$$

$$\boldsymbol{\tau} = \mathbf{r} \times \mathbf{f}_e \tag{5}$$

where $\mathbf{r}$ points to the place where the force acts. The above derivation is valid for a body moving in free space. The transformation of the force and torque to the joint torque is achieved using the principle of virtual work, see (Spong and Vidyasagar, 1989). The following formula for the joint torques can be derived

$$\boldsymbol{\tau}_e = J_0(\mathbf{q})^T \begin{bmatrix} \mathbf{R}_0^{b4} & \mathbf{0} \\ \mathbf{0} & \mathbf{R}_0^{b4} \end{bmatrix} \mathbf{f} \quad \text{with } \mathbf{f} = \begin{bmatrix} \mathbf{f}_l \\ \boldsymbol{\tau} \end{bmatrix} \tag{6}$$

where R_0^{b4} transforms the force expressed in the segment body frame to the base frame, $J_0(\mathbf{q})$ is the Jacobian which transforms the joint speeds to the linear/rotational speeds in the base frame and $\boldsymbol{\tau}_e$ is a vector with joint torques/forces.

2.3 Hydraulic Thrust Cylinders

The model consists of a collision model and a hydraulic model. The collision model must detect the collision between the hydraulic thrust cylinders and the formwork ring, which means that the model must detect whether the tips are contacting the segment body (body4). If there is a collision, the magnitude of the force can be calculated with the scheme explained in Section 2.2.

The shape of a segment is illustrated in Fig. 6. The segment (body4) is a part of a hollow cylinder. For a controlled collision, the y coordinate of tip of the hydraulic thrust cylinders must be between the inner radius r_1 and the outer radius r_2, and within the range φ, see Fig. 6. The kinematic transformation of two hydraulic thrust cylinders tips to the C_{b4} frame is used to detect collision.

When the actuators come into contact with the segment, this will result in a force and torque as explained in Fig. 5. Only the normal force acting on the segment is simulated. The error between the segment and the thrust cylinders is at most 3 degrees, satisfying the assumption to neglect the tangential force. The force and torque in the C_{b4} frame, are calculated with the following equations. Let H_{a1}^{b4} and H_{a2}^{b4} be the homogeneous transformations from the thrust cylinder frames (C_{a1}, C_{a2}) to C_{b4}. The two force and torque vectors can be calculated as follows

$$\mathbf{f}_{a1} = |f_{a1}|[0\ 1\ 0]^T \quad \mathbf{r}_{a1} = (H_{b4}^{a1}[0\ 0\ 0\ 1]^T)_{123}$$
$$\boldsymbol{\tau}_{a1} = \mathbf{r}_{a1} \times \mathbf{f}_{a1} \tag{7}$$

$$\mathbf{f}_{a2} = |f_{a2}|[0\ 1\ 0]^T \quad \mathbf{r}_{a2} = (H_{b4}^{a2}[0\ 0\ 0\ 1]^T)_{123}$$
$$\boldsymbol{\tau}_{a2} = \mathbf{r}_{a2} \times \mathbf{f}_{a2}. \tag{8}$$

where the subscript 123 refers to the first three elements. These forces and torques in the body frame can be transformed to the joint torques by means of the principle of virtual work (6).

3 HYBRID CONTROLLER

The manipulator in free space is position controlled, but when it is in contact with the environment, the lateral motion and rotation become constrained. When the segment is pushed by the hydraulic thrust cylinders, stable contact between the thrust cylinders and the segment is maintained by controlling the contact force (Fig. 7). When the segment comes into contact with the segment ring, it aligns to the desired orientation and position. During the transition between the modes it is required that the contact forces are limited. An impedance controller gives the manipulator the desired compliant behavior (Hogan, 1985).

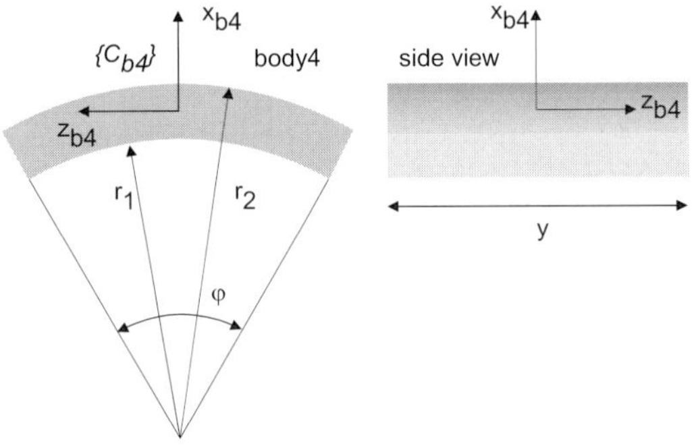

Figure 6: The dimensions of the segment (body4).

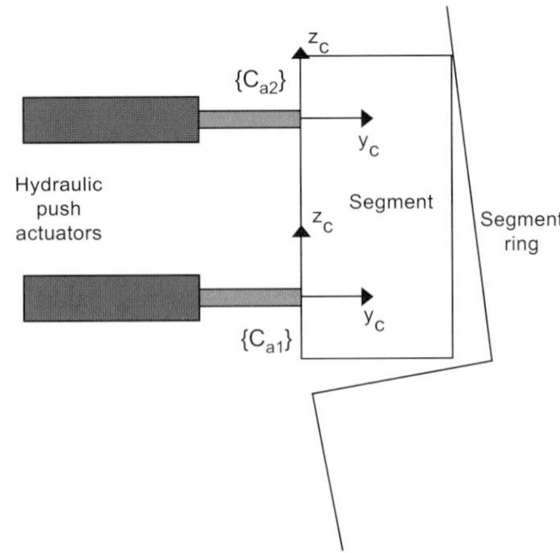

Figure 7: Top view of the segment.

A control scheme known as a hybrid position force controller of (Raibert and Craig, 1981) allows to specify for each DOF the type of control. The frame in which the segment is controlled is positioned such that the direction, in which the position is controlled, is perpendicular to the direction in which the force is controlled. This frame is called the constraint frame. In our case the constraint frame is aligned with the two thrust cylinder frames (see Fig. 7). In this situation, the yaw, y, and the pitch, z, DOFs are constrained. We choose for the y degree of freedom a force controller to assure contact. For the other position constrained degrees of freedom an impedance controller is chosen which guarantees compliant behavior.

The controller consists of four blocks: feedback linearization, transformation of sensor data (kinematics, Jacobian and coordinate transform), selection matrix and the controllers (see Fig. 8).

The coordinate-transform block transforms the data $\mathbf{f}$, $\mathbf{x}$, R, $\mathbf{v}$ and ω from the base frame to the constraint frame and vice versa, where $\mathbf{f}$ is the measured force, $\mathbf{x}$ is the position of the segment, R is the rotation matrix which defines the orientation of the segment, $\mathbf{v}$ is the linear speed of the segment and ω is the angular speed of the segment. The inputs of the force controller are $\mathbf{f}_d$ and $\dot{\mathbf{f}}_d$, which are the desired force and it's derivative. The inputs of the position controller are $\mathbf{x}_d$, $\dot{\mathbf{x}}_d$ and $\ddot{\mathbf{x}}_d$, which are the desired position, velocity and acceleration respectively. The vector $\mathbf{a}$ is the resolved acceleration vector and τ is the torque/force vector for the manipulator.

3.1 Feedback Linearization

The following controller is used for feedback linearization (Siciliano and Villani, 1999)

$$\tau = \hat{D}(\mathbf{q})\mathbf{u} + \hat{C}(\mathbf{q},\dot{\mathbf{q}})\dot{\mathbf{q}} + \hat{F}_v\dot{\mathbf{q}} + \hat{\phi}(\mathbf{q}) + J_0^{b4}(\mathbf{q})^T \mathbf{f} \tag{9}$$

where $\mathbf{u}$ is the new acceleration control input. Substituting (9) into (1) and assuming perfect modeling results in the following system

$$\ddot{\mathbf{q}} = \mathbf{u} \tag{10}$$

The goal of the controller is to control the manipulator in the constraint frame to its desired values.

The Jacobian transforms the joint velocities to the linear and rotational speeds. The Jacobian is given by

$$\begin{bmatrix} \mathbf{v}_0^{b4} \\ \omega_0^{b4} \end{bmatrix} = J_0^{b4}(\mathbf{q})\dot{\mathbf{q}} \tag{11}$$

Differentiating (11) gives

$$\begin{bmatrix} \dot{\mathbf{v}}_0^{b4} \\ \dot{\omega}_0^{b4} \end{bmatrix} = \ddot{\mathbf{x}} = J_0^{b4}(\mathbf{q})\ddot{\mathbf{q}} + \dot{J}_0^{b4}(\mathbf{q})\dot{\mathbf{q}} \tag{12}$$

which represents the relationship between the joint acceleration and the segment's linear and rotational speed. A new control input can now be chosen as

$$\mathbf{u} = (J_0^{b4}(\mathbf{q}))^{-1}\left(\mathbf{a} - \dot{J}_0^{b4}(\mathbf{q})\dot{\mathbf{q}}\right). \tag{13}$$

Substitute (13) into (10) and the result into (12), assuming perfect modeling and perfect force measurement, leads to the following total plant model

$$\ddot{\mathbf{x}} = \mathbf{a} = \mathbf{a}_p + \mathbf{a}_f \tag{14}$$

where $\mathbf{a}$ is the resolved acceleration in the base frame and $\mathbf{a}_p$, $\mathbf{a}_f$ are the outputs of the position controller and force controller in the base frame. The vector $\mathbf{a}_p$ can be partitioned into linear acceleration $\mathbf{a}_l$ and angular acceleration $\mathbf{a}_o$. Likewise, $\mathbf{a}_f$ can be partitioned into linear acceleration $\mathbf{a}_{fl}$ and angular acceleration $\mathbf{a}_{ft}$:

$$\mathbf{a}_p = \begin{bmatrix} \mathbf{a}_l \\ \mathbf{a}_o \end{bmatrix} \qquad \mathbf{a}_f = \begin{bmatrix} \mathbf{a}_{fl} \\ \mathbf{a}_{ft} \end{bmatrix}. \tag{15}$$

3.2 Transformation of Sensor Data

The different positions and speeds are measured in the joints of the manipulator. The force is measured using a force sensor, located in the manipulator head. To control the manipulator in Cartesian space these measurements must be transformed to a Cartesian frame. This is done with the kinematic matrix and the Jacobian matrix. The frame in which the manipulator is controlled is the constraint frame.

3.3 Selection Matrix

The DOF for position control and for force control can be selected. This is done by multiplying the force/torque vector by a diagonal selection matrix S and the position/velocity vector by $I - S$, with I the identity matrix. For a DOF that is position controlled $S_i = 0$ and for a force controlled DOF $S_i = 1$. For a manipulator moving in free space the selection matrix is a zero matrix, meaning position control.

To make sure that the switch between the two situations is in a controlled manner, the switch function of the selection matrix must be smooth. The time of switching in unknown, so the switch should be triggered by the contact force (Zhang et al., 2001).

$$\Phi_i(f_i) = \begin{cases} f_i - f_{ref1}^i, & \text{if } f_i \geq 0 \\ f_{ref2}^i - f_i, & \text{if } f_i < 0 \end{cases} \tag{16}$$

The smooth switch function can now be defined as

$$S_i = \begin{cases} 0, & \text{if } \Phi_i(f_i) < 0 \\ 1 - sech(\alpha\Phi_i(f_i)), & \text{otherwise} \end{cases} \tag{17}$$

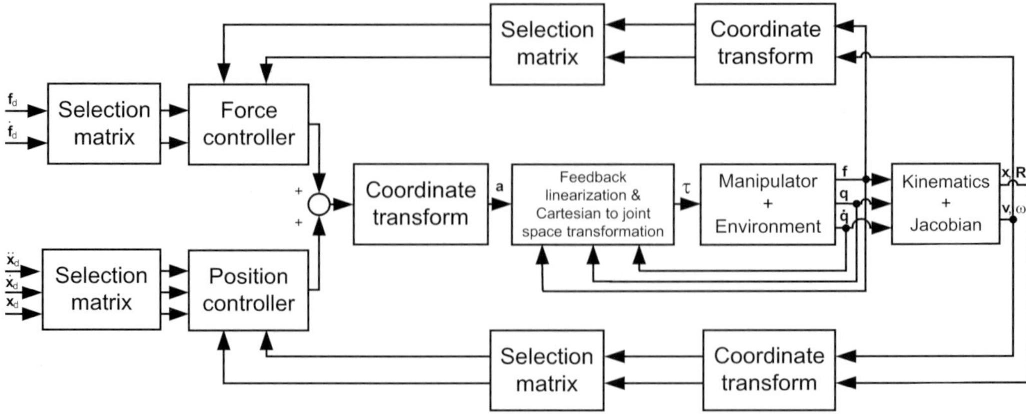

Figure 8: Block diagram of the hybrid position force controller.

with

$$sech(x) = \frac{2}{e^x + e^{-x}} \qquad (18)$$

The value α determines the steepness of the curve, and f^i_{ref1}, f^i_{ref2} are the upper and lower threshold values respectively. The steepness α should be chosen such that the $S_i = 1$, when the desired contact force is reached. The force in the y DOF (in the constraint frame) is chosen to switch from position control to force/impedance control.

3.4 Position Control

For the linear and decoupled system (14), the following control can be used in order to give the system the desired dynamics (Siciliano and Villani, 1999).

$$\mathbf{a}_l = \ddot{\mathbf{x}}_d + D_l(\dot{\mathbf{x}}_d - \dot{\mathbf{x}}) + P_l(\mathbf{x}_d - \mathbf{x}) \qquad (19)$$

$$\mathbf{a}_o = \dot{\boldsymbol{\omega}}_d + D_o(\boldsymbol{\omega}_d - \boldsymbol{\omega}) + P_o R^c_{b4}\epsilon^d_{b4} \qquad (20)$$

This is a PD controller for the position controlled DOF, where R^c_{b4} transforms the quarternion error ϵ^d_{b4} from the C_{b4} frame to the constraint frame. The matrices D_l and D_o are the diagonal damping matrices, and P_l and P_o are the proportional gain matrices. Substituting (19) and (20) into (14) assuming perfect modeling gives the total plant model

$$\ddot{\mathbf{x}} = \overline{S} \begin{bmatrix} \ddot{\mathbf{x}}_d + D_l(\dot{\mathbf{x}}_d - \dot{\mathbf{x}}) + P_l(\mathbf{x}_d - \mathbf{x}) \\ \dot{\boldsymbol{\omega}}_d + D_o(\boldsymbol{\omega}_d - \boldsymbol{\omega}) + P_o R^e_2\epsilon^{de}_e \end{bmatrix} \qquad (21)$$

3.5 Force Control

The force controller has an inner position/orientation loop within the force loop. For the position loop, the following PD-controller is chosen (Siciliano and Villani, 1999).

$$\mathbf{a}_f = -D_f\dot{\mathbf{x}} + P_{fp}(\mathbf{x}_c - \mathbf{x}) \qquad (22)$$

where $\mathbf{x}_c - \mathbf{x}$ is the difference between $\mathbf{x}_c$ the output of the force loop and the measured position/orientation $\mathbf{x}$ and P_{fp} is the proportional matrix gain and D_f is damping gain. For the force loop a proportional-integral (PI) control is chosen

$$\mathbf{x}_c = P^{-1}_{fp}\left(P_f(\mathbf{f}_d - \mathbf{f}) + P_I\int_0^t(\mathbf{f}_d - \mathbf{f})d\varsigma\right) \qquad (23)$$

where P_f is the proportional gain, P_I is the integral gain, $\mathbf{f}_d$ is the desired force and $\mathbf{f}$ is the measured force. The P^{-1}_{fp} in (23) has been introduced to make the resulting force and moment control action in (22) independent of the choice of the proportional gain on the position and orientation.

3.6 Impedance Control

For the impedance controlled DOF the following control law can be chosen (Siciliano and Villani, 1999)

$$\mathbf{a}_f = \begin{bmatrix} \ddot{\mathbf{x}}_d + M^{-1}_{li}(D_{li}(\dot{\mathbf{x}}_d - \dot{\mathbf{x}}) + P_{li}(\mathbf{x}_d - \mathbf{x}) - \mathbf{f}_l) \\ \dot{\boldsymbol{\omega}}_d + M^{-1}_{oi}(D_{oi}(\boldsymbol{\omega}_d - \boldsymbol{\omega}) + P_{oi}R^c_{b4}\epsilon^d_{b4} - \mathbf{f}_t) \end{bmatrix} \qquad (24)$$

with M_{li} and M_{oi} positive definite matrix gains, leading to

$$M_{li}(\ddot{\mathbf{x}}_d - \ddot{\mathbf{x}}) + D_{li}(\dot{\mathbf{x}}_d - \dot{\mathbf{x}}) + P_{li}(\mathbf{x}_d - \mathbf{x}) = \mathbf{f}_l \qquad (25)$$

$$M_{oi}(\dot{\boldsymbol{\omega}}_d - \dot{\boldsymbol{\omega}}) + D_{oi}(\boldsymbol{\omega}_d - \boldsymbol{\omega}) + P_{oi}R^c_{b4}\epsilon^d_{b4} = \mathbf{f}_t \qquad (26)$$

Equations (25) and (26) allow specifying the manipulator dynamics through the desired impedance, where the translational impedance is specified by M_{li}, D_{li} and P_{li} and the rotational impedance is specified by M_{oi}, D_{oi} and P_{oi}.

3.7 Simulation Results

The rotation around the spine is fixed during fine positioning at $q_0 = 0$. The segment ring has a different orientation than the segment; it is rotated around the yaw axis by 0.05 radians. In order to take model uncertainties into account, errors are introduced in the model. For the dynamic parameters such as inertia an error of 10% is introduced and for length and positions an error of 1% is introduced. As it is impossible to simulate all combination of parameter errors, the errors are randomly chosen for every simulation.

In the position loop, the P-gain is chosen $P_l = 10I$. This corresponds with a natural frequency of 3.2 rad/s and for the D-gain $D_l = 6.3I$ and this corresponds to a damping ratio of 1. For the pitch and roll axes the same gains $P_o = 10I$ and $D_o = 6.3$ are used. The desired force input is 500 N. The position loop in the force controller has also a P-gain of $P_{fp} = 10I$ and a D-gain of $D_f = 6.3I$. The P-gain of the force controller is $P_{fp} = 5 \cdot 10^{-3}I$ and the I-gain is set $P_I = 5 \cdot 10^{-3}I$. The parameters of the impedance-controlled DOFs are given by $M_{li} = 100I$, $D_{li} = 250I$ and $P_{li} = 500I$ for the z DOF and $M_{oi} = 300I$, $D_{oi} = 50000I$ and $P_{oi} = 100I$ for rotational DOF.

At t=0 s the manipulator is in position control, and there is no contact between the segment and the environment. The thrust cylinders start moving and get in contact with the segment at t=4.8 s. The manipulator switches to force/impedance control and moves further with the same speed as the hydraulic thrust cylinders (see Fig. 9). The y position of the environment is at $y = 0.25$ m. When the segment contacts the environment at $t = 8$ s, the segment aligns with the environment as shown in Fig. 10. This contact gives rise to a torque τ_x of 800 Nm in the head, which is measured by the force sensor (Fig. 11). This torque is caused by the damping term in the impedance controller. The rotational speed is 0.0017 rad/s the damping term is 50000, the torque is then $0.0017 \times 50000 = 833$ Nm.

The simulations are repeated 50 times, each time the parameters in the controller's internal model are chosen randomly. In Fig. 11 and Fig. 12, the measurements of the force sensor are plotted. The magnitude of the force in the y DOF is not influenced; the time of impact varies due to the parameter changes. The torque around the z axis is influenced by the parameter change.

4 CONCLUSIONS

A dynamic model of the manipulator and its environment has been successfully derived and implemented in Matlab/Simulink. The control requirements have

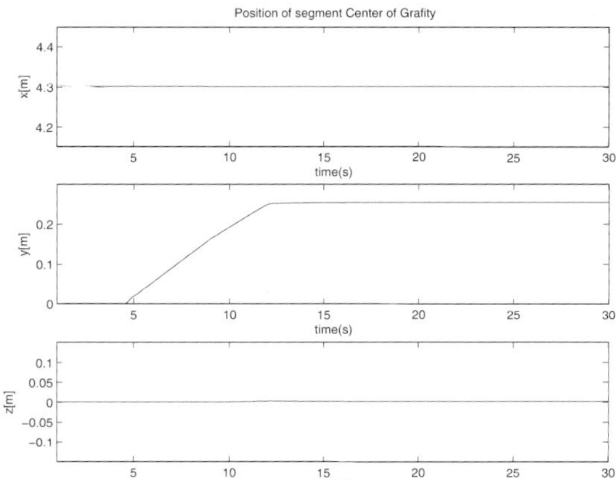

Figure 9: Position of the segment.

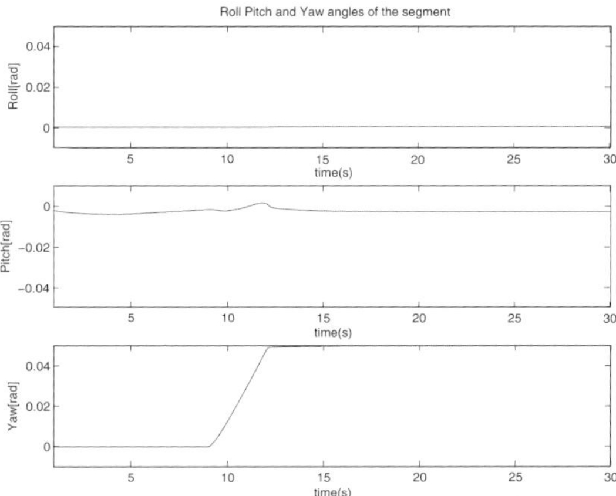

Figure 10: Orientation of the segment.

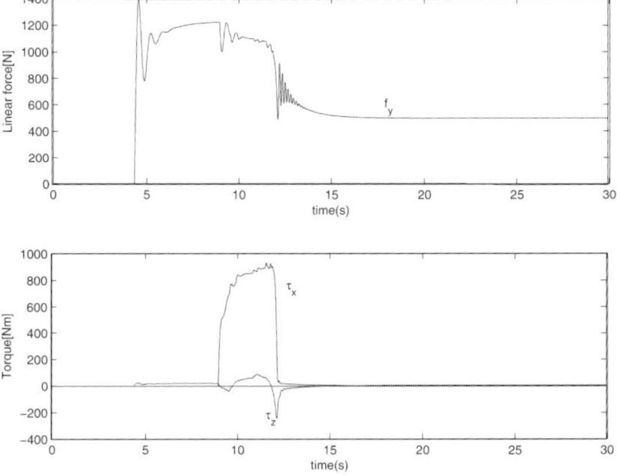

Figure 11: Force measurement in the segment force sensor (one simulation run).

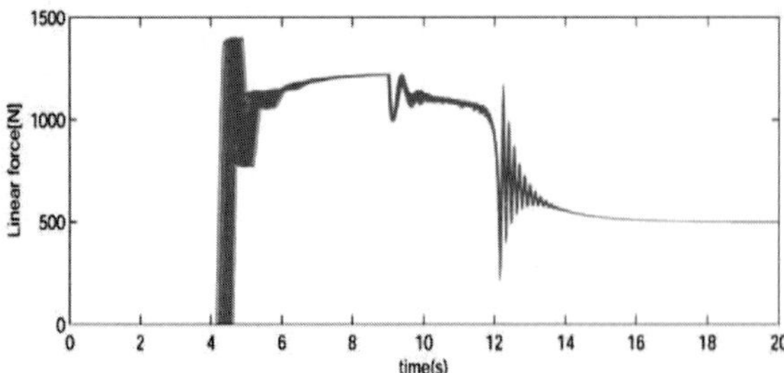

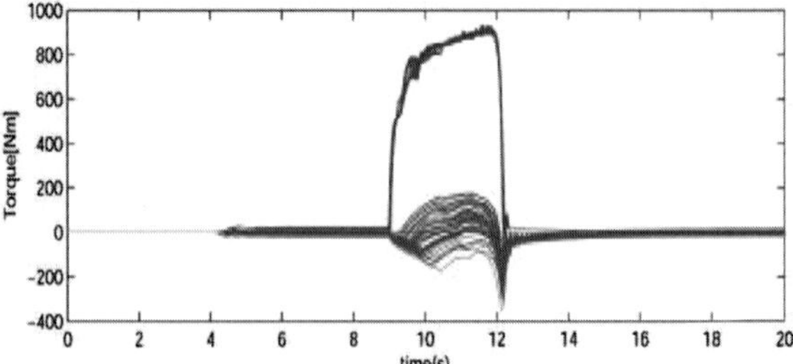

Figure 12: Force measurement in the segment force sensor (50 simulation runs).

been met by using a combination of a force controller, an impedance controller and a position controller. The position precision is 0.01 m and for the orientation the precision is 0.005 rad (0.3 degrees)

The hybrid controller is based on feedback linearization to make the system linear and decoupled. The feedback linearization is an inverse model of the manipulator. If the dynamic parameters in the feedback linearization are within 10% accuracy and the dimensional parameter are within 1% accuracy, a stable hybrid controller is obtained in simulations.

The position loops in the hybrid controller consist of a PD-controller. The P and the D gains are tuned by applying the following rules. When the natural frequency ω and the damping ratio ζ are known, then the P-gain is defined by ω^2 and the D-gain is defined by $2\zeta\omega$.

It is assumed that the manipulator head is equipped with a six DOF force sensor. If this is not possible in the mechanical design, the possibility must be investigated of estimating the force, by measuring the pressures in the hydraulic actuators.

REFERENCES

Hogan, N. (1985). Impedance control: An approach to manipulation: Part i - theory, part ii - implementation, part iii - applications. *ASME Journal of Dynamic Systems, Measurement, and Control*, 107: 1–24.

K. Kosuge et al. (1996). Assembly of segments for shield tunnel excavation system using task-oriented force control of parallel link mechanism. *IEEE*.

Raibert, M. and Craig, J. (1981). Hybrid position/force control of manipulators. *J of Dyn Sys, Meas, and Cont*.

Siciliano, B. and Villani, L. (1999). *Robot Force Control*. Kluwer Academic Publishers.

Spong, M. W. and Vidyasagar, M. (1989). *Robot Dynamics and Control*. John Wiley & Sons.

Tanaka, Y. (1995). Automatic segment assembly robot for shield tunneling machine. *Microcomputers in Civil Engineering*, 10:325–337.

Vukobratović, M. (1994). Contact control concepts in manipulation robotics-an overview. In *IEEE Transactions on industrial electronics, vol. 41, no. 1*, pages 12–24.

Zhang, H., Zhen, Z., Wei, Q., and Chang, W. (2001). The position/ force control with self-adjusting select-matrix for robot manipulators. In *Proceedings of the 2001 IEEE International Conference on Robotics & Automation*. Department of Automatic Control, National University of Defense Technology.

MOCONT LOCATION MODULE: A CONTAINER LOCATION SYSTEM BASED ON DR/DGNSS INTEGRATION

Joseba Landaluze, Victoria del Río, Carlos F. Nicolás, José M. Ezkerra, Ana Martínez

IKERLAN Research Centre, Arizmendiarrieta 2, 20500 Arrasate/Mondragon, The Basque Country (Spain)
Email: jlandaluze@ikerlan.es

Keywords: DR/DGNSS, DR, GPS, DGPS, inertial navigation system, location system.

Abstract: The problem of identifying and adequately positioning containers in the terminal yard during handling with Reach Stackers still remains to be solved in an appropriate manner, while this is extremely important in making the identification and positioning operations automatic. A precise knowledge in the Terminal Operating System (TOS) of such data in Real Time would have considerable economic impact on the logistic treatment of operations. The MOCONT system sets out to provide a solution to this lack. In particular, the MOCONT Location Module establishes the position of the container in the yard while it is being handled by a Reach Stacker. This system is based on the integration of the Differential Global Navigation Satellite System (DGNSS) with a Dead Reckoning (DR) inertial system. This article presents the general characteristics of the MOCONT Location Module, its structure, and the structure of data fusion, besides some results obtained experimentally.

1 INTRODUCTION

The problem of localising the containers in the yard of container terminals had been partially solved for terminals equipped with huge machines, such as Rubber Tyre (RTG) and Rail Mounted Gantry Cranes (RNG) or even Straddle Carriers, but is still a problem for those terminals equipped with Reach Stackers or Front Loaders. Moreover, the automatic identification of containers on board the handling machine is still a problem remaining to be solved.

The European projects MOCONT ("Monitoring the yard in container terminals", IST-1999-10057) and MOCONT-II ("Monitoring the yard in container terminals – Trials", IST-2001-34186), aimed at providing a system to track the containers in the yard in Real Time. The projects aimed at developing a system that automatically identified containers and localised them when moved by small machines, called Reach Stackers.

This paper presents the Location Module of the MOCONT system, developed in said European projects, and more particularly, the development of the inertial navigation system or Dead Reckoning (DR) system. The Location Module is based on the integration of a Differential Global Navigation Satellite System (DGNSS) and an inertial DR system. Data are integrated by means of a Kalman Filter. This system reckons the position of the vehicle at all times, improving estimates supplied by the DGNSS. The exact position of the container is determined by means of a transformation of coordinates from the vehicle body since the length and angle of inclination of the boom are known. The system is particularly useful when the GNSS does not supply quality data or is interrupted, due to few satellites being accessible, multipath phenomena or due to working inside container canyons or in dark areas. The DR inertial system is able to continue estimating vehicle position with precision for short time intervals, with bounded errors, until GNSS signals with sufficient quality are available.

In the literature, different structures of inertial systems appear for land vehicles. The most common case is the use of a sensor for rotation speed of yaw angle and odometric information obtained from vehicle wheels (Aono, 1999; Ramjattan, 1995). Other redundant sensors are often used to help, each being different depending on the type of vehicle and application in question (Aono, 1999; Zhang, 1999). In some cases, especially in vehicles for agricultural purposes, a digital compass or a geomagnetic direction sensor are used instead of a yaw angle speed gyroscope (Benson, 1998; Zhang, 1999). In order to avoid errors, which may be introduced by odometric sensors on wheels due to slipping, Sukkarieh (1999) proposes an Inertial Measurement Unit (IMU), comprising three accelerometers, three

151

J. Braz et al. (eds.), Informatics in Control, Automation and Robotics I, 151–158.
© 2006 *Springer. Printed in the Netherlands.*

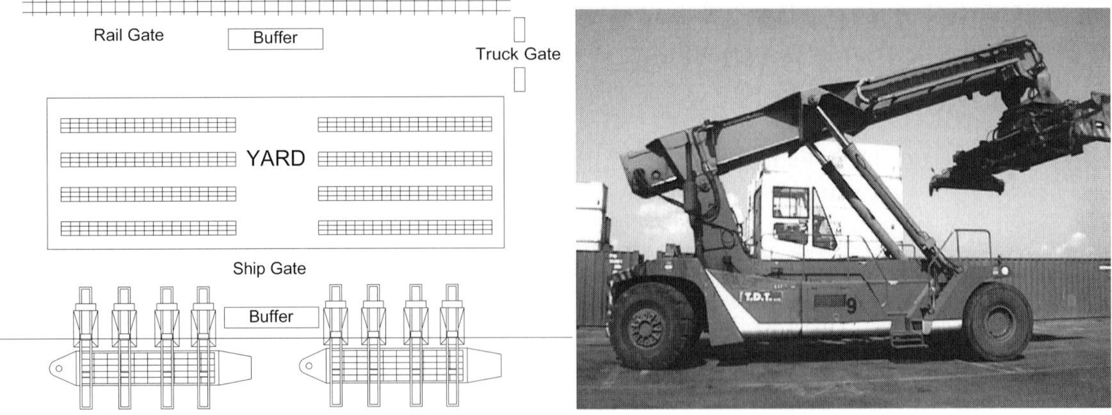

Figure 1: Overall scheme of a container terminal and Reach Stacher container handling machine.

gyroscopes and two pendular gyroscopes, by which vehicle acceleration and yaw and tilt rotation speeds are obtained. Rogers (1999) presents an inertial system consisting of a low cost fibre optic rate gyroscope and of a radar ground speed sensor.

Due to the specifications of the MOCONT project, according to which it was not possible to change the structure of the Reach Stacker machines and nor was it possible to install encoders in the wheels to obtain odometric information, initially a structure based on a triad of accelerometers and a triad of solid state gyroscopes was selected. Although this set up may be valid for on-road vehicles, once field data were obtained for a Reach Stacker in normal work tasks, it was concluded that this set up was not valid for such a machine; due to the low speeds and considerable degree of vibration to which this machine is subject during operations, the accelerometers did not provide information on machine movements on the yard, and only the yaw rate ant tilt rate gyroscope signals could be used (Landaluze, 2003). Therefore, a sensor structure similar to that proposed in (Rogers, 1999) was chosen. Finally, since a subcentimetric receiver was replaced by a submetric receiver, and taking into account the "non-collocation" between the Reach Stacker chassis sensors and the GPS antenna, fitted on the highest point of the Reach Stacker boom, sensoring was completed with a digital compass.

This paper firstly presents an overview of the MOCONT project, followed by an explanation of the structure of the Location Module in the MOCONT system. Then follows a description of the DR inertia system and the Kalman Filter implemented. Lastly, some experimental results are shown, as well as a statistical evaluation of the results obtained by the Location Module of the MOCONT system in the course of the MOCONT-II project.

2 OVERVIEW OF THE MOCONT SYSTEM

The MOCONT project was presented as a new landmark in the application of telematics in intermodal transport, especially in the control of container terminals. The main objective of the project was to develop a system to identify the position of containers in the yard in Real Time. Although said follow-up problem had already been partially solved in the case of large loading and unloading cranes (Rubber Tyre RTD, Rail Mounted Gantry Crane RMG, Straddle Carriers), this was and is a problem in terminals with Reach Stacker machines. It was in these machines, therefore, where the project comes to bear (Figure 1).

The Reach Stacker is an off-road machine used to handle containers in the terminal yard. It is characterised by a small body with an extensible boom (the arm) mounted over of the operator cabin. It is equipped with a spreader, namely the handling device used to pick and keep containers by the machine itself. The Reach Stacker can stack up to the fourth height (up to the fifth height in case of empty containers).

Taking into account the main objective of the project, the MOCONT system should perform the following operations:

- Identify the container, reading the identification code when picked up by each Reach Stacker.
- Follow each movement of the container in the terminal yard while being handled by the Reach Stacker, recording (i.e., the position of the container in the yard – row, column, height) where the container is picked up or released.
- Inform on the position of the container and its identification without the intervention of human operators.

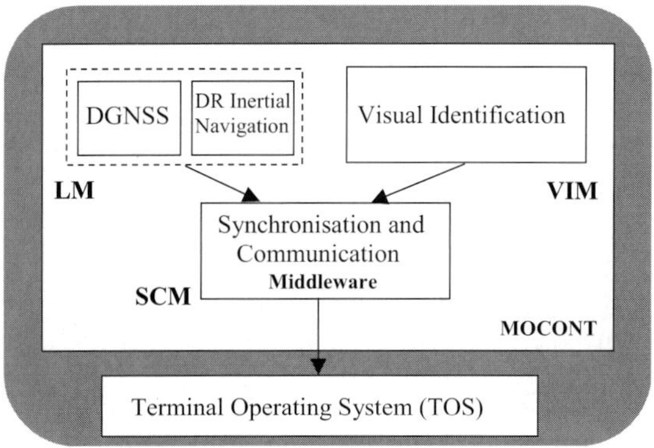

Figure 2: MOCONT functional architecture.

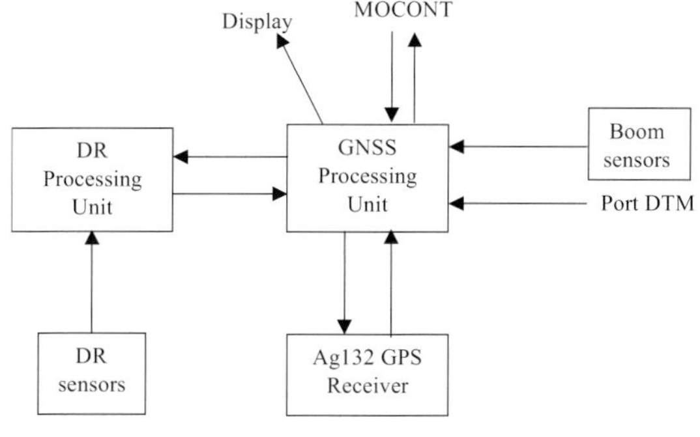

Figure 3: Scheme and flow data of the Location Module.

- In Real Time, update the position of any container handled by a Reach Stacker in the Terminal Operating System (TOS).

In order to obtain the objectives proposed, the MOCONT system incorporates three different modules (Figure 2): the Location Module, the Visual Identification Module and the Synchronisation and Communication Module. The Location Module determines the position of the container in the terminal yard; the Visual Identification Module reads the container's identification code; the Synchronisation and Communication Module acquires the identification and position data on the container and informs the TOS.

3 STRUCTURE OF THE LOCATION MODULE

The Location Module consists of two subsystems: the DR subsystem and the DGNSS subsystem. This last one has two different parts: the GPS receiver and the GNSS Processing Module.

In the final MOCONT Location Module Trimble's Ag132 GPS receiver is used at the heart of the GNSS and, therefore, of the location system, for the positioning of the Reach Stacker. The Ag132 GPS receiver combines high-performance GNSS reception with radio-beacon DGNSS capability in a single durable waterproof housing, ideal for use in the yard environment. The receiver uses differential GNSS to provide sub-metre accuracy.

Differential GNSS requires two or more receivers. One receiver, called the reference or base station, is located at a known point to determine the GNSS measurement errors. This could be housed on the roof of the main administration buildings, to allow easy access and constant monitoring. An unlimited number of AgGPS receivers, sometimes

called rovers, collect GNSS data at unknown locations onboard each Reach Stacker. Over a radio band, the reference station broadcasts correction values, which are applied to the Ag132 GPS receiver positions. Errors common at both the reference and rover receivers and then removed from the solution.

The performance of the Ag132 GPS receiver is improved by direct GNSS augmentation with height aiding. Height aiding improves the solution by enhancing satellite visibility, and reducing the positioning challenge from a three-dimensional to a two dimensional problem. Using a DTM of the port and the current location of the Reach Stacker, an interpolation algorithm provides an accurate measure of the current ground height. With knowledge of the Reach Stacker geometry, the boom extension and boom inclination, the height of the GNSS antenna on board the vehicle, and indeed the height of the container carried by the Reach Stacker, can be continually computed.

In addition, the Location Module provides complimentary DR augmentation for periods when GNSS positioning with height aiding is not possible. The DR subsystem consists of a Processing Unit and some DR sensors, by means of which the Reach Stacker position is continuously estimated. The GNSS Processing Module continually provides the DR subsystem with the current position from the Ag132 GPS receiver (in projected UTM coordinates) and some indication of the quality of that position fix (by means of a covariance matrix of the computed parameters). In return the DR subsystem continually updates the GNSS Processing Module with the best estimate of the current position.

The GNSS Processing Module will then pass the position information to the driver and the rest of the MOCONT system.

The GNSS Processing Module, which interfaces with the Ag132 GPS receiver, the DR subsystem,

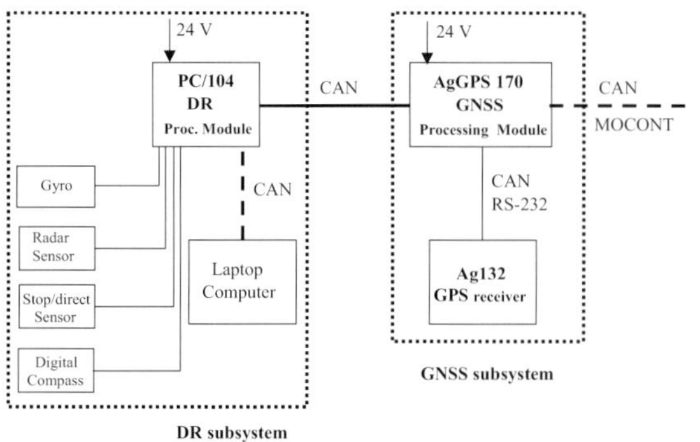

Figure 4: Elements of the Location Module.

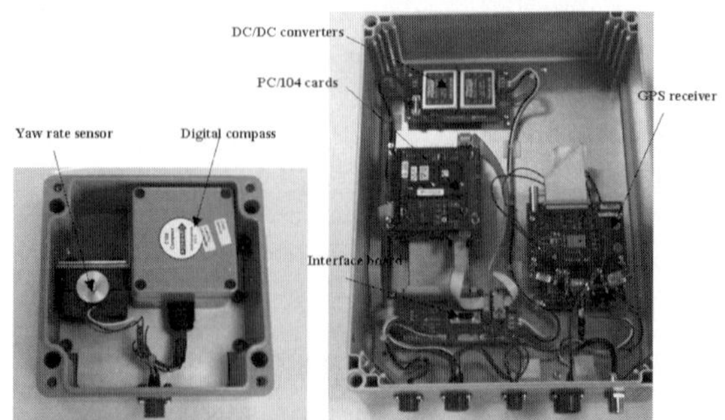

Figure 5: Heading box and Location Module box.

the boom sensors and the Synchronisation and Communication Module, has been developed using the robust and compact AgGPS 170 Field Computer. The AgGPS 170 is designed to withstand the environmental extremes that are typical of the container port environment.

A scheme of the Location Module, with its components and the flow of data, is shown in Figure 3 and Figure 4. As can be observed, the communication between the two subsystems of the Location Module is by means of a CAN bus. The communication between the GNSS subsystem and the Synchronisation and Communication Module is also by another CAN bus (Figure 4). The values measured by the boom sensors are supplied by the MOCONT middle-ware through this bus.

4   DR SUBSYSTEM DESIGN

4.1 Description of the DR Subsystem

Although initially different sensorings were tested for the DR subsystem, a sensor structure similar to that proposed by Rogers (1999) was finally selected. Likewise, replacing the MS750 subcentimetric GPS receiver in the final structure of the Location Module with an Ag132 submetric receiver, made it necessary to complete sensoring in the DR subsystem with a digital compass. As a result of these changes, the final structure used for the DR subsystem was as follows:

- The DR Processing Module, which consisted of a sandwich of three PC/104 boards: a CPU based on a 233 MHz Pentium Processor, an I/O data acquisition board and a 2-channel CAN communication board.
- The DR sensor set. This included the following sensors: a solid state gyroscope to measure the speed of rotation around the yaw axis; a radar

technology Ground Speed Sensor, which provided the forward/backward speed of the vehicle; a stop/direction sensor, in order to detect if the vehicle is stopped or not, as well as the movement direction of the vehicle, forward or backward; a digital compass, to measure the vehicle heading angle.

Figure 4 shows the main elements of the Location Module and the structure of the DR subsystem. A laptop computer was used to monitor and configure the DR subsystem, as well as to collect raw data. Most of the elements of the DR subsystem were included in two boxes, the Location Module box and the Heading box, as they appear in Figure 5.

4.2 DR/DGNSS Integration

Although different data fusion algorithms were tested, a kinematical Kalman Filter was finally chosen due to its simplicity and the good results obtained. For the navigation equations, it is assumed that the vehicle is moving on a tangent-plane, as it was a point, so the positioning involves locating the vehicle in cardinal directions: N-S-E-W. Figure 6 shows the local level geographic navigation and the

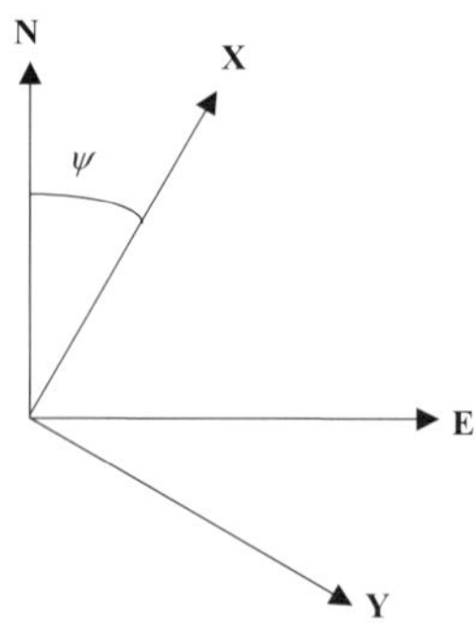

Figure 6: Local level geographic and body frames.

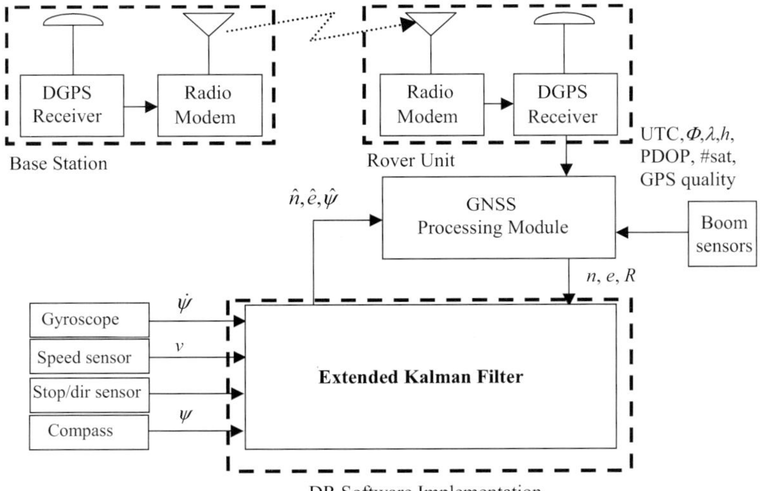

Figure 7: DR/DGNSS integration scheme.

body reference frames. It is assumed that the only transformation between these two frames is via the heading angle ψ.

The DR navigation system implements a distance travelled (integrated velocity) and a travel direction. The distance travelled is referenced to the body frame, which is then transformed into a local level geographic navigation frame through the angle ψ. This implementation assumes that other vehicle attitudes, i.e., roll and pitch, are sufficiently small as to be ignored.

The body referenced velocity is represented as a nominal velocity v, defined such that the X-component is along the primary direction of travel, plus velocity errors as a result of speed sensor scale factors ε_v.

The computed body referenced velocity is represented as

$$v_B = v + v \cdot \varepsilon_v \qquad (1)$$

where v is the measured or estimated vehicle velocity.

Assuming that the body to navigation frame transformation and body-referenced velocity are approximately constant over a small time interval, the sampling time, it can be written:

$$\dot{n} = v_B \cdot \cos\psi = v \cdot \cos\psi + v \cdot \varepsilon_v \cdot \cos\psi$$
$$\dot{e} = v_B \cdot \sin\psi = v \cdot \sin\psi + v \cdot \varepsilon_v \cdot \sin\psi \qquad (2)$$

where $\dot{n}$ and $\dot{e}$ are the velocities in the local frame and v is the measured vehicle velocity.

The heading angle rate could be expressed as follows:

$$\dot{\psi} = \alpha_\psi \cdot V_\psi + \varepsilon_\psi \cdot V_\psi + b_\psi \qquad (3)$$

where

α_ψ : gyroscope gain

ε_ψ : gyroscope scale factor error

b_ψ : gyroscope bias

V_ψ : measured gyroscope voltage

The speed sensor scale factor error ε_v, the gyroscope scale factor error ε_ψ and the gyroscope bias b_ψ are modelled as random-walk processes. From equations (1), (2) and (3) the continuous-time state-space realisation for the DR/DGPS is deduced:

$$\dot{\underline{x}} = f(\underline{x}, \underline{u}) + \underline{w}$$
$$\underline{y} = h(\underline{x}) + \underline{v} \qquad (4)$$

where $\underline{u} = \begin{bmatrix} v & V_\psi \end{bmatrix}$, $\underline{x} = \begin{bmatrix} n & e & \varepsilon_v & \psi & \varepsilon_\psi & b_\psi \end{bmatrix}'$ and $\underline{w}$ and $\underline{v}$ are random variables. The augmented state equations for the DR subsystem can be stated in direct form or in terms of residual errors, and therefore the structure of the Extended Kalman Filter can be deduced, the measurement vectors being:

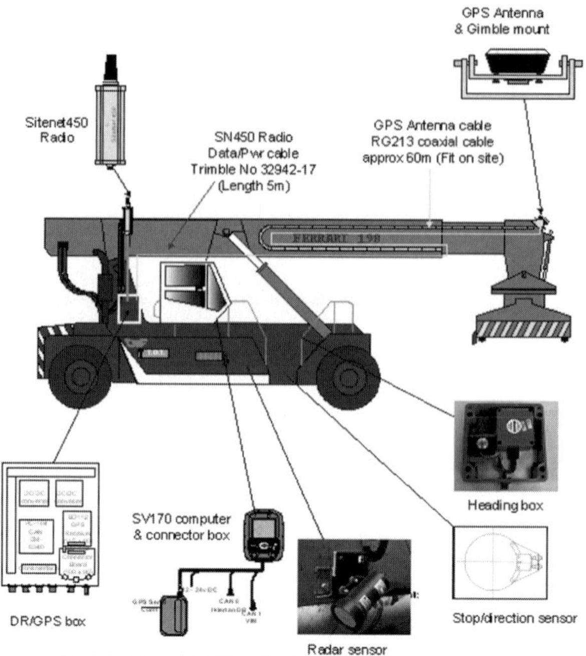

Figure 8: Installation of all elements of the MOCONT Location Module on a Reach Stacker.

$$\underline{u}(k) = \begin{bmatrix} v \\ V_\psi \end{bmatrix} \text{ with a frequency of 200 Hz}$$

$$\underline{z}(k) = \begin{bmatrix} n \\ e \\ \psi \end{bmatrix} \text{ with a frequency of 2 Hz}$$

Figure 7 shows the DR/DGNSS integration scheme. As outputs of the prediction steps of the Extended Kalman Filter, a vehicle position estimate is obtained with a frequency of 200 Hz, although the correction of state estimate is applied with a frequency of 2 Hz.

5   EXPERIMENTAL RESULTS

Figure 8 shows how all the MOCONT Location Module elements are fitted in a Reach Stacker. In the course of the MOCONT project, a system prototype was tested at the Terminal Darsena Toscana (TDT) in Livorno (Italy). In the MOCONT-II project, however, the system was implemented on 8 prototypes tested intensively over a six-month period at the TDT and at the VTE terminal in Genoa (Italy), taking a considerable number of data for statistical analysis.

The main objective of the Location Module is to locate precisely the container in the terminal yard.

The yard is the surface of the terminal dedicated to the container storage. It is subdivided into modules, each one composed of corridors (carriageways used to move containers within a module and between different modules) and groups of slots. A group of slot is referred to as a lane. Lanes are numbered using capital letters, starting from A. One slot is uniquely identified within a lane by its yard coordinates: row, column and height (Figure 1). Therefore, the container position in the corresponding slot should be accurately estimated. During the project MOCONT-II more than 5600 container position messages were collected. Results presented in the Final Public Report (MOCONT-II, 2004) led to the conclusion that the performance of the MOCONT Location Module was 99.7% of correct localisation resulting from the wide set of trials carried out.

The advantages of the DR subsystem are highlighted during the work inside container canyons, and generally, in the work near high stacks of containers which, on the other hand, are the most important moments for the correct position identification as this is when the Reach Stacker is picking up or releasing a container in a given slot.

Figure 9 shows typical results of a Reach Stacker handling a container. It corresponds to data collected in Livorno, where the cases of container canyons were quite common. The figure shows the GNSS estimate for the position of the vehicle chassis and the estimate conducted by the DR subsystem, which

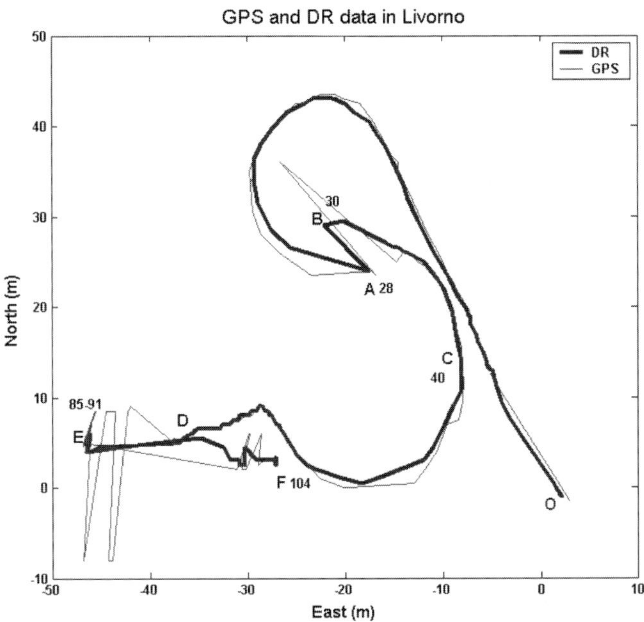

Figure 9: Example of experimental results at the TDT terminal in Livorno.

is transmitted to the GNSS on a frequency of 2 Hz. North and east relative position components, with respect to the starting point O, appear. The vehicle moves forward from the origin O to point A (instant 28 s). It then moves backward to point B (instant 30 s) from where it moves forward once again to point E (instant 85 s), after having passed through points C (instant 40 s) and D. It is at point E until instant 91 s, when it then moves backward to point F (instant 104 s). Covariance is not very good (always greater than 1), but also, between points C and D, it is approximately 3. As shown in the graph of Figure 9, when the covariance of the GNSS estimate is relatively low (about 1), the DR estimate continues to be in line with that of the GNSS. When the value of the covariance is large, particularly in the D-E-F stretch, the DR estimate is based mainly on the information provided by its own sensors.

In Figure 10 a detail of results obtained in Genoa is shown. In Genoa true container canyons rarely appeared, but sometimes the influence of high stacks of containers was evident, as in the case shown in the figure. The movement of the Reach Stacker starts at the point O, point considered as the origin of coordinates. The vehicles moves forward to point A (instant 12 s) and it is stopped at that point until instant 74 s, when it then moves backward to point B (instant 78 s). The Reach Stacker is at point B until instant 106 s. Then it goes forward to point C (instant 110 s) and after 17 s at that point the vehicle

moves backward to point D (instant 130 s). After 11 s it continues going backward to point E (instant 153 s), changes movement direction and moves forward to point F (instant 188 s) and then it continues its travel. As it can be deduced from the figure, the Reach Stacker performed operations with containers at points A and C. A container was picked up from a slot at point A and then it was released in other slot at point C.

The GNSS data covariance was very bad from instant 6 s until instant 78 s and from instant 106 s until instant 110 s (east covariance value higher than 3.5 and north covariance value higher than 9). Therefore, from point A to point C the GNSS estimates have poor quality, as it can be observed in Figure 10, but DR estimates show very well the movement carried out.

6 CONCLUSIONS

Analysis of a large number of experimental data obtained in the course of the MOCONT-II project has proven the success of the MOCONT Location Module in tasks involving tracking the position of containers in terminal yards while these are being handled by the Reach Stackers, recording the slot (row, column, tier), where the container is picked up or released. Integration of the Differential Global

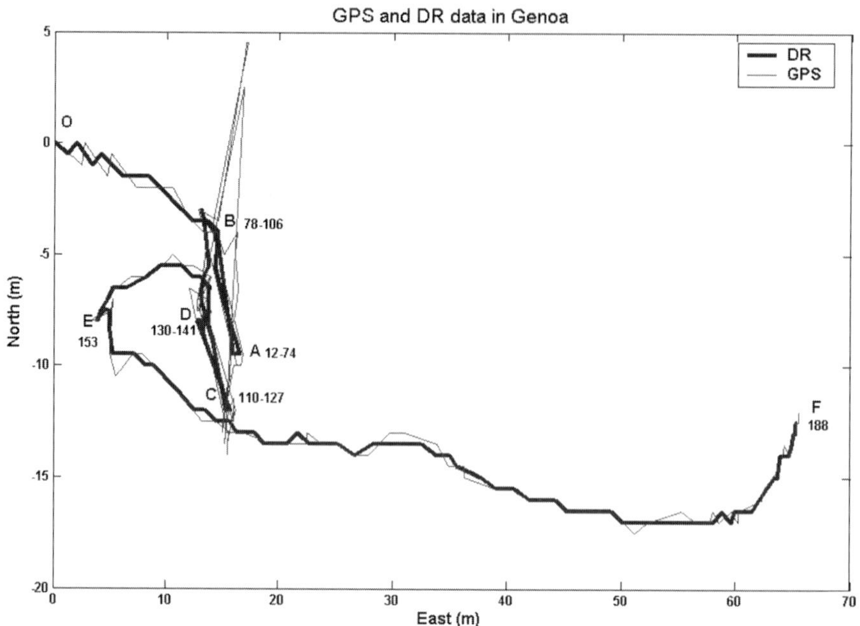

Figure 10: Example of experimental results at the VTE terminal in Genoa.

Navigation Satellite System (DGNSS) with a Dead Reckoning (DR) inertial system has been demonstrated to be effective in estimating the position of the vehicle and of the container. Positioning is effective even when working in container canyons, on the condition that it be for a limited movement time of some 40 s. Apart from the general characteristics of the MOCONT system, the structure of the MOCONT Location Module has been presented as well as the data fusion diagram. Likewise, typical experimental data and the final evaluation of effectiveness are also presented.

ACKNOWLEDGEMENTS

The material in this paper has been partially funded by the European Union under the scope of the Information Society Technologies programme (research projects MOCONT: IST-1999-10057 and MOCONT-II: IST-2001-34186).

REFERENCES

Aono, T., K. Fujii and S. Hatsumoto. 1999. Positioning of a Vehicle con Undulating Ground Using GPS and Internal Sensors. *T. SICE, vol. 35, no. 8, pp. 1004/1001.*

Benson, E.R., T.S. Stombaugh, N. Noguchi, J.D. Will and J.F. Reid. 1998. An evaluation of a geomagnetic direction sensor for vehicle guidance in precision agriculture applications. *ASAE paper 983203.*

Landaluze, J., V. del Río, A. Milo, J. M. Ezkerra and A. Martínez. 2003. Desarrollo de un sistema de localización DR/DGPS para vehículos "off-road". *5ª Semana Geomática. Barcelona, 11-14 febrero.*

MOCONT-II. 2004. *Final Public Report.* www.mocont.org.

Ramjattan, A.N. and P.A. Cross. 1995. A Kalman Filter Model for an Integrated Land Vehicle Navigation System. *Amy 1995, vol. 48 (no.2), pp. 293-302.*

Rogers, R.M. 1999. Integrated DR/DGPS using Low Cost Gyro and Speed Sensor. *ION National Technical Meeting.*

Sukkarieh, S., E.M. Nebot and H.F. Durrant-Whyte. 1999. A High Integrity IMU/GPS Navigation Loop for Autonomous Land Vehicle Applications. *IEEE Transactions on Robotics and Automation, vol. 15, no. 3.*

Zhang, Q, J.F. Reid and N. Noguchi. 1999. Agricultural Vehicle Navigation Using Multiple Guidance Sensors. *FSR'99 International Conference on Field and Service Robotics.*

PARTIAL VIEWS MATCHING USING A METIIOD BASED ON PRINCIPAL COMPONENTS

Santiago Salamanca Miño
Universidad de Extremadura (UEX)
Escuela de Ingenirías Industriales. Badajoz. Spain
Email: ssalaman@unex.es

Carlos Cerrada Somolinos
Universidad Nacional de Educación a Distancia (UNED)
Escuela Técnica Superior de Ingeniería Industrial. Madrid. Spain
Email: ccerrada@ieec.uned.es

Antonio Adán Oliver
Universidad de Castilla la Mancha (UCLM)
Escuela Superior de Ingeniería Informática. Ciudad Real. Spain
Email: Antonio.Adan@uclm.es

Miguel Adán Oliver
Universidad de Castilla la Mancha (UCLM)
Escuela Universitaria de Ingeniería Técnica Agrícola. Ciudad Real. Spain
Email: Miguel.Adan@uclm.es

Keywords: Matching, recognition, unstructured range data, 3D computer vision.

Abstract: This paper presents a method to estimate the pose (position and orientation) associated to the range data of an object partial view with respect to the complete object reference system. A database storing the principal components of the different partial views of an object, which are generated virtually, is created in advance in order to make a comparison between the values computed in a real view and the stored values. It allows obtaining a first approximation to the searched pose transformation, which will be afterwards refined by applying the Iterative Closest Point (ICP) algorithm.

The proposed method obtains very good pose estimations achieving very low failure rate, even in the case of the existence of occlusions. The paper describes the method and demonstrates these conclusions by presenting a set of experimental results obtained with real range data.

1   INTRODUCTION

Relative pose of the partial view of an object in a scene with respect to the reference system attached to the object can be determined by using matching techniques. This work is concerned with the problem of matching 3D range data of a partial view over the 3D data of the complete object. Resolution of this problem is of utmost practical interest because it can be used in applications like industrial robotics, mobile robots navigation, visual inspection, etc.

A standard way of dealing with this problem is to generating a model from the data, which allows extracting and representing some information associated to the source data. There are two basic classes of representation (Mamic and Bennamoun, 2002): object based representations and view based representations.

In the first class, models are created by extracting representative features of the objects. This type can be divided into four major categories: boundaries representations, generalized cylinders, surface representations and volumetric representations, being the third one the mostly used. In this case, a surface is fitted from the range data and then certain features are extracted from the fitted surface. Spherical representations belong to this category, being the Simplex Angle Image (SAI) representation (Higuchi et al., 1994; Hebert et al., 1995; Adán et al., 2001b; Adán et al., 2001a) an important example of this type of surface representation. In general terms, object based representations are not the most suitable ones for application in partial views matching.

Concerning to the view based representations they try to generate the model as a function of the diverse appearances of the object from different points of view. There exist a lot of techniques that belong to

J. Braz et al. (eds.), Informatics in Control, Automation and Robotics I, 159–166.

this class (Mamic and Bennamoun, 2002). Let us remark the methods based in principal components (PC) (Campbell and Flynn, 1999; Skocaj and Leonardis, 2001), which use them in the matching process as discriminant parameters to reduce the initial range images database of an object generated from all possible viewpoints.

The method presented in this paper can be classified halfway of the two classes because the appearance of the object from all the possible points of view are not stored and managed. Instead of that, only some features of each view are stored and handled. More specifically, just three critical distances are established from each point of view. These distances are determined by means of the principal components computation. A first approach to the transformation matrix between the partial view and the complete object can be obtained from this computation. The definitive transformation is finally achieved by applying a particular version of the Iterative Closest Point (ICP) algorithm (Besl and McKay, 1992) to the gross transformation obtained in the previous stage. A comparative study of different variants of the ICP algorithm can be seen in (Rusinkiewicz and Levoy, 2001).

The paper is organized as follows. A general description of the overall method is exposed in the next section. The method developed to obtain the database with the principal components from all the possible viewpoints are described is section 3. Section 4 is devoted to present the matching algorithm built on top of the principal components database. A set of experimental results is shown in section 5 and conclusions are stated in section 6.

2 OVERALL DESCRIPTION OF THE METHOD

As it has been mentioned, the first stage of the proposed method is based on the computation of the principal components from the range data corresponding to a partial view. Let us call X to this partial view. Principal components are defined as the set of eigenvalues and eigenvectors $\{(\lambda_i, \tilde{\mathbf{e}}_i) \mid i = 1, \ldots, m\}$ of the $\mathbf{Q}$ covariance matrix:

$$\mathbf{Q} = \mathbf{X}_c \mathbf{X}_c^T \qquad (1)$$

where $\mathbf{X}_c$ represents the range data translated with respect to the geometric center.

In the particular case we are considering, $\mathbf{X}_c$ is a matrix of dimension $n \times 3$ and there are three eigenvectors that point to the three existing principal directions. The first eigenvector points to the spatial direction where the data variance is maximum. From a geometrical point of view, and assuming that the

range data is homogeneously distributed, it means that the distance between the most extreme points projected over the first direction is the maximum among all possible $\mathbf{X}_c$ couple of points. The second vector points to another direction, normal to the previous one, in which the variance is maximum for all the possible normal directions. It means again that the distance between the most extreme points projected over the second direction is the maximum among all the possible directions normal to the first vector. The third vector makes a right-handed reference system with the two others. The eigenvalues represent a quantitative measurement of the maximum distances in each respective direction.

From the point of view of its application to the matching problem it is important to remark firstly that the eigenvalues are invariant to rotations and translations and the eigenvectors invariant to translations, and secondly that the frame formed by the eigenvectors represents a reference system fixed to the own range data. The first remark can be helpful in the process of determining which portion of the complete object is being sensed in a given partial view, i.e. the recognition process. The second one gives an initial estimation of the searched transformation matrix that matches the partial view with the complete object in the right pose. In fact, it is only valid to estimate the rotation matrix because the origins of the reference systems do not coincide.

To implement the recognition process it is necessary to evaluate, in a previous stage, all the possible partial views that can be generated for a given complete object. We propose a method that considers a discretized space of the viewpoints around the object. Then a *Virtual Partial View (VPV)* is generated for all the discretized viewpoints using a z-buffer based technique. Principal components of each one of these VPV are then computed and stored, such that they can be used in the recognition process.

Initial information of the possible candidate zones to matching a sample partial view to the complete object can be extracted by comparing the eigenvalues of the sample with the stored values. Nevertheless, this is only global information and it only concerns to orientation estimation. A second process must be implemented in order to obtain a fine estimate of the final transformation matrix. We use the ICP algorithm applied to the set of candidates extracted in the previous process. The transformation matrix estimated from the eigenvectors is used as the initial approximation required by the ICP algorithm. The final matching zone is selected as the one that minimizes the ICP error. Figure 1 shows a schematic block diagram of the entire developed method. More implementation details are explained in next sections.

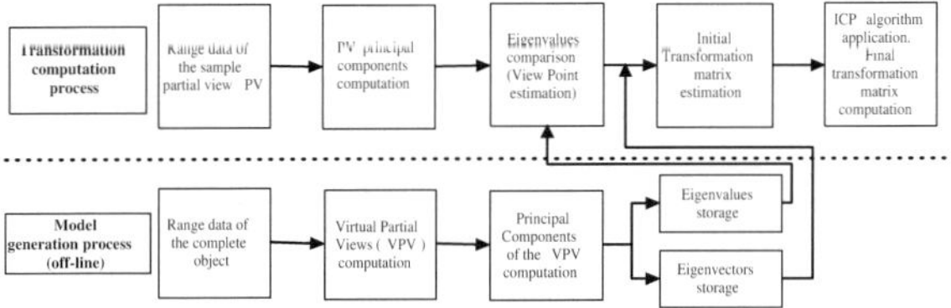

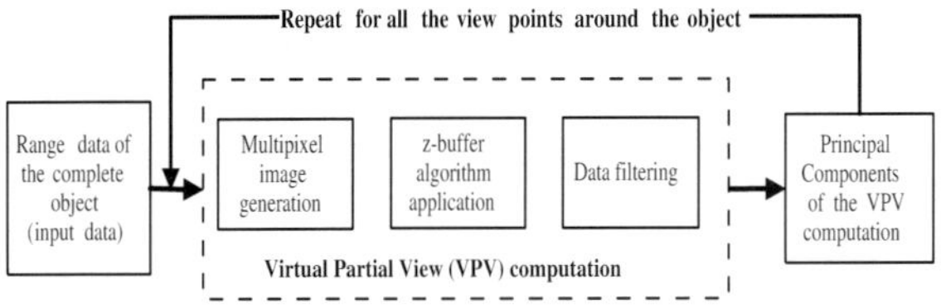

Figure 1: Block Diagram of the proposed method to find the best match of a given range data partial view with the complete object.

Figure 2: Block diagram of the process followed to generate the principal components database.

3 PRINCIPAL COMPONENTS DATABASE GENERATE

A database storing the principal components of all the possible partial views that can be generated from the range data of a given complete object must be computed in an off-line previous stage. Figure 2 shows the steps followed in this process.

Range data of the complete object is firstly translated to its geometric center and then normalized to the unit value. Therefore, we will start to work with the following normalized range data:

$$\mathbf{M}_n = \frac{\mathbf{M} - \mathbf{c}}{\max\left(\|\mathbf{M} - \mathbf{c}\|\right)} \qquad (2)$$

where $\mathbf{M}$ is the range data of the complete object, $\mathbf{c}$ is the geometric center, max is the maximum function and $\| \cdot \|$ is the Euclidean distance. At this point, as can be observed in Figure 2, the most important step of the method is the computation of the virtual partial views (VPV) described in the next subsection.

3.1 Virtual Partial Views Computation

A VPV can be defined as a subset of range data points $\mathbf{O} \subset \mathbf{M}_n$ virtually generated from a specific viewpoint VP, and that can be approximated to the real partial view obtained with a range sensor from the same VP.

Notice that in this definition it is necessary to consider, apart from the object range data itself, the viewpoint from which to look at the object. In order to

take into account all the possible viewpoints, a discretization of the visual space around the object must be considered. A tessellated sphere circumscribed to the object is used for that and each node of the sphere can be considered as a different and homogeneously distributed viewpoint from where to generate a VPV (see Figure 3).

For a specific VP its corresponding virtual partial view is obtained by applying the z-buffer algorithm. This algorithm is widely used in 3D computer graphics applications and it allows defining those mesh patches that can be viewed from a given viewpoint. Specifically, only the facets with the highest value of the z component will be visible, corresponding the Z-axis with the viewing direction.

This method is designed to apply when there is information about the surfaces to visualize, but not when just the rough 3D data points are available. Some kind of data conversion must be done previous to use the algorithm as it is. We have performed this conversion by generating the named *multipixel matrix* of the range data. This matrix can be obtained as follows. First a data projection over a plane normal to the viewing direction is performed:

$$\mathbf{M}'_n = \mathbf{U}\mathbf{M}_n \qquad (3)$$

where U is the matrix representing such projection. This transformation involves a change from the original reference system $S = \{O, X, Y, Z\}$ to the new one $S' = \{OX'Y'Z'\}$ where the Z' component, denoted as z', directly represents the depth value. From these new data an image can be obtained by discretizing the $X'Y'$ plane in as many pixels as required, and by assigning the z' coordinate of each point as the image value. Notice that in this process several points can be associated to the same pixel if they have the

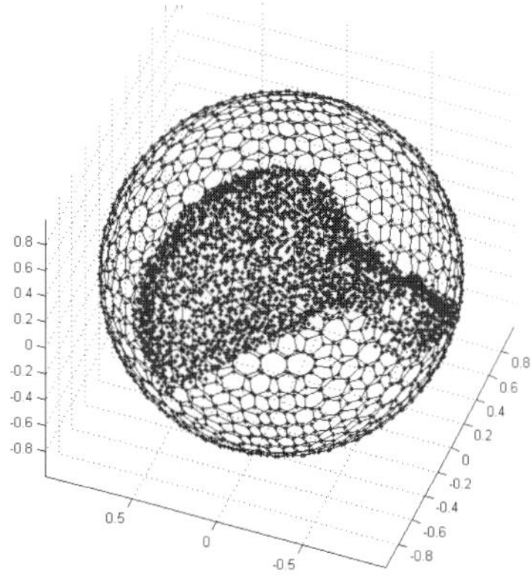

Figure 3: Visual space discretization around the complete object range data. Each sphere node constitutes a different viewpoint from which a VPV is estimated.

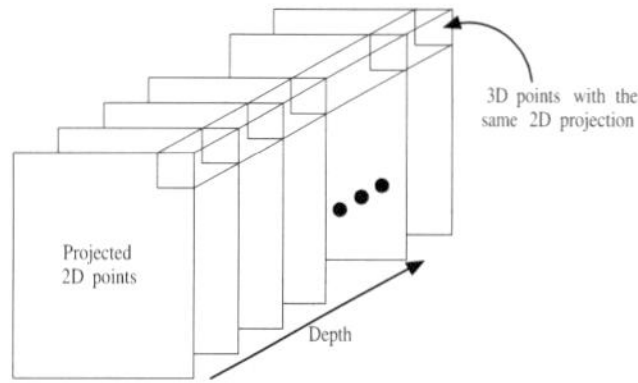

Figure 4: Multipixel matrix representation.

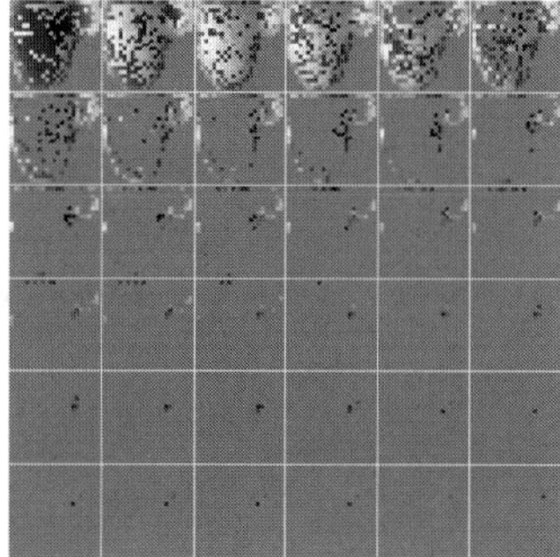

Figure 5: Visualization of each one of the planes of the multipixel matrix as a compacted intensity image.

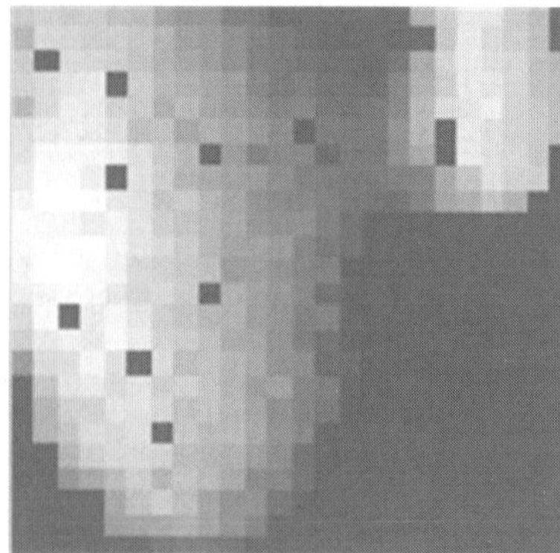

Figure 6: Intensity image associated to the maximum value of z' at each pixel.

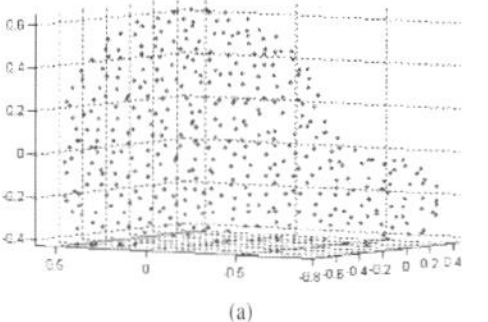 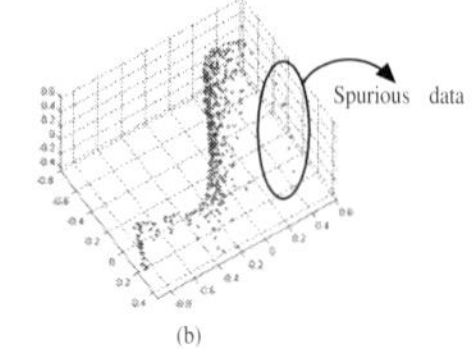

Figure 7: Range data points corresponding to a Virtual Partial View. They are shown from two different points of view to improve their visualization. In (b) the existence of spurious data are more evident.

same (x',y') coordinate values. To avoid depth information loss in these cases, the image values are stored in a three dimensional matrix. For that reason the created structure is denoted as multipixel matrix.

Figure 4 shows a typical scheme of this matrix. On the other hand, figure 5 shows as a grey-scaled image and in a compacted manner all the two-dimensional matrices that conform this structure. Pixels next to white color represent higher z' values. Figure 6 shows an image obtained by selecting for each pixel the maximum of its associated z' values. If the data points corresponding to these selected z' values are denoted as $\mathbf{O}' \subset \mathbf{M}'_n$, then the VPV can be obtained by applying the inverse of the $\mathbf{U}$ transformation defined in equation (3), i.e.:

$$\mathbf{O} = \mathbf{U}^{-1}\mathbf{O}' \qquad (4)$$

The set of range data points corresponding to a VPV obtained after the application of the described process to a sample case can be seen in the figure 7 (a) and (b). It can be observed that results are generally acceptable but some spurious data appear. These values were already noticeable from figure 6 where they look like a salt-pepper noise effect. Due to this fact, a median filter (see figure 8) is applied to the image before evaluating the equation (4), in order to obtain the definitive range data points of the searched VPV (figure 9).

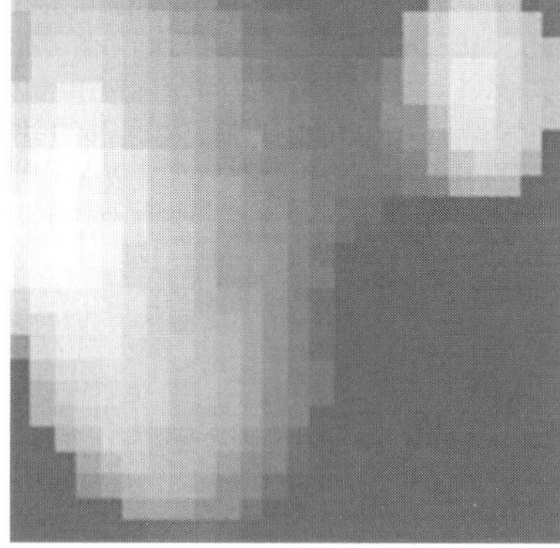

Figure 8: Image obtained after applying a median filter to the image shown in figure 6.

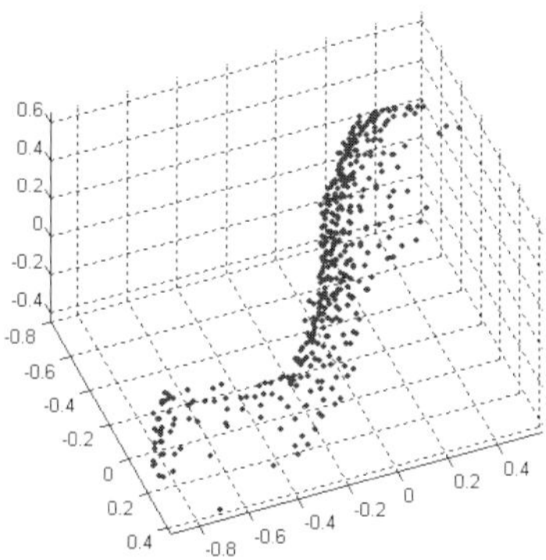

Figure 9: Definitive range data of a VPV. It can be observed, comparing with the figure 7(b), that the noise has been removed.

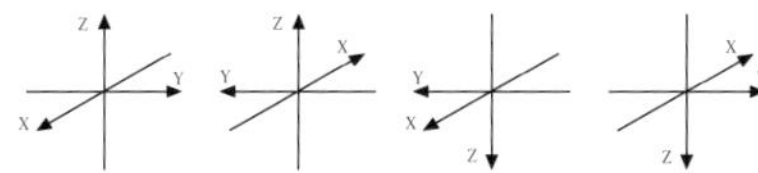

Figure 10: Four ambiguous frames for direction combinations.

3.2 Principal Components of VPV Computation

Once the range data points of a VPV have been determined the associated principal components can be computed from them. They will be the eigenvectors and the eigenvalues of the covariance matrix at (1), where $\mathbf{X}_c$ is now $\mathbf{O}$.

Due to the fact that the eigenvectors only provide information about the line vectors but not about their directions, some uncertainties can appear in the ulterior matching process. For example, using right handed coordinate systems like in figure 10, if no direction is fixed in advance there are four ambiguous possibilities for frame orientation. It can be also seen in the figure 10 that when one of the line directions is fixed only two possibilities appear, being related one to the other by a rotation of π radians around the fixed axis. For that reason, before storing the VPV eigenvectors, following steps are applied:

1. Verify if the third eigenvector forms an angle smaller than $\pi/2$ with the viewing direction. If not, we take the opposite direction for this eigenvector.

2. The first eigenvector direction is taken to build together with the two others a right-handed frame.

After these steps, principal components of the VPV are stored in the database for posterior use in the matching process. Figure 11 shows the definitive principal components of a given VPV.

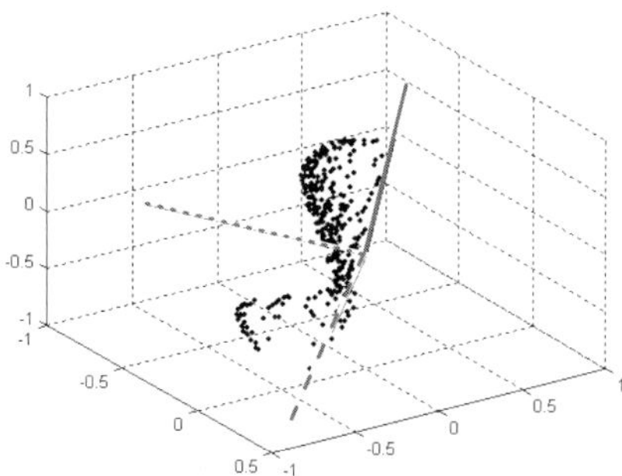

Figure 11: Definitive eigenvectors computed from the range data of a VPV. The first one is represented in red, second in green and the third in blue.

4 MATCHING PROCESS

The developed matching process is divided in four stages (see figure 1):

1. Principal components computation of the acquired partial view.

2. eigenvalues comparison and initial candidates zones selection.

3. Initial transformation estimation by using the eigenvectors.

4. ICP algorithm application to determine the final transformation.

The method developed to carry out the three first stages is described in the next subsection. Then we will explain how the ICP algorithm is applied to obtain the final result.

4.1 Initial Transformation Matrix Computation

The first thing to do is to compute the eigenvectors and eigenvalues of the acquired range partial view. To maintain equivalent initial conditions than in the VPV computation we must apply the same steps applied to the complete object range data: normalization with respect to the geometric center of the complete object, $\mathbf{M}$; multipixel matrix generation; z-buffer algorithm application and, finally, data filtering. In this case the main objective is to try the handled data being the most similar possible to those used in the principal component computation of the VPV.

After that, the principal components are computed and the definitive vector directions are established following the steps described in subsection 3.2. Then the eigenvalues of the acquired view are compared with the eigenvalues of all the stored VPV by evaluating the following error measurement:

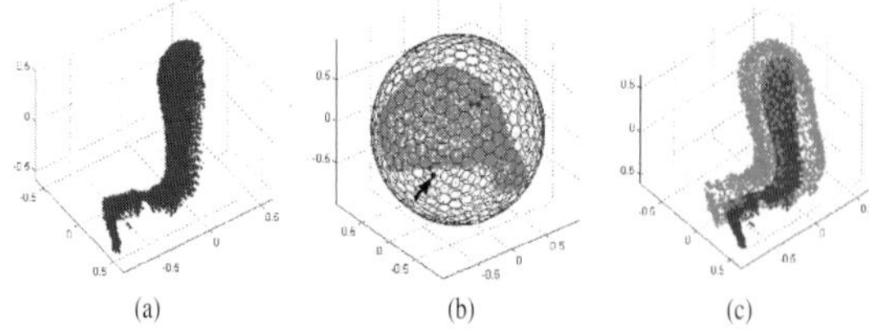

(a)   (b)   (c)

Figure 12: Results of the comparing eigenvectors algorithm. (a) Visualization of the acquired range data partial view. (b) Five selected candidate viewpoints from e_λ error computation. (c) Rotated data results for the first candidate.

$$e_\lambda = ||\Lambda^v - \Lambda^r|| \qquad (5)$$

where $\Lambda^v = \{\lambda_1^v, \lambda_2^v, \lambda_3^v\}$ is the vector formed by the eigenvalues of a VPV, $\Lambda^r = \{\lambda_1^r, \lambda_2^r, \lambda_3^r\}$ is the vector formed by the eigenvalues of the real partial view and $|| \cdot ||$ is the Euclidean distance.

Notice that a given VPV can achieve the minimum value of e_λ and not being the best candidate. This is because the feature we are using is a global one. For that reason we take a set of selected candidates to compute the possible initial transformation. Specifically, we are selecting the five VPV candidates with less error.

These initial transformations do not give the exact required rotation to coupling the eigenvector because the original data are not normalized with respect the same geometric center. Mathematically, what we are computing is the rotation matrix $\mathbf{R}$ such that applied to the eigenvectors of the real partial view, $\mathbf{E}^r$, gives a set of eigenvectors coincident with the VPV ones, $\mathbf{E}^v$. Because both $\mathbf{E}^r$ and $\mathbf{E}^v$ represent orthonormal matrices, the searched rotation matrix can be computed from the next expression:

$$\mathbf{R} = \mathbf{E}^r (\mathbf{E}^v)^{-1} = \mathbf{E}^r (\mathbf{E}^v)^T \qquad (6)$$

Because of the ambiguity of the two possible directions existing in the eigenvectors definition, the final matrix can be the directly obtained in (6) or another one obtained after rotating an angle of π radians around the third eigenvector.

Figure 12 shows the results of the comparing eigenvectors algorithm. In (a) the real partial view is shown. The five selected candidates with less e_λ error values are remarked over the sphere in part (b). An arrow indicates the first candidate, i.e. the corresponding to the minimum e_λ value. The approximate $\mathbf{R}$ matrix obtained as a result of equation (6) evaluation can be observed in (c). In this case the result corresponds to the VPV associated to the first candidate viewpoint. Definitive matrix will be obtained after a refinement process by means of the ICP algorithm application.

Summarizing, the resulting product of this phase is a transformation matrix $\mathbf{T}$ whose sub matrix $\mathbf{R}$ is obtained from equation (6) and whose translation vector is $\mathbf{t} = [0, 0, 0]^T$.

4.2 ICP Algorithm Application

The Iterative Closest Point (ICP) (Besl and McKay, 1992) is an algorithm that minimizes the medium quadratic error

$$e(k) = \frac{1}{n} \sum ||\mathbf{P} - \mathbf{P}'||^2 \qquad (7)$$

among the n data points of an unknown object $\mathbf{P}'$, called scene data, and the corresponding data of the database object $\mathbf{P}$, called model data. In our particular case, the scene data are the range data of the real partial view, $\mathbf{X}_c$, normalized and transformed by the $\mathbf{T}$ matrix, and the model data are the subset of points of the complete object $\mathbf{M}_n$ that are the nearest to each of the scene points. The latest will change at each iteration step.

Once the model data subset in a given iteration step k are established, and assuming that the error $e(k)$ is still bigger than the finishing error, it is necessary to determine the transformation matrix that makes minimum the error $e(k)$. Solution for the translation part, the t vector, is obtained from the expression (Forsyth and Ponce, 2002):

$$\mathbf{t} = \frac{1}{n} \sum_{i=1}^{n} \mathbf{r}_i - \frac{1}{n} \sum_{i=1}^{n} \mathbf{r}_i' \qquad (8)$$

where r_i and r_i' are the coordinates of the model data points and scene data points respectively.

With respect to the rotation part, the rotation matrix $\mathbf{R}$ that minimizes the error, we have used the Horn approximation (Horn, 1988), in which $\mathbf{R}$ is formed by the eigenvectors of the matrix $\mathbf{M}$ defined as:

$$\mathbf{M} = \mathbf{P} (\mathbf{P}')^T \qquad (9)$$

This approximation gives a closed-form solution for that computation, which accelerates significantly the ICP algorithm.

The algorithm just described is applied to the five candidates selected in the previous phase. The chosen final transformation matrix will be that one for which the finishing ICP error given by expression (7) is the smaller one.

Figure 13 shows the results for the same data points of figure 11 after application of the ICP algorithm. Partial view after applying the final transformation matrix matched over the object and the complete object range data are plotted together. Plots from two different points of view are shown to improve the visualization of the obtained results.

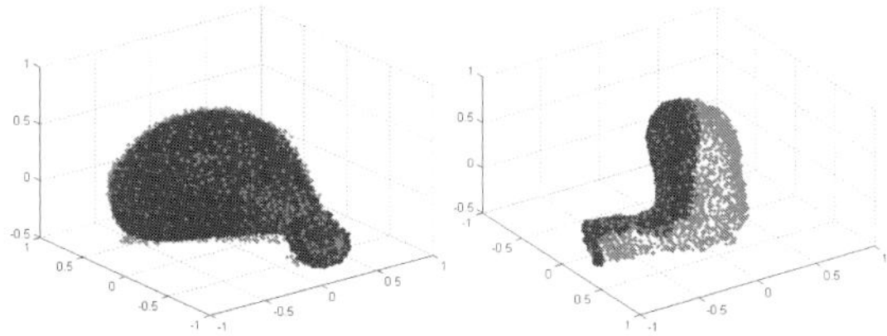

Figure 13: Final results of the overall matching algorithm visualized from two different points of view.

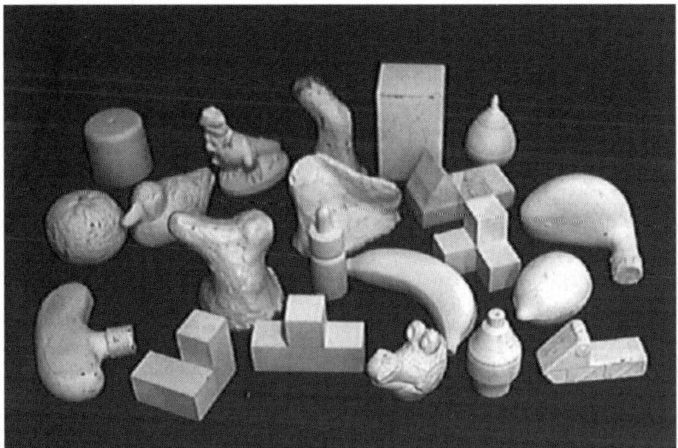

Figure 14: Set of objects used to test the presented algorithm.

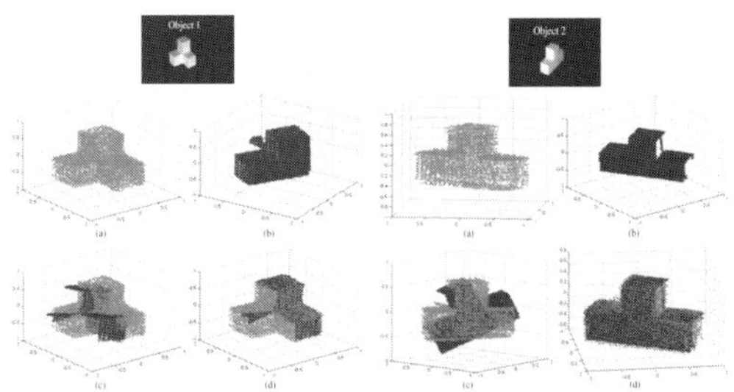

Figure 15: Some results over polyhedral objects of the proposed method.

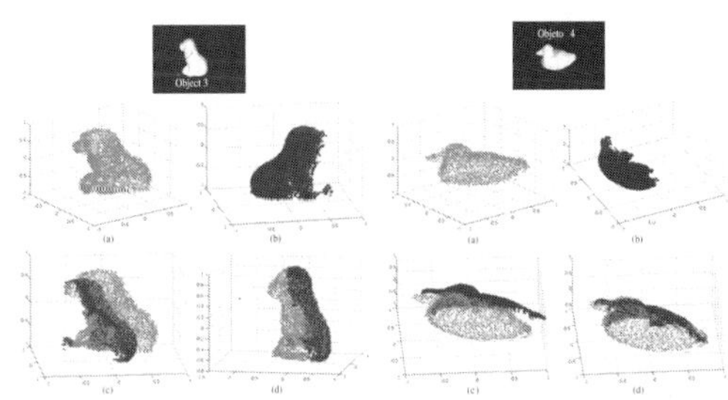

Figure 16: Some results over free form objects of the proposed method.

5   EXPERIMENTAL RESULT

The algorithm proposed in the present work has been tested over a set of 21 objects. Range data of these objects have been acquired by means of a GRF-2 range sensor which provides an average resolution of approximately 1 mm. Real size of the used objects goes from 5 to 10 cm. height and there are both polyhedral shaped and free form objects (see figure 14).

We have used a sphere with 1280 nodes to compute the principal components for the complete object. Some tests have been made with a mesh of 5120 nodes, but the associated increment of the computation time does not bring as a counterpart any improvement in the obtained matching results.

With respect to the multipixel matrix, sizes around 22x22 pixels have empirically given good results. The matrix size can be critical in the method functionality because high value implies poor VPV generation, and low value involves a reduction in the number of range data points that conform the VPV, making unacceptable the principal components computation.

Once the principal components database was generated for all the considered objects we have checked the matching algorithm. Three partial views have been tested for each object, making a total of 63 studied cases. The success rate has been the 90,67%, what demonstrates the validity of the method.

The average computation time invested by the algorithm has been 75 seconds, programmed over a Pentium 4 at 2.4 GHz. computer under Matlab environment. The time for the phase of eigenvalues comparison and better candidates' selection is very small, around 1 sec. The remaining time is consumed in the computation of the principal components of the real partial view and, mainly, in the application of the ICP algorithm to the five candidates: five times for the direct computation of the R matrix from equation (6), and another five times due to the direction ambiguity existing in the eigenvectors definition described in subsection 3.2.

Figures 15 and 16 show the results obtained with several polyhedral and free-form objects respectively. Apart from the intensity image of the object, subplot (a) presents the range data of the complete object, subplot (b) shows the partial view set of points, subplot (c) shows the data transformed after eigenvalues comparison, and subplot (d) contains the final results after ICP algorithm application. Some plots have been rotated to enhance data visualization.

Finally it is important to remark that the developed algorithm can handle partial views with auto-occlusions. Figure 17 shows an example of this case.

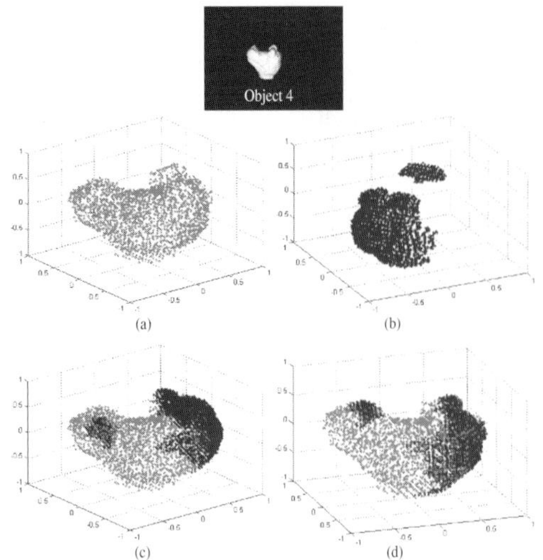

Figure 17: Results of the method in a free form object with existence of auto occlusions.

6 CONCLUSIONS

A method to find the best matching of a range data partial view with the complete object range data has been presented in this work. The method takes advantage of the principal components concept. The eigenvalues comparison allows determining the most probable matching zones among the partial view and the complete object. The corresponding eigenvectors are used to compute an initial transformation matrix. Applying the ICP algorithm refines this one and the definitive transformation is obtained.

For the comparison purposes a database of principal components must be generated in advance. A procedure has been designed and described to virtually generate all the possible partial views of a given complete object and then to compute the associated principal components. The procedure is based on the z-buffer algorithm.

The method has been tested over a database containing 21 objects, both polyhedral and free form, in a 63 case study (three different views for each object). The success rate has been the 90,67%. The method has proven its robustness to auto-occlusions. Some improvements can still to be made in a future concerning to the candidate selection step. Several candidates appear at the same zone and they could be grouped into the same one for the following step of ICP application.

Finally it is important to remark that the presented method has very good performance in the shown matching problem but it can also be applied in recognition applications with similar expected performance.

ACKNOWLEDGEMENTS

This work has been developed under support of the Spanish Government through the CICYT DPI2002-03999-C02-02 contract.

REFERENCES

Adán, A., Cerrada, C., and Feliu, V. (2001a). Automatic pose determination of 3D shapes based on modeling wave sets: a new data structure for object modeling. *Image and Vision Computing*, 19(12): 867–890.

Adán, A., Cerrada, C., and Feliu, V. (2001b). Global shape invariants: a solution for 3D free-form object discrimination/identification problem. *Pattern Recognition*, 34(7):1331–1348.

Besl, P. and McKay, N. (1992). A method for registration of 3-D shapes. *IEEE Transactions on Pattern Analysis and Machine Intelligence*, 14(2):239–256.

Campbell, R. J. and Flynn, P. J. (1999). Eigenshapes for 3D object recognition in range data. In *Proc. of the IEEE Conference on Computer Vision and Pattern Recognition*, volume 2, pages 2505–2510.

Forsyth, D. A. and Ponce, J. (2002). *Computer Vision: A Modern Approach*. Prentice Hall.

Hebert, M., Ikeuchi, K., and Delingette, H. (1995). A spherical representation for recognition of free-form surfaces. *IEEE Transactions on Pattern Analysis and Machine Intelligence*, 17(7):681–690.

Higuchi, K., Hebert, M., and Ikeuchi, K. (1994). Merging multiple views using a spherical representation. In *IEEE CAD-Based Vision Workshop*, pages 124–131.

Horn, B. K. (1988). Closed form solutions of absolute orientation using orthonormal matrices. *Journal of the Optical Society A*, 5(7):1127–1135.

Mamic, G. and Bennamoun, M. (2002). Representation and recognition of 3D free-form objects. *Digital Signal Processing*, 12(1):47–76.

Rusinkiewicz, S. and Levoy, M. (2001). Efficient variants of the ICP algorithm. In *Proceeding of the Third International Conference on 3D Digital Imaging and Modeling (3DIM01)*, pages 145–152, Quebec, Canada.

Skocaj, D. and Leonardis, A. (2001). Robust recognition and pose determination of 3-D objects using range image in eigenspace approach. In *Proc. of 3DIM'01*, pages 171–178.

TOWARDS A CONCEPTUAL FRAMEWORK-BASED ARCHITECTURE FOR UNMANNED SYSTEMS

Norbert Oswald

European Aeronautic Defence and Space Company - EADS,
Military Aircraft, 81663 Munich, Germany
norbert.oswald@m.eads.net

Keywords: Software architecture, autonomous system, modelling.

Abstract: Future unmanned aerial systems demand capabilities to perform missions automatically to the greatest possible extent. Missions like reconnaissance, surveillance, combat, or SEAD usually consist of recurring phases and contain resembling or identical portions such as autonomous flight control, sensor processing, data transmission, communication or emergency procedures. To avoid implementing many similar singular solutions, a systematic approach for the design of an unmanned avionic software architecture is needed. Current approaches focus on a coarse system design, do not integrate off-the-shelf middleware, and do not consider the needs for having on-board intelligence.

This paper presents a reference software architecture to design and implement typical missions of unmanned aerial vehicles based on a CORBA middleware. The architecture is composed of identical components and rests upon the peer-to-peer architectural style. It describes the internal structure of a single component with respect to autonomous requirements and provides a framework for the rapid development and implementation of new components. The framework separates functionality and middleware by hiding ORB specific statements from components. Experimental tests simulating a reconnaissance mission using various ORB implementations indicate the benefits of having an architectural design supporting multi-lingual multi-process distributed applications.

1 INTRODUCTION

As a result of technological advances in many disciplines like flight control, data and signal processing, sensor engineering, communication links, and integrated modular avionics, the development of unmanned aerial systems is currently of great interest in the domain of military aircrafts. Such systems raise the possibility to conduct military operations in a more efficient and less risky fashion than before but require robustness and reliability. Because of its dynamic, stochastic, and largely unknown environment, the execution of missions needs software systems that are able to act autonomously especially in situations were no remote control is possible.

Missions like reconnaissance, surveillance, combat, suppression of enemy air defence, or air-to-air contain a number of components that can be recycled. From the software point of view a mission typically consists of a mission independent and a mission dependent part. Tasks of the mission independent part like takeoff, landing, or autonomous flight recur for various missions. But also in the independent part resembling or identical portions such as data transmission, communication or emergency procedures occur. This suggests to design and build a unique software architecture facilitating to cover demands of most missions. To do so, one has to incorporate features of today's avionic systems. Characteristic for avionic applications are their heterogeneous and distributed environment with various platforms as well as existing and approved multi-lingual legacy code. As redesign, porting and testing of software to new platforms is costly and time-consuming, one needs a software architecture that integrates existing code written for particular platforms, in various languages, for different operating systems as well as COTS products. Desirable would be an open platform that enables distributed computing, dynamic algorithm plug-in, and a rapid algorithm switching, also incorporating aspects such as real-time, fault tolerance and security.

When dealing with the development of such a software architecture one has to consider the work done so far from three subject areas, the military sector, the

J. Braz et al. (eds.), Informatics in Control, Automation and Robotics I, 167–177.
© 2006 *Springer. Printed in the Netherlands.*

artificial intelligence community and the field of architecture description languages (ADL). In the military sector, there is still a primary focus on the design of physical platforms, low-level control systems and sensing. Recently, there has been some work started in order to build so-called open source platforms. OCP is a joint venture from Boeing, Georgia Institute of Technologies and Honeywell Laboratories to build an open control platform designed to give future UAVs more capable flight control systems and integrates interdisciplinary systems that enables reconfigurable control (Schrage and Vachtsevanos, 1999) (Wills et al., 2001). A prototype of OCP was lately tested on a UAV helicopter. JAUS is another example of a joint venture to design an architecture for future unmanned systems. The main focus of this project is based upon building an architecture for various unmanned systems that tries to become independent from technological trends (JAUS, 2004) (Hansen, 2003). A third initiative is the avionics architecture description language (AADL) which is expected to pass as a standard soon. The AADL has be designed to support a model-based architecture-driven development of large-scale systems with an emphasis on concepts that address performance-critical embedded systems concerns like timing and performance (Feiler et al., 2003) (Dissaux, 2004).

Building of software architectures has a long tradition in the artificial intelligence community. The first important developments go back to the early 80s with the deliberative architectures from (Nilsson, 1980) that are characterised through predictable and predetermined communication between layers. Reactive behaviour-based architectures introduced by (Brooks, 1986) that emphasised parallelism over hierarchy appeared later. Intelligent behaviour in such an architecture emerges from the interactions of multiple behaviours and the real world. To combine the benefits of both approaches and thus enabling reactive behaviour and planning capabilities a multitude of hybrid architectures have been introduced over the intervening years. An overview about various architectures in AI applications can be found e.g. in (Kortenkamp et al., 1998). Currently, distributed agent architectures are under investigation. A sample architecture is the Open Agent Architecture, a domain-independent framework for integrating heterogeneous agents in a distributed environment based on the interagent communication language (Cheyer and Martin, 2001).

Architecture description languages have become an area of intense research in the software engineering community. They define software architectures as reusable and transferable abstractions of software systems (Clements et al., 2002), composed of components, connectors, and configurations to define locus, interaction, and topology (Garlan and Shaw, 1993). In general, software architectures are used to describe the structure of software systems. Support exists from a number of architectural description languages like Adage (Coglianese and Szymanski, 1993), C2 (Medvidoc et al., 1996) (Medvidoc, 1999), Meta-H (Binns and Vestal, 1993), or ACME (Garlan et al., 2000) to name but a few. xADL, another ADL, has been evaluated by a number of projects like AWACS aircraft software system or Mission Data System (Dashofy et al., 2002). It can be used in a flexible manner to develop new styles and architectures on a coarse description level, suited even for families of architectures. An extensive comparison of the various description languages can be found in (Medvidovic and Taylor, 2000). Although used in a variety of applications, each ADL has its particular justification but so far none of them has accomplished as being a standard.

At present, most of the propagated approaches lack on the one hand of support for integrating off-the-shell middleware (Nitto and Rosenblum, 1999) and consider on the other hand only the design of coarser-grained architectural elements. Instead of ignoring the results that practitioners have achieved in the definition of middlewares, the design of a software architecture should incorporate both, the benefits of top-down and bottom-up approaches. Some research concerning the integration of middleware has already been started (e.g. (Dashofy et al., 1999)).

To build systems with capabilities for self-dependent acting requires a fine-grained design. This regards in particular the internal architecture of single architectural elements like components or connectors. To support this, the present article describes an integrated view to the design of a reference software architecture for the domain of unmanned aerial vehicles regarding coarse-grained and fine-grained aspects with respect to autonomous requirements. Components are considered as architectural elements that encapsulate functionality in a blackboard style. That means, the way of providing a service is of no consideration as long as a reasonable result is returned and each component therefore possesses mechanisms to ensure the required quality of service.

2 DESCRIPTION OF THE SOFTWARE ARCHITECTURE

The reference software architecture is an abstract model that may be substantiated for a multiplicity of missions. It consists of identical constructed components, so-called autonomous entities (AEs), covering roles and functionalities, connectors that contain various ORB middleware implementations, and configurations that describe potential connection structures.

The design of the architecture follows a peer-to-peer approach meaning that peers respectively AEs may exist independently from each other. A priori, no defined hierarchies exist, but resulting from the selected AE during setup inherent hierarchies might appear. The proposed hybrid architecture combines behaviour-based and functional-based aspects and thus provides deliberative and reactive system behaviour. The proposed architecture assists coopera-

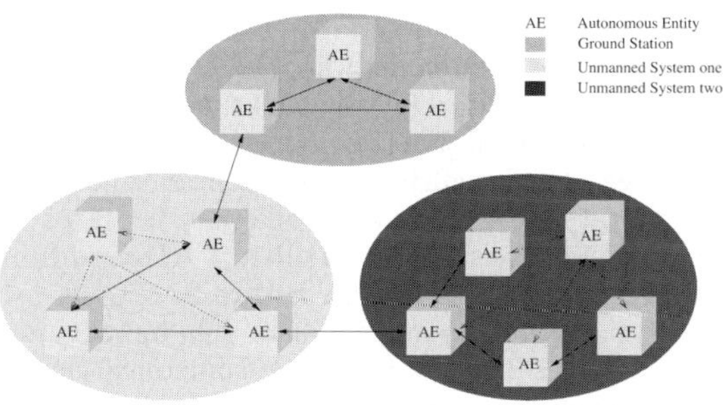

Figure 1: Design of the software architecture.

tion not only between AEs of a single system but also in a group of unmanned systems. Figure 1 shows the design and configuration of the software architecture in a single system as well as in a more complex context, being a part of a compound systems, e.g. with a ground station and another unmanned system.

2.1 Architectural Elements

In the following, the architectural elements are identified and described at domain level. This is done in a non-formal way but with respect to standard ADLs.

2.1.1 Components

Components respectively AEs are major ingredients including the functionalities that sum up to the system capabilities. They work concurrently, synchronously, or asynchronously and adopt certain roles according to the embedded functionality. AEs are identified through the process of decomposing missions in order to locate recurring conceptual formulations. At this level of abstraction, the internal representation of an AE is up to the blackboard style known from artificial intelligence technologies. To grant access, AEs provide fixed interfaces for their services to the general public. These so-called service interfaces are a generalisation of their embedded functionality, that is to say, they contain no explicit implementation details.

At design time, own service interfaces are made available but the precise implementation of an AE's functionality is unknown. Because of that, there exist no information about possibly required service interfaces from other AEs. Thus, there exist no a priori knowledge about connections between AEs at this level.

2.1.2 Connectors

Connectors are used to model the communication between AEs in ADL. In the above software architecture, the CORBA middleware is separated from the AEs and embedded into the connectors by hiding all ORB specific activities. If using exclusively TCP/IP-based communication, it is possible to join the connectors to constitute a communication network based on CORBA middleware. To do so with heterogeneous mediums, some effort to implement pluggable protocols is required (see e.g. (Ryan et al., 2000)). Resolving known services to the locations of their providing AEs require either a central component similar to the yellow pages or numerous communications between AEs performing sophisticated protocols. In terms of efficiency, a Service Broker is used that manages the connectors by resolving their requests and keeping their services and thus assists a component to find a particular service in the network. An AE accesses a service simply by calling *connect(name)*. The connector forwards this call to the Service Broker which resolves the request and returns a CORBA object. The latter is narrowed by the connector and returned to the client in form of a usual object reference. This procedure applies too for the access of service interfaces beyond the particular system.

2.1.3 Configuration

The configuration is used to describe how a system is built-up. As a consequence of using the framework that hides the middleware, the coupling between AEs and connectors during design time requires only one identical interface, namely *connect(name)* with *name* specifying the name of the service interface. Thus, an explicit modelling of the relationship between AEs and connectors is reduced to a single recurring description.

As a matter of fact, a linking between AEs occurs through claiming of general public service interfaces. A concrete configuration will be chosen only at the beginning of a session. An operator selects from a set of available AEs the one, that provide the required mission functionality. The connection structure for the concrete mission emerges not until the system was instantiated. Nevertheless, dynamically added AEs to the running system may be integrated into the connection structure if a dependency to functionalities of other AEs exists.

2.2 Middleware-Based Framework

Off-the-shelf middleware provides useful mechanisms to enable communication among several possibly distributed components together with a number of services like transactional communication or event-based interactions. In order to use these benefits without having dependencies onto the chosen middleware, a framework was developed, that encapsulates ORB specific functionality into the connectors. Essential elements in the framework are the structure of services and the Service Broker.

2.2.1 Structure of a Service

Basic element of the architecture is the service, as communication between AE is mainly based upon the use of services. Services are defined by IDL interfaces describing how to access a particular func-

```
<component name="Navigator">
 <interfaces>
  <implements>
   <type>IDL:AE_navigatorI/INavigator</type>
   <optional>
    <parameter name="Planner">
     <type>IDL:AE_navigator/IPlanner</type>
    </parameter>
    <parameter name="Time limit"></parameter>
   </optional>
   <attribute name="latency" value="100" />
  </implements>
 </interfaces>
 <dependencies>
  <depends>IDL:AE_flightControllerI/IFlightController
  </depends>
 </dependencies>
 <subcomponents>
  <subcomponent>IDL:AE_navigator/IPlanner</subcomponent>
 </subcomponents>
</component>
```

Figure 2: Describing a service to an AE.

tionality. To access a service, the name of the service interface has to be known. At run-time, an AE must know about the service interface it provides and about the service interfaces it requires. These informations are covered in XML notation for each AE. Figure 2 shows a sample description of the service interface *INavigator* provided by the component *Navigator*. The *implements* tag indicates that *INavigator* is the service interface of the AE. In order to be able to provide a service, an AE may depend upon additional service providers. Such dependencies are tagged by *dependencies*. In the above example, the AE needs

optionally an implementation for the interface *IPlanner* and depends on an AE that provides the service interface *IFlightController*.

To separate implementation details from the definition of the service interface, a service uses only a single parameter of type string. That means, the implementation part is hidden in a textual context, needing a description on how to interpret it. The parameters can be required or optional like *Planner*, and inform clients of which information they need to provide in order to use the service. The parameters may just be simple data or they may be complete object references. Attributes tell the client something about the properties of a service. For example, a service may have a latency attribute which tells the client how slow to respond the service is likely to be. This allows the client to make an intelligent selection from the available services and the best to be chosen. All the attributes and parameters that might be requested by a service are predifined in the component definition file in order to maintain consistent semantics.

2.2.2 Service Broker

The Service Broker enables the access to a particular service. It's functionality is similar to the one the Broker Pattern described in (Douglass, 2002) provides. The Service Broker fulfils two roles: firstly, that of a name server. It allows service interfaces to be mapped to a predefined name, which is known to other components in the system at design-time. This name can then be resolved to a concrete object reference, allowing components to find each other at runtime. Secondly, the Service Broker allows component to be registered by the interfaces they implement. For each interface, the Service Broker maintains a list of references to components currently running on the network which implement the interface. Other components can query the Service Broker to find this list.

Figure 3 shows the class diagram for the Service Broker. Each time a new component starts it registers with the Service Broker and submits a service description of itself by calling *addService()*. This description complies with the XML notation of figure 2. When an AE needs to find a service provider with a particular service interface, it asks the Service Broker by calling *connect(INavigator)*, which in that case returns a reference to the component that provides *INavigator*. Establishing a connection is a task of the connector for which the latter calls *findService()*. Because the Service Broker possesses a reference to the service description, the callee receives a description of the required service interface too.

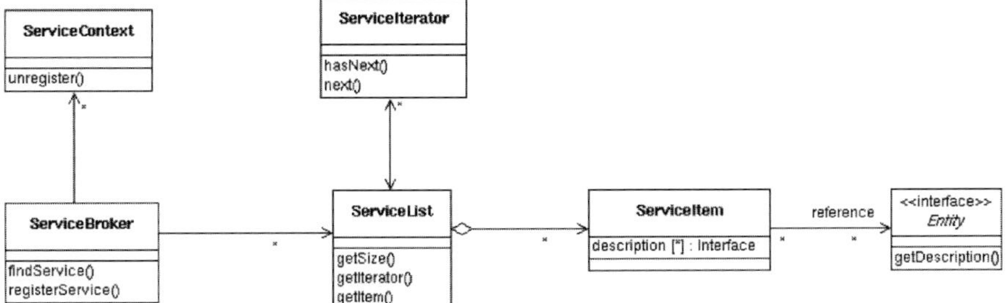

Figure 3: Service Broker.

3 DESIGN OF AN AUTONOMOUS ENTITY

The coarse-grained design of the architecture is build from a set of identical AE without specifying their internal structure. However, the architectural design of an AE is of major importance with regard to the building of an unmanned system because that requires to have autonomous capabilities. While (Maes, 1994) demands from an autonomous system to be adaptive, robust and effective, we use a similar interpretation but with different terms. A component is called to be an autonomous entity, when the following four major characteristics are fulfilled. It must be able

- to perceive,
- to plan,
- to decide, and
- to act.

Perceiving and acting are parts of today's avionics systems. Some of these systems already include planning and deciding capabilities without human intervention, but the definitive decision-making process during a mission is still incumbent upon the pilot. To automize this in order to build the capability of self-dependent acting requires to have several executable options. This means, that instead of using a single algorithm, an AE should comprise of a set of algorithms that solve the same task. To push that claim, we distinguish between two kinds of components in the software architecture. An AE on the one hand provides autonomous capabilities and builds a visible brick in the architecture. On the other hand, there are so-called concrete service provider (CSP) that quote implementations of basic functionality. These components are encapsulated into single AEs and hidden to public access.

3.1 Structure of an AE

Robustness and reliability requirements force unmanned systems to allocate mechanisms that enable support for deciding and planning tasks. To do so, we divide an AE into a $Head$ and a $Body$ and distinguish between allocating basic functionality and functionality that enable intelligent support. The $Body$ is responsible for providing the basic functionality of the AE and thus covers the execution of the functionality only. The $Head$ is responsible for building the AE's capability to act autonomously and covers planning, modelling and decision aspects. Both parts of an AE can be regarded as building blocks, with the $Head$ being on top of the $Body$.

3.1.1 The Head

The $Head$ provide means that enables an AE to act independently. The ability to effectively control the behaviour of an AE indeed depends on the allocated functionality for the $Head$. In case there is no intelligent support available for the $Head$, an AE acts like a conventional component without any decision, planning or learning capabilities simply based on the functionality of the $Body$. To provide responsibility, the $Head$ requires of intelligent support that allow for plausible decisions in dynamic situations. Such a decision support which can be considered as a building block inside the $Head$ is needed in various cases, among other things for

1. the selection of algorithms to use,

2. the evaluation of calculated results,

3. the synthesis of results, and

4. the handling of exceptional circumstances.

The first task of the $Head$ is a central one with respect to administrate a set of CSPs. This exercise occurs each time when there exists more than one implementation for a problem. Decisions on what algorithm to select are required for the $Body$, the *evaluation* and the *synthesis*. Certainly, one could also select the *selection*, but for this essay, we assume having loaded a selection algorithm at design-time. The second task mentions methods which are required to inspect the plausibility of calculated results. If provisionally results appear, the $Head$ has to decide how to proceed in order to guarantee a service. This might lead to a change in the selection of the native algorithm, improvements of a learning component, an update of model knowledge, or instructions to other AEs. The third task regards a situation where several algorithms work in parallel. To build a result for the client, data have to be fused in an appropriated way. Therefore, the $Head$ comprises of a number of CSPs allocating fusion functionality. Task four is of major importance to the $Head$ because of the AE's claim to act autonomously which requires robustness and reliability. The $Head$ has to try all feasibilities like using several implementations in order to execute a given task reliable. This task does not consider replication aspects as these should be hidden to the components and reside inside the connectors. The above list of tasks for the $Head$ shows minimum requirements and may be extended on demand without affecting the architecture due to their modular design.

3.1.2 The Body

The task of the $Body$ is to provide the functionality, that an AE has registered with the Service Broker. Therefore, the $Body$ implements the propagated service interface according to figure 4. The service interface *ServiceInterface* is visible to all other AEs. Usually, the $Body$ core itself contains no service implementations. These are provided from the CSP introduced in 3.2. That means, to solve requests that belong to the same problem domain, a $Body$ has a number of CSPs at it's disposal. This indirect access to the real implementation offers a number of advantages, it allows

- to administrate several identical sources,

- to group services of the same type, and

- to provide distributed processing.

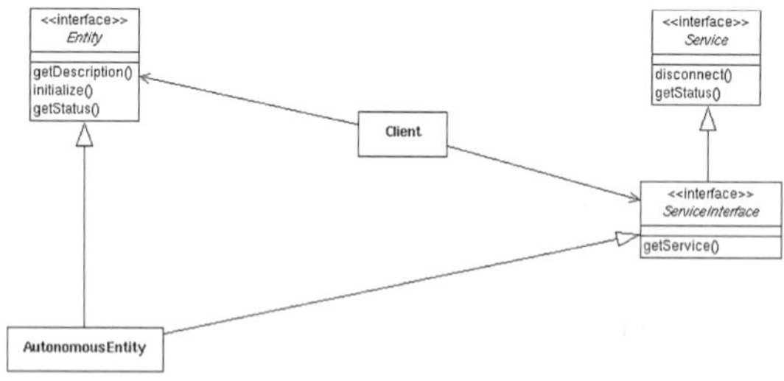

Figure 4: Provision of a service by an AE.

Although the CSP wraps the native interface of an algorithm to the service interface, the required input parameter may differ. As described in the service structure 2.2, some of the input parameters are fixed others are optional, that means they will be collected

at run-time. To meet all requirements, the $Body$ resolves all variable parameters needed to run a CSP.

The $Body$ communicates with the $Head$, whenever a *decision*, an *evaluation* or a *synthesis* is required. Also, the $Body$ may appear in the role of a client to other AEs, if the current task requires additional services. That means, the $Body$ communicates with other $Head$s until a reference to another service provider, CSP or $Body$ has been selected.

3.2 Provider of Concrete Services

While an AE provides services without containing a concrete implementation, a CSP provides an implementation of a single algorithm. CSPs are independent components of the software architecture that belong typically to one particular AE. They can be accessed only indirectly by accessing an AE's service interface.

```
<component name="AStar">
  <interfaces>
    <implements>AE_navigator/IPlanner</implements>
  </interfaces>
</component>
```

Figure 5: CSP providing a service.

Services from CSP are provided according to figure 6. A CSP provides an implementation of the interface *ManagedServiceInterface*. The CSP registers that service interface, which is tagged by *implements*, with the Service Broker as well as a description on how to use that particular algorithm. In the example of figure 5 the component *AStar* provides the interface *IPlanner*. As *ManagedServiceInterface* is not a general public service interface, it is invisible to all AEs except the one that depends on that concrete service marked with the *dependency* tag.

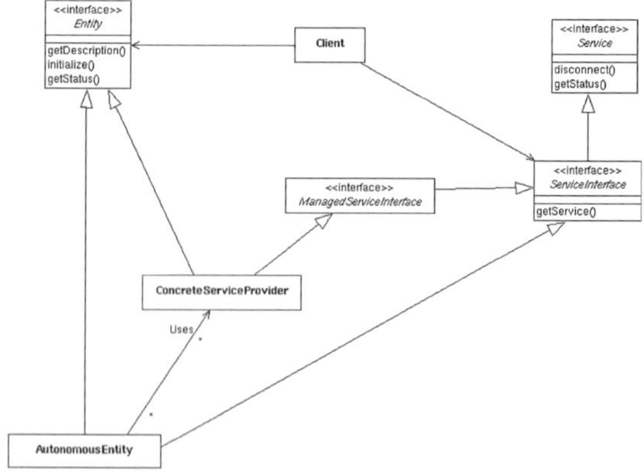

Figure 6: Provision of a service by an CSP.

As already mentioned, there are a number of algorithms in the aeronautics domain, that belong to the same problem domain. Although they solve the same task, algorithms differ in their native interface, their behaviour, or may have different constraints. Importing a native interface from legacy code will usually not meet the service interface. Thus, the CSP wraps the native interface of the legacy code to the service interface.

3.3 Retrieving CSPs

An AE provides neither particular basic functionality nor particular intelligent support and thus requires of having a number of CSP. Each CSP implements external functionality that belongs either to the $Head$ or to the $Body$. As CSPs are part of an AE the latter can be considered as an application independent skeleton or frame that provides, among other things, a mechanism for a purposive access to one or several CSPs. To do so, we have built the *Algorithm Selection* pattern shown in figure 7. The scope of this pattern is to return a list of references to implementations of algorithms that shall be executed. It constitutes a proposal for the first task of the $Head$ introduced in 3.1.1. The *Algorithm Selection* pattern is used both from the $Body$ to find a CSP that provides the required basic functionality and from the $Head$ to retrieve CSPs that allocate intelligent support like *evaluation* and *synthesis*.

The *Algorithm Selection* pattern is a collection of other patterns, namely Factory, Strategy, Adapter, Iterator and Proxy. It works in the following manner: a client, either a $Head$ or a $Body$, tries to resolve references for a given task. First, it calls the Service Broker to return an appropriated list of currently available CSPs on the net together with their constraints. From that list, it then decides with the pre-selected decision algorithm what CSP to use and returns one or several references.

3.4 Provision of Services

After having introduced the construction of an AE, interactions to provide a service are focused now by means of the sequence diagram shown in figure 8. An AE in its role as a client tries to access a particular service by calling *connect(AE_X)*. The connector of the receiving AE then calls the *initialize()* method of the $Head$ of the AE, named as *AE_XHead* in the diagram. At that stage, the $Head$ builds a reference to its $Body$ and returns that reference to the client. The client now calls the $Body$ directly with one of the provided service interfaces. When doing that, the $Body$ either executes directly a pre-selected CSP or it asks the $Head$ first to select CSPs according to the *Algorithm Selection* pattern. The result of *AlgorithmXCSP*

is returned to the $Body$ which calls the $Head$ for evaluation. If several results were calculated it might be required to use a fusion algorithm. The $Body$ asks the $Head$ to take care of the fusion. In both cases, the result of these calculations is returned to the $Body$. The latter returns the result to the client. For *evaluate()* and *synthesize()* the selection of algorithms follows according to the *Algorithm Selection* pattern. To finish a connection or to choose other selection algorithms the client calls *disconnect()*.

The selection of CSPs according to the *Algorithm Selection* pattern works as explained in Fig. 9. After having received *SelectAlgorithm* the $Head$ calls *select()* to the selection algorithm. There, a *findService()* call is made to the Service Broker. This retrieves a list of currently available algorithms respectively CSPs on the net. From that list, the selection algorithm chooses one or more appropriate CSPs and return those to the $Head$. The $Head$ then initialises the chosen CSPs and returns those references to the $Body$.

4 EXPERIMENTAL RESULTS

To test the framework we have designed an engineering platform for autonomous systems (EPAS). This platform builds the hardware and software infrastructure to develop, implement and test the behaviour of constructed software architectures. It consists on the one hand of a number of different hardware components, operating systems and various ORB implementations. On the other hand, there exists a repository of already constructed AEs and CSPs. Although AEs in the repository have default settings, they can be adapted to current mission purposes simply by reassigning the set of CSPs for each AE. The status quo of EPAS is shown in the table of figure 10. In total, four operating systems, seven ORB implementations and three languages are supported currently together with a number of pre-designed AE bricks as well as two simulators.

To rapidly construct software architectures, we have developed a graphical user interface, the so-called System Designer as shown in figure 11. According to mission requirements, a user selects AE from the repository shown as boxes in the System Designer, or he creates new components. For each AE the user then assigns the CSPs that resolve the proposed service interfaces. A minimal design is complete when all dependencies are resolved. Currently, the are no intelligent decision algorithms involved. Their behaviour is simulated by a simple but flexible decision technique. So far, no system aspects were considered for the developed software architecture. To do so, the System Designer allows to do a

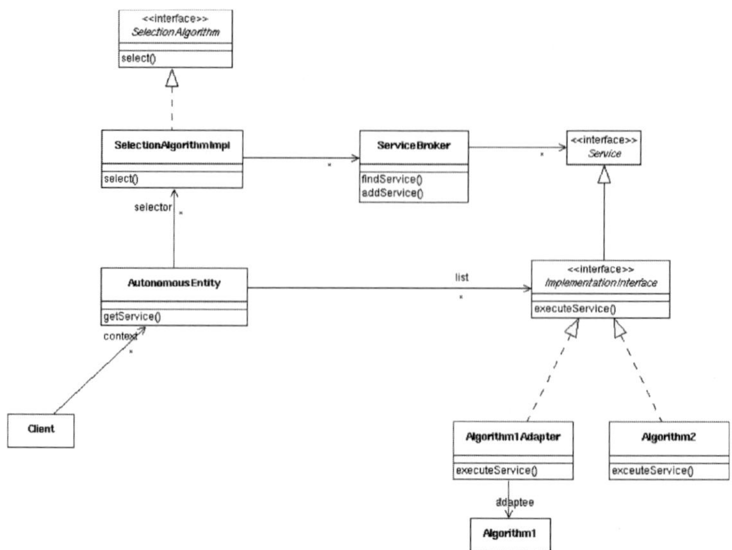

Figure 7: *Algorithm Selection* pattern.

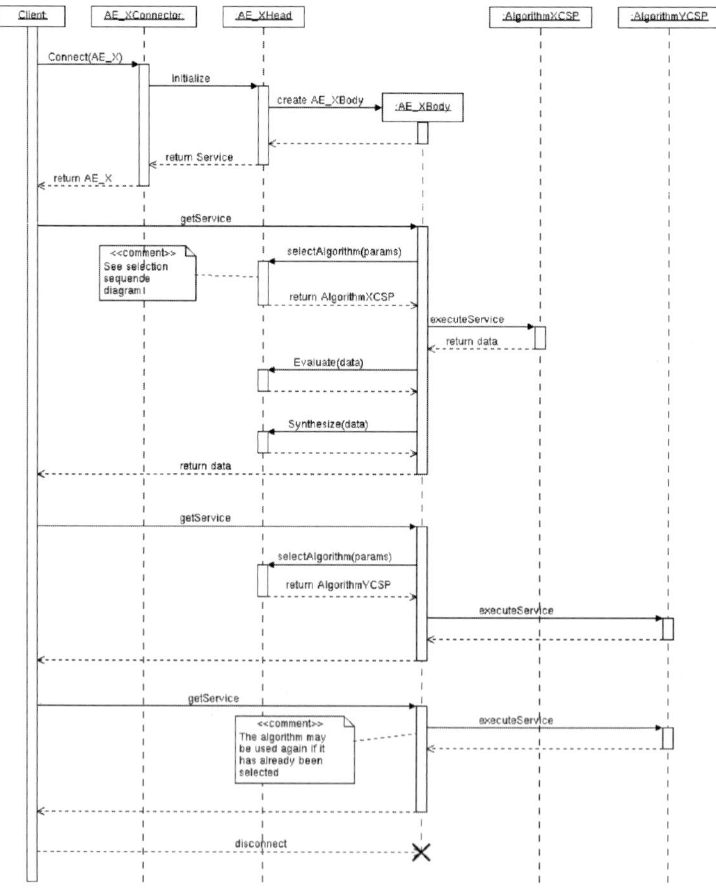

Figure 8: Sequence diagram for service provision.

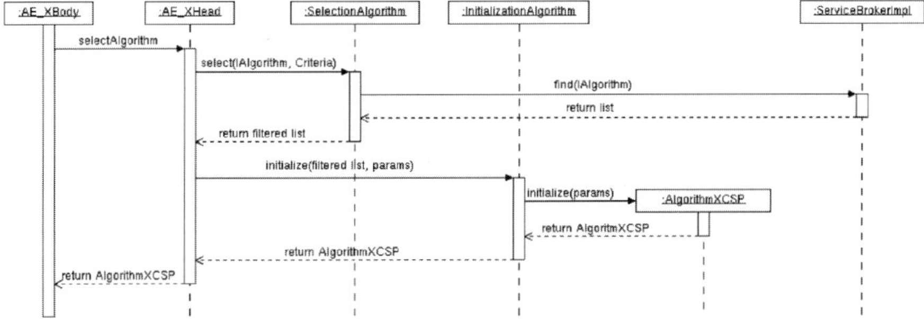

Figure 9: Sequence diagram for algorithm selection.

Object	Operating System					Language			Device							Corba Provider							Name	
	Win 2000	Solaris	Linux	VxWorks	LynxOs	C++	Java	Ada	Kandinsky (GS)	Dix	Klee (FCS)	Picasso (FT)	Janssen	Degas	tm g-ws2	JacORB	TAO	ORBacus Java	MICO	ORBacus C++	e*ORB	ORBExpress	TAO	ORBacus
Ground Station			X				X		X							X		X						X
NHP			X				X			X						X		X						X
Navigator			X				X			X						X		X						X
Mission Manager			X				X			X						X		X						X
EO Manager			X			X				X		X					X			X				X
Radar Warner			X			X	X						X				X		X	X				X
Flight Controller			X				X			X								X						X
Collision Manager			X			X											X			X				
Object Tracker					X	X								X							X			X
Flight Simulator	X					X					X						X						X	
Threat Simulator			X			X							X						X					X
EGI Manager				X				X							X							X	X	
AD Manager				X				X							X							X	X	

Figure 10: Infrastructure provided by EPAS.

configuration for each AE and each CSP. Configuration parameters include aspects like what ORB to use and on what machine to run a service. These informations are stored in an XML-based configuration file. From the chosen configuration a start-up script is build that sets up the whole system. At first, the Service Broker is started. Then, components are started as independent peers without any interactions except that they register their services with the Service Broker. So far, there exist no connection structures in the real application. The resolving of dependencies for an AE starts with the connect() call. Components check dependencies on demand and may resolves them by querying the Service Broker.

5 CONCLUSIONS

We introduced a reference software architecture based on a component framework to design and implement typical missions of unmanned aerial vehicles. The framework separates functionality and middleware by encapsulating CORBA into connectors and thus supports, at no expenses for the developer, using the heterogeneous avionics environment with various platforms, multi-lingual software, and different operating systems. Furthermore, the framework provides a unique architectural skeleton for a single autonomous entity to meet the requirements for self-dependent acting. A software architecture for a particular mission emerges from the composition of autonomous entities

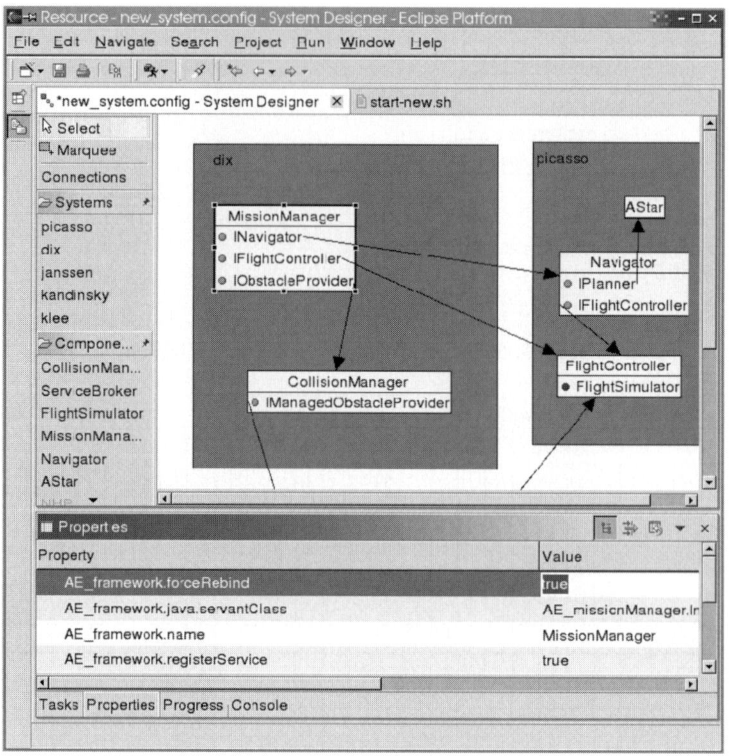

Figure 11: System Designer.

covering different functionality and interacting analogue to peers. Advantages of the presented approach are

- cost saving over a variety of missions because of reusable design technologies and fast turnaround times,

- reuse of existing and tested software,

- rapid system analysis of the composed mission architecture,

- dramatic reduction in mission completion time, and

- using of COTS and open source components.

Experiments in an appropriate engineering platform for autonomous systems containing various ORB implementations showed the rapid development process to build and test a software architecture. By means of a system designer, autonomous entities can be customised, combined and distributed at design-time in a flexible manner.

REFERENCES

Binns, P. and Vestal, S. (1993). Formal real-time architecture specification and analysis. In *In Tenth IEEE Workshop on Real-Time Operating Systems and Software*, New York.

Brooks, R. (1986). A layered control system for a mobile robot. *3rd Symposium. MIT Press*, pages 367–372.

Cheyer, A. and Martin, D. (2001). The open agent architecture. *Journal of Autonomous Agents and Multi-Agent Systems*, 4(1): 143–148. OAA.

Clements, P., Kazman, R., and Klein, M. (2002). Evaluating software architectures: Methods and case studies. Technical report, SEI, Series in Software Engineering.

Coglianese, L. and Szymanski, R. (1993). Dssa-adage: An environment for architecture-based avionics development. In *AGARD'93*.

Dashofy, E., Hoek, A., and Taylor, R. (2002). An infrastructure for the rapid development of xml-based architecture description languages. In *The 24th International Conference on Software Engineering*, Orlando.

Dashofy, E., Medvidoc, N., and Taylor, R. (1999). Using off-the-shelf middleware to implement connectors in distributed software architectures. In *International Conference on Software Engineering*, Los Angeles.

Dissaux, P. (2004). Using the aadl for mission critical software development. In *ERTS conference*, Toulouse.

Douglass, B. (2002). *Real-Time Design Pattern*. Addison Wesley.

Feiler, P., Lewis, B., and Vestal, S. (2003). The sae avionics architecture description language (aadl) standard: A basis for model-based architecture-driven embedded systems engineering. In *RTAS 2003*, Washington. vorhanden.

Garlan, D., Monroe, R., and Wile, D. (2000). Acme: Architectural description of component-based systems. Technical report, CMU, Software Engineering Institute.

Garlan, D. and Shaw, M. (1993). An introduction to software architecture. In Ambriola, V. and Tortora, G., editors, *Advances in Software Engineering and Knowledge Engineering*.

Hansen, S. (2003). Integration of autonomous system components using the jaus architecture. In *AUVSI Unmanned Systems 2003*, Baltimore.

JAUS (2004). Joint architecture for unmanned systems. http://www.jauswg.org.

Kortenkamp, D., Bonassao, R., and Murphy, R., editors (1998). *Artificial Intelligence and Mobile Robots*. MIT Press.

Maes, P. (1994). Mmodelling adaptive autonomous agents. *Artificial Life*, 1(1): 135–162.

Medvidoc, N. (1999). *Architecture-based Specifi-cation-Time Software Evolution*. PhD thesis, University of California.

Medvidoc, N., Oreizy, P., Robbins, J., and Taylor, R. (1996). Using object-oriented typing to support architectural design in the c2 style. In *SIGSOFT'96*. ACM Press.

Medvidovic, N. and Taylor, R. N. (2000). A classification and comparison framework for software architecture description languages. *Software Engineering*, 26(1): 70–93.

Nilsson, N. (1980). *Principles of Artificial Intelligence.* Tioga Press.

Nitto, E. D. and Rosenblum, D. (1999). Exploiting ADLs to specify architectural styles induced by middleware infrastructures. In *Int. Conf. on Software Engineering*, pages 13–22.

Ryan, C., Kuhns, F., Schmidt, D., Othman, O., and Parsons, J. (2000). The design and performance of a pluggable protocol framework for real-time distributed object computing middleware. In ACM/IFIP, editor, *Proceedings of the Middelware 2000 Conference*.

Schrage, D. P. and Vachtsevanos, G. (1999). Software enabled control for intelligent UAVs. In *1999 IEEE International Conference on Control Applications*.

Wills, L., Kannan, S., Sanders, S., Guler, M., Heck, B., Prasad, J., Schrage, D., and Vachtsevanos, G. (2001). An open platform for reconfigurable control. *IEEE Control Systems Magazine*, 21(3).

A INTERPOLATION-BASED APPROACH TO MOTION GENERATION FOR HUMANOID ROBOTS

Koshiro Noritake, Shohei Kato and Hidenori Itoh

Dept. of Intelligence and Computer Science
Nagoya Institute of Technology
Gokiso-cho, Showa-ku, Nagoya 466-8555, Japan
Email: {noritake, shohey, itoh}@ics.nitech.ac.jp

Keywords: Humanoid robot, motion generation, Tai Chi Chuan, balance checking.

Abstract: This paper proposes a static posture based motion generation system for humanoid robots. The system generates a sequence of motion from given several postures, and the motion is smooth and stable in the balance. We have produced all the motions of Tai Chi Chuan by the system. Motion generation for humanoids has been studied mainly based on the dynamics. Dynamic based method has, however, some defects: e.g., numerous parameters which can not be always prepared, expensive computational cost and no guarantee that the motions are stable in balance. We have, thus, studied less dependent-on-dynamics approach. A motion is described as a sequence of postures. Our system figure out if we need extra postures to insert for stability. This method enables humanoid robot, HOAP-1 to do Tai Chi Chuan.

1 INTRODUCTION

In recent years, robotics has greatly developed, especially, research for humanoid robots has attracted much attention (e.g,(Nishiwaki et al., 2002), (Sugihara et al., 2002), (Huang et al., 2001), (Yamaguchi et al., 1993), (Li et al., 1993), (Kagami et al., 2001), (Kuffner et al., 2001), (Kuffner et al., 2002), (Kuwayama et al., 2003)). Existing methods of motion generation for humanoid robots are mostly based on the dynamic control theory and the optimization technique. These methods are often specialized in some particular motions, such as walk and standing, which are simple, symmetric or cyclic. This presents an obstacle to general-purpose. These methods may require the mastery of dynamic for use. The methodologies based on the dynamics often require highly expensive computational cost, and the motion control for unconstraint motions is still hard problem. Mechanical characteristic of humanoid robot is an increase in DOFs. There is, however, few studies for motion control such that the DOFs are fully utilized.

In this research, we, thus, take an intelligent software approach to motion control with useful interface and application for various motions. In this paper, we propose a motion generation system for humanoid robots. Our system generates a sequence of motion from given several postures, and the motion is smooth and stable in the balance.

We have produced all the motions of Tai Chi Chuan by the system. All motions have been performed by a humanoid robot.

2 HUMANOID ROBOT AND THE TARGET MOTIONS

2.1 Humanoid Robot

In this paper, we consider the motion control of a humanoid robot, HOAP-1 (Humanoid for Open Architecture Platform) produced by Fujitsu(Murase et al., 2001), shown in Figure 1. The total weight is 6 (kg) and the height is 48 (cm). HOAP-1 has 20 DOFs in total, 6 in each leg and 4 in each arm. The link structure is shown in Figure 2. The sensor architecture of HOAP-1 is consisted of 4 pressure sensors on each sole and angular rate and acceleration sensors mounted in breast. HOAP-1 is controlled with RT-Linux OS on itself or computer connected with USB.

2.2 Tai Chi Chuan

We consider Tai Chi Chuan as the target motion for humanoid robots. Tai chi has several styles. In this paper, we adopted Tai Chi 48. Tai Chi motions have various movement of entire body. Tai Chi motions are

179

J. Braz et al. (eds.), Informatics in Control, Automation and Robotics I, 179–185.

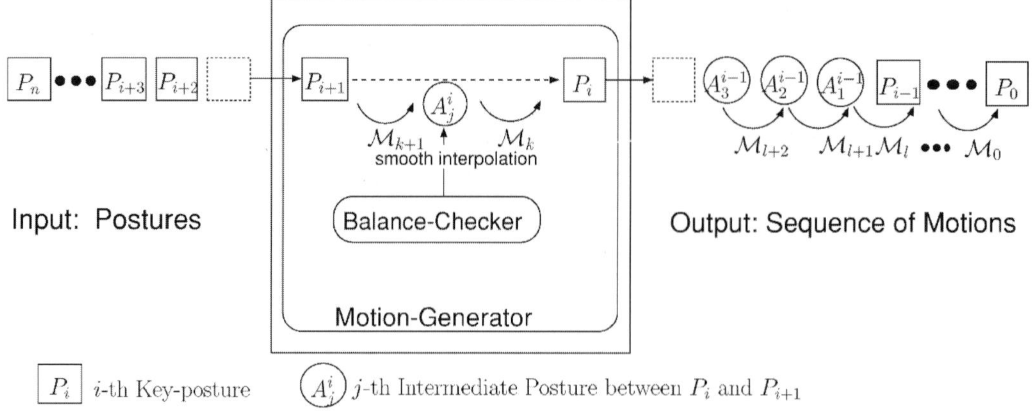

Figure 3: Data flow in our motion generation system.

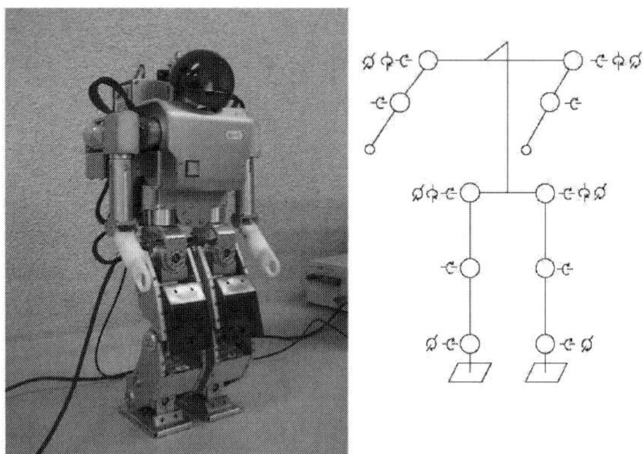

Figure 1: HOAP-1.

Figure 2: The link structure of HOAP-1.

performed slowly, softly and with smooth and even transitions between them. Tai Chi motions require sophisticated balance control for robots. In addition to stability or energy consumption, Tai Chi Chuan requires concinnous forms. Unlike to walking, Tai Chi motions should not be changed markedly in appearance even for the stability.

3 MOTION GENERATION SYSTEM

3.1 Interpolation-based Motion Generation

The section describes a motion generation system for humanoid robots. Figure 3 shows the out-line of the system. Input to the system is a sequence of postures $P = P_0, P_1, \cdots, P_n$, (called *key-postures*). Each key-posture has characteristic form of a motion to generate. In this system, we suppose that each key-posture is statically stable in balance. This supposition is supported by the key-posture adjustment, such that the center of mass (COM) of the upper body is positioned just above the ankle of the supporting leg. Output from the system is a motion sequence $M = \mathcal{M}_0, \mathcal{M}_0, \cdots, \mathcal{M}_m$.

The basic function of motion generation is the smooth interpolation between two postures, P_i and P_{i+1}. The interpolated motions, needless to say, should be stable in balance for humanoid robots. *Balance Checker* evaluates the distance between P_i and P_{i+1} in balance space, and then, inserts a few intermediate postures (corresponding to A_j^i in Figure 3) if the distance is over a threshold. It is, in general, difficult to calculate the distance of postures in balance space. We have, thus, made a rough approximation described in the following section.

3.2 Classification of the Postures in Balance Space

Tai Chi Chuan, our target motion, has wide variation in posture. Interpolation-based motion generation is, however, vulnerable to wide or quick variation in postures. It is, thus, important to consider some relation between two postures to be interpolated. The admissibility of the interpolated motion may be estimated by the relation. In this research, we consider the distance in balance space as the admissibility.

To calculate the distance of postures in balance space is hard problem. In this research, we give this problem an audacious consideration that postures of

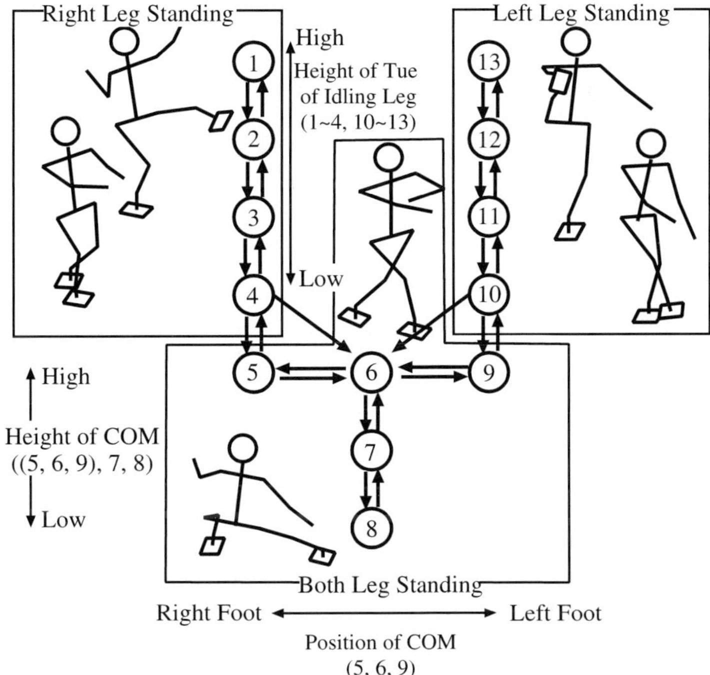

Figure 4: The directed graph of 13 posture classes.

Tai Chi motion are classified into 13 groups (*posture class*, or simply *class*) and these groups are connected by 26 arcs in the balance space. The balance space is represented by a directed graph shown in Figure 4. The classification is based on the relative position of COM and the relation of the supporting and idling legs. Firstly, postures are classified into three classes: both leg standing, right leg standing and left leg standing. Single leg standing is classified into four classes according to the height of the idling leg in relation to the supporting leg (see Figure 4):

- lower than the ankle (#4 and #10),

- higher or equal than the ankle and lower than the knee (#3 and #11),

- higher or equal than the knee and lower than the hip (#2 and #12),

- higher or equal than the hip (#1 and #13).

Both leg standing is classified into five classes according to the position of the COM in relation to the COM of the erect posture (see Figure 4):

- the vertical component is less than 80% (#8),

- the vertical component is more or equal than 80% and less than 90% (#7),

- the vertical component is more or equal than 90% and,

- the horizontal component leans to right more than 20% (#5),

- the horizontal component leans to left more than 20% (#9),

- the horizontal component is within 20% from the center of the both feet (#6).

The classification enables our system to approximate the distance of two postures. Let a and b be arbitrary postures to be interpolated, and let g_a to g_b be the classes which a and b belong to, respectively. The distance between a and b is given by the distance between g_a and g_b: the length of the path from g_a to g_b. We, therefore, can estimate the admissibility of the interpolated motion from a to b by the distance between g_a and g_b.

The balance checker in our system utilizes the directed graph in Figure 4 for the decision to insert an intermediate posture or not. The directed graph give a constraint for the stability in balance on the motion generator. The constraint is that two postures to be interpolated should belong to one class or adjoining two classes. For example, interpolation from a posture in class #6 to a posture in class #4 has the risk of tumble. A posture in class #5 should be inserted between them.

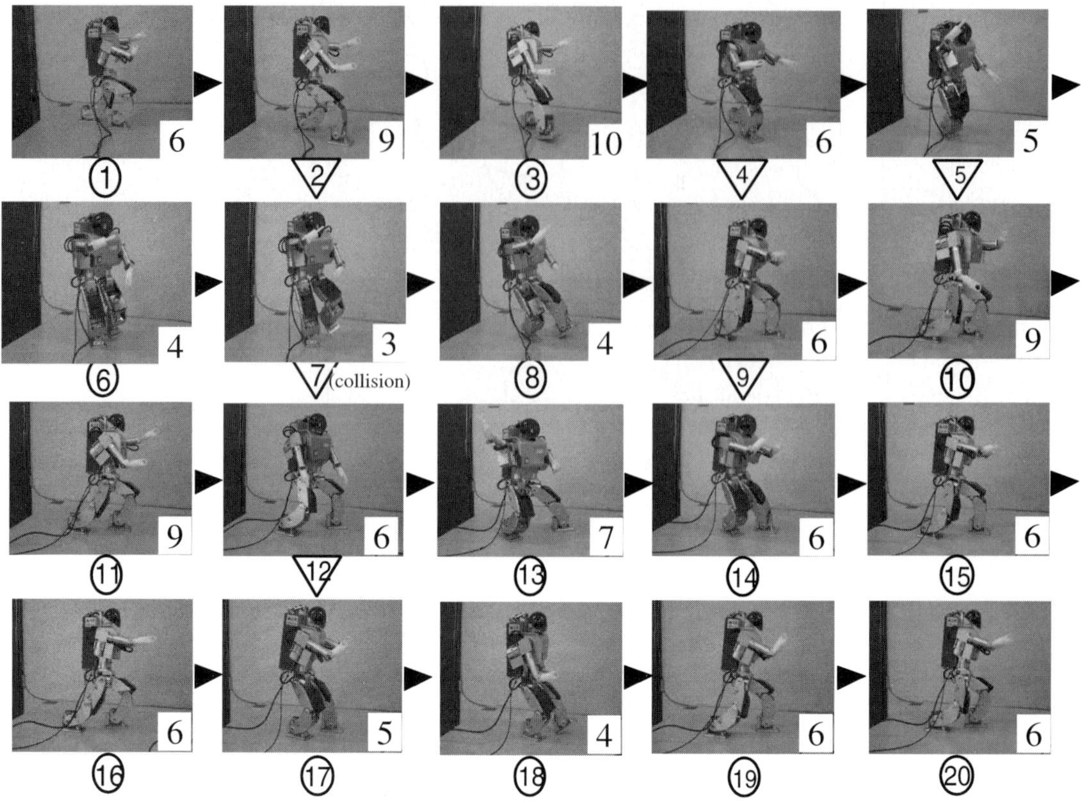

Figure 5: The snapshot of Tai Chi #7 motion by HOAP-1.

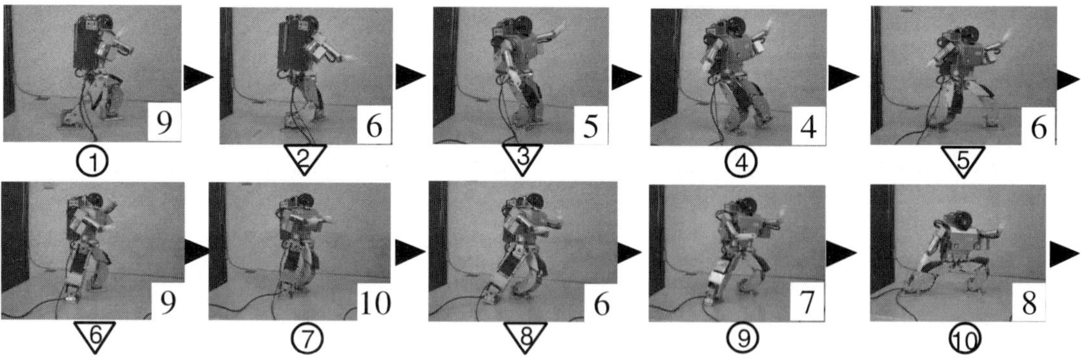

Figure 6: The snapshot of Tai Chi #17 motion by HOAP-1.

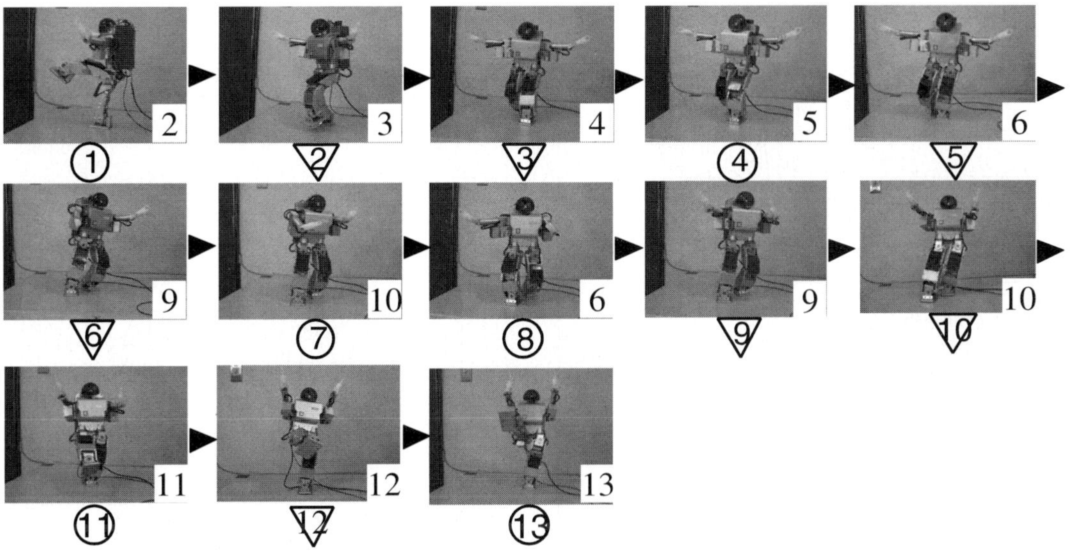

Figure 7: The snapshot of Tai Chi #44 motion by HOAP-1.

4　EXPERIMENT

We have produced all the motions of Tai Chi Chuan without tipping over. Firstly for data input to our motion generation system, we have prepared key-postures for 48 Tai Chi motions, from a tutorial book (Defang, 1999). The system has, secondly, generated the control sequences of servomotors for all the motions. In this particular examples, all motions are well performed by HOAP-1. Table 1 and 2 show the listing of Tai Chi motions and the numerical relation between input and output postures by our system.

4.1　Performance Results

Figure 5, 6 and 7 show the snapshots of Tai Chi #07, #17 and #44 motion by HOAP-1, respectively. Tai Chi #7 composed of the basic walking motion, called *shang bu*. This is easy one in Tai Chi, however, it occurs tipping over without our system. Tai Chi #17 composed of the motion keeping robot's head low, called *pu hu*. This motion HOAP is imbalance in backward and forward movement. Tai Chi #44, which contains high kick called *bai lian*, requires high skill balance control of robot. This is one of the most difficult motions. HOAP-1 can support its weight with one leg skillfully, although the imbalance becomes very high. In these figures, the number in the lower right of the each snapshot means the class which the posture belongs to. The transition of classes satisfies all constraints by the directed graph shown in Figure 4.

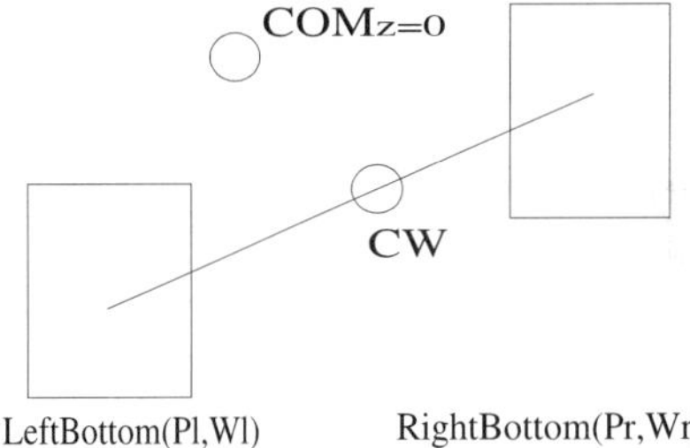

Figure 8: Components for balance quantification.

4.2　Effectiveness of Balance Checker

To verify the effectiveness of our system, we have evaluate the balancing performance for the motions. In this paper, we suppose the quantity of balance as follows. Let COM be the center of mass of whole robot, let COM_z be the vertical component of COM, and let $COM_{z=0}$ be the projection of COM on the floor. Further, let P_l and P_r be the position of the left and right sole, and let W_l and W_r be the weight on the left and right sole, respectively (see Figure 8). The balance is quantified as the following equation:

$$imbalance = |COM_{z=0} - CW| \cdot COM_z, \quad (1)$$

where

$$CW = \frac{Pr \cdot Wr + Pl \cdot Wl}{Wr + Wl}. \quad (2)$$

Figure 9, 10 and 11 show the trajectories *imbalance* of two Tai Chi by HOAP-1: our method and without *Balance-Checker* for Tai Chi #7, #17 and

Table 1: Key Postures of Tai Chi Generated by Our System.

#	name	postures after balance checker	postures by tutorial book
(0)	qi shi	4	4
1	bai he liang chi	5	4
2	zuo lou xi ao bu	4	4
3	zuo dan bian	20	10
4	zuo pi pa shi	4	3
5	lu ji shi	26	15
6	zuo ban lan chui	13	7
7	zuo peng lu ji an	19	11
8	xie shen kao	8	4
9	zhou di chui	11	7
10	dao juan gong	29	12
11	zhuan shen tui zhang	18	13
12	you pi pa shi	4	3
13	lou xi zai chui	9	6
14	bai she tu xin	9	6
15	pai jiao fu hu	31	18
16	zuo pie shen chui	11	5
17	chuan quan xia shi	9	4
18	du li cheng zhang	14	6
19	you dan bian	20	10
20	you yun shou	25	12
21	you zuo fen zong	13	7
22	gao tan ma	5	3
23	you deng jiao	10	6
24	shuang feng guan er	5	3

Table 2: Key Postures of Tai Chi (continued).

#	name	postures after balance checker	postures by tutorial book
25	zuo deng jiao	8	4
26	yan shou liao quan	7	3
27	hai di zhen	6	3
28	shan tong bei	5	2
29	you zuo fen jiao	21	9
30	lou xi ao bu	15	7
31	shang bu qin da	10	4
32	ru feng shi bi	7	4
33	zuo yun shou	25	11
34	you pie hen chui	11	5
35	zuo you chuan suo	33	16
36	tui bu chuan zhang	5	2
37	xu bu ya zhang	6	2
38	du li tuo zhang	2	1
39	ma bu kao	9	3
40	zhuan shen da lu	9	5
41	liao zhang xia shi	11	7
42	shang bu qi xing	4	2
43	du li kua hu	7	4
44	zhuan shen bai lian	12	4
45	wan gong she hu	6	4
46	you ban lan chui	16	7
47	you peng lu ji an	19	11
48	shi zi shou	5	3
(49)	shou shi	3	3

#44 motions, respectively. The result indicates that our system can generate a stable Tai Chi motion for HOAP-1, while the robot loses the balance without our balance checker. In the figures, the numbers over the graph correspond to the frame labeled the same number in snapshots shown in Figure 5, 6 and 7. All motions without balance check tipped over, and after that, the imbalance is not accurate because sensors are not calibrated.

5   CONCLUSION

In this paper, we proposed a motion generation system for humanoid robot. The motion generated by our system is smooth and stable in the balance. Humanoid robot, HOAP-1 with our system has performed whole 48 Tai Chi motions in good balance. Our system performs various and unrestricted motions for humanoid robots without hard problem in dynamics. Our system still has some constraints, that is, platform dependent, interpolation interval, and at least one sole on the floor.

The future work will focus on three phases. Firstly, we will deal with the automatic choreographing the postures to insert. Secondly, we will dedicate to the more generalized control of the motion generation. Especially, it is important to let system free from interpolating interval. Our system often makes motions sprawly in time-axis. This is caused by our policy that stability is more important than speed. There are, however, scenes that speed is important, too. We will respond to this tradeoff. Thirdly, we will dedicate to some interpolation methods specialized to Tai Chi Chuan. The value of the Tai Chi motion is not only stability and speed, but also aesthetic sense and smoothness of the motions. In this phase, we will firstly focus on finding ordinality in the rhythm of Tai Chi Chuan.

ACKNOWLEDGEMENTS

This work was supported in part by Artificial Intelligence Research Promotion Foundation and Ministry of Education, Science, Sports and Culture, Grant–in–Aid for Scientific Research under grant #14780307.

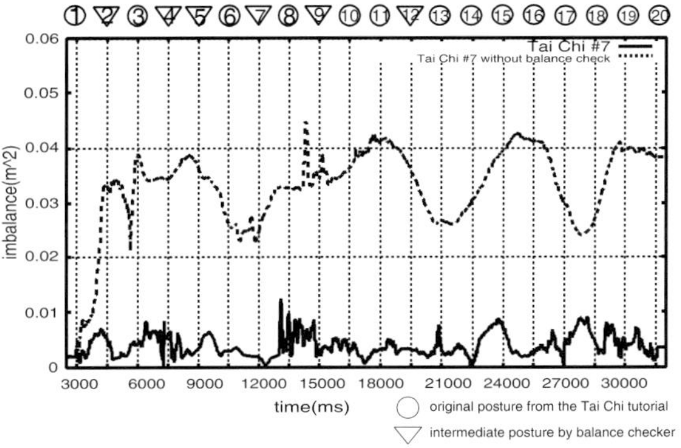

Figure 9: The imbalance of Tai Chi #7.

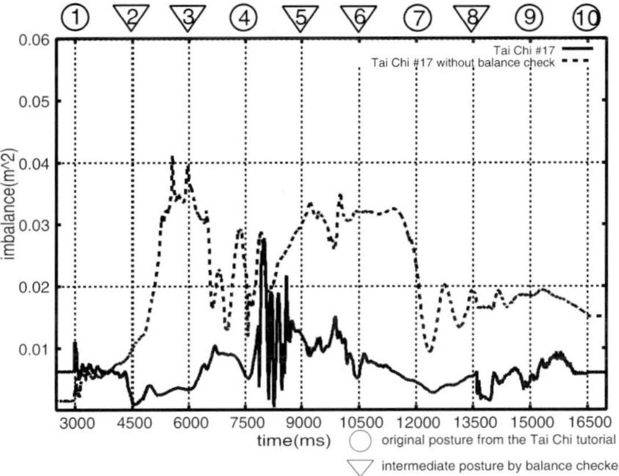

Figure 10: The imbalance of Tai Chi #17.

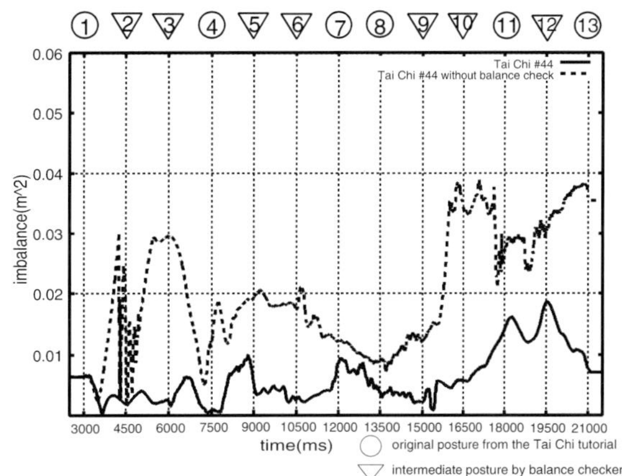

Figure 11: The imbalance of Tai Chi #44.

REFERENCES

Defang, L. (1999). *Introduction to Tai Chi consist of 48 motions*. BAB Japan.

Huang, Q., Yokoi, K., Kajita, S., Kaneko, K., Arai, H., Koyachi, N., and Tanie, K. (2001). Planning walking patterns for a biped robot. *IEEE Transactions on robotics and automation*, 17(3).

Kagami, S., Kanehiro, F., Tamiya, Y., Inaba, M., and Inoue, H. (2001). Autobalancer: An online dynamic balance compensation scheme for humanoid robots. In *Robotics: The Algorithmic Perspective, Workshop on Algorithmic Foundations on Robotics*, pages 329–340.

Kuffner, J. J., Nishiwaki, K., Kagami, S., Inaba, M., and Inoue, H. (2001). Motion planning for humanoid robots under obstacle and dynamic balance constraints. In *Proc. of the IEEE Int'l Conf. on Robotics and Automation (ICRA'2001)*, pages 692–698.

Kuffner, J. J., Nishiwaki, K., Kagami, S., Kuniyoshi, Y., Inaba, M., and Inoue, H. (2002). Self-collision detection and prevention for humanoid robots. In *Proc. of the IEEE Int'l Conf. on Robotics and Automation (ICRA'2002)*, pages 2265–2270.

Kuwayama, K., Kato, S., Seki, H., Yamakita, T., and Itoh, H. (2003). Motion control for humanoid robots based on the concept learning. In *Proc. of International Symposium on Micromechatoronics and Human Science*, pages 259–263.

Li, Q., Takanishi, A., and Kato, I. (1993). Learning control for a biped walking robot with a trunk. In *Proc. of the IEEE/RSJ International Conference on Intelligent Robot and Systems*, page 1771.

Murase, Y., Yasukawa, Y., Sakai, K., and et al. (2001). Design of a compact humanoid robot as a platform. In *Proc. of the 19-th conf. of Robotics Society of Japan*, pages 789–790. (in Japanese), http://pr.fujitsu.com/en/news/2001/09/10.html.

Nishiwaki, K., Kagami, S., Kuniyoshi, Y., Inaba, M., and Inoue, H. (2002). Online generation of humanoid walking motion based on a fast generation method of motion pattern that follows desired zmp. In *Proc. of the 2002 IEEE/RSJ International Conference on Intelligent Robots and Systems*, pages 2684–2689.

Sugihara, T., Nakamura, Y., and Inoue, H. (2002). Realtime humanoid motion generation through zmp manipulation based on inverted pendulum control. In *Proc. of IEEE International Conference on Robotics and Automation (ICRA2002)*, volume 2, pages 1404–1409, Washington D.C., U.S.A.

Yamaguchi, J., Takanishi, A., and Kato, I. (1993). Development of biped walking robot compensating for three-axis moment by trunk motion. In *Proc. of the IEEE/RSJ International Conference on Intelligent Robot and Systems*, page 561.

REALISTIC DYNAMIC SIMULATION OF AN INDUSTRIAL ROBOT WITH JOINT FRICTION

Ronald G.K.M. Aarts, Ben J.B. Jonker
University of Twente, Faculty of Engineering Technology
Enschede, The Netherlands
Email: R.G.K.M.Aarts@utwente.nl, J.B.Jonker@utwente.nl

Rob R. Waiboer
Netherlands Institute for Metals Research
Delft, The Netherlands
Email: R.R.Waiboer@utwente.nl

Keywords: Realistic closed-loop trajectory simulation, Industrial robot, Perturbation method, Friction modelling.

Abstract: This paper presents a realistic dynamic simulation of the closed-loop tip motion of a rigid-link manipulator with joint friction. The results from two simulation techniques are compared with experimental results. A six-axis industrial Stäubli robot is modelled. The LuGre friction model is used to account both for the sliding and pre-sliding regime. The manipulation task implies transferring a laser spot along a straight line with a trapezoidal velocity profile.

Firstly, a non-linear finite element method is used to formulate the dynamic equations of the manipulator mechanism. In a closed-loop simulation the driving torques are generated by the control system. The computed trajectory tracking errors agree well with the experimental results. Unfortunately, the simulation is very time-consuming due to the small time step of the discrete-time controller.

Secondly, a perturbation method has been applied. In this method the perturbed motion of the manipulator is modelled as a first-order perturbation of the nominal manipulator motion. Friction torques at the actuator joints are introduced at the stage of perturbed dynamics. A substantial reduction of the computer time is achieved without loss of accuracy.

1 INTRODUCTION

Robotic manipulators for laser welding must provide accurate path tracking performance of 0.1 mm and less at relatively high tracking speed exceeding 100 mm/s. High speed manipulator motions are accompanied by high-frequency vibrations of small magnitudes whereas velocity reversals in the joints lead to complicated joint friction effects. These dynamic phenomena may significantly effect the weld quality, since laser welding is sensitive to small variations in the processing conditions like welding speed as well as to disturbances of the position of the laser spot with respect to the seam to be welded. To study the applicability of industrial robotic manipulators as in Fig. 1 for laser welding tasks, a framework for realistic dynamic simulations of the robot motion is being developed. Such a simulation should predict the path tracking accuracies, thereby taking into account the dynamic limitations of the robotic manipulator such as joint flexibility, friction and limits of the drive system like the maximum actuator torques. Obviously the accuracy of the simulation has to be sufficiently high with respect to the allowable path deviations.

Unfortunately, closed-loop dynamic simulations with a sufficient degree of accuracy are computationally expensive and time consuming especially when static friction, high-frequency elastic modes and a digital controller operating with small discrete time steps are involved. In an earlier paper we discussed a so-called perturbation method (Jonker and Aarts, 2001) which proved to be accurate and computationally efficient for simulations of robotic manipulators with flexible links. In this paper a similar perturbation scheme is presented for the closed-loop dynamic simulation of a rigid-link manipulator with joint friction. The extension to an industrial robot with elastic joints is still a topic of ongoing research.

We consider the motion of a six axes industrial robot (Stäubli RX90B), Fig. 1. The unknown robot manipulator parameters, such as inertias of the manipulator arms, the stiffness and the pretension of a gravity compensation spring are determined by means of parameter identification techniques. The friction characteristics are identified separately. The controller model is based on manufacturer's data and has been verified experimentally using system identification techniques. Details of the identification tech-

J. Braz et al. (eds.), Informatics in Control, Automation and Robotics I, 187–194.

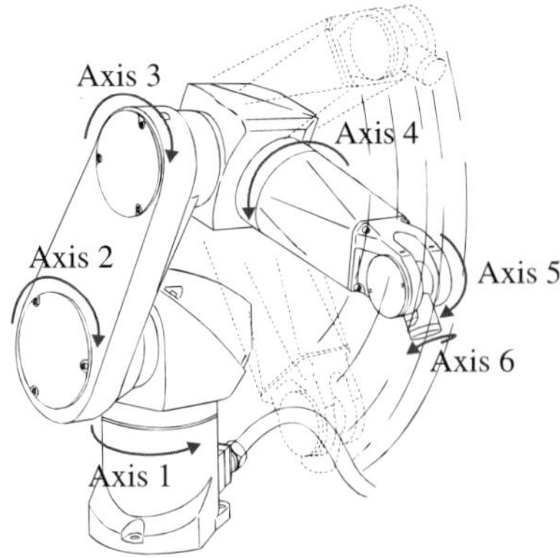

Figure 1: The six axes Stäubli RX90B Industrial robot.

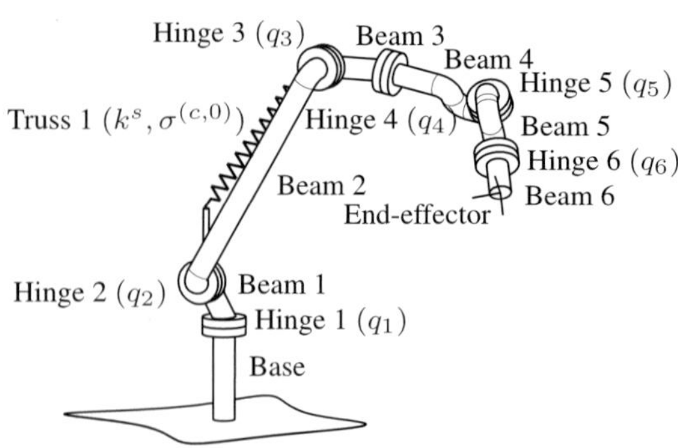

Figure 2: The 6 DOF Finite Element Model.

niques are outside the scope of this paper and will be outlined elsewhere (Waiboer, 2004).

The present paper will focus on two simulation techniques and a comparison of numerical and experimental results. At first, a non-linear finite element method (Jonker, 1990) is used to formulate the dynamic equations of the manipulator mechanism, section 2. The finite element formulation and the proposed solution method are implemented in the program SPACAR (Jonker and Meijaard, 1990). An interface to MATLAB is available and the closed-loop simulations are carried out using SIMULINK's graphical user interface. A driving system is added to the manipulator model, section 3, and the influence of joint friction is taken into account, section 4. The LuGre friction model (Wit et al., 1995) has been used as it accounts for both the sliding and pre-sliding regimes. Finally the closed-loop model is assembled by including the control system, section 5.

The second approach is the so-called perturbation method, section 6. The differences between the actual manipulator motion and the nominal (desired) motion are modelled as first order perturbations of that nominal motion. The computation of the perturbed motion is carried out in two steps. In the first step, the finite element method permits the generation of locally linearised models (Jonker and Aarts, 2001). In the second step, the linearised model can simulate the perturbed motion of the manipulator accurately as long as it remains sufficiently close to the nominal path. The perturbation method is computationally more efficient as the non-linear model only has to solved during the first step in a limited number of points along the nominal trajectory. This is the solution of an inverse dynamic analysis along a known path, so with only algebraic equations. The friction torques at the actuator joints and the control system are introduced at the stage of perturbed dynamics in the second step. In section 7, the simulation results will be compared with the experimental results. Also results acquired with the perturbation method will be compared with those obtained from a full non-linear dynamic simulation.

2 FINITE ELEMENT REPRESENTATION OF THE MANIPULATOR

In the non-linear finite element method, a manipulator mechanism is modelled as an assembly of finite elements interconnected by joint elements such as hinge elements and (slider) truss elements. This is illustrated in Fig. 2, where the Stäubli robot with six degrees of freedom is modelled by three different types of elements. The gravity compensation spring is modelled as a slider-truss element. The manipulator arms are modelled by beam elements. Finally, the joints are represented by six cylindrical hinge elements, which are actuated by torque servos. The manipulator mechanism is assembled by allowing the elements to have nodal points in common. The configuration of the manipulator is then described by the vector of nodal coordinates x, some of which may be the Cartesian coordinates, while others describe the orientation of orthogonal triads rigidly attached at the element nodes. The motion of the manipulator mechanism is described by relative degrees of freedom which are the actuator joint angles denoted by the vector q. By means of the geometric transfer functions $\mathcal{F}^{(x)}$ and $\mathcal{F}^{(e,c)}$, the nodal coordinates x and the elongation of the gravity compensating spring $e^{(c)}$ are expressed as functions of the joint angles q

$$x = \mathcal{F}^{(x)}(q) \tag{1}$$

and

$$e^{(c)} = \mathcal{F}^{(e,c)}(q). \tag{2}$$

Differentiating the transfer functions with respect to time gives

$$\dot{x} = D\mathcal{F}^{(x)}\dot{q} \tag{3}$$

and

$$\dot{e}^{(c)} = D\mathcal{F}^{(e,c)}\dot{q}, \tag{4}$$

where the differentiation operator D represents partial differentiation with respect to the degrees of freedom. The acceleration vector $\ddot{x}$ is obtained by differentiating (3) again with respect to time,

$$\ddot{x} = D\mathcal{F}^{(x)}\ddot{q} + (D^2\mathcal{F}^{(x)}\dot{q})\dot{q}. \tag{5}$$

The geometric transfer functions $\mathcal{F}^{(x)}$ and $\mathcal{F}^{(e,c)}$ are determined numerically in an iterative way (Jonker, 1990).

The inertia properties of the concentrated and distributed mass of the elements are described with the aid of lumped mass matrices. Let $M(x)$ be the global mass matrix, obtained by assembling the mass matrices of the individual elements and let $f(x, \dot{x}, t)$ be the global force vector including gravity forces and the velocity dependent inertia forces. Then the equations of motion described in the dynamic degrees of freedom are given by

$$\bar{M}\ddot{q} + D\mathcal{F}^{(x)T}\left[M(D^2\mathcal{F}^{(x)}\dot{q})\dot{q} - f\right] \\ + D\mathcal{F}^{(e,c)T}\sigma^{(c)} = \tau, \tag{6}$$

where $\bar{M} = D\mathcal{F}^{(x)T}MD\mathcal{F}^{(x)}$ is the reduced mass matrix and $\sigma^{(c)}$ is the total stress in the gravity compensating spring. The vector τ represents the joint driving torques. The constitutive relation for the gravity compensating spring is described by

$$\sigma^{(c)} = \sigma^{(c,0)} + k^s e^{(c)}, \tag{7}$$

where $\sigma^{c,0}$ and k^s denote the pre-stress and the stiffness coefficients of the spring, respectively.

This finite element method has been implemented in the SPACAR software program (Jonker and Meijaard, 1990).

3   THE DRIVING SYSTEM

The robot is driven by servo motors connected via gear boxes to the robot joints. The inputs for the drive system are the servo currents i_j and the outputs are the joint torques τ_j, where the subscript $j = 1..6$ denotes either the motor or the joint number. A schematic model of the drives of joints 1 to 4 is shown in Fig. 3. The relations between the vector of motor angles φ, the vector of joint angles q and their time derivatives are given by

$$\varphi = Tq, \tag{8}$$
$$\dot{\varphi} = T\dot{q}, \tag{9}$$
$$\ddot{\varphi} = T\ddot{q}, \tag{10}$$

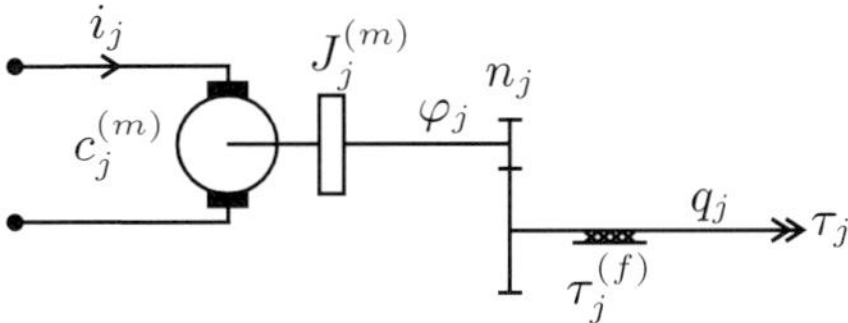

Figure 3: Schematic representation of the drives for joints 1 to 4.

where T is the transmission model for the joints. For joints $j = 1..5$ only the respective gear ratio n_j of the joint plays a role. The drives for the wrist of the robot, drives 5 and 6, are coupled. This causes an extra term leading to

$$T = \begin{bmatrix} n_1 & 0 & 0 & 0 & 0 & 0 \\ 0 & n_2 & 0 & 0 & 0 & 0 \\ 0 & 0 & n_3 & 0 & 0 & 0 \\ 0 & 0 & 0 & n_4 & 0 & 0 \\ 0 & 0 & 0 & 0 & n_5 & 0 \\ 0 & 0 & 0 & 0 & n_6 & n_6 \end{bmatrix}. \tag{11}$$

The friction torque in each joint is denoted $\tau_j^{(f)}$ with $j = 1..6$. Due to the coupling in the wrist, an additional friction torque $\tau_7^{(f)}$ is identified on the axis of motor 6. The vector of joint friction torques is then defined as

$$\tau^{(f)} = \left[\tau_1^{(f)}, \tau_2^{(f)}, \tau_3^{(f)}, \tau_4^{(f)}, \cdots \right. \\ \left. \tau_5^{(f)} + \tau_7^{(f)}, \tau_6^{(f)} + \tau_7^{(f)}\right]^T. \tag{12}$$

The joint friction modelling is continued in section 4.

The drives are equipped with permanent magnet, three-phase synchronous motors, yielding a linear relation between motor currents and torque. The vector of joint driving torques τ is then given by

$$\tau = T^T C^{(m)} i - T^T J^{(m)} \ddot{\varphi} - \tau^{(f)}, \tag{13}$$

where the matrices $C^{(m)}$ and $J^{(m)}$ are diagonal matrices with the motor constants $c_j^{(m)}$ and rotor inertias $J_j^{(m)}$, respectively. Substitution of (10) and (13) into (6) and some rearranging yields

$$\bar{M}^{(n)}\ddot{q} + D\mathcal{F}^{(x)T}\left[M(D^2\mathcal{F}^{(x)}\dot{q})\dot{q} - f\right] \\ + D\mathcal{F}^{(e,c)T}\sigma^{(c)} = \tau^{(n)}, \tag{14}$$

where the mass matrix $\bar{M}^{(n)}$ is defined by

$$\bar{M}^{(n)} = \bar{M} + T^T J^{(m)} T, \tag{15}$$

as the rotor inertias $T^T J^{(m)} T$ obviously have to be added to the reduced mass matrix $\bar{M}$ in the equations of motion (6). Furthermore, the vector of net joint torques is defined as

$$\tau^{(n)} = T^T C^{(m)} i - \tau^{(f)}. \tag{16}$$

The inertia properties and spring coefficients have been found by means of parameter identification techniques (Waibocr, 2004).

4   JOINT FRICTION MODEL

For feed-forward dynamic compensation purposes and robot inertia identification techniques it is common to model friction in robot joints as a torque $\tau^{(f)}$ which consists of a Coulomb friction torque and an additional viscous friction torque for non-zero velocities (Swevers et al., 1996; Calafiore et al., 2001). These so-called static or kinematic friction models are valid only at sufficiently high velocities because they ignore the presliding regime. At zero velocity they show a discontinuity in the friction torque which gives rise to numerical integration problems in a forward dynamic simulation. To avoid this problem we apply the LuGre friction model (Wit et al., 1995), that accounts for both the friction in the sliding and in the presliding regime.

In the LuGre friction model there is an internal state z that describes the average presliding displacement, as introduced by Heassig et al. (Haessig and Friedland, 1991). The state equations with a differential equation for the state z and an output equation for the friction moment $\tau^{(f)}$ are

$$\dot{z} = \dot{q} - \frac{|\dot{q}|}{g(\dot{q})} z, \tag{17}$$

$$\tau^{(f)} = c^{(0)} z + c^{(1)} \dot{z} + c^{(2)} \dot{q}. \tag{18}$$

Note that a subscript $j = 1 \ldots 7$ should be added to all variables to distinguish between the separate friction torques in the robot model, but it is omitted here for better readability. For each joint friction model with $j = 1..6$ the input velocity equals the joint velocity $\dot{q}_j$. For the extra friction model $\tau_7^{(f)}$ the input velocity is defined as the sum of the joint velocities of joints 5 and 6.

In general, the friction torque in the pre-sliding regime is described by a non-linear spring-damper system that is modelled with an equivalent stiffness $c^{(0)}$ for the position–torque relationship at velocity reversal and a micro-viscous damping coefficient $c^{(1)}$. At zero velocity, the deformation of the non-linear spring torque is related to the joint (micro) rotation q.

The viscous friction torque in the sliding regime is modelled by $c^{(2)} \dot{q}$, where $c^{(2)}$ is the viscous friction coefficient. In addition, at a non-zero constant velocity $\dot{q}$, the internal state z, so the average deformation, will approach a steady state value equal to $c^{(0)} g(\dot{q})$. The function $g(\dot{q})$ can be any function that represents the constant velocity behaviour in the sliding regime. In this paper the Stribeck model will be used, which models the development of a lubricating film between the contact surfaces as the relative velocity increases from zero. The Stribeck model is given by

$$c^{(0)} g(\dot{q}) = \tau^{(c)} + (\tau^{(s)} - \tau^{(c)})e^{-(|\dot{q}|/\omega_s)^\delta}, \tag{19}$$

where $\tau^{(c)}$ is known as Coulomb friction torque, $\tau^{(s)}$ is the static friction torque and ω_s and δ are shaping parameters. The values for the static friction torque $\tau^{(s)}$ and Coulomb friction torque $\tau^{(c)}$ may be different for positive and negative velocities and are therefore distinguished by the subscripts $+$ and $-$, respectively.

For each friction torque in the robot model, the parameters describing the sliding regime of the LuGre friction model are estimated separately using dedicated joint torque measurements combined with non-linear parameter optimisation techniques (Waiboer, 2004). The parameters describing the presliding regime are approximated by comparing the pre-sliding behaviour in simulation with measurements.

5   CLOSED-LOOP ROBOT MODEL

The SPACAR model of the manipulator mechanism, the controller, the friction models and the robot drives are assembled into a complete model of the closed-loop robot system, Fig. 4.

The CS7B controller used in the Stäubli RX90B is an industrial PID controller based on the Adept V+ operating- and motion control system. The controller includes six SISO controllers, one for each servo-motor. A trajectory generator computes the set-points $\ddot{q}^{(r)}$, $\dot{q}^{(r)}$ and $q^{(r)}$ at a rate of 1 kHz. The actual joint position q is compared with the set-point $q^{(r)}$ and the error ϵ is fed into the PI-block of the controller, which includes the proportional part and the integrator part. The acceleration set-points $\ddot{q}^{(r)}$ and velocity set-points $\dot{q}^{(r)}$ are used for acceleration feed-forward and friction compensation, both included in the FF-block. The friction compensation includes both Coulomb and viscous friction. The velocity feedback $\dot{q}$ is fed into the D-block, representing the differential part of the control scheme. The block C represents the current loop, including the power amplifier. Note that the position loop runs at $f_s^{(1)} = 1$ kHz and the velocity loop and current loop run at $f_s^{(2)} = 4$ kHz. The transfer functions of the different blocks have been identified and implemented in a SIMULINK block scheme of the robot controller (Waiboer, 2004).

The current vector i is fed into the drive model where the joint torques are computed. The LuGre friction models compute the friction torques from the joint velocities $\dot{q}$. The net joint torque $\tau^{(n)}$ in (14) is the input for the non-linear manipulator model (SPACAR). The output of the manipulator model contains the joint positions, and velocities, denoted by q and $\dot{q}$, respectively.

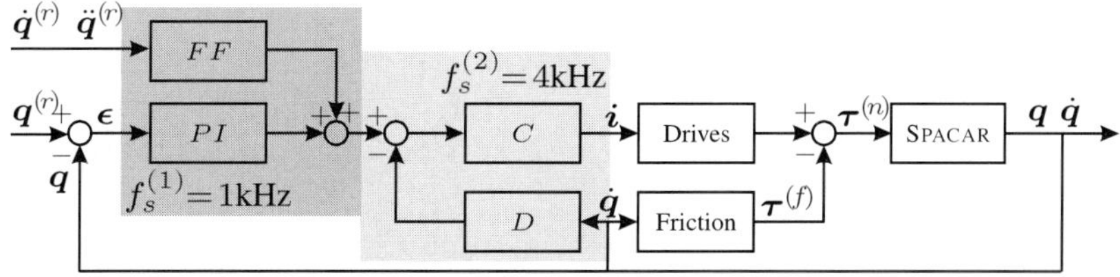

Figure 4: Assembly of the closed-loop robot model for a simulation with the non-linear finite element method (SPACAR block).

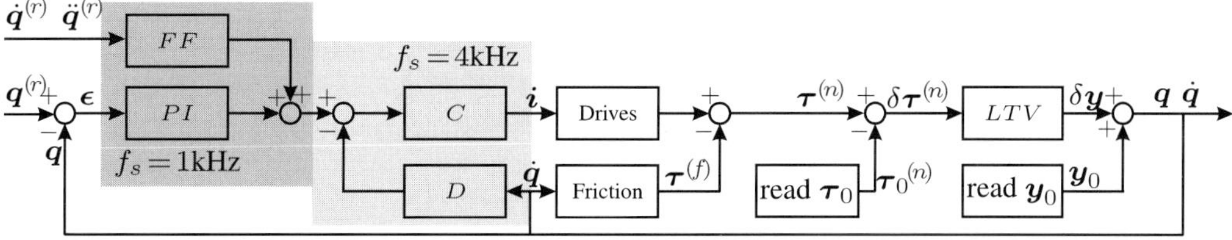

Figure 5: Perturbation scheme with the linearised equations of motion (LTV block).

6 PERTURBATION METHOD

In the presented closed-loop model, both the SPACAR model and the friction model are continuous in time, while the controller sections are discrete and running at 1 kHz and 4 kHz, respectively. In SIMULINK this implies that the SPACAR model is simulated at a time step equal to 0.25 ms or even smaller if this is required by SIMULINK's integrator. Due to the fact that the position, computed within the SPACAR simulation model, is obtained in an iterative way, the computation effort is quite elaborate. This results in long simulation times. In order to overcome this drawback, the perturbation method (Jonker and Aarts, 2001) is applied.

In the perturbation method, deviations from the (nominal) desired motion $(q_0, \dot{q}_0, \ddot{q}_0)$ due to joint friction and errors of the control system are modelled as first-order perturbations $(\delta q, \delta \dot{q}, \delta \ddot{q})$ of the nominal motion, such that the actual variables are of the form:

$$q = q_0 + \delta q, \tag{20}$$
$$\dot{q} = \dot{q}_0 + \delta \dot{q}, \tag{21}$$
$$\ddot{q} = \ddot{q}_0 + \delta \ddot{q}, \tag{22}$$

where the prefix δ denotes a perturbation. Expanding (1) and (3) in their Taylor series expansion and disregarding second and higher order terms results in the linear approximations

$$\delta x = \mathrm{D}\mathcal{F}_0 \cdot \delta q, \tag{23}$$
$$\delta \dot{x} = \mathrm{D}\mathcal{F}_0 \cdot \delta \dot{q} + (\mathrm{D}^2\mathcal{F}_0 \cdot \dot{q}_0) \cdot \delta q. \tag{24}$$

The subscript 0 refers to the nominal trajectory.

The nominal motion is computed first and is described by the non-linear manipulator model. The non-linear equations of motion (14) for the nominal motion become

$$\bar{\mathrm{M}}_0^{(n)} \ddot{q}_0$$
$$+ \mathrm{D}_q\mathcal{F}_0^T \left[\mathrm{M}_0(\mathrm{D}^2\mathcal{F}_0 \cdot (\dot{q}_0)) \cdot (\dot{q}_0) - f \right] \tag{25}$$
$$+ \mathrm{D}\mathcal{F}_0^{(e,c)} \sigma_0^{(c)} = \tau_0^{(n)},$$

where $\sigma_0^{(c)} = \sigma^{(c,0)} + k^s e_0^{(c)}$. The right hand side vector $\tau_0^{(n)}$ represents the nominal net input torque necessary to move the manipulator along the nominal (desired) trajectory. It is determined from an inverse dynamic analysis based on (25). Note that $\tau_0^{(n)}$ is computed without the presence of friction. Obviously, the desired nominal motion $(\ddot{q}_0, \dot{q}_0, q_0)$ equals the reference motion $(\ddot{q}^{(r)}, \dot{q}^{(r)}, q^{(r)})$ of the closed loop system.

Next the perturbed motion is described by a set of linear time-varying equations of motion, which are obtained by linearising the equations of motion around a number of points on the nominal trajectory. The resulting equations of motion for the perturbations of the degrees of freedom δq are

$$\bar{\mathrm{M}}_0^{(n)} \delta \ddot{q} + \mathrm{C}_0 \delta \dot{q} + \bar{\mathrm{K}}_0 \delta q = \delta \tau^{(n)}, \tag{26}$$

where $\bar{\mathrm{M}}_0^{(n)}$ is the reduced system mass matrix as in (14), C_0 is the velocity sensitivity matrix. Matrix $\bar{\mathrm{K}}_0$ is a shorthand notation for the combined stiffness matrices

$$\bar{\mathrm{K}}_0 = \mathrm{K}_0 + \mathrm{N}_0 + \mathrm{G}_0, \tag{27}$$

where $\mathbf{K}_0$ denotes the structural stiffness matrix, $\mathbf{N}_0$ and $\mathbf{G}_0$ are the dynamic stiffening matrix and the geometric stiffening matrix, respectively. The matrices $\bar{\mathbf{M}}_0^{(n)}$, $\mathbf{K}_0$ and $\mathbf{G}_0$ are symmetric, but $\mathbf{C}_0$ and $\mathbf{N}_0$ need not to be symmetrical. Expressions for these matrices are given in (Jonker and Aarts, 2001).

Fig. 5 shows the block scheme of the closed-loop perturbation model. Note that the non-linear manipulator model (SPACAR) is replaced by the LTV-block which solves (26). Its input $\delta\boldsymbol{\tau}^{(n)}$ is computed according to

$$\delta\boldsymbol{\tau}^{(n)} = \boldsymbol{\tau}^{(n)} - \boldsymbol{\tau}_0^{(n)}, \tag{28}$$

where $\boldsymbol{\tau}^{(n)}$ represents the vector of the net joint torques as defined by (16). The perturbed output vector $\delta\boldsymbol{y}$ is added to the output vector $\boldsymbol{y}_0$ of the nominal model, yielding the actual output $\boldsymbol{y}$. This output vector will be discussed below.

Within the SIMULINK framework the LTV block represents a linear time-varying state space representation of a system

$$\begin{aligned}\dot{\boldsymbol{x}}_{\mathrm{ss}} &= \mathbf{A}_{\mathrm{ss}}\boldsymbol{x}_{\mathrm{ss}} + \mathbf{B}_{\mathrm{ss}}\boldsymbol{u}_{\mathrm{ss}} \\ \boldsymbol{y}_{\mathrm{ss}} &= \mathbf{C}_{\mathrm{ss}}\boldsymbol{x}_{\mathrm{ss}}\end{aligned} \tag{29}$$

where $\boldsymbol{u}_{\mathrm{ss}}$, $\boldsymbol{y}_{\mathrm{ss}}$ and $\boldsymbol{x}_{\mathrm{ss}}$ are the input, output and state vectors, respectively, and $\mathbf{A}_{\mathrm{ss}}$, $\mathbf{B}_{\mathrm{ss}}$ and $\mathbf{C}_{\mathrm{ss}}$ are the time-varying state space matrices. Equation (26) is written in this representation by defining the state and input vectors

$$\begin{aligned}\boldsymbol{x}_{\mathrm{ss}} &= [\delta\dot{\boldsymbol{q}}, \delta\boldsymbol{q}]^T, \\ \boldsymbol{u}_{\mathrm{ss}} &= \delta\boldsymbol{\tau}^{(n)}.\end{aligned} \tag{30}$$

The state space matrices $\mathbf{A}_{\mathrm{ss}}$ and $\mathbf{B}_{\mathrm{ss}}$ are then computed from the matrices in (26) using a straightforward procedure :

$$\begin{aligned}\mathbf{A}_{\mathrm{ss}} &= \begin{bmatrix} \mathbf{0} & \mathbf{I} \\ -\bar{\mathbf{M}}_0^{(n)^{-1}}\mathbf{K}_0 & -\bar{\mathbf{M}}_0^{(n)^{-1}}\mathbf{C}_0 \end{bmatrix}, \\ \mathbf{B}_{\mathrm{ss}} &= \begin{bmatrix} \mathbf{0} \\ \bar{\mathbf{M}}_0^{(n)^{-1}} \end{bmatrix}.\end{aligned} \tag{31}$$

The output matrix $\mathbf{C}_{\mathrm{ss}}$ depends on the output vector $\boldsymbol{y}_{\mathrm{ss}}$ that is defined by the user. It may consist of (first derivatives of) the degrees of freedom $\delta\boldsymbol{q}$ present in $\boldsymbol{x}_{\mathrm{ss}}$ or (first derivatives of) nodal coordinates $\delta\boldsymbol{x}$ that are computed from $\boldsymbol{x}_{\mathrm{ss}}$ using (23) and (24). E.g. the coordinates $\delta\boldsymbol{x}$ in $\boldsymbol{y}_{\mathrm{ss}}$ are computed using

$$\mathbf{C}_{\mathrm{ss}} = [\ \mathbf{D}\mathcal{F}_0 \quad \mathbf{0}\ ]. \tag{32}$$

For a practical implementation of the perturbation method the total trajectory time T is divided into intervals. The state space matrices and the vectors $\boldsymbol{\tau}_0$ and $\boldsymbol{y}_0$ are computed during a preprocessing run at a number N of discrete time steps $t = t_i$ $(i = 0, 1, 2, ..., N)$ and stored in a file. During the

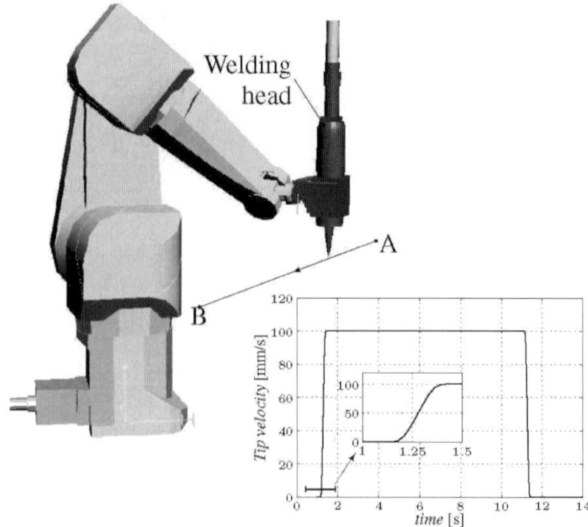

Figure 6: The simulated straight line path and the path velocity.

simulation runs the matrices and vectors are then read back from this file. Between two time steps the coefficients of the state space matrices are interpolated linearly. Cubic interpolation is used for the vectors $\boldsymbol{\tau}_0$ and $\boldsymbol{y}_0$.

7   SIMULATION RESULTS

In Fig. 6 an experiment is defined in which a straight line motion of the robot tip has to be performed in the horizontal plane with a velocity profile as presented in the insert. The welding head remains in the vertical position as shown in Fig. 6. The joint set-points are computed with the inverse kinematics module of the SPACAR software. The joint velocities are shown in Fig. 7. The figures show clearly the non-linear behaviour of the robot. Note the velocity reversal of joints 2 and 3 nearly halfway the trajectory.

First, the trajectory will be simulated with the non-linear model. The simulation has been performed in SIMULINK using the `ode23tb` integration scheme, which is one of the stiff integrators in SIMULINK. The stiff integration scheme is needed due to the high equivalent stiffness in the pre-sliding part of the friction model (18).

The path performance as predicted by the model will be compared to measurements carried out on the actual robot. In both the measurements and simulations, the path position is based on joint axis positions and the nominal kinematic model. In Fig. 8, the horizontal and vertical path deviations are shown. The results show that the agreement between the simulated prediction and the measurement is within 20 μm

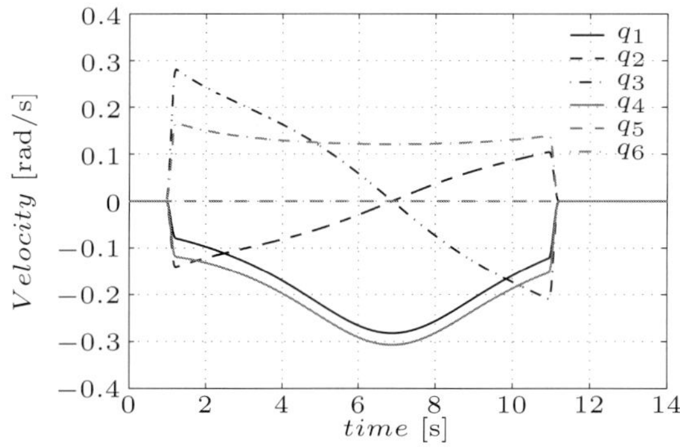

Figure 7: The joint velocity.

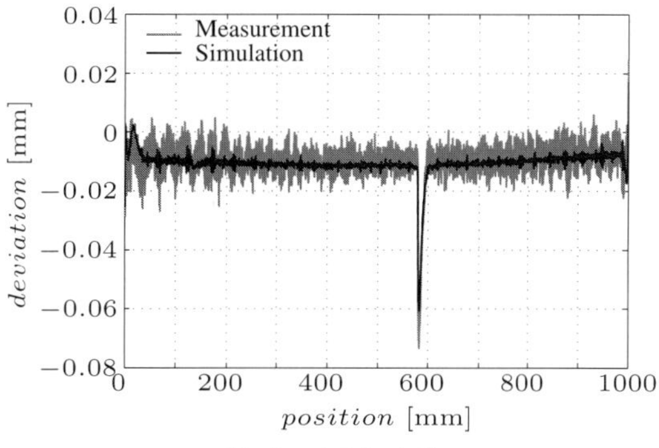

a. Horizontal deviation

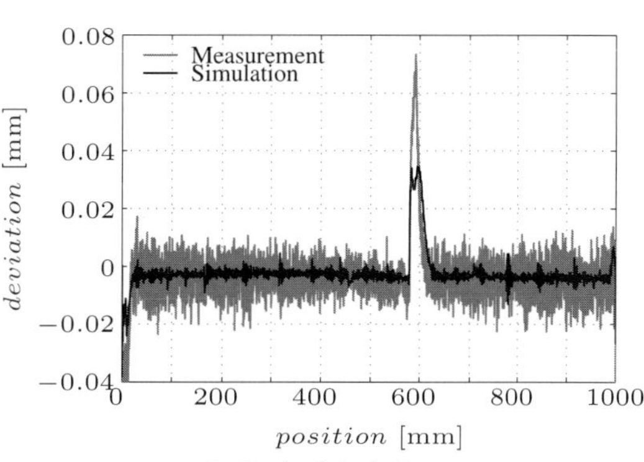

b. Vertical deviation

Figure 8: Simulated and measured path deviation.

which is well within the range of accuracy required for laser welding. At about 600 mm on the trajectory, at the velocity reversal in joints 2 and 3, there is a jump of the friction force, resulting in a relatively large path error.

During the experiment the motor currents were recorded. In Fig. 9 the measured motor currents of joint 3 is compared with the simulated motor current as an example. The simulation shows good agreement with the measurements. It can be observed that the motor currents are predicted quite accurately in the sliding regime and near the velocity reversal. However, the model is unable to predict the stick-torques with a high level of accuracy. This can be explained by the fact that the stiction torque can be anywhere between $\tau_-^{(s)}$ and $\tau_+^{(s)}$ over a very small position range of tenths of μrad, due to the high equivalent stiffness $c^{(0)}$ in (18).

The required simulation time with the non-linear simulation model was about 70 minutes on a 2.4 GHz Pentium 4 PC, which indicates that it is computationally quite intensive. This is caused by the small time steps at which the manipulator configuration, see (1) and (2), needs to be solved by an iterative method. To overcome this drawback, the perturbation approach will be used next.

For the perturbation method the number of points along the trajectory N at which the system is linearised is taken to be 400. The simulation results of the perturbation method and the non-linear simulation have been compared. Fig 10 shows the path accuracy along the trajectory. There is no noticeable difference between the results at this scale. The error stays well below 10 μm and is mostly less than 2 μm. Also the agreement between the simulated motor currents for both the non-linear simulation and the perturbation method is good, especially when the manipulator is in motion. At rest, the torques in the stick or pre-sliding regime shows some differences as before. A significant reduction of simulation time by about a factor 10 has been achieved.

8   CONCLUSIONS

In this paper a realistic dynamic simulation of a rigid industrial robot has been presented. The following components are essential for the closed-loop simulation:

- A finite element-based robot model has been used to model the mechanical part of the robot. Robot identification has been applied to find the model parameters accurately.

- Furthermore, joint friction has been modelled using the LuGre friction model.

- Finally, a controller model has been identified and implemented.

The complete closed-loop model has been used for the simulation of the motion of the robot tip along a prescribed trajectory. The simulation results show good agreement with the measurements. It was found that a closed-loop simulation with the non-linear robot model was very time consuming due to the small time step imposed by the discrete controller. Application of the perturbation method with the linearised model yields a substantial reduction in computer time without any loss of accuracy. The non-linear finite

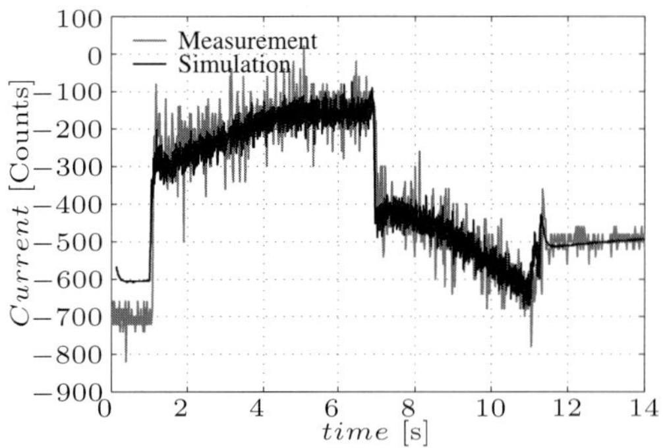

Figure 9: Measured and simulated motor current for joint 3.

element model and the perturbation method can relatively easy be extended to account for flexibilities in the robot.

ACKNOWLEDGEMENTS

This work was carried out under the project number MC8.00073 in the framework of the Strategic Research programme of the Netherlands Institute for Metals Research in the Netherlands (http://www.nimr.nl/).

The authors acknowledge the work carried out on the identification of the controller, inertia parameters and friction parameters by W. Holterman, T. Hardeman and A. Kool, respectively.

REFERENCES

Calafiore, G., Indri, M., and Bona, B. (2001). Robot dynamic calibration: Optimal excitation trajectories and experimental parameter estimation. *Journal of Robotic Systems*, 18(2):55–68.

Haessig, D. and Friedland, B. (1991). On the modeling and simulation of friction. *Transactions of the ASME, Journal of Dynamic Systems, Measurement and Control*, 113(3):354–362.

Jonker, J. (1990). A finite element dynamic analysis of flexible manipulators. *International Journal of Robotics Research*, 9(4):59–74.

Jonker, J. and Aarts, R. (2001). A perturbation method for dynamic analysis and simulation of flexible manipulators. *Multibody System Dynamics*, 6(3):245–266.

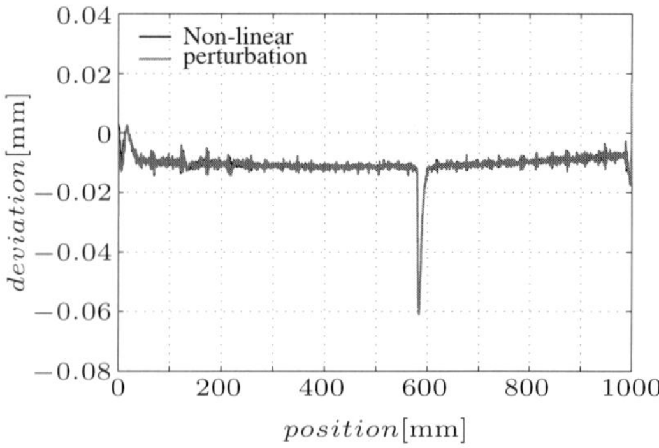

a. Horizontal deviation

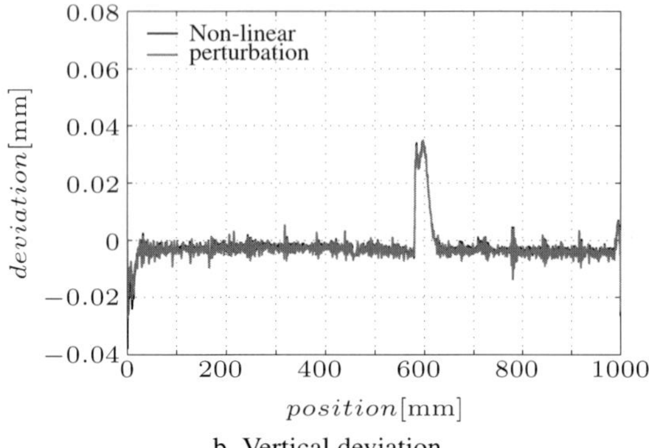

b. Vertical deviation

Figure 10: Simulated path deviation (non-linear versus perturbation method).

Jonker, J. and Meijaard, J. (1990). Spacar-computer program for dynamic analysis of flexible spatial mechanisms and manipulators. In Schiehlen, W., editor, *Multibody systems handbook*, pages 123–143. Springer-Verlag, Berlin.

Swevers, J., Ganseman, C., Schutter, J. D., and Brussel, H. V. (1996). Experimental robot identification using optimised periodic trajectories. *Mechanical Systems and Signal Processing*, 10(5):561–577.

Waiboer, R. R. (to be published, 2004). *Dynamic robot simulation and identification – for Off-Line Programming in robotised laser welding*. PhD thesis, University of Twente, The Netherlands.

Wit, C. d., Olsson, H., Åström, K., and Lischinsky, P. (1995). A new model for control of systems with friction. *IEEE Transactions on Automatic Control*, 40(3):419–425.

A NEW PARADIGM FOR SHIP HULL INSPECTION USING A HOLONOMIC HOVER-CAPABLE AUV

Robert Damus, Samuel Desset, James Morash, Victor Polidoro, Franz Hover, Chrys Chryssostomidis
Sea Grant Autonomous Underwater Vehicles Lab, Massachusetts Institute of Technology, Cambridge, MA, USA
Email: auvs@mit.edu

Jerome Vaganay, Scott Willcox
Bluefin Robotics Corporation, Cambridge, MA, USA
Email: vaganay@bluefinrobotics.com, swillcox@bluefinrobotics.com

Keywords: Autonomous Underwater Vehicle, hovering, feature-relative navigation, inspection.

Abstract: The MIT Sea Grant AUV Lab, in association with Bluefin Robotics Corporation, has undertaken the task of designing a new autonomous underwater vehicle, a holonomic hover-capable robot capable of performing missions where an inspection capability similar to that of a remotely operated vehicle is the primary goal. One of the primary issues in this mode of operating AUVs is how the robot perceives its environment and thus navigates. The predominant methods for navigating in close proximity to large iron structures, which precludes accurate compass measurements, require the AUV to receive position information updates from an outside source, typically an acoustic LBL or USBL system. The new paradigm we present in this paper divorces the navigation routine from any absolute reference frame; motions are referenced directly to the hull. We argue that this technique offers some substantial benefits over the conventional approaches, and will present the current status of our project.

1 INTRODUCTION AND EXISTING CAPABILITIES

The majority of existing autonomous underwater vehicles (AUVs) are of a simple, torpedo-like design. Easy to build and control, the torpedo-shaped AUV has proven useful in many applications where a vehicle needs to efficiently and accurately survey a wide area at low cost. As the field of underwater robotics continues to grow, however, new applications for AUVs are demanding higher performance: in maneuvering, precision, and sensor coverage. In particular, the ability to hover in place and execute precise maneuvers in close quarters is now desirable for a variety of AUV missions. Military applications include hull inspection and mine countermeasures, while the scientific community might use a hovering platform for monitoring coral reefs, exploring the crevices under Antarctic ice sheets, or close-up inspection in deep-sea archaeology. An autonomous hovering platform has great potential for industrial applications in areas currently dominated by work-class remotely operated vehicles (i.e., tethered, ROVs): subsea rescue, intervention, and construction, including salvage and wellhead operations.

Frequent hull inspection is a critical maintenance task that is becoming increasingly important in these security-conscious times. Most ships (whether civilian or military) are only inspected by hand, in dry-dock, and thus rarely - certainly not while they are in active service. Standards do exist for UWILD (Underwater Inspection in Lieu of Drydock), but divers have typically performed underwater inspections, a time-consuming, hazardous job. Additionally, there is a high probability of divers missing something important, because it is so difficult for a human being to navigate accurately over the hull of a ship, with their hands, and often in poor visibility. With a loaded draft on the order of 30m and a beam of 70m for a large vessel, debilitating mines can be as small as 20cm in size, and in this scale discrepancy lies the primary challenge of routine hull inspection.

J. Braz et al. (eds.), Informatics in Control, Automation and Robotics I, 195–200.
© 2006 *Springer. Printed in the Netherlands.*

The simplest inspection is a visual examination of the hull surface. Underwater however, (particularly in harbors and at anchor in coastal waters) a visual inspection must be performed very close to the ship. The health of a ship's skin may also be judged by measuring plating thickness, or checking for chemical evidence of corrosion. For security purposes, a sonar image may be adequate because of larger target size. For instance, the US Customs Service currently uses a towfish sidescan sonar to check hulls (Wilcox, 2003).

Some military vessels are now using small, free-swimming ROVs for in-situ inspection (Harris & Slate, 1999). This method eliminates the safety hazard of diver work, but retains the disadvantage of uncertain navigation and human load. The only commercial hull inspection robot, at the time of this writing, is the Imetrix *Lamp Ray*. *Lamp Ray* is a small ROV designed to crawl over the hull surface. The ROV is deployed from the vessel under inspection; the vehicle swims in and closes with the hull under human control, then holds itself in place using front-mounted thrusters for suction. The operator then drives the ROV over the hull surface on wheels. This limits the survey to flat areas of the hull; more complex geometry around e.g. sonar domes, propeller shafts, etc. must still be visually inspected with a free-swimming ROV. The *Cetus II* AUV is an example of a free-swimming autonomous system that has also conducted ship hull surveys (Trimble & Belcher, 2002). Using altimeters to maintain a constant relative distance from the hull, and the AquaMap long baseline navigation system (DesertStar, Inc.), *Cetus II* records globally-referenced position information, and this (with depth and bearing to the hull) is the primary navigation sensor used to ensure and assess full coverage. The AquaMap system uses a transponder net deployed in the vicinity of the ship being inspected (see URL in References); clearly, a long baseline acoustic system could be used for any vehicle.

Our vehicle program has three unique aspects to address the needs of ship hull inspection: development of a *small* autonomous vehicle optimized for *hovering*, and of a *hull-relative navigation* procedure, wherein dependence on a deployed acoustic navigation system is avoided. The data product this vehicle will produce is a high-resolution sonar mosaic of a ship hull, using the DIDSON imaging sonar (University of Washington's Applied Physics Laboratory) as a nominal payload (Belcher et al., 2003).

2 PHYSICAL VEHICLE OVERVIEW

The hovering AUV (HAUV, Figure 1) has eight hubless, bi-directional DC brushless thrusters, one main electronics housing, and one payload module. The symmetrical placement of the large number of thrusters makes the vehicle agile in responding to wave disturbances, and capable of precise flight maneuvers, such as orbiting targets for inspection or hovering steadily in place. The vehicle is intended to operate in water depths ranging from the Surf Zone (SZ) through Very Shallow Water (VSW) and beyond, up to depths of 100 meters; and to perform in waves up to Sea State Three.

Onboard non-payload instruments include a Doppler velocity log (DVL), inertial measurement unit (IMU), depth sensor, and acoustic modem for supervisory control. While we do carry a magnetic compass, this cannot be expected to work well in close proximity to a metal hull. As noted above, the nominal payload at this writing is the DIDSON imaging sonar. Both the DIDSON and the DVL are mounted on independent pitching servos at the front of the vehicle, because the DIDSON produces good imagery at an incidence angle greater than 45 degrees, while the DVL needs to maintain a normal orientation to the hull. The DVL can also be pointed down for a bottom-locked velocity measurement.

The vehicle is strongly passively stable, with a gravity-buoyancy separation of about 3cm. It has approximate dimensions of 100cm long, 80cm wide, and 30cm tall; it displaces about 45kg. Of this weight, about 12kg are for a 1.5kWh battery.

3 OUR APPROACH TO HULL NAVIGATION

We have chosen to attack this problem from a feature-relative navigation standpoint, as this has some advantages compared to current approaches. Our basic strategy is to measure tangential velocity relative to the hull being inspected using a Doppler velocity log (DVL), and to servo a desired distance from the hull, and orientation, using the individual ranges from acoustic beams.

The immediate impact of this functionality is the elimination of support gear for the robot itself; no localized network setup like LBL is needed. This reduces complexity and provides a simple, quick deployment where the robot can operate unattended; our long-term goal is that the mission focus could

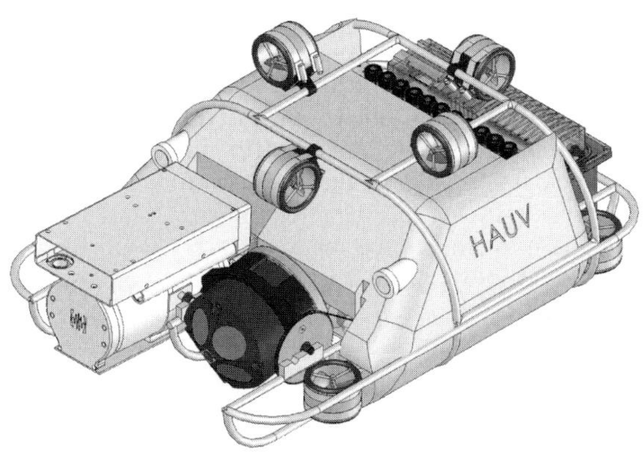

Figure 1: The HAUV, showing DIDSON (light brown) and DVL (dark blue) on the front, yellow flotation in the mid-body, and a large battery at the stern. Thruster locations are reconfigurable; the main electronics housing is underneath the foam.

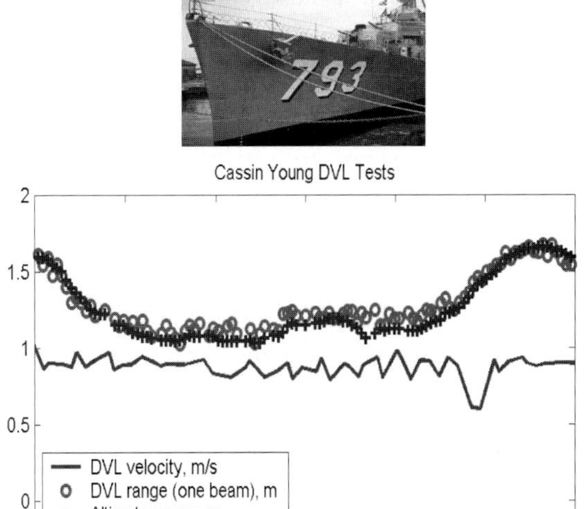

Figure 2: DVL performance when towed along the hull of the *USS Cassin Young*.

shift towards analyzing the data collected. The lack of a shipboard system presence also means the craft can be deployed quickly to respond to developing situations below the waterline.

As a second benefit, the proposed feature-relative control schemes should work when the ship being inspected is fixed within a close berth (where LBL navigation could be poor), anchored and moving slowly about its mooring, or moving freely at very low speed, e.g., adrift.

The key technical point to note about navigating relative to a fixed hull surface is that the vehicle is constrained absolutely in the DOF normal to the hull, but not tangentially. A featureless hull is a poor candidate for visual or sonar image serving, and the use of DVL velocity measurements for positioning invokes an obvious drift error over time.

3.1 Suitability of the DVL for this Task

The DVL (RD Instruments; see URL in References) comprises four narrow beam transducers, arranged uniformly at a spread angle of 30 degrees, and operating broadband in the frequency range of 1200kHz. The Doppler shift is measured for each beam, and an average sensor-relative tangential velocity vector is computed. We also have available the four ranges from the individual transducers: the device provides range by using the return times from each sensor and the speed of sound in water. Complete (four-transducer) measurements are available at a bandwidth of 3-8Hz, depending on signal quality and range.

We performed a series of tests with the DVL, with the specific goal of determining suitability for the hull-relative inspection task. Specifically, we have considered: a) what is the drift rate of the integrated velocities? b) What is the noise characteristic of the independent range measurements? c) What is the effect of a metal hull, with biofouling? d) Does the DIDSON acoustic imaging system interfere with the DVL?

- On a cement and glass wall at the MIT Ocean Engineering Testing Tank, the position error in integrating velocity was confirmed to be about 0.5 percent of distance traveled. The error goes up substantially when the sensor is oriented more than 30 degrees from normal to the hull.

- We performed field tests along the hull of the *USS Cassin Young*, at the Navy Shipyard in Charlestown, Massachusetts. As shown in Figure 2, the range and velocity measurements are well behaved.

- We performed controlled tests at the Testing Tank, with simultaneous operation of the DIDSON and the DVL. DIDSON images (at 5fps) show the DVL pings as a faint flash, but the image is by no means unusable. Conversely, there is a slight degradation of the DVL's velocity performance. The drift rate approximately doubles, but remains below 1cm per meter of distance traveled, which is sufficiently low enough to satisfy our concept of operations.

Figure 3: The horizontal slice method; the vehicle makes passes at constant depth.

3.2 Two Approaches Using "Slicing"

The DVL can be used to servo both orientation and distance to the hull (through the four independent range measurements) and to estimate the distance traveled, with reasonable accuracy. When coupled with an absolute depth measurement, two plausible inspection scenarios emerge for the majority of a large ship's surface: vertical and horizontal "slicing." For the purposes of this paper, we confine our discussion to the large, relatively smooth surface of the hull sides, bottom, and bow. As with other existing automated inspection methods, the stern area with propellers, rudders, shafting and bosses cannot be easily encompassed within our scheme.

In the case of horizontal slicing (Figures 3 and 4), paths in the horizontal plane are performed. The absolute depth provides bounded cross-track error measurement, while the integrated velocity provides the along-track estimate of position. This along-track position, with depth, is recorded for each image.

Defining the end of a track at a given depth is a sensing challenge to which we see several possible approaches. First, there may be landmarks, such as weld lines, protuberances, or sharp edges as found near the bow or stern areas. These landmarks, especially if they occur at many depths, can be used to put limits on the search area, and to re-zero the integrated velocity error. Certainly prior knowledge of the ship's lines and these features can be incorporated into the mapping strategy at some level.

On the other hand, the complete absence of features is workable also: operate at a given depth until the integrated velocity safely exceeds the circumference of the vessel, then move to another depth. When an object of interest is detected, immediate surfacing must occur in this scenario since location along the hull would be poorly known.

The horizontal slice method is very good for the sides and bow of a vessel. Many vessels, for example, large crude carriers (LCC's) have flat bottoms, which must also be inspected. Here, aside from the fact that the vehicle or the imaging sensor and DVL must be reoriented to look up, there is no cross-track error available, since the depth is roughly constant. Long tracks parallel to the hull centerline would be subject to accrued errors on the order of several meters. The vertical slice approach (Figure 5) addresses this problem, by making paths down the sides of the hull and then underneath, in a plane normal to the hull centerline. Once at the centerline, options are to turn around and come back up on the same side, or to continue all the way under the hull to surface on the other side, after a 180-degree turn in place (which must be constructed based on rate gyro information only). In either case, the important property here is that the path length is limited, so that the cross-track errors are limited, and overlap can be applied as necessary. For instance, using a vertical path length of 130m implies a cross-track error on the order of 65cm, which is easily covered by overlapping images with field of view several meters, assuming no systematic bias.

Convex or concave, two-axis curvature of the hull also requires some overlap. For instance, in the extreme case of a spherical hull and the vertical survey, like ribbons around a ball, the imaged path lines converge at the bottom. These cases will require further study and mission design at a high level.

3.3 Role of Low- and Mid-Level Control

Dynamically, the vehicle is equipped with high-performance thrusters so as to operate in shallow waters, waves, and in proximity to hulls. The primary sensor we have available, the DVL, however, is a comparatively low bandwidth device, which cannot provide robust measurements for direct control – the noise properties may be unpredictable, timing may vary, and missed data are not uncommon. Furthermore, loss of contact with the hull can occur in regular operation, and even be exploited as a landmark.

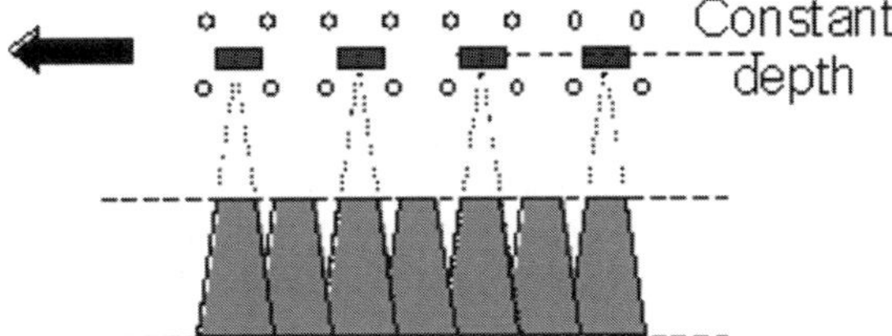

Figure 4: Operation during horizontal survey, looking at the side of the vessel. The vehicle is shown in blue, with the four DVL footprints in yellow on the hull. The DIDSON images (green) are taken looking downward as the vehicle moves.

In waves, the depth sensor also fails as a high-bandwidth navigation sensor. As a consequence of these facts, the vehicle has to be capable of short-term autonomous navigation, through a high-end inertial measurement unit, and an integrated low-level control system. The division of control can be stated as follows: The low-level controller depends only on the core sensors of the IMU, while a mid-level layer incorporates the DVL and depth sensor, and a high-level controller manages the mission and desired pathlines. This multi-level control system is to be of the inner-outer loop type, with the DVL and depth sensor providing setpoints for higher-bandwidth inner loops. As in most cases of inner-outer design, the outer loop bandwidth should be at least 3-5 times slower than the inner loop.

Consider for example the case of yaw control relative to the hull. At the innermost level, a yaw rate servo runs at maximum update frequency and closed-loop bandwidth, employing a model-based estimator, i.e., a Kalman Filter for handling vehicle dynamics and sensor channels that are coupled due to gravity. The mid-level control has coupling, due to the fact that the DVL is like a velocity sensor on a moment arm, so that yaw and sway at the wall are kinematically coupled. This is one of many concepts from visual servoing that are appropriate here (e.g., Hutchison et al., 1996). Figure 6 gives an illustration of hull servoing using nested low- and mid-level control, and DVL data.

4 SUMMARY

Doppler velocimetry with ranging facilitates a new feature-relative approach for autonomous ship hull inspection, one which allows several intuitive strategies that can account for the majority of the hull surface. The use of landmarks and ship's lines, as well as survey techniques for complex stern arrangements are still open questions.

Support is acknowledged from the Office of Naval Research (Dr. T.F. Swean) under Grant N00014-02-1-0946, and from NOAA and the Sea Grant College Program, Grant NA 16RG2255.

Figure 5: Vertical slice survey; the vehicle makes depth passes with zero sway velocity.

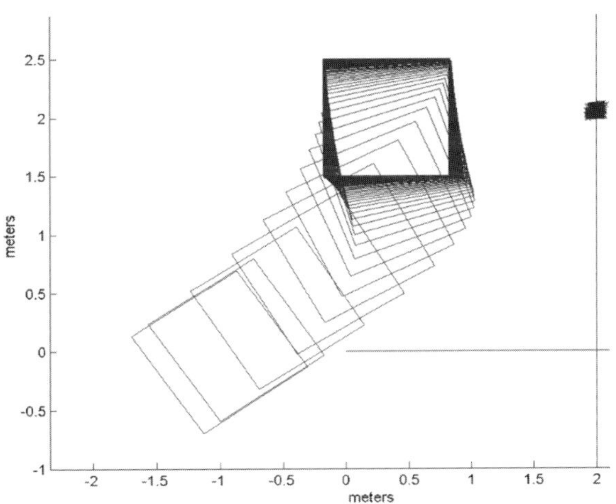

Figure 6: Example of low- (PID) and mid-level (LQG) coupled control in the yaw-sway hull positioning problem. Vehicle initially is at a 42 degree bearing, 3m range; final position is zero degrees bearing, 1.7m range. The controller keeps the tangential velocity small while reorienting, so that the excursion of the DVL "pointer" on the wall (line on right hand side) is 12cm.

REFERENCES

Belcher, E., B. Matsuyama, and G. Trimble, 2003. *Object Identification with Acoustic Lenses.* http://www.apl.washington.edu/programs/DIDSON/Media/object_ident.pdf.

Harris, S.E. and E.V. Slate, 1999. *Lamp Ray: Ship Hull Assessment for Value, Safety and Readiness.* Proc. IEEE/MTS Oceans.

Hutchinson, S., G.D. Hager, and P.I. Corke, 1996. A tutorial on visual servo control. IEEE Trans. Robotics and Automation, 12:651-670.

RD Instruments DVL, http://www.dvlnav.com.

Ship Hull Inspections with AquaMap http://www.desertstar.com/newsite/positioning/shiphul l/manuals/Ship%20Hull%20Inspections.pdf.

Trimble, G. and E. Belcher, 2002. *Ship Berthing and Hull Inspection Using the CetusII AUV and MIRIS High-Resolution Sonar,* Proc. IEEE/MTS Oceans. http://www.perrymare.com/presentations/Oceans%202 002%20Homeland%20Defense.pdf. *See also:* http://web.nps.navy.mil/~brutzman/Savage/Submersib les/UnmannedUnderwaterVehicles/CetusFlyerMarch2 001.pdf.

Wilcox, T., 2003. Marine Sonic Technologies Ltd., Personal communication.

DIMSART: A REAL TIME - DEVICE INDEPENDENT MODULAR SOFTWARE ARCHITECTURE FOR ROBOTIC AND TELEROBOTIC APPLICATIONS

Jordi Artigas, Detlef Reintsema, Carsten Preusche, Gerhard Hirzinger

Institute of Robotics and Mechatronics, DLR (German Aerospace Center)
Oberpfaffenhofen, Germany
Email: firstname.lastname@dlr.de

Keywords: Telepresence, Distributed Control, Indpendency, Robotic, Telerobotic, DIMSART, RTLinux, VxWorks.

Abstract: In this paper a software architecture for robotic and telerobotic applications will be described. The software is device and platform independent, and is distributed control orientated. Thus, the package is suitable for any real time system configuration. The architecture allows designers to easily build complex control schemes for any hardware device, easily control and manage them, and communicate with other devices with a plug-in/plug-out modular concept. The need to create a platform where control engineers/designers could freely implement their algorithms, without needing to worry about the device driver and programming related issues, further motivated this project. Implementing a new control algorithm with the software architecture described here, requires that the designer simply follow a template where the necessary code is reduced to only those functions having to do with the controller. We conducted several teleoperation schemes, one of which will be presented here as a configuration example.

1 INTRODUCTION

Control methods are nowadays totally related to soft computing techniques. From this relationship a new area in software engineering is emerging, which explores the interplay between the control theory and software engineering worlds. It is in this research direction that the authors found the need of building a robotic control software architecture. Among other things, the architecture should facilitate the development of robotic and telerobotic control schemes by defining beneficial constraints on the design and implementation of the specific application, without being too restrictive. Keeping this goal in mind, the DIMSART has been developed by the Telepresence group of the Institute of Robotics and Mechatronics to provide a practical, convenient and original solution.

1.1 Telepresence Environment

The focus of the DIMSART is the development of Telepresence systems. Telepresence is an extension of the telerobotics concept in which a human operator is coupled with as much sensory information as possible to a remote environment through a robot, in order to produce an intuitive and realistic interaction with that environment. The range of senses can encompass vision, tactile, auditory, and even smell and taste senses. Our interest is focused on the haptic channel, therefore the type of information which is sent is motion and force data. From the control point of view, such systems are often referred to as *bilateral control* schemes (see fig.1), because two controls are simultaneously performed in the telepresence global closed loop, one on the master side (controlling the master device), and one on the slave side (controlling the slave device).

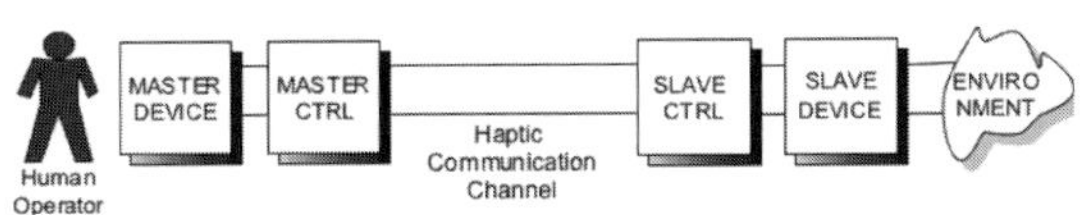

Figure 1: Telepresence Scheme.

A telepresence system developed by the Telepresence research group from DLR Oberpfaffenhofen will be used in this article to show the needs and requirements of the architecture, and thus, motivate the development of the DIMSART to the reader. Its extrapolation to "mono-lateral" robotics applications will be straightforward.

J. Braz et al. (eds.), Informatics in Control, Automation and Robotics I, 201–208.
© 2006 *Springer. Printed in the Netherlands.*

1.2 Control and Software engineering Interplay

A not less significant goal is the creation of an harmony between the computer science world and the control engineering world. Not rarely, control engineers are faced with obstacles arising from the programming work required for controlling and driving hardware devices. The software architecture described here allows a control engineer to forget about issues related to device drivers, real time programming, concurrencies or thread programming. Thus, he/she only needs to concentrate on the specific control algorithm. As it will be seen, by using this software platform, the designer is only faced with the creation a "module", for which a template-based methodology is used. This goal will result in a programming time saving for the control designer and thus the consequence of investing the whole energy in the control design itself.

1.3 Existing Architectures

The generalized needs of facilitating efficient robotic system designs and implementations resulted in the development of several software architectures. In (Ève Coste-Manière and Redi Simmons, 2000) these general needs are formalized. Furthermore, the importance of making the right architecture choice is noted, and some architectures are compared and contrasted.

Some architectures for robotic systems are very complete packages which consider the overall structure for controlling a robot, including all possible levels of competence in a robotic system (sensing, mapping, modeling, planning, AI, task execution, motor control, etc...) either in a hierarchical layered style (Albu et al., 1988), in a behavioral one (Brooks, 1986), or in an hybrid one (Borrelly et al., 1998), (Schneider et al., 1998), (RTI, 2004), (Volpe et al., 2001). However these kind of architectures are more appropriate for autonomous robot vehicles where a control at high level or, alternatively, a *Decision* Layer, as mentioned in (Volpe et al., 2001), is of high importance.

Other architectures are not so concerned with layered structures and emphasize instead the real time operation as a sequence of control components executed at a lower level. These architectures tend to be simpler but more flexible (Scholl, 2001), (Stasse and Kuniyoshi, 2000).

The DIMSART architecture shines for its simplicity, flexibility and portability. It can be included to the second group of architectures, but it is more focused in the automatic control level. Furthermore, it can be easily embedded in other architectures or systems with almost any Linux/Unix operating system type: Linux, Solaris, RTLinux, VxWorks.

The outline of this paper is as follows. We first present, in section 2, a chapter dedicated to general robotic control concepts with a particular scheme example which will be used in the subsequent sections to introduce the DIMSART architecture. We then describe the software platform and its main parts in section 3. Section 4 introduces a complete experimental setup as a DIMSART configuration example. Finally, in section 5, some concluding remarks and future lines are given.

2 ROBOT CONTROL

We will make use of a telepresence control scheme example to focus our interest in three aspects: distributed control, the data flow and the definition of the acting regions of our framework. Also in this chapter, a more abstract view of a general robotic control scheme will be introduced, and later it will be specified within the mentioned example.

2.1 Wave Variables Scheme as Bilateral Control Example

In fig.2 a block diagram of a Wave Variable control scheme can be seen. The Wave Variables Theory is a common approach to minimize the degradative effects of time-delayed communication channels in telepresence systems. For detailed information about this theory refer to (Niemeyer, 1996) and (Artigas, 2003) . It is not the aim of this paper to detail the control theory behind the scheme. Rather, it is intended to be used as a reference point for the DIMSART approach. We will refer to this example in some of the following sections to give support to the theoretical explanations of the architecture.

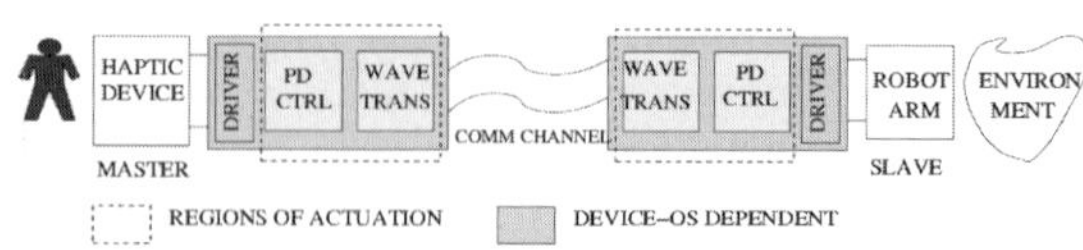

Figure 2: Global control software is decoupled from hardware device, driver and communication.

Distributed Control

In telepresence scenarios the concept of distributed control becomes an important issue. Although from the hardware point of view both master and slave devices can be quite different, from the control point of view they are not so dissimilar. The main idea of distributed control is to divide the global control task of the system in n identical or quasi identical local control components. The nature of a bilateral control is to distribute the control task between both sides, master and slave. The control component, -henceforth referred to as *module*-, will have to be sufficiently powerful to support

1. The existing differences between master and slave robots/environments characteristics (for instance, controller constants, input/output variables, algorithm differences, etc...)

2. Possible different OS platforms. For example, the master could be controlled by a RTLinux machine, and the slave by a VxWorks one.

In our bilateral control scheme, the control task is distributed between master and slave sides through the *Wave Transformer* and *PD Controller* blocks.

Main Operation and Data Flow

The two blocks on each side of the system in fig.2, *PD Controller* and *Wave Transformer*, can be viewed, from the software point of view, as a chain of algorithms with similar characteristics sequentially called. This reasoning leads to an object-oriented direction, in which a Module Template class can be constructed and from which different objects (the control elements) can be defined. Modularity will facilitate a *Top-down* design methodology, as well as code reuse.

Defining Operating Boundaries

By defining the operating boundaries shown in fig.2, the independence from the hardware driver and the communication channel will be preserved, or, in other words, the control task will be uncoupled (from the software point of view) from the rest of the system. Portability and independence are direct consequences. That is, portability to other robotic systems can be achieved, independently of the robot and communication channel.

2.2 General Robot Control setup

Fig.3 skews the components of a general robotic system. On the lowest level, we find the sensors and actuators of the haptic/robot device. The driver, which is in charge of initializing and closing the device, and

reading and writing data from/to it, is part of what we call the *frame*. The *frame* encompasses elements located between the low level control layer, the hardware I/O layer and the network communication layer. The high level control layer deals with non-real time control such as task planning or behavioral control. Our framework will be focused on the low level control layer and its relationship with the *frame*.

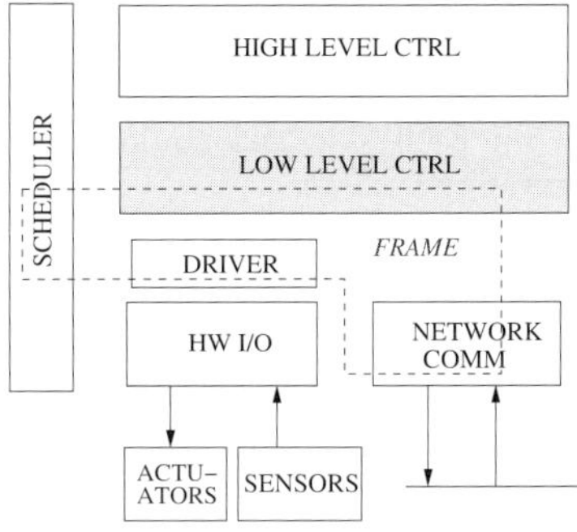

Figure 3: Components of a general robot control setup.

Some of the tasks of the *frame* include the communication between the software architecture, the hardware device and the network, and the real time reading, processing and writing scheduling. The following code exemplifies the master's main operation of the example depicted in fig.2 for a mono-thread *frame*:

```
dev_open(&Phantom);
dev_init(&Phantom);

main_interrupt_thread(arg){ /*called every 1 ms*/
      /* Some inis*/
   Read_Comm(in_waves);
   Robot_read(local_pos);
  /*---------- Control ---------*/
      exec_PDctrl(local_pos,
                  prev_des_pos,
                  out_force);
      exec_WaveTrans(in_waves,
                  out_waves,
                  des_pos);
      /* context save:*/
      prev_des_pos=des_pos;
  /*--------------------------*/
   Write_Comm(out_data);
   Robot_Command(force);
}
dev_close(&Phantom);
```

Often, the low level controlling task is performed by the same *frame* in the real time main interrupt. However, defining the boundaries indicated in fig.2, and thus isolating the control task from the rest of the system, would bring significant benefits. The DIM-

SART architecture provides the needed mechanisms for this job.

2.3 Bilateral Control Scheme with DIMSART Embedded

At this point, we are ready to concretely define specific requirements for the architecture, as well as its location in a robotic scheme. Fig.4 shows the introduced bilateral control example with a DIMSART on each side. Each DIMSART performs the control for each side and could be running in different operating systems.

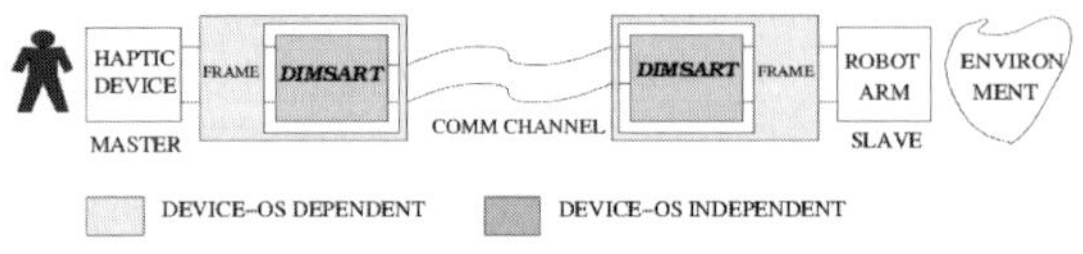

Figure 4: Bilateral control scheme with two DIMSART: one on the master side, one on the slave side.

Requirements

1. **OS independency**.
2. **Device independency**. This implies that the architecture can be set for any DoF[1], I/O data types and sampling rate.
3. **Modular**. Permits a Top-down control scheme design methodology. Flexibility upon the design.
4. **Dynamic**. Only one compilation must be needed to construct any control scheme configuration.
5. Must allow **distributed control**.

3   ARCHITECTURE OVERVIEW

The DIMSART can be defined as a real time device independent software architecture for distributed control in robotic and telerobotic applications. The central point of the DIMSART consist of a dynamic Data Base. Around the Data Base, there is a *frame* and modules. These two kind of elements interact with the Data Base by writing data to it or reading from it. A module, which implements a specific control algorithm, gets its input from the Data Base and writes its output to it. The device driver (*frame*) also reads and writes from and to the Data Base in a similar manner. The modules are sequentially called by a `Module Engine` and transform the input data to produce their output. The following subsections describe each element mentioned above. Fig.5 is a block

---
[1]DoF: Degrees of Freedom

diagram of the DIMSART overview. Furthermore, in a higher layer, a GUI has been developed to configure the robotic control scheme. The user can choose from a list of read-to-use modules which ones to activate, and configuration parameters can also be set for each module.

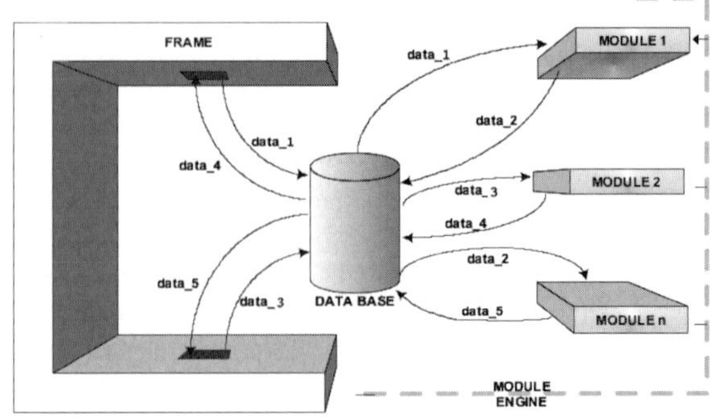

Figure 5: DIMSART Concept Diagram. "data_x" stands for data types.

3.1   Modules

In this framework, a Module is a piece of software intended to perform real time operations. As already mentioned, the module is the data processing unit of the DIMSART. The range of possible functionalities of a module is quite wide. Some examples are control algorithms such as P, PD, PID controllers for robots or haptic devices; simulation of an environment such as a virtual wall; a magnitude converter such as a data normalizer for sending data through a network; a numerical integrator or derivator such as a velocity-to-position converter; a wave transformer as the one shown in fig.2.

There are two types of data with which the Module interacts: internal local data, which stores configuration parameters of the module and is located in the same module, and the data to process, which is stored in the Data Base. The real time main operations of a module can be synthesized in three steps:

1) Read: read data from the Data Base.
2) Compute: process the data to perform the control.
3) Write: write the output data to the Data Base.

The type or types of data which the module extracts from the Data Base, and later the types to be written, are defined in its activate function. There, the module tells the Data Base the data type or types which need to be read, and the type or types which are to be written.

Moreover, as it will be seen in section 3.3, the `Module Engine` provides a mechanism for communicating with the active modules. By means of a `Command` interface, the user will be able to send commands from the *frame* space to a specific module. These commands are internally defined in the module, and they are thought to set the configuration parameters of the module (in the previous example, a `Command` could be used to set the constants of the *PD Controller*, or to specify the configuration of the *Wave Transformer*).

Once the DIMSART is loaded in a robotic system, any control engineer can easily implement a module by simply following a module pattern in a pure "write your code here" style. The aim of the DIMSART is to create a list of modules, from which control engineers will be able to build complex control schemes by choosing and combining modules from this list. Thus, a chain or sequence of modules will be created to perform the control task of a robotic application. As it will be seen in the following subsections, after a module "library" is created, the user will be able to activate and deactivate modules in a plug-in/ plug-out methodology.

Module Template

As already stated, every control block has similar characteristics. The Module Template formalizes the base of every module in an object oriented way. The class[2] `struct Module` is declared here with a group of functions and some attributes (see fig.6).

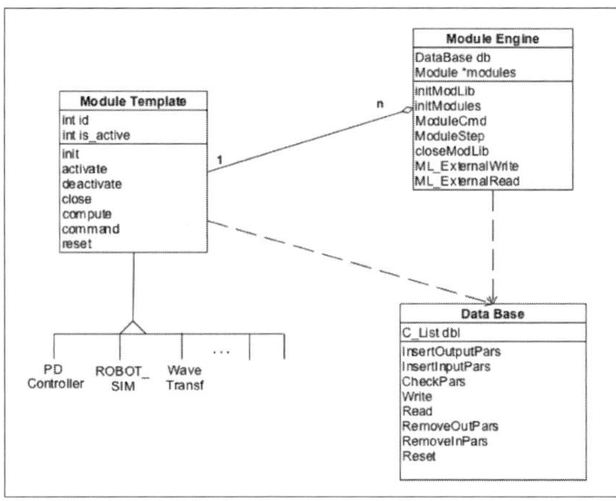

Figure 6: Class-like Diagram of the DIMSART.

3.1.1 Initialization and Activation

Initialization of a module means that its internal memory is allocated, and its internal variables and constants are initialized. This process takes place for every module belonging to the `modules` list (attributed in the `Module Engine`, see fig.6), during the initialization process of the DIMSART.

On the other hand, only the desired modules for the specific control scheme will be activated. Activate means to insert the input and output data types needed by the module under consideration into the Data Base. Thus the Data Base will dynamically grow by means of each module activation. The `Module Engine`, or more precisely, the `Module_Step` function (which will be called at every time step), will act by calling the *Compute* function of each active module.

It is important to note the differences between the tasks of the initialization and activation processes. In order to make the DIMSART compatible with most operating systems, the conception of these initial processes is based on a kernel/modules[3] model of operating systems like Unix or Linux (Corbet, 2001). A kernel module in RTLinux, for instance, allocates its memory during its insertion process. Once the RTLinux module is inserted it should not allocate more memory.

3.2 The Data Base

The core of the Data Base is a dynamic list which stores the incoming data from each active module, and the data coming from the *frame* (which comes from the hardware device). Furthermore, the Data Base incorporates a set of mechanisms for the data interaction between the active modules and the dynamic list. Its construction is performed during the initialization of the `Module Engine` and at each module activation processes.

During the initialization the dynamic list is created according to a *Device Descriptor* of the hardware device. In this descriptor, characteristics such as the DoF and input and output data types are enclosed. After the initialization process, the Data Base is created. At this point, each time a module is activated new data fields are created in accordance with the data types needed by the module. Modules can be inserted or removed at any time during system operation (which in turn will insert or remove input and output types into the Data Base).

---

[2] Simulated class in C code

[3] Unix/Linux kernel modules

Types Matching Check

The matching check function performs a test to validate the coherency of the relationship between the input and output types of a constructed scheme. Fig.7 presents a closer view of the master side of the example in section 2.1. It shows the data interactions between the blocks in the master side. The *PD Controller*, for instance, requires position and desired position as data inputs and outputs desired force. After the scheme is constructed, the *Types matching check* ensures that some other module provides the required data. In our example, the *Wave Transformer* provides desired position to the *PD Controller*, and the *frame* position.

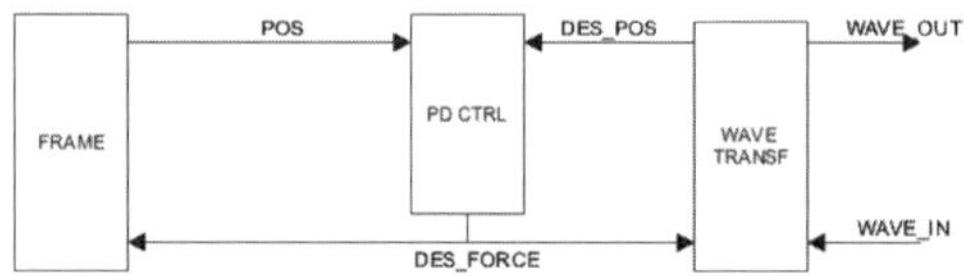

Figure 7: Detailed view of the control scheme in the master side.

3.3   The Module Engine

The `Module Engine` is the software layer between the driver and the Data Base and modules. Through it, a set of capabilities are provided to control and schedule the engine of the DIMSART from the *frame*. A description of the main functionalities and attributes follows. Refer to fig.6 to locate each function and attribute described here.

Frame Descriptor: The *frame* provides to the `Module Engine` a descriptor of the robot in the `Module Engine` initialization function. This descriptor contains information about the DoF of the device, number of input and output data types, and which types are needed by the device. With this information, the `Module Engine` initializes the Data Base.

List of Modules: During the initialization step, a list of modules is created with all the included modules in the DIMSART software architecture from which, the user chooses which ones to activate.

Module Step: This is the beating heart of the `Module Engine`. This is the function that the main loop of the *frame* calls at every time step. The Module Step sequentially calls the `Compute` function of each activated Module. This is how each one of the activated modules performs its real time computation.

Module Command: An interface to send commands to the DIMSART from the *frame* space is also provided in the `Module Engine`. A command can be sent to the `Module Engine`, or to a specific module as seen in section 3.1. By sending commands to the `Module Engine`, the user will be able to set the configuration of the control scheme. Some of the most representative commands are the activation and deactivation of a module, the reset of a module, or the initialization or the close of the DIMSART.

Frame Communication: The hardware driver or a user space (the *frame*) communicates with the Data Base in a similar way as the modules do. These communication mechanisms are also provided in the `Module Engine`.

3.4   The Frame

The *frame*, as indicated in section 2.2, is a space which joins elements belonging to the hardware device communication, to the network communication, and to the DIMSART architecture. The main functions of a *frame* are outlined here.

- Initialization of the hardware device.
- Initialization of the DIMSART.
- Interaction with the Data Base.
- Call of the `Module Step` (see sec. 3.3) in every time step.
- Close the DIMSART.
- Close the hardware device.

The initialization of the DIMSART needs to know the device characteristics. This is done through the *Device Descriptor* (defined in the *frame*), in which the following data is enclosed: number of degrees of freedom of the device; input data types needed by the *frame*, which is the data to be commanded to the device; and output data types, which is the data to be processed by the active modules in the DIMSART.

The interaction with the Data Base is the read and write mechanisms for the *frame*. Two "external" functions are provided for this purpose. To review the above concepts, the following code corresponding to the example presented in section 2.1, shows how a device driver with a DIMSART embedded should look like.

```
init_funct(DevDescr){
  dev_open(&Phantom);
  dev_init(&Phantom);
  initModLib(&DevDescr); /*ini DataBase*/
  initModules(&initarg); /*ini all mods*/
  cmd.ModNr=MOD_ENGINE;
  cmd.ParamNr=MOD_ON;
```

```
  cmd.Value=PDctrl;
  ModuleCmd(&cmd); /*Activates PD ctlr*/
   /*Other mod activations go here*/.
}
main_interrupt_thread(arg){
   /* Some inis */
   Read_Comm(inwaves);
   Robot_read(pos);
   DB_ext_write(pos,inwaves);
   Module_Step(); /*compute active mods*/
   DB_ext_read(force,outwaves);
   Write_Comm();
   Robot_Command(force);
}
```

DB_ext_write : Writes the device output data to the Data Base (*position* in the example).

Module_Step : function which schedules all active modules (in the previous example the *compute* function of the modules *PD Controller* and *Wave Transformer* are called).

DB_ext_read : Reads the device input data from the Data Base (*desired force* in the example.)

The data coming from the robot is no longer in the driver, but instead in a data container, from which other elements will be able to read. Thus, a software boundary between the *frame* and the DIMSART is defined.

4 EXAMPLE

Fig.8 shows the complete scheme for the above presented Wave Variables scheme example, and table 1 its Commands configuration. The system is distributed within two computers, each one running a DIMSART architecture. The computer governing the master is equipped with a RTLinux OS. The slave runs under Linux. On the master side there is a *frame* driving a PHANToM 3DoF (Massie and Salisbury, 1994) as a master device with a DIMSART configuration. The communication between master and slave is performed through UDP sockets . On the slave side, a *user frame* governing a DIMSART is used to simulate a slave robot and a virtual environment. This specific scheme was built for testing the performance of the Wave Variables control scheme as well as for verifying the modular container approach of the DIMSART.

As it can be seen, the communication between the two sides is performed through some dedicated modules. These two modules provide the interface between the DIMSART and the communication channel.

Table 1: Master and slave Commands configuration. Dest is destination. m&s means master and slave. MOD_ON, for instance, activates the module.

Dest.	Module	Command	Value
m&s	COMM_RXMOD	CRX_ADD_DATA	CH2_WAVES
m&s	MOD_ENG	MOD_ON	COMM_RXMOD
m&s	WTRANSF_MOD	B_PARAM	10
master	WTRANSF_MOD	CONF_MODE	MASTER
slave	WTRANSF_MOD	CONF_MODE	SLAVE
m&s	MOD_ENG	MOD_ON	WTRANSF_MOD
master	PDCTRL_MOD	K_PARAM	3000
slave	PDCTRL_MOD	K_PARAM	5000
m&s	MOD_ENG	MOD_ON	PDCTRL_MOD
master	TIMEDELAY_MOD	TD_PARAM	10
master	MOD_ENG	MOD_ON	TIMEDELAY_MOD
m&s	COMM_TXMOD	CTX_ADD_DATA	CH1_TD_WAVES
m&s	MOD_ENG	MOD_ON	COMM_TXMOD
m&s	MOD_ENG	MATCHING_CHECK	/

5 CONCLUDING REMARKS AND REMAINING ASPECTS

The DIMSART has been presented in this paper. It provides a solution to embed a robotic or telerobotic control scheme in a hardware device-OS-communication channel configuration. Independence, distributed control, dynamics and flexibility are values provided to the architecture. The system performance depends on several parameters which may vary in every particular scheme. The number of degrees of freedom, the number of active modules and the length of the active modules influence the system performance. The system has been tested in several robotic and telerobotic contexts and has shown good performance. One of them has been presented in this paper as an example and as a tool to describe the DIMSART architecture to the reader. A very representative application is the ROKVISS[4], a Telepresence experiment in which a force feedback joystick on Earth is to be controlled with the DIMSART architecture, and a 2 DoF robot, which will be mounted on the ISS is also to be controlled with the same architecture. For detailed information about the ROKVISS experiment see (Preusche et al., 2003).

Future work for DIMSART will include several aspects. "Online" module compilation and the extension of the Data Base to multiple data formats and dimensions are issues which are already being developed. Future lines could extend the DIMSART to high level control and multi-thread architectures.

---

[4]Robotic Component Verification on the International Space Station (ISS)

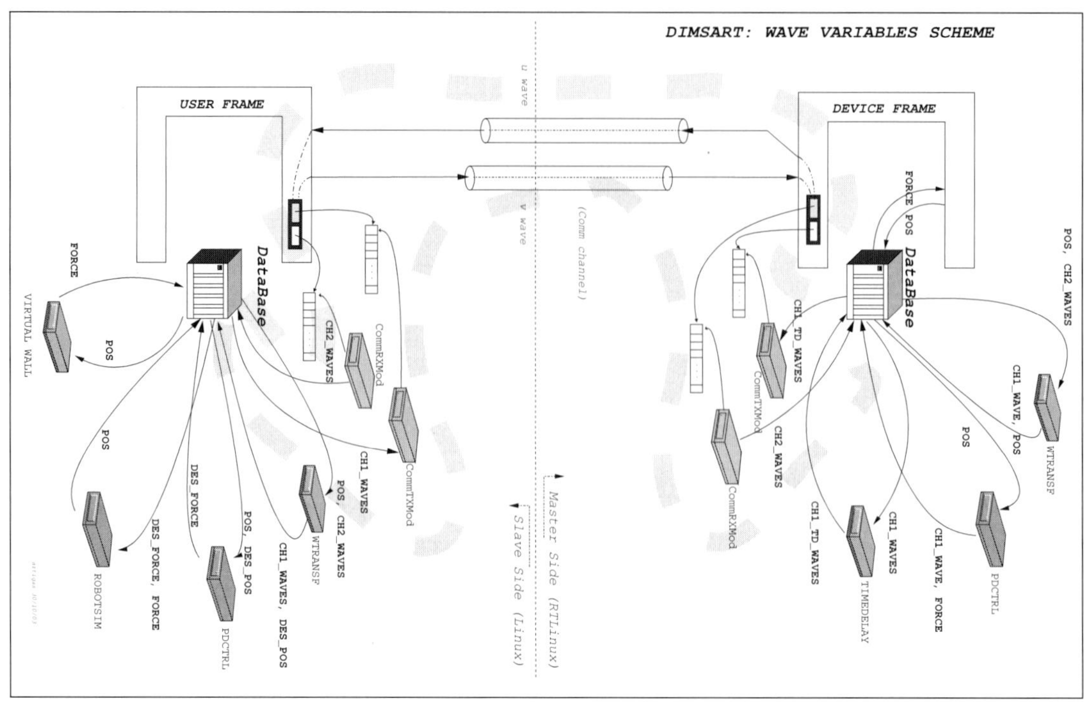

Figure 8: A real bilateral DIMSART setup.

REFERENCES

Albu, J., Lumia, R., and McCain, H. (1988). Hierarchical control of intelligent machines applied to space station telerobots. *Transactions on Aerospace and Electronic Systems*, 24(24):535–541.

Artigas, J. (2003). Development and implementation of bilateral control using the wave variables therory in the rokviss experiment. Internal publication, DLR (German Aerospace Center) - Insitute of Robotics and Mechatronics.

Borrelly, J.-J., Éve Coste-Manière, Espiau, B., Kapellos, K., Pissard-Gibollet, R., Simon, D., and Turro, N. (1998). The orccad architecture. *The International Journal of Robotics Research*, 17(4): 338–359.

Brooks, R. A. (1986). A robust layered control system for a mobile robot. *IEEE Journal of Robotics and Automation*, 2(1):14–23.

Corbet, A. R. J. (2001). *Linux Device Drivers, 2nd Edition*. Number 0-59600-008-1.

Éve Coste-Manière and Redi Simmons (2000). Architecture, the backbone of robotic systems. In *Proceedings of the 2000 IEEE International Conference on Robotics and Automation*, San Francisco, CA.

Massie, T. and Salisbury, J. (1994). The phantom haptic interface: A device for probing virtual objects. In *Proceedings of the ASME International Mechanical Engineering Congress and Exhibition*, pages 295–302, Chicago.

Niemeyer, G. (1996). *Using Wave Variables in Time Delayed force Reflecting Teleoperation*. PhD thesis, Massachussetts Institute of Technology.

Preusche, C., Reintsema, D., Landzettel, K., and Hirzinger, G. (2003). Rokviss - towards telepresence control in advanced space missions. In *Humanoids 2003 - The Third IEEE International Conference on Humanoid Robots*, Munich, Karlsruhe (Germany).

RTI (2004). Constellation. www.rti.com/products/ constellation/index.html.

Schneider, S. A., Chen, V. W., Pardo-Castellote, G., and Wang, H. H. (1998). Controlshell: A software architecture for complex electromechanical systems. *The International Journal of Robotics Research*, 17(4):360–380.

Scholl, K.-U. (2001). MCA2 (Modular Controller Architecture). Software platform. http://mca2.sourceforge.net.

Stasse, O. and Kuniyoshi, Y. (2000). Predn: Achieving efficiency and code re-usability in a programming system for complex robotic applications. In *Proceedings of the 2000 IEEE International Conference on Robotics and Automation*, San Francisco, CA.

Volpe, R., Nesnas, I., Estlin, T., Mutz, D., Petras, R., and Das, H. (2001). The claraty architecture for robotic autonomy. In *2001 Aerospace Conference IEEE Proceedings*, pages 1/121–1/132.

PART 3

Signal Processing, Systems Modeling and Control

ON MODELING AND CONTROL OF DISCRETE TIMED EVENT GRAPHS WITH MULTIPLIERS USING (MIN, +) ALGEBRA

Samir Hamaci
Jean-Louis Boimond
Sébastien Lahaye
LIS
62 avenue Notre Dame du Lac - Angers, France
Email: [hamaci, boimond,lahaye]@istia.univ-angers.fr,

Keywords: Discrete timed event graphs with multipliers, timed weighted marked graphs, dioid, (min, +) algebra, Residuation, just-in-time control.

Abstract: Timed event graphs with multipliers, also called timed weighted marked graphs, constitute a subclass of Petri nets well adapted to model discrete event systems involving synchronization and saturation phenomena. Their dynamic behaviors can be modeled by using a particular algebra of operators. A just in time control method of these graphs based on Residuation theory is proposed.

1 INTRODUCTION

Petri nets are widely used to model and analyze discrete-event systems. We consider in this paper timed event graphs[1] with multipliers (TEGM's). Such graphs are well adapted for modeling synchronization and saturation phenomena. The use of multipliers associated with arcs is natural to model a large number of systems, for example when the achievement of a specific task requires several units of a same resource, or when an assembly operation requires several units of a same part. Note that TEGM's can not be easily transformed into (ordinary) TEG's. It turns out that the proposed transformation methods suppose that graphs are strongly connected under particular server semantics hypothesis (*single* server in (Munier, 1993), or *infinite* server in (Nakamura and Silva, 1999)) and lead to a duplication of transitions and places.

This paper deals with just in time control, *i.e.,* fire input transitions *at the latest* so that the firings of output transitions occur at the latest before the desired ones. In a production context, such a control input minimizes the work in process while satisfying the customer demand. To our knowledge, works on this tracking problem only concern timed event graphs without multipliers (Baccelli et al., 1992, §5.6), (Cohen et al., 1989), (Cottenceau et al., 2001).

TEGM's can be handled in a particular algebraic structure, called *dioid*, in order to do analogies with conventional system theory. More precisely, we use an algebra of operators mainly inspired by (Cohen et al., 1998a), (Cohen et al., 1998b), and defined on a set of operators endowed with pointwise minimum operation as addition and composition operation as multiplication. The presence of multipliers in the graphs implies the presence of inferior integer parts in order to preserve integrity of discrete variables used in the models. Moreover, the resulting models are non linear which prevents from using a classical transfer approach to obtain the just in time control law of TEGM's. As alternative, we propose a control method based on "backward" equations.

The paper is organized as follows. A description of TEGM's by using recurrent equations is proposed in Section 2. An algebra of operators, inspired by (Cohen et al., 1998a), (Cohen et al., 1998b), is introduced in Section 3 to model these graphs by using a state representation. In addition to operators γ, δ usually used to model discrete timed event graphs (without multipliers), we add the operator μ to allow multipliers on arcs. The just in time control method of TEGM's is proposed in Section 4 and is mainly based on Residuation theory (Blyth and Janowitz, 1972). After recalling basic elements of this theory, we recall the residuals of operators γ, δ, and give the residual of operator μ which involves using the superior integer part. The just in time control is expressed as the great-

---

[1]Petri nets for which each place has exactly one upstream and one downstream transition.

211

J. Braz et al. (eds.), Informatics in Control, Automation and Robotics I, 211–216.

est solution of a system of "backward" equations. We give a short example before concluding.

2 RECURRENT EQUATIONS OF TEGM's

We assume the reader is familiar with the structure, firing rules, and basic properties of Petri nets, see (Murata, 1989) for details.

Consider a TEGM defined as a valued bipartite graph given by a five-tuple (P, T, M, m, τ) in which:

- P represents the finite set of *places*, T represents the finite set of *transitions*;

- a *multiplier M* is associated with each *arc*. Given $q \in T$ and $p \in P$, the multiplier M_{pq} (respectively, M_{qp}) specifies the weight (in $\mathbb{N}$) of the arc from transition q to place p (respectively, from place p to transition q) (a zero value for M codes an absence of arc);

- with each place is associated an *initial marking* (m_p assigns an initial number of tokens (in $\mathbb{N}$) in place p) and a *holding time* (τ_p gives the minimal time a token must spend in place p before it can contribute to the enabling of its downstream transitions).

We denote by $^\bullet q$ (resp. $q^\bullet$) the set of places upstream (resp. downstream) transition q. Similarly, $^\bullet p$ (resp. $p^\bullet$) denotes the set of transitions upstream (resp. downstream) place p.

In the following, we assume that TEGM's function according to the *earliest firing* rule: a transition q fires as soon as all its upstream places $\{p \in {}^\bullet q\}$ contain enough tokens (M_{qp}) having spent at least τ_p units of time in place p (by convention, the tokens of the initial marking are present since time $-\infty$, so that they are immediately available at time 0). When the transition q fires, it consumes M_{qp} tokens in each upstream place p and produces $M_{p'q}$ tokens in each downstream place $p' \in q^\bullet$.

Remark 1 We disregard without loss of generality *firing times* associated with transitions of a discrete event graph because they can always be transformed into holding times on places (Baccelli et al., 1992, §2.5).

Definition 1 (Counter variable) With each transition is associated a *counter variable*: x_q is an increasing map from $\mathbb{Z}$ to $\mathbb{Z} \cup \{\pm\infty\}$, $t \mapsto x_q(t)$ which denotes the cumulated number of firings of transition q up to time t.

Assertion 1 The counter variables of a TEGM (under the earliest firing rule) satisfy the following *transition to transition* equation:

$$x_q(t) = \min_{p \in {}^\bullet q,\, q' \in {}^\bullet p} \lfloor M_{qp}^{-1}(m_p + M_{pq'} x_{q'}(t - \tau_p)) \rfloor.$$

$$(1)$$

Note the presence of inferior integer part to preserve integrity of Eq. (1). In general, a transition q may have several upstream transitions ($\{q' \in {}^{\bullet\bullet}q\}$) which implies that its associated counter variable is given by the $\min$ of *transition to transition* equations obtained for each upstream transition.

Example 1 The counter variable associated with transition q described in Fig. 1 satisfies the following equation:

$$x_q(t) = \lfloor a^{-1}(m + b\, x_{q'}(t - \tau)) \rfloor.$$

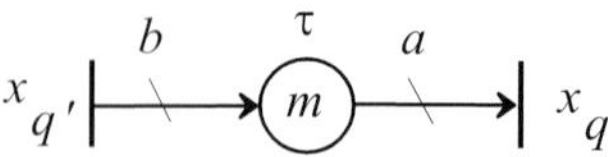

Figure 1: A simple TEGM.

Example 2 Let us consider TEGM depicted in Fig. 2. The corresponding counter variables satisfy the following equations:

$$\begin{cases} x_1(t) &= \min(3 + x_3(t-2), u(t)), \\ x_2(t) &= \min(\lfloor \frac{2x_1(t-2)}{3} \rfloor, \lfloor \frac{6 + 2x_3(t-2)}{3} \rfloor), \\ x_3(t) &= 3x_2(t-1), \\ y(t) &= x_2(t). \end{cases}$$

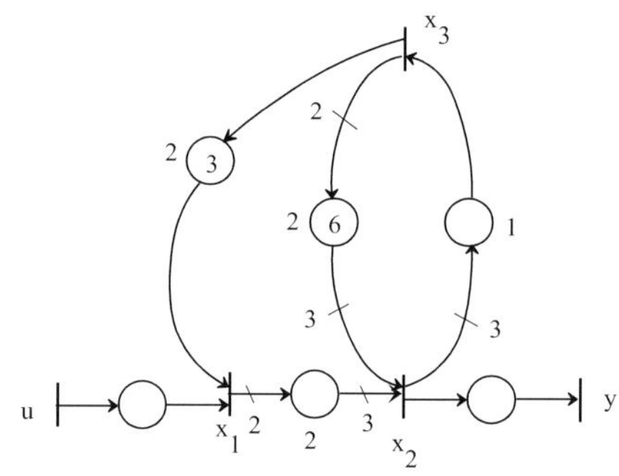

Figure 2: A TEGM.

3 DIOID, OPERATORIAL REPRESENTATION

Let us briefly define dioid and algebraic tools needed to handle the dynamics of TEGM's, see (Baccelli et al., 1992) for details.

Definition 2 (Dioid) A *dioid* $(\mathcal{D}, \oplus, \otimes)$ is a semiring in which the addition $\oplus$ is idempotent ($\forall a,\ a \oplus a = a$). Neutral elements of $\oplus$ and $\otimes$ are denoted ε and e respectively.

A dioid is *commutative* when the product $\otimes$ is commutative. Symbol $\otimes$ is often omitted.

Due to idempotency of $\oplus$, a dioid can be endowed with a natural order relation defined by $a \preceq b \Leftrightarrow b = a \oplus b$ (the least upper bound of $\{a,b\}$ is equal to $a \oplus b$).

A dioid $\mathcal{D}$ is *complete* if every subset A of $\mathcal{D}$ admits a least upper bound denoted $\bigoplus_{x \in A} x$, and if $\otimes$ left and right distributes over infinite sums.

The greatest element noted $\top$ of a complete dioid $\mathcal{D}$ is equal to $\bigoplus_{x \in \mathcal{D}} x$. The greatest lower bound of every subset X of a complete dioid always exists and is noted $\bigwedge_{x \in X} x$.

Example 3 • The set $\mathbb{Z} \cup \{\pm\infty\}$, endowed with min as $\oplus$ and usual addition as $\otimes$, is a complete dioid noted $\overline{\mathbb{Z}}_{\min}$ with neutral elements $\varepsilon = +\infty$, $e = 0$ and $\top = -\infty$.

• Starting from a 'scalar' dioid $\mathcal{D}$, let us consider $p \times p$ matrices with entries in $\mathcal{D}$. The sum and product of matrices are defined conventionally after the sum and product of scalars in $\mathcal{D}$:

$$(A \oplus B)_{ij} = A_{ij} \oplus B_{ij} \text{ and } (A \otimes B)_{ij} = \bigoplus_{k=1}^{p} A_{ik} \otimes B_{kj}.$$

The set of square matrices endowed with these two operations is also a dioid denoted $\mathcal{D}^{p \times p}$.

Counter variables associated with transitions are also called *signals* by analogy with conventional system theory. The set of signal is endowed with a kind of module structure, called *min-plus semimodule*; the two associated operations are:

• pointwise minimum of time functions to add signals:

$$\forall t, (u \oplus v)(t) = u(t) \oplus v(t) = \min(u(t), v(t));$$

• addition of a constant ($\in \overline{\mathbb{Z}}_{\min}$) to play the role of external product of a signal by a scalar:

$$\forall t, \forall \lambda \in \mathbb{Z} \cup \{\pm\infty\}, (\lambda.u)(t) = \lambda \otimes u(t) = \lambda + u(t).$$

A modeling method based on operators is used in (Cohen et al., 1998a), (Cohen et al., 1998b), a similar approach is proposed here to model TEGM's. Let us recall the definition of *operator*.

Definition 3 (Operator, linear operator) An *operator* is a map from the set of signals to the set of signals. An operator H is *linear* if it preserves the min-plus semimodule structure, *i.e.*, for all signals u, v and constant λ,

$$H(u \oplus v) = H(u) \oplus H(v) \text{ (additive property)},$$

$$H(\lambda \otimes u) = \lambda \otimes H(u) \text{ (homogeneity property)}.$$

Let us introduce operators γ, δ, μ which are central for the modeling of TEGM's:

1. Operator γ^ν represents a shift of ν units in counting ($\nu \in \mathbb{Z} \cup \{\pm\infty\}$) and is defined as $\gamma^\nu x(t) = x(t) + \nu$. It verifies the following relations:
$$\begin{cases} \gamma^\nu \oplus \gamma^{\nu'} = \gamma^{\min(\nu, \nu')}, \\ \gamma^\nu \otimes \gamma^{\nu'} = \gamma^{\nu+\nu'}. \end{cases}$$

2. Operator δ^τ represents a shift of τ units in dating ($\tau \in \mathbb{Z} \cup \{\pm\infty\}$) and is defined as $\delta^\tau x(t) = x(t - \tau)$. It verifies the following relations:
$$\begin{cases} \delta^\tau \oplus \delta^{\tau'} = \delta^{\max(\tau, \tau')}, \\ \delta^\tau \otimes \delta^{\tau'} = \delta^{\tau+\tau'}. \end{cases}$$

3. Operator μ_r represents a scaling of factor r and is defined as $\mu_r x(t) = \lfloor r \times x(t) \rfloor$ in which $r \in \mathbb{Q}^+$ (r is equal to a ratio of elements in $\mathbb{N}$). It verifies the following relation: $\mu_r \oplus \mu_{r'} = \mu_{\min(r, r')}$. Note that $\mu_r \otimes \mu_{r'}$ can be different from $\mu_{(r \times r')}$.

See Fig.3.a-3.c for a graphical interpretation of operators γ, δ, μ respectively. We note that operators γ, δ are linear while operator μ is only additive. We have the following properties:

1. $\gamma^\nu \delta^\tau = \delta^\tau \gamma^\nu, \mu_r \delta^\tau = \delta^\tau \mu_r$ (commutative properties),

2. Let $a, b \in \mathbb{N}$, we have $\mu_{a^{-1}} \mu_b = \mu_{(a^{-1}b)}$.

Let us introduce dioid $\mathcal{D}_{\min}[\![\delta]\!]$. First, we denote by $\mathcal{D}_{\min}$ the (noncommutative) dioid of finite sums of operators $\{\mu_r, \gamma^\nu\}$ endowed with pointwise min ($\oplus$) and composition ($\otimes$) operations, with neutral elements $\varepsilon = \mu_{+\infty} \gamma^{+\infty}$ and $e = \mu_1 \gamma^0$. Thus, an element in $\mathcal{D}_{\min}$ is a map $p = \bigoplus_{i=1}^{k} \mu_{r_i} \gamma^{\nu_i}$ such that

$$\forall t \in \mathbb{Z}, p(x(t)) = \min_{1 \leq i \leq k} (\lfloor r_i(\nu_i + x(t)) \rfloor).$$

Operator δ is considered separately from the other operators in order to allow the definition of a dioid of formal power series. With each value of time delay τ (*i.e.*, with each operator δ^τ) is associated an element of $\mathcal{D}_{\min}$. More formally, we define a map $g : \mathbb{Z} \to \mathcal{D}_{\min}, \tau \mapsto g(\tau)$ in which $g(\tau) = \bigoplus_{i=1}^{k_\tau} \mu_{r_i^\tau} \gamma^{\nu_i^\tau}$.

Such an application can be represented by a formal power series in the indeterminate δ. Let the series $G(\delta)$ associated with map g defined by:

$$G(\delta) = \bigoplus_{\tau \in \mathbb{Z}} g(\tau) \delta^\tau.$$

The set of these formal power series endowed with the two following operations:

$$\begin{aligned} F(\delta) \oplus G(\delta) : (f \oplus g)(\tau) &= f(\tau) \oplus g(\tau) \\ &= \min(f(\tau), g(\tau)), \\ F(\delta) \otimes G(\delta) : (f \otimes g)(\tau) &= \bigoplus_{i \in \mathbb{Z}} f(i) \otimes g(\tau - i) \\ &= \inf_{i \in \mathbb{Z}} (f(i) + g(\tau - i)), \end{aligned}$$

is a dioid noted $\mathcal{D}_{\min}[\![\delta]\!]$ with neutral elements $\varepsilon = \mu_{+\infty} \gamma^{+\infty} \delta^{-\infty}$ and $e = \mu_1 \gamma^0 \delta^0$.

Elements of $\mathcal{D}_{\min}[\![\delta]\!]$ allow modeling the transfer between two transitions of a TEGM. A signal x can be also represented by a formal series of $\mathcal{D}_{\min}[\![\delta]\!]$ $(X(\delta) = \bigoplus_{\tau \in \mathbb{Z}} x(\tau)\,\delta^\tau)$, simply due to the fact that it is also equal to $x \otimes e$ (by definition of neutral element e of $\mathcal{D}_{\min}$). For example, the graph depicted in Fig. 1 is represented by equation $X_q(\delta) = \mu_{a-1}\gamma^m\delta^\tau\mu_b X_{q'}(\delta)$ where $X_q(\delta)$ and $X_{q'}(\delta)$ denote elements of $\mathcal{D}_{\min}[\![\delta]\!]$ associated with transitions q and q' respectively.

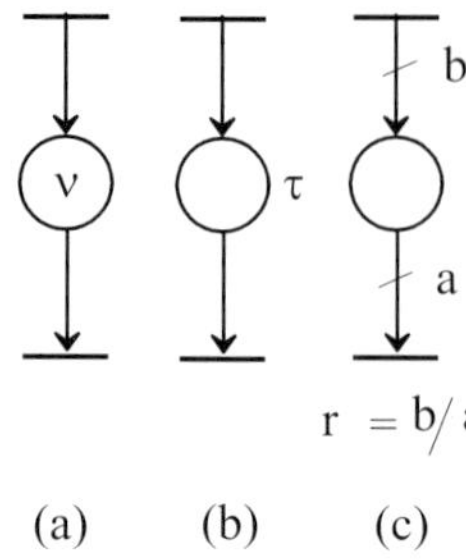

Figure 3: Graphs corresponding to operators γ, δ, μ.

In the following, matrices or scalars with elements in dioid $\mathcal{D}_{\min}[\![\delta]\!]$ are denoted by upper case letters, *i.e.*, X is a shorter notation for $X(\delta)$.

Let us extend the product notation to compose matrices of operators with vectors of signals (with compatible dimensions). Given a matrix of operators A and a vector of signals X with elements in $\mathcal{D}_{\min}[\![\delta]\!]$, we set $(AX)_i \stackrel{def}{=} \bigoplus_j A_{ij}(X_j)$.

Assertion 2 The counter variables of a TEGM satisfy the following state equations:

$$\begin{cases} X &= AX \oplus BU, \\ Y &= CX \oplus DU, \end{cases} \tag{2}$$

in which state X, input U and output Y vectors are composed of signals, entries of matrices A, B, C, D belong to dioid $\mathcal{D}_{\min}[\![\delta]\!]$.

Example 4 TEGM depicted in Fig. 2 admits the following state equations:

$$\begin{cases} \begin{pmatrix} X_1 \\ X_2 \\ X_3 \end{pmatrix} = \begin{pmatrix} \varepsilon & \varepsilon & \gamma^3\delta^2 \\ \mu_{1/3}\delta^2\mu_2 & \varepsilon & \mu_{1/3}\gamma^6\delta^2\mu_2 \\ \varepsilon & \delta\mu_3 & \varepsilon \end{pmatrix} \begin{pmatrix} X_1 \\ X_2 \\ X_3 \end{pmatrix} \oplus \begin{pmatrix} e \\ \varepsilon \\ \varepsilon \end{pmatrix} U, \\[2em] Y = (\varepsilon \quad e \quad \varepsilon) \begin{pmatrix} X_1 \\ X_2 \\ X_3 \end{pmatrix} \oplus \varepsilon U. \end{cases} \tag{3}$$

4 JUST IN TIME CONTROL

4.1 Residuation Theory

Laws $\oplus$ and $\otimes$ of a dioid are not reversible in general. Nevertheless *Residuation* is a general notion in lattice theory which allows defining "pseudo-inverses" of some *isotone* maps (f is isotone if $a \preceq b \Rightarrow f(a) \preceq f(b)$). Let us recall some basic results on this theory, see (Blyth and Janowitz, 1972) for details.

Definition 4 (Residual of map) An isotone map $f :$ $\mathcal{D} \to \mathcal{C}$ in which $\mathcal{D}$ and $\mathcal{C}$ are ordered sets is *residuated* if there exists an isotone map $h : \mathcal{C} \to \mathcal{D}$ such that $f \circ h \preceq Id_\mathcal{C}$ and $h \circ f \succeq Id_\mathcal{D}$ ($Id_\mathcal{C}$ and $Id_\mathcal{D}$ are identity maps on $\mathcal{C}$ and $\mathcal{D}$ respectively). Map h, also noted $f^\sharp$, is unique and is called the *residual* of map f.

If f is residuated then $\forall y \in \mathcal{C}$, the least upper bound of subset $\{x \in \mathcal{D} \mid f(x) \preceq y\}$ exists and belongs to this subset. This *greatest* "subsolution" is equal to $f^\sharp(y)$.

Let $\mathcal{D}$ be a complete dioid and consider the isotone map $L_a : x \mapsto a \otimes x$ from $\mathcal{D}$ into $\mathcal{D}$. The greatest solution to inequation $a \otimes x \preceq b$ exists and is equal to $L_a^\sharp(b)$, also noted $\frac{b}{a}$. Some results related to this map and used later on are given in the following proposition.

Proposition 1 *((Baccelli et al., 1992, §4.4, 4.5.4, 4.6))*
Let maps $L_a : \mathcal{D} \to \mathcal{D}, x \mapsto a \otimes x$ and $L_b : \mathcal{D} \to \mathcal{D}, x \mapsto b \otimes x$.

1. $\forall a, b, x \in \mathcal{D}$,
   $L_{ab}^\sharp(x) = (L_a \circ L_b)^\sharp(x) = (L_b^\sharp \circ L_a^\sharp)(x)$.
   More generally, if maps $f : \mathcal{D} \to \mathcal{C}$ and $g : \mathcal{C} \to \mathcal{B}$ are residuated, then $g \circ f$ is also residuated and $(g \circ f)^\sharp = f^\sharp \circ g^\sharp$.

2. $\forall a, x \in \mathcal{D}, x \succeq a\,x \Leftrightarrow x \preceq \frac{x}{a}$.

3. Let $A \in \mathcal{D}^{n \times p}, B \in \mathcal{D}^{n \times q}, \frac{B}{A} \in \mathcal{D}^{p \times q}$
   and $(\frac{B}{A})_{ij} = \bigwedge_{l=1}^n \frac{B_{lj}}{A_{li}}, 1 \leq i \leq p, 1 \leq j \leq q$.

Proposition 2 The residuals of operators γ, δ, μ are given by:
$\gamma^{\nu^\sharp} : \{x(t)\}_{t \in \mathbb{Z}} \mapsto \{x(t) - \nu\}_{t \in \mathbb{Z}}$ in which $\nu \in \mathbb{Z} \cup \{+\infty\}$,
$\delta^{\tau^\sharp} : \{x(t)\}_{t \in \mathbb{Z}} \mapsto \{x(t + \tau)\}_{t \in \mathbb{Z}}$ in which $\tau \in \mathbb{Z}$,
$\mu_r^\sharp : \{x(t)\}_{t \in \mathbb{Z}} \mapsto \{\lceil \frac{1}{r} \times x(t) \rceil\}_{t \in \mathbb{Z}}$ in which $r \in \mathbb{Q}^+$ ($\lceil \alpha \rceil$ stands for the superior integer part of real number α).

Proof
Expressions of residuals of operators γ, δ are classical (Baccelli et al., 1992, Chap. 4), (Menguy, 1997, Chap. 4).

Relatively to residuation of operator μ, let us express that $\mu_r = P\mu'_r I$ in which
$I : (\mathbb{Z} \cup \{\pm\infty\})^{\mathbb{Z}} \to (\mathbb{R} \cup \{\pm\infty\})^{\mathbb{Z}}$,
$\{x(t)\}_{t\in\mathbb{Z}} \mapsto \{x(t)\}_{t\in\mathbb{Z}}$,
$\mu'_r : (\mathbb{R} \cup \{\pm\infty\})^{\mathbb{Z}} \to (\mathbb{R} \cup \{\pm\infty\})^{\mathbb{Z}}$,
$\{x(t)\}_{t\in\mathbb{Z}} \mapsto \{r \times x(t)\}_{t\in\mathbb{Z}}$
and $P : (\mathbb{R} \cup \{\pm\infty\})^{\mathbb{Z}} \to (\mathbb{Z} \cup \{\pm\infty\})^{\mathbb{Z}}$,
$\{x(t)\}_{t\in\mathbb{Z}} \mapsto \{\lfloor x(t) \rfloor\}_{t\in\mathbb{Z}}$.
Operator I is residuated, its residual is defined by
$I^\sharp : (\mathbb{R} \cup \{\pm\infty\})^{\mathbb{Z}} \to (\mathbb{Z} \cup \{\pm\infty\})^{\mathbb{Z}}$,
$\{x(t)\}_{t\in\mathbb{Z}} \mapsto \{\lceil x(t) \rceil\}_{t\in\mathbb{Z}}$.
Operator P is residuated, its residual is defined by
$P^\sharp : (\mathbb{Z} \cup \{\pm\infty\})^{\mathbb{Z}} \to (\mathbb{R} \cup \{\pm\infty\})^{\mathbb{Z}}$,
$\{x(t)\}_{t\in\mathbb{Z}} \mapsto \{x(t)\}_{t\in\mathbb{Z}}$, we have $P^\sharp = I$.
Residuations of I and P are proven directly from Def. 4. Indeed $I, P, I^\sharp$ and $P^\sharp$ are isotone, moreover, $\forall t \in \mathbb{Z}$,
$\forall x \in (\mathbb{R} \cup \{\pm\infty\})^{\mathbb{Z}}, I\,I^\sharp(x(t)) = I(\lceil x(t) \rceil) \preceq x(t)$
and $\forall x \in (\mathbb{Z} \cup \{\pm\infty\})^{\mathbb{Z}}, I^\sharp I(x(t)) = \lceil x(t) \rceil = x(t);$
$\forall x \in (\mathbb{Z} \cup \{\pm\infty\})^{\mathbb{Z}}, PP^\sharp(x(t)) = PI(x(t)) = \lfloor x(t) \rfloor = x(t)$ and $\forall x \in (\mathbb{R} \cup \{\pm\infty\})^{\mathbb{Z}}, P^\sharp P(x(t)) = IP(x(t)) = \lfloor x(t) \rfloor \succeq x(t)$.
Residual of operator μ' is classical, it is defined by
$\mu'^\sharp_r : (\mathbb{R} \cup \{\pm\infty\})^{\mathbb{Z}} \to (\mathbb{R} \cup \{\pm\infty\})^{\mathbb{Z}}$,
$\{x(t)\}_{t\in\mathbb{Z}} \mapsto \{\frac{1}{r} \times x(t)\}_{t\in\mathbb{Z}}$. Hence, we can deduce the residuation of operator μ. We have $\mu^\sharp_r = (P\mu'_r I)^\sharp = I^\sharp \mu'^\sharp_r P^\sharp$ thanks to Prop. 1.1, *i.e.*,
$\forall x \in (\mathbb{Z} \cup \{\pm\infty\})^{\mathbb{Z}}, \mu^\sharp_r x(t) = \lceil \mu'^\sharp_r I(x(t)) \rceil = \lceil \frac{1}{r} \times x(t) \rceil$.

4.2 Control Problem Statement

Let us consider a TEGM described by Eqs. (2). The just in time control consists in firing input transitions (u) *at the latest* so that the firings of output transitions (y) occur at the latest before the desired ones. Let us define reference input z as the *counter* of the desired outputs: $z_i(t) = n$ means that the firing numbered n of the output transition y_i is desired at the latest at time t. More formally, the just in time control noted u_{opt} is the *greatest* solution (with respect to the order relation $\preceq$) to Eqs. (2) such that $y \preceq z$ (with respect to the usual order relation $\leq$, u_{opt} is the lowest control such that $y \geq z$).

Its expression is deduced from the following result based on Residuation theory.

Proposition 3 Control u_{opt} of TEGM described by Eqs. (2) is the greatest solution (with respect to the order relation $\preceq$) to the following equations:
$$\begin{cases} \xi & = \frac{\xi}{A} \wedge \frac{Z}{C}, \\ U & = \frac{\xi}{B} \wedge \frac{Z}{D}. \end{cases}$$

ξ is the greatest solution of the first equation and corresponds to the latest firings of state transition X ($\xi \succeq X$).

Proof We deduce from Eqs. (2) that state X and output Y are such that $\begin{cases} X & \succeq AX \quad (i) \\ X & \succeq BU \quad (2i) \end{cases}$ and
$\begin{cases} Y & \succeq CX \\ Y & \succeq DU \end{cases}$. Moreover, we look for control U such that $Y \preceq Z$ which leads to $\begin{cases} Z & \succeq CX \quad (3i) \\ Z & \succeq DU \quad (4i) \end{cases}$.
The greatest solution to Eq. (3i) is equal to $\frac{Z}{C}$. Hence we deduce thanks to Prop. 1.2 that the greatest solution noted ξ verifying Eqs. (i) and (3i) is equal to $\xi = \frac{\xi}{A} \wedge \frac{Z}{C}$ (sizes of ξ and X are equal). So the greatest solution verifying Eqs. (2i) and (4i) (in which ξ replaces X) is equal to $\frac{\xi}{B} \wedge \frac{Z}{D}$.

For example let us consider the TEGM depicted in Fig. 2 and modeled by Eqs. (3). Let us give the expression of the just in time control, which leads to calculating the greatest solution of the following equations:
$$\begin{cases} \begin{pmatrix} \xi_1 \\ \xi_2 \\ \xi_3 \end{pmatrix} = \begin{pmatrix} \dfrac{\xi_2}{\mu_{(1/3)}\delta^2\mu_2} \\ \dfrac{\xi_3}{\delta\mu_3} \wedge Z \\ \dfrac{\xi_1}{\gamma^3\delta^2} \wedge \dfrac{\xi_2}{\mu_{(1/3)}\gamma^6\delta^2\mu_2} \end{pmatrix}, \\ U = \xi_1. \end{cases}$$
Let us express these equations in usual counter setting. The recursive equations are backwards in time numbering and are supposed to start at some finite time, noted t_f, which means that system is only controlled until this time. So let us consider the following "initial conditions":
$$z(t) = z(t_f) \text{ and } \xi(t) = \xi(t_f), \ \forall t > t_f.$$
For all $t \in \mathbb{Z}$, we have:
$$\begin{cases} \begin{pmatrix} \xi_1(t) \\ \xi_2(t) \\ \xi_3(t) \end{pmatrix} = \begin{pmatrix} \lceil 3/2 \times \xi_2(t+2) \rceil \\ \max(\lceil 1/3 \times \xi_3(t+1) \rceil, z(t)) \\ \max(\xi_1(t+2) - 3, \lceil 3/2 \times \xi_2(t+2) - 3 \rceil) \end{pmatrix}, \\ u(t) = \xi_1(t). \end{cases}$$
Reference input z and output y are represented in Fig.4, z is such that $z(t) = z(t_f), \ \forall t > t_f = 15$. Control u is represented in Fig.5 and is as late as possible so that desired behavior of output transition is satisfied ($y \preceq z$). Moreover, control u is such that components of ξ are greater than or equal to those of x ($x \preceq \xi$).

5 CONCLUSION

Most works on dioid deal with discrete timed event graphs without multipliers. We aim at showing here the efficiency of dioid theory to also just in time control TEGM's without additional difficulties. The proposed method is mainly based on Residuation theory and the control is the greatest solution of "backward" equations. A possible development of this work would consist in considering hybrid systems or more complex control objectives.

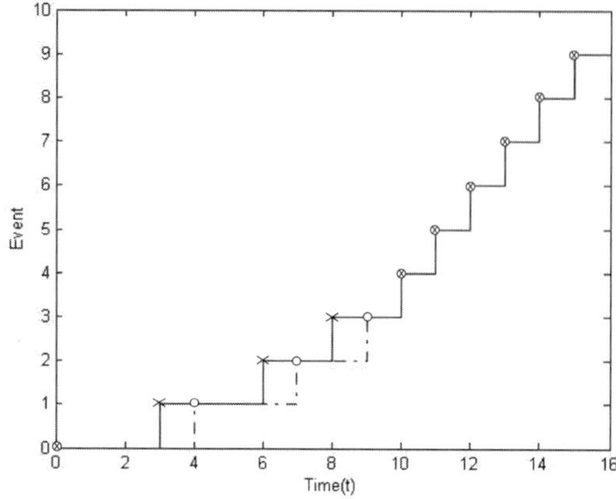

Figure 4: Output y (thin line) and reference input z (dotted line).

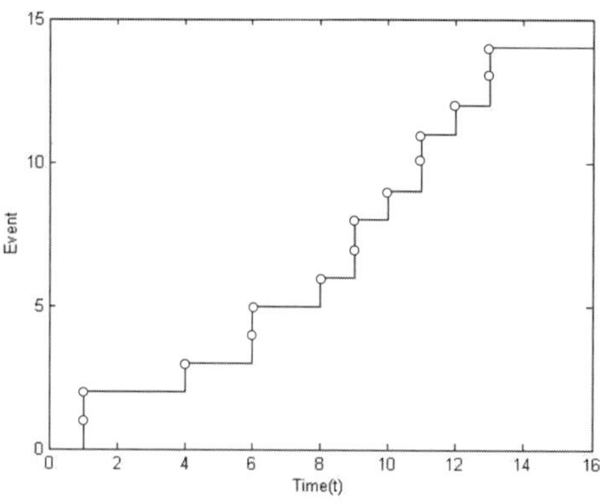

Figure 5: Control u.

REFERENCES

Baccelli, F.,Cohen, G., Olsder, G., and Quadrat, J.-P. (1992). *Synchronization and Linearity: n lgebra for Discrete vent Systems*. Wiley.

Blyth, T. and Janowitz, M. (1972). *"esiduation Theory"*. Pergamon Press, Oxford.

Cohen, G., Gaubert, S., and Quadrat, J.-P. (1998a). *"l-gebraic System nalysis of Timed Petri Nets"*. pages 145-170, In: *Idempotency*,J. Gunawardena Ed., Collection of the Isaac Newton Institute, Cambridge University Press. ISBN 0-521-55344-X.

Cohen, G., Gaubert, S., and Quadrat, J.-P. (1998b). *"Timed-Event Graphs with Multipliers and Homogeneous Min-Plus Systems"*. *I T C*, 43(9):1296–1302.

Cohen, G., Moller, P., Quadrat, J.-P., and Viot, M. (1989). *"Algebraic Tools for the Performance Evaluation of Discrete Event Systems"*. *I Proceedings*, 77(1):39−58.

Cottenceau, B., Hardouin, L., Boimond, J.-L., and Ferrier, J.-L. (2001). *"Model Reference Control for Timed Event Graphs in Dioids"*. *utomatica*, 37:1451−1458.

Menguy, E. (1997). *"Contribution à la commande des systèmes linéaires dans les dioïdes"*. PhD thesis, Université d'Angers, Angers, France.

Munier, A. (1993). Régime asymptotique optimal d'un graphe d'événements temporisé généralisé : Application à un problème d'assemblage. In *IPO- PII*, volume 27(5), pages 487−513.

Murata, T. (1989). *"Petri Nets: Properties, Analysis and Applications"*. *I Proceedings*, 77(4):541–580.

Nakamura, M. and Silva, M. (1999). Cycle Time Computation in Deterministically Timed Weighted Marked Graphs. In *I - T* , pages 1037−1046.

MODEL PREDICTIVE CONTROL FOR HYBRID SYSTEMS UNDER A STATE PARTITION BASED MLD APPROACH (SPMLD)

Jean Thomas, Didier Dumur

Supélec, F 91192 Gif sur Yvette cedex, France   Phone : +33 (0)1 69 85 13 75
Email: jean.thomas@supelec.fr, didier.dumur@supelec.fr

Jean Buisson, Herve Guéguen

Supélec, F 35 511 Cesson-Sévigné cedex, France   Phone : +33 (0)2 99 84 45 43
Email: jean.buisson@supelec.fr, herve.gueguen@supelec.fr

Keywords: Hybrid dynamical systems, Mixed Logical Dynamical systems (MLD), Piecewise Affine systems (PWA), Model Predictive Control (MPC)

Abstract   This paper presents the State Partition based Mixed Logical Dynamical (SPMLD) formalism as a new modeling technique for a class of discrete-time hybrid systems, where the system is defined by different modes with continuous and logical control inputs and state variables, each model subject to linear constraints. The reformulation of the predictive strategy for hybrid systems under the SPMLD approach is then developed. This technique enables to considerably reduce the computation time (with respect to the classical MPC approaches for PWA and MLD models), as a positive feature for real time implementation. This strategy is applied in simulation to the control of a three tanks benchmark.

1 INTRODUCTION

Hybrid systems become an attractive field of research for engineers as it appears in many control applications in industry. They include both continuous and discrete variables, discrete variables coming from parts described by logic such as for example on/off switches or valves. Various approaches have been proposed for modeling hybrid systems (Branicky et al., 1998), like Automata, Petri nets, Linear Complementary (LC), Piecewise Affine (PWA) (Sontag, 1981), Mixed Logical Dynamical (MLD) models (Bemporad, and Morari, 1999).

This paper examines a class of discrete-time hybrid systems, which consists of several models with different dynamics according to the feasible state space partition. Each model is described with continuous and logical states and control inputs. Consequently, the dynamic of the system depends on the model selected in relation to linear constraints over the states and on the inputs values.

On the other hand, model predictive control (MPC) appears to be an efficient strategy to control hybrid systems. Considering the previous particular class of hybrid systems, implementing MPC leads to a problem including at each prediction step the states and inputs vectors (both continuous and discrete variables), the dynamic equation and linear constraints, for which a quadratic cost function has to be optimized. Two classical approaches exist for solving this optimization problem.

First, all possible logical combinations can be studied at each prediction time, which leads solving a great number of QPs. Each of these QPs is related to a particular scenario of logical inputs and modes. This is the PWA approach. The number of QPs can be reduced by reachability considerations (Pena et al., 2003).

The second moves the initial problem through the MLD formalism to a single general model used at each prediction step. This MLD formalism introduces many auxiliary logical and continuous variables and linear constraints. At each prediction step, all the MLD model variables have to be solved (even if some of them are not active). However, the MLD transformation allows utilizing the Branch and Bound (B&B) technique (Fletcher and Leyffer, 1995), reducing the number of QPs solved.

This paper develops a technique which aims at implementing MPC strategy for the considered class of hybrid systems, as a mixed solution of the two classical structures presented before. It is based on a

J. Braz et al. (eds.), Informatics in Control, Automation and Robotics I, 217–224.
© 2006 *Springer. Printed in the Netherlands.*

new modeling technique, called State Partition based MLD approach (SPMLD) formalism, combining the PWA and MLD models. The complexity of this formalism is compared to that obtained with the usual PWA and MLD forms, which can also model this class of hybrid systems as well.

The paper is organized as follows. Section 2 presents a short description of the PWA and MLD hybrid systems. General consideration about model predictive control (MPC) and its classical application to PWA and MLD systems are summarized in Section 3. Section 4 develops the State Partition based MLD approach (SPMLD) and examines the application of MPC to hybrid systems under this formalism. Section 5 gives an application of this strategy to water level control of a three tanks benchmark. Section 6 gives final conclusions.

2 HYBRID SYSTEMS MODELING

2.1 Mixed Logical Dynamical Model

The MLD model appears as a suitable formalism for various classes of hybrid systems, like linear hybrid or constrained linear systems. It describes the systems by linear dynamic equations subject to linear inequalities involving real and integer variables, under the form (Bemporad and Morari, 1999)

$$\mathbf{x}_{k+1} = \mathbf{A}\mathbf{x}_k + \mathbf{B}_1\mathbf{u}_k + \mathbf{B}_2\boldsymbol{\delta}_k + \mathbf{B}_3\mathbf{z}_k$$
$$\mathbf{y}_k = \mathbf{C}\mathbf{x}_k + \mathbf{D}_1\mathbf{u}_k + \mathbf{D}_2\boldsymbol{\delta}_k + \mathbf{D}_3\mathbf{z}_k \quad (1)$$
$$\mathbf{E}_2\boldsymbol{\delta}_k + \mathbf{E}_3\mathbf{z}_k \leq \mathbf{E}_1\mathbf{u}_k + \mathbf{E}_4\mathbf{x}_k + \mathbf{E}_5$$

where

$$\mathbf{x} = \begin{pmatrix} \mathbf{x}_c \\ \mathbf{x}_l \end{pmatrix} \in \Re^{n_c} \times \{0,1\}^{n_l} , \quad \mathbf{u} = \begin{pmatrix} \mathbf{u}_c \\ \mathbf{u}_l \end{pmatrix} \in \Re^{m_c} \times \{0,1\}^{m_l} ,$$

$$\mathbf{y} = \begin{pmatrix} \mathbf{y}_c \\ \mathbf{y}_l \end{pmatrix} \in \Re^{p_c} \times \{0,1\}^{p_l} , \quad \boldsymbol{\delta} \in \{0,1\}^{r_l} , \quad \mathbf{z} \in \Re^{r_c}$$

are respectively the vectors of continuous and binary states of the system, of continuous and binary (on/off) control inputs, of output signals, of auxiliary binary and continuous variables.

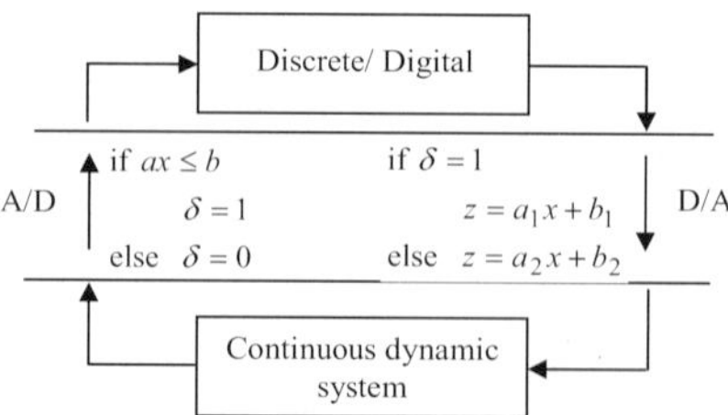

Figure 1: MLD model structure.

The auxiliary variables are introduced when translating propositional logic into linear inequalities

as described in Figure 1. All matrices appearing in (1) can be obtained through the specification language HYSDEL (Hybrid System Description Language), see (Torrisi et al., 2000).

2.2 Piecewise Affine Model

Another framework for discrete time hybrid systems is the PWA model (Sontag, 1981), defined as

$$S^i : \begin{cases} \mathbf{x}_{k+1} = \mathbf{A}^i\mathbf{x}_k + \mathbf{B}^i\mathbf{u}_k + \mathbf{f}^i \\ \mathbf{y}_k = \mathbf{C}^i\mathbf{x}_k + \mathbf{g}^i \end{cases} , \text{ for: } \begin{bmatrix} \mathbf{x}_k \\ \mathbf{u}_k \end{bmatrix} \in \chi_i \quad (2)$$

where $\{\chi_i\}_{i=1}^s$ is the polyhedral partition of the state and input spaces (s being the number of subsystems within the partition). Each χ_i is given by

$$\chi_i = \left\{ \begin{bmatrix} \mathbf{x}_k \\ \mathbf{u}_k \end{bmatrix} \middle| \ \mathbf{Q}^i \begin{bmatrix} \mathbf{x}_k \\ \mathbf{u}_k \end{bmatrix} \leq \mathbf{q}^i \right\} \quad (3)$$

where $\mathbf{x}_k, \mathbf{u}_k, \mathbf{y}_k$ denote the state, input and output vector respectively at instant k. Each subsystem S^i defined by the 7-uple $\left(\mathbf{A}^i, \mathbf{B}^i, \mathbf{C}^i, \mathbf{f}^i, \mathbf{g}^i, \mathbf{Q}^i, \mathbf{q}^i\right)$, $i \in (1,2,\cdots,s)$ is a component of the PWA system (2). $\mathbf{A}^i \in \Re^{n \times n}, \mathbf{B}^i \in \Re^{n \times m}, \mathbf{C}^i \in \Re^{p \times n}, \mathbf{Q}^i \in \Re^{p_i(n+m)}$ and $\mathbf{f}^i, \mathbf{g}^i, \mathbf{q}^i$ are suitable constant vectors or matrices, where n, m, p are respectively the number of states, inputs, outputs, and p_i is the number of hyperplanes defining the i-polyhedral. In this formalism, a logical control input is considered by developing an affine model for each input value (1/0), defining linear inequality constraints linking the model with the relevant input value.

It has been shown in (Bemporad et al., 2000), that MLD and PWA models are equivalent, which enables transformation from one model to the other. A MLD model can be transferred to a PWA model with the number of subsystems inside the polyhedral partition equal to all possible combination of all the integer variables of the MLD model (Bemporad *et al.*, 2000) (a technique for avoiding empty region is presented in (Bemporad, 2003))

$$s = 2^{n_l + m_l + r_l} \quad (4)$$

3 MODEL PREDICTIVE CONTROL

Model predictive control (MPC) has proved to efficiently control a wide range of applications in industry, including systems with long delay times, non-minimum phase, unstable, multivariable and constrained systems.

The main idea of predictive control is the use of a plant model to predict future outputs of the system. Based on this prediction, at each sampling period, a sequence of future control values is elaborated through an on-line optimization process, which maximizes the tracking performance while satisfying constraints. Only the first value of this optimal sequence is applied to the plant according to the 'receding' horizon strategy (Dumur and Boucher, 1998).

Considering the particular class of hybrid systems previously described, implementing MPC leads to a problem including at each prediction step the states vector, the inputs vector (both continuous and discrete), the dynamic equation and linear constraints, for which a quadratic cost function has to be optimized. Two classical approaches exist for solving this optimization problem, the Branch and Bound technique that can be used with the MLD formalism and the enumeration of all possible logical system combinations at each prediction time corresponding to all particular scenarios of logical inputs and modes used with the PWA formalism.

3.1 Model Predictive Control for the MLD Systems

For a MLD system of the form (1), the following model predictive control problem is considered. Let k be the current time, $\mathbf{x}_k$ the current state, $(\mathbf{x}_e, \mathbf{u}_e)$ an equilibrium pair or a reference trajectory value, $k + N$ the final time, find $\mathbf{u}_k^{k+N-1} = (\mathbf{u}_k \ \cdots \ \mathbf{u}_{k+N-1})$ the sequence which moves the state from $\mathbf{x}_k$ to $\mathbf{x}_e$ and minimizes

$$\min_{\mathbf{u}_k^{k+N-1}} J(\mathbf{u}_k^{k+N-1}, \mathbf{x}_k) = \sum_{i=0}^{N-1} \left\| \mathbf{u}_{k+i} - u_e \right\|_{Q_1}^2$$

$$+ \left\| \boldsymbol{\delta}_{k+i/k} - \boldsymbol{\delta}_e \right\|_{Q_2}^2 + \left\| \mathbf{z}_{k+i/k} - \mathbf{z}_e \right\|_{Q_3}^2 \qquad (5)$$

$$+ \left\| \mathbf{x}_{k+i+1/k} - \mathbf{x}_e \right\|_{Q_4}^2 + \left\| \mathbf{y}_{k+i/k} - \mathbf{y}_e \right\|_{Q_5}^2$$

subject to (1), where N is the prediction horizon, $\boldsymbol{\delta}_e$, $\mathbf{z}_e$ are the auxiliary variables of the equilibrium point or the reference trajectory value, calculated by solving a MILP problem for the inequality equation of (1). $\mathbf{x}_{k+i/k}$ denotes the predicted state vector at time $k + i$, obtained by applying the input sequence $\mathbf{u}_k^{k+N-1}$ to model (1) starting from the current state $\mathbf{x}_k$ (same for the other input and output variables), $\mathbf{Q}_i = \mathbf{Q}_i' > 0$, for $i = 1,4$, and $\mathbf{Q}_i = \mathbf{Q}_i' \geq 0$, for $i = 2,3,5$.

The optimization procedure of (5) leads to MIQP problems with the following optimization vector

$$\boldsymbol{\chi} = [\mathbf{u}_k, \cdots, \mathbf{u}_{k+N-1}, \boldsymbol{\delta}_k, \cdots, \boldsymbol{\delta}_k$$
$$+N-1, \mathbf{z}_k, \cdots, \mathbf{z}_{k+N-1}]^{\mathrm{T}} \qquad (6)$$

The number of binary optimization variables is $L = N(m_l + r_l)$. In the worst case, the maximum number of solved QP problems is

$$\text{No of QPs} = 2^{L+1} - 1 \qquad (7)$$

So the main drawback of this MLD formalism remains the computational burden related to the complexity of the derived Mixed Integer Quadratic Programming (MIQPs) problems. Indeed MIQPs problems are classified as NP-complete, so that in the worst case, the optimization time grows exponentially with the problem size, even if branch and bounds methods (B&B) may reduce this solution time (Fletcher and Leyffer, 1995).

3.2 Model Predictive Control for the PWA Systems

Considering the PWA system under the form (2), assuming that the current state $\mathbf{x}_k$ is known the model predictive control requires solving at each time step (Pena et al., 2003).

$$\min_{\mathbf{u}_k^{k+N-1}} J(\mathbf{u}_k^{k+N-1}, \mathbf{x}_k) = \sum_{i=1}^{N} q_{ii} \left\| \mathbf{y}_{k+i/k} - \mathbf{w}_{k+i} \right\|^2$$

$$+ \sum_{i=0}^{N-1} r_{ii} \left\| \mathbf{u}_{k+i} \right\|^2 \qquad (8)$$

$$\text{s.t.} : \mathbf{u}_{\min} \leq \mathbf{u}_{k+i} \leq \mathbf{u}_{\max}$$

where N is the prediction horizon, $\mathbf{w}_{k+i}$ is the output reference, and $\mathbf{y}_{k+i/k}$ denotes the predicted output vector at time $k + i$, obtained by applying the input sequence $\mathbf{u}_k^{k+N-1} = (\mathbf{u}_k \ \cdots \ \mathbf{u}_{k+N-1})$ to the system starting from the current state $\mathbf{x}_k$. q_{ii}, r_{ii} are the elements of Q, R weighting diagonal matrices.

In order to solve this equation the model applied at each instant has to be determined and all potential sequences of subsystems $I = \{I_k, I_{k+1}, \cdots, I_{k+N-1}\}$ have to be examined, where I_{k+i} is one sub-region among the s subsystems at prediction time i for $i = 1,2,\cdots,N-1$. As for each model the value of the logical variable is fixed, the MPC problem is solved by a QP for each potential sequence. As the current state $\mathbf{x}_k$ is known, the starting region according to the state partition is identified. But the initial sub-region related to the current input control is not known as it appears in the domain definition (3). Similarly, the next steps subsystems are also unknown, depending on the applied control sequence. In general, all potential sequences of subsystems I have to be examined, which increases the computation burden. If no constraints are considered, the number of possible sequences for a prediction horizon N is $m_p s^{N-1}$, where m_p is the number of all possible sub-regions at instant k

according to the input space partition. In order to solve the MPC problem of (8), the number of quadratic programming problems to be solved is

$$\text{No QPs} = m_p \, s^{N-1} \tag{9}$$

4 MPC FOR STATE PARTITION BASED MLD (SPMLD) FORMALISM

4.1 The SPMLD Formalism

The SPMLD model is a mixed approach where in each region of the feasible space a simple MLD model is developed. Starting from the MLD model, the auxiliary binary variables are divided into two groups $\boldsymbol{\delta} = [\boldsymbol{\delta}_1 \ \boldsymbol{\delta}_2]^T$. Where $\boldsymbol{\delta}_2 \in \{0,1\}^{r_{l2}}$ is chosen in order to include the δ variables that are not directly depending on the state variables (the inequalities that define δ variables are not depending on $\mathbf{x}$), and $\boldsymbol{\delta}_1 \in \{0,1\}^{r_{l1}}$ depending on $\mathbf{x}$. This partition will be further justified. The SPMLD model is then developed by giving $\boldsymbol{\delta}_1$ a constant value: for each possible combination of $\boldsymbol{\delta}_1$ a sub-region is defined with the corresponding $\mathbf{Q}^i, \mathbf{q}^i$ constraints as in (3). As some logical combinations may not be feasible, the number of sub-regions of the polyhedral partition is

$$s \leq 2^{r_{l1}} \tag{10}$$

Consequently, this model requires a smaller number of sub-regions than the classical PWA model for the same modeled system. Each sub-region has its own dynamic described in the same way as (1) but with a simpler MLD model that represents the system behavior in this sub-region and includes only the active variables in this sub-region. This partition always implies a reduction in the size of $\mathbf{z}$ and/or $\mathbf{u}$. For example, some control variables may not be active in sub-regions and the auxiliary continuous variables $\mathbf{z}$ depending on the $\boldsymbol{\delta}_1$ variables may disappear or become fixed as $\boldsymbol{\delta}_1$ is fixed where:

$$z = \delta f(x) \rightarrow z = \begin{cases} 0 & \text{if } \delta = 0 \\ f(x) & \text{if } \delta = 1 \end{cases} \tag{11}$$

Consequently, simplified sub-regions models can be derived, an example of this simplification is given in the application section.

The system is thus globally modeled as

$$\mathbf{x}_{k+1} = \mathbf{A}^i \mathbf{x}_k + \mathbf{B}_1^i \mathbf{u}_k + \mathbf{B}_2^i \boldsymbol{\delta}_k + \mathbf{B}_3^i \mathbf{z}_k$$
$$\mathbf{y}_k = \mathbf{C}^i \mathbf{x}_k + \mathbf{D}_1^i \mathbf{u}_k + \mathbf{D}_2^i \boldsymbol{\delta}_k + \mathbf{D}_3^i \mathbf{z}_k \tag{12}$$
$$\mathbf{E}_2^i \boldsymbol{\delta}_k + \mathbf{E}_3^i \mathbf{z}_k \leq \mathbf{E}_1^i \mathbf{u}_k + \mathbf{E}_4^i \mathbf{x}_k + \mathbf{E}_5^i$$

$\mathbf{A}^i, \mathbf{B}_{1-3}^i, \mathbf{C}^i, \mathbf{D}_{1-3}^i, \mathbf{E}_{1-5}^i$ are the matrices of the ith MLD model defining the dynamics into that sub-region. $\mathbf{Q}^i, \mathbf{q}^i$ constraints has to be included in (12).

4.2 Reformulation of the MPC Solution

At this stage, the MPC technique developed for the PWA formalism must be rewritten to fit the new SPMLD model. Consider the initial subsystem I_k

$$\mathbf{x}_{k+1} = \mathbf{A}^k \mathbf{x}_k + \mathbf{B}_1^k \mathbf{u}_k + \mathbf{B}_2^k \boldsymbol{\delta}_k + \mathbf{B}_3^k \mathbf{z}_k$$
$$\mathbf{y}_k = \mathbf{C}^k \mathbf{x}_k + \mathbf{D}_1^k \mathbf{u}_k + \mathbf{D}_2^k \boldsymbol{\delta}_k + \mathbf{D}_3^k \mathbf{z}_k \tag{13}$$

with $\quad -\mathbf{E}_1^k \mathbf{u}_k + \mathbf{E}_3^k \mathbf{z}_k \leq \mathbf{E}_4^k \mathbf{x}_k + \mathbf{E}_5^k - \mathbf{E}_2^k \boldsymbol{\delta}_k$

Where $\mathbf{A}^k$ will now denote for simplification purposes the $\mathbf{A}^i$ matrix of model i at instant k (the same notations are used for $\mathbf{B}_{1-3}^k, \mathbf{C}^k, \mathbf{D}_{1-3}^k, \mathbf{E}_{1-5}^k$). For a given sequence over the prediction horizon N i.e. for $I = \{I_k, I_{k+1}, \cdots, I_{k+N-1}\}$, the system is recursively defined as follows

$$\overline{\mathbf{x}} = \mathbf{F}_x \mathbf{x}_k + \mathbf{H}_x \overline{\mathbf{u}} + \mathbf{P}_x \overline{\boldsymbol{\delta}} + \mathbf{G}_x \overline{\mathbf{z}}$$
$$\overline{\mathbf{y}} = \mathbf{F}_y \mathbf{x}_k + \mathbf{H}_y \overline{\mathbf{u}} + \mathbf{P}_y \overline{\boldsymbol{\delta}} + \mathbf{G}_y \overline{\mathbf{z}} \tag{14}$$

Where $\overline{\mathbf{x}} = [\mathbf{x}_{k+1} \ \mathbf{x}_{k+2} \ \cdots \ \mathbf{x}_{k+N}]^T$,
$\overline{\mathbf{u}} = [\mathbf{u}_k \ \cdots \ \mathbf{u}_{k+N-1}]^T, \overline{\mathbf{y}} = [\mathbf{y}_k \ \cdots \ \mathbf{y}_{k+N-1}]^T$
$\overline{\mathbf{z}} = [\mathbf{z}_k \ \cdots \ \mathbf{z}_{k+N-1}]^T, \overline{\boldsymbol{\delta}} = [\boldsymbol{\delta}_k \ \cdots \ \boldsymbol{\delta}_{k+N-1}]^T$

$$\mathbf{F}_x = \begin{bmatrix} \mathbf{A}^k \\ \mathbf{A}^{k+1}\mathbf{A}^k \\ \vdots \\ \mathbf{A}^{k+N-1}\cdots\mathbf{A}^k \end{bmatrix}, \mathbf{F}_y = \begin{bmatrix} \mathbf{C}^k \\ \mathbf{C}^{k+1}\mathbf{A}^k \\ \vdots \\ \mathbf{C}^{k+N-1}\mathbf{A}^{k+N-2}\cdots\mathbf{A}^k \end{bmatrix}$$

$$\mathbf{H}_x = \begin{bmatrix} \mathbf{B}_1^k & 0 & \cdots & 0 \\ \mathbf{A}^{k+1}\mathbf{B}_1^k & \mathbf{B}_1^{k+1} & \cdots & 0 \\ \vdots & \vdots & \ddots & \vdots \\ \mathbf{A}^{k+N-1}\cdots\mathbf{A}^{k+1}\mathbf{B}_1^k & \mathbf{A}^{k+N-1}\cdots\mathbf{A}^{k+2}\mathbf{B}_1^{k+1} & \cdots & \mathbf{B}_1^{k+N-1} \end{bmatrix}$$

$$\mathbf{G}_x = \begin{bmatrix} \mathbf{B}_3^k & 0 & \cdots & 0 \\ \mathbf{A}^{k+1}\mathbf{B}_3^k & \mathbf{B}_3^{k+1} & \cdots & 0 \\ \vdots & \vdots & \ddots & \vdots \\ \mathbf{A}^{k+N-1}\cdots\mathbf{A}^{k+1}\mathbf{B}_3^k & \mathbf{A}^{k+N-1}\cdots\mathbf{A}^{k+2}\mathbf{B}_3^{k+1} & \cdots & \mathbf{B}_3^{k+N-1} \end{bmatrix}$$

$$\mathbf{P}_x = \begin{bmatrix} \mathbf{B}_2^k & 0 & \cdots & 0 \\ \mathbf{A}^{k+1}\mathbf{B}_2^k & \mathbf{B}_2^{k+1} & \cdots & 0 \\ \vdots & \vdots & \ddots & \vdots \\ \mathbf{A}^{k+N-1}\cdots\mathbf{A}^{k+1}\mathbf{B}_2^k & \mathbf{A}^{k+N-1}\cdots\mathbf{A}^{k+2}\mathbf{B}_2^{k+1} & \cdots & \mathbf{B}_2^{k+N-1} \end{bmatrix}$$

$$\mathbf{H}_y = \begin{bmatrix} \mathbf{D}_1^k & 0 & \cdots & 0 \\ \mathbf{C}^{k+1}\mathbf{B}_1^k & \mathbf{D}_1^{k+1} & \cdots & 0 \\ \vdots & \vdots & \ddots & \vdots \\ \mathbf{C}^{k+N-1}\mathbf{A}^{k+N-2}\cdots\mathbf{A}^{k+1}\mathbf{B}_1^k & \mathbf{C}^{k+N-1}\mathbf{A}^{k+N-2}\cdots\mathbf{A}^{k+2}\mathbf{B}_1^{k+1} & \cdots & \mathbf{D}_1^{k+N-1} \end{bmatrix}$$

$$\mathbf{G}_y = \begin{bmatrix} \mathbf{D}_3^k & 0 & \cdots & 0 \\ \mathbf{C}^{k+1}\mathbf{B}_3^k & \mathbf{D}_3^{k+1} & \cdots & 0 \\ \vdots & \vdots & \ddots & \vdots \\ \mathbf{C}^{k+N-1}\mathbf{A}^{k+N-2}\cdots\mathbf{A}^{k+1}\mathbf{B}_3^k & \mathbf{C}^{k+N-1}\mathbf{A}^{k+N-2}\cdots\mathbf{A}^{k+2}\mathbf{B}_3^{k+1} & \cdots & \mathbf{D}_3^{k+N-1} \end{bmatrix}$$

$$\mathbf{P}_y = \begin{bmatrix} \mathbf{D}_2^k & 0 & \cdots & 0 \\ \mathbf{C}^{k+1}\mathbf{B}_2^k & \mathbf{D}_2^{k+1} & \cdots & 0 \\ \vdots & \vdots & \ddots & \vdots \\ \mathbf{C}^{k+N-1}\mathbf{A}^{k+N-2}\cdots\mathbf{A}^{k+1}\mathbf{B}_2^k & \mathbf{C}^{k+N-1}\mathbf{A}^{k+N-2}\cdots\mathbf{A}^{k+2}\mathbf{B}_2^{k+1} & \cdots & \mathbf{D}_2^{k+N-1} \end{bmatrix}$$

Then the MPC optimization problem (8) leads to the following cost function

$$\mathbf{J} = \begin{bmatrix} \overline{\mathbf{u}} & \overline{\mathbf{z}} & \overline{\boldsymbol{\delta}} \end{bmatrix} \mathbf{H}_o \begin{bmatrix} \overline{\mathbf{u}} \\ \overline{\mathbf{z}} \\ \overline{\boldsymbol{\delta}} \end{bmatrix} + 2\mathbf{f}_o \begin{bmatrix} \overline{\mathbf{u}} \\ \overline{\mathbf{z}} \\ \overline{\boldsymbol{\delta}} \end{bmatrix} + \mathbf{g}_o \tag{15}$$

where $\mathbf{H}_o = \begin{bmatrix} \mathbf{H}_y^T \overline{\mathbf{Q}}\mathbf{H}_y + \overline{\mathbf{R}} & \mathbf{H}_y^T \overline{\mathbf{Q}}\mathbf{G}_y & \mathbf{H}_y^T \overline{\mathbf{Q}}\mathbf{P}_y \\ \mathbf{G}_y^T \overline{\mathbf{Q}}\mathbf{H}_y & \mathbf{G}_y^T \overline{\mathbf{Q}}\mathbf{G}_y & \mathbf{G}_y^T \overline{\mathbf{Q}}\mathbf{P}_y \\ \mathbf{P}_y^T \overline{\mathbf{Q}}\mathbf{H}_y & \mathbf{P}_y^T \overline{\mathbf{Q}}\mathbf{G}_y & \mathbf{P}_y^T \overline{\mathbf{Q}}\mathbf{P}_y \end{bmatrix}$

$$\mathbf{f}_o = \begin{bmatrix} \mathbf{x}_k^T \mathbf{F}_y^T \overline{\mathbf{Q}}\mathbf{H}_y - \overline{\mathbf{w}}^T \overline{\mathbf{Q}}\mathbf{H}_y \\ \mathbf{x}_k^T \mathbf{F}_y^T \overline{\mathbf{Q}}\mathbf{G}_y - \overline{\mathbf{w}}^T \overline{\mathbf{Q}}\mathbf{G}_y \\ \mathbf{x}_k^T \mathbf{F}_y^T \overline{\mathbf{Q}}\mathbf{P}_y - \overline{\mathbf{w}}^T \overline{\mathbf{Q}}\mathbf{P}_y \end{bmatrix}^T$$

$$\mathbf{g}_o = [\mathbf{x}_k^T \mathbf{F}_y^T \overline{\mathbf{Q}}\mathbf{F}_y \mathbf{x}_k - 2\overline{\mathbf{w}}^T \overline{\mathbf{Q}}\mathbf{F}_y \mathbf{x}_k + \overline{\mathbf{w}}^T \overline{\mathbf{Q}}\overline{\mathbf{w}}]$$

$$\overline{\mathbf{R}} = \mathrm{diag}[r_{ii}], \quad \left(\overline{\mathbf{R}} = \overline{\mathbf{R}}^T \succ 0\right)$$

$$\overline{\mathbf{Q}} = \mathrm{diag}[q_{ii}], \quad \left(\overline{\mathbf{Q}} = \overline{\mathbf{Q}}^T \succ 0\right)$$

This technique allows choosing different weighting factors for each sub-region according to its priority.

The constraints over the state and input domains for each sub-region are included in the inequality equation of the MLD model of that sub-region using the HYSDEL program. The MPC optimization problem (15) is solved subject to the constraints

$$\begin{bmatrix} \mathbf{M}_1 & \mathbf{M}_3 & \mathbf{M}_2 \end{bmatrix} \begin{bmatrix} \overline{\mathbf{u}} \\ \overline{\mathbf{z}} \\ \overline{\boldsymbol{\delta}} \end{bmatrix} \leq \mathbf{N} \tag{16}$$

where $\mathbf{N} = \begin{bmatrix} \mathbf{E}_4^k \mathbf{x}_k + \mathbf{E}_5^k \\ \mathbf{E}_4^{k+1}\mathbf{A}^k \mathbf{x}_k + \mathbf{E}_5^{k+1} \\ \vdots \\ \mathbf{E}_4^{k+N-1}\mathbf{A}^{k+N-2}\cdots\mathbf{A}^k \mathbf{x}_k + \mathbf{E}_5^{k+N-1} \end{bmatrix}$

$$\mathbf{M}_1 = \begin{bmatrix} -\mathbf{E}_1^k & 0 & \cdots & 0 \\ -\mathbf{E}_4^{k+1}\mathbf{B}_1^k & -\mathbf{E}_1^{k+1} & \cdots & 0 \\ \vdots & \vdots & \ddots & 0 \\ -\mathbf{E}_4^{k+N-1}\mathbf{A}^{k+N-2}\cdots\mathbf{A}^{k+1}\mathbf{B}_1^k & -\mathbf{E}_4^{k+N-1}\mathbf{A}^{k+N-2}\cdots\mathbf{A}^{k+2}\mathbf{B}_1^{k+1} & \cdots & -\mathbf{E}_1^{k+N-1} \end{bmatrix}$$

$$\mathbf{M}_2 = \begin{bmatrix} \mathbf{E}_2^k & 0 & \cdots & 0 \\ -\mathbf{E}_4^{k+1}\mathbf{B}_2^k & \mathbf{E}_2^{k+1} & \cdots & 0 \\ \vdots & \vdots & \ddots & 0 \\ -\mathbf{E}_4^{k+N-1}\mathbf{A}^{k+N-2}\cdots\mathbf{A}^{k+1}\mathbf{B}_2^k & \cdots & -\mathbf{E}_4^{k+N-1}\mathbf{B}_2^{k+N-2} & \mathbf{E}_2^{k+N-1} \end{bmatrix}$$

$$\mathbf{M}_3 = \begin{bmatrix} \mathbf{E}_3^k & 0 & \cdots & 0 \\ -\mathbf{E}_4^{k+1}\mathbf{B}_3^k & \mathbf{E}_3^{k+1} & \cdots & 0 \\ \vdots & \vdots & \ddots & 0 \\ -\mathbf{E}_4^{k+N-1}\mathbf{A}^{k+N-2}\cdots\mathbf{A}^{k+1}\mathbf{B}_3^k & \cdots & -\mathbf{E}_4^{k+N-1}\mathbf{B}_3^{k+N-2} & \mathbf{E}_3^{k+N-1} \end{bmatrix}$$

The number of binary optimization variables, with $\boldsymbol{\delta}_1$ known and constant, is given by the relation

$$L = \sum_{i=0}^{N-1} m_{lj}^i + r_{l2j}^i, \qquad j \in \{1,2,\cdots,s\} \tag{17}$$

Where m_{lj}^i, r_{l2j}^i are the number of modeled logical control and $\boldsymbol{\delta}_2$ elements respectively in the jth sub-region at prediction time i.

Therefore, if the sequence I is fixed, the problem can be solved minimizing (15) subject to the constraints of (16). But, as only I_k is known (where $\mathbf{x}(k)$ is considered as known, and $\boldsymbol{\delta}_1(k)$ depends on $\mathbf{x}(k)$), all possible sequences as in Figure 2 have to be solved. So the number of possible sequences is s^{N-1}. The optimal solution is provided by the resolution of these s^{N-1} MIQPs. In order to find the solution more quickly, these problems are not solved independently and the optimal value of the criterion provided by the solved MIQP is used to bind the result of the others. It can then be used by the B&B algorithm to cut branches.

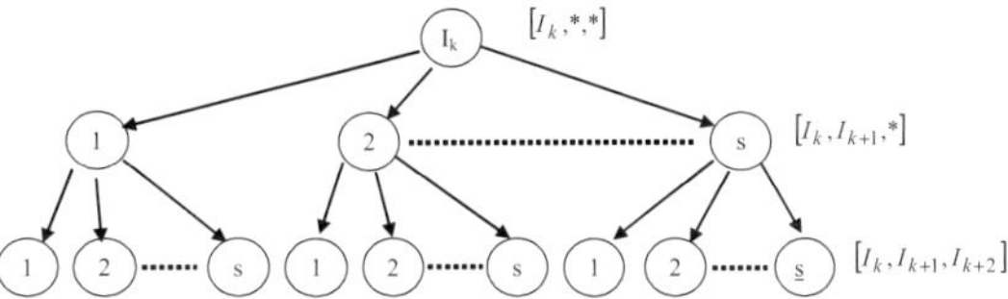

Figure 2: State transition graph of the MPC optimization problem for a system under the SPMLD form ($N = 3$).

4.3 Compared Computational Burden

The global complexity of the MPC resolution with systems under the SPMLD form is reduced. First the related number of subsystems s is smaller than that with the classical PWA model. Then, the B&B technique considerably decreases the number of solved QPs. The index sequence I imposes the successive values of $\boldsymbol{\delta}_1$ over the prediction horizon and then the succession of region on the state space partition the state has to reach. This leads to non feasible solutions in many sub-problems, effectively reducing the number of solved QPs according to the B&B technique. This is why we partition $\boldsymbol{\delta}$ vector.

First, the SPMLD technique is faster than the classical MLD because for a known sequence of index I, only $2^{\eta_1(N-1)}$ simple B&B trees with only $\mathbf{u}, \boldsymbol{\delta}_2, \mathbf{z}$ optimization variables have to be solved; i.e. smaller number of optimization binary variables L (17), and simpler MLD models as previously explained. Moreover, as explained, the optimization algorithm just has to look for the control sequence that could force the system to follow the index I and optimize the cost function with respect to all the associated constraints. In many root nodes at level F (Figure 3), this leads to non-feasible solution (more frequent than in classical MLD approach), due to non feasible sequence whatever the value of the control inputs are, thus that MIQP will then quickly be eliminated. Furthermore, if there is a solution for a B&B tree at level F, it is considered as an upper bound for the global optimized solution for all the following B&B trees, which reduces the number of solved QPs according to B&B technique.

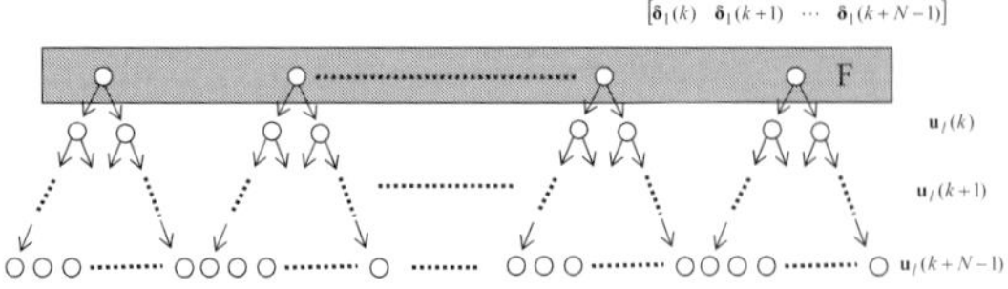

Figure 3: B&B trees for optimization with SPMLD.

Then, the SPMLD technique is obviously faster than the classical PWA technique. First the initial index I_k is completely known as the space partition only depends on $\mathbf{x}(k)$ and not on $\mathbf{u}$ (as in the PWA model where m_p possible subsystems at instant k have to be examined). In addition, SPMLD models use the B&B technique, which considerably reduces the number of solved QPs while in classical PWA systems all the QPs must be solved.

4.4 Further Improvements of the Optimization Time

Two different techniques can be considered to reduce the computation load for real time applications. The first one introduces the control horizon N_u, which reduces the number of unknown future control values, i.e. $\mathbf{u}(k+i)$ is constant for $i \geq N_u$. This decreases the number of binary optimization variables (17) and the optimization time

$$L_{N_u} = \sum_{i=0}^{N_u-1} m_{lj}^i + \sum_{i=0}^{N-1} r_{l2j}^i, \quad j \in \{1,2,\cdots,s\} \quad (18)$$

The second one, called the reach set, aims at determining the set of possible regions that can be reached from the actual region in next few sampling times (Kerrigan, 2000). That is, all sequences that cannot be obtained are not considered.

5 APPLICATION

5.1 Description of the Benchmark

The proposed control strategy is applied on the three tanks benchmark used in (Bemporad et al., 1999). The simplified physical description of the three tanks system proposed by COSY as a standard benchmark for control and fault detection problems is presented in Figure 4 (Dolanc et al., 1997).

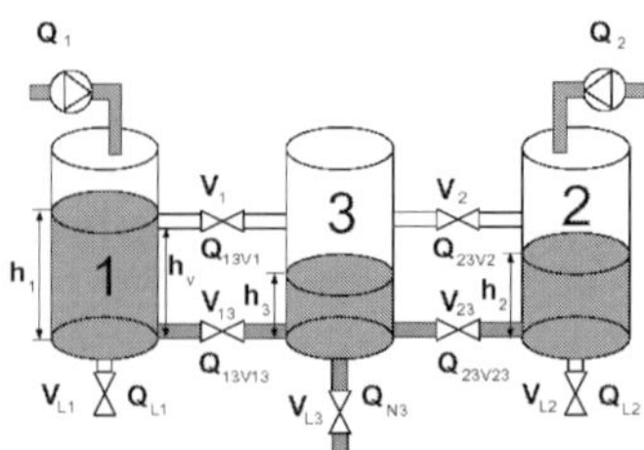

Figure 4: COSY three tanks benchmark system.

The system consists of three tanks, filled by two independent pumps acting on tanks 1 and 2, continuously manipulated from 0 up to a maximum flow Q_1 and Q_2 respectively. Four switching valves V_1, V_2, V_{13} and V_{23} control the flow between the tanks, those valves are assumed to be either completely opened or closed ($V_i = 1$ or 0 respectively). The V_{L3} manual valve controls the nominal outflow of the middle tank. It is assumed in further simulations that the V_{L1} and V_{L2} valves are always closed and V_{L3} is open. The liquid levels to be controlled are denoted h_1, h_2 and h_3 for each tank respectively. The conservation of mass in the tanks provides the following differential equations

$$\dot{h}_1 = \frac{1}{A}(Q_1 - Q_{13V1} - Q_{13V13})$$

$$\dot{h}_2 = \frac{1}{A}(Q_2 - Q_{23V2} - Q_{23V23}) \quad (19)$$

$$\dot{h}_3 = \frac{1}{A}(Q_{13V1} + Q_{13V13} + Q_{23V2} + Q_{23V23} - Q_N)$$

where the Qs denote the flows and A is the cross-sectional area of each of the tanks. A MLD model is derived as developed in (Bemporad et al., 1999), introducing the following variables

$$\mathbf{x} = [h_1 \ h_2 \ h_3]'$$
$$\mathbf{u} = [Q_1 \ Q_2 \ V_1 \ V_2 \ V_{13} \ V_{23}]'$$
$$\boldsymbol{\delta} = [\delta_{01} \ \delta_{02} \ \delta_{03}]' \quad (20)$$
$$\mathbf{z} = [z_{01} \ z_{02} \ z_{03} \ z_1 \ z_2 \ z_{13} \ z_{23}]'$$

where
$$[\delta_{0i}(t) = 1] \leftrightarrow [h_i(t) \geq h_v] \quad i = 1,2,3$$
$$z_{0i}(t) = \delta_{0i}(t)(h_i(t) - h_v) \quad i = 1,2,3$$
$$z_i(t) = V_i(z_{0i}(t) - z_{03}(t)) \quad i = 1,2$$
$$z_{i3}(t) = V_{i3}(h_i(t) - h_3) \quad i = 1,2$$

5.2 Application of MPC for the SPMLD Formalism

In this system, $\boldsymbol{\delta}_1 = \boldsymbol{\delta}$ since the three introduced auxiliary binary variables depend on the states, thus $r_{l1} = r_l$ and the number of sub-systems is

$$s = 2^{r_{l1}} = 8 \quad (21)$$

Inside each sub-region, a simple MLD model is developed, that takes into account only the system dynamics in this sub-region. In some sub-regions a reduction in the size of $\mathbf{u}$ and $\mathbf{z}$ appears; for example in the sub-region where $\boldsymbol{\delta}_1' = \begin{bmatrix} 0 & 0 & 0 \end{bmatrix}$ it clearly appears that the two valves V_1 and V_2 of the input vector are not in progress, as the liquid level in this region is always less than the valves level. Consequently, the continuous auxiliary variables $\{z_{0i}\}_{i=1,2,3}$ and $\{z_i\}_{i=1,2}$ corresponding to the flows that pass through the upper pipes are useless. It results from this a simple model with:

$$\mathbf{x} = [h_1 \ h_2 \ h_3]', \ \mathbf{u} = [Q_1 \ Q_2 \ V_{13} \ V_{23}]'$$
$$\mathbf{z}_1 = [\, z_{13} \ z_{23}]' \tag{22}$$

Let us consider now the following specification: starting from zero levels (the three tanks being completely empty), the objective of the control strategy is to reach the liquid levels $h_1 = 0.5\,\text{m}$, $h_2 = 0.5\,\text{m}$ and $h_3 = 0.1\,\text{m}$. The MPC technique for a SPMLD model has been implemented in simulation to reach the level specification with $N = 2$. The results are presented on Figure 5 for the tanks levels and on Figure 6 for the control signals.

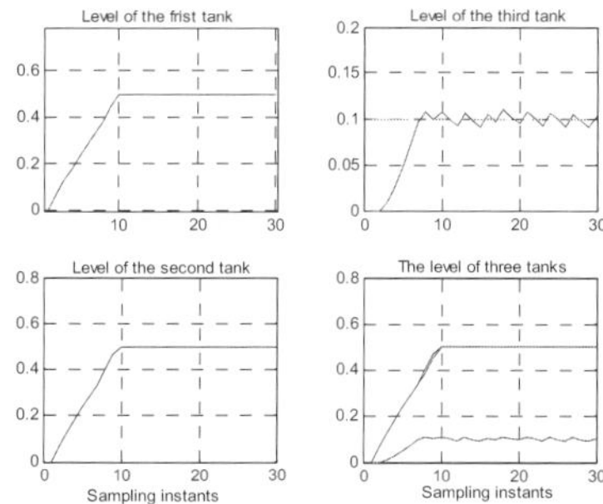

Figure 5: Water levels in the three tanks.

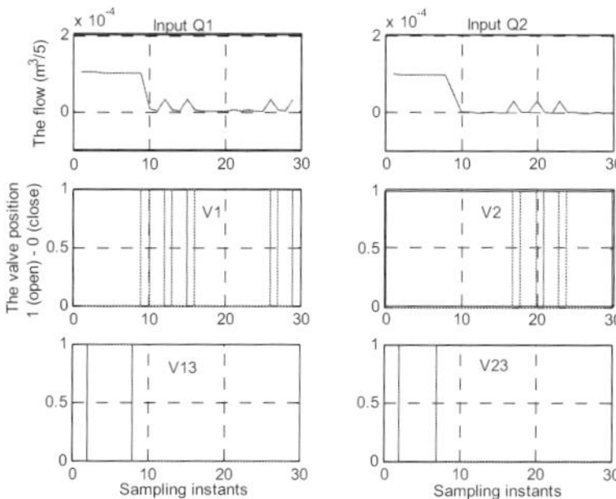

Figure 6: Controlled variables.

The level of the third tank oscillates around 0.1 as $h_3 = 0.1$ does not correspond to an equilibrium point. Consequently, the system opens and closes the two valves V_1 and V_2 to maintain the level in the third tank around the desired level of 0.1m.

5.3 Comparison of the Approaches

As a comparison purpose between the SPMLD model, the classical MLD model and the classical PWA model strategies, the same previous level specification has been considered with $N = 2$. The MLD model described in (Bemporad et al., 1999) has been used for the three tanks modeled by (20); this MLD model transfers to a PWA model with $s = 128$ subsystems (with 28 empty regions). The classical PWA model has not been developed as it needs 100 sub-models and is in fact not required to compare complexity. For that comparison, looking at the number of QPs that have to be solved during optimization is sufficient.

Table 1 illustrates for $N = 2$ the total time required to reach the specification level, the total number of QPs solved, and the maximum time and maximum QPs to find the optimized solution at each iteration. It can be seen that the difference between the SPMLD technique and the other classical techniques is quite large, the SPMLD model allowing real time implementation and avoiding exponential explosion of the algorithm (the sampling time of the three tanks benchmark is 10 s.). All data given above were obtained using the MIQP Matlab code (Bemporad and Mignone, 2000), on a 1.8 MHz PC with 256 Mo of ram. Same comparisons are presented with $N = 3$ in table 2.

Table 1. Comparison of performances obtained with the SPMLD model, the classical MLD model and the classical PWA model for $N = 2$.

Approach	No of QPs solved	Max. No. QPs / step	Total time	Max. time / step
Classical PWA	8 800	1 600	*	*
Classical MLD	11 130	2 089	822.97 s	138.97 s
SPMLD	832	218	15.28 s	3.90 s

Table 2. Comparison of performances obtained with the SPMLD, MLD and PWA models for $N = 3$.

Approach	No of QPs solved	Max. No. QPs / step	Total time	Max. time / step
Classical PWA	880 000	160 000	*	*
Classical MLD	25 606	6 867	5243.6 s	1 147.80 s
SPMLD	3 738	1 054	137.54 s	37.65 s

This table shows that no real time implementation is possible with $N = 3$ for the SPMLD form, although the maximum time per iteration is much smaller in this case. But it must be noticed that the results in table 1 and 2 for the SPMLD model are achieved without applying techniques described in section 4.4. For example using a prediction horizon $N = 3$ and a control horizon $N_u = 1$ leads to the following results enabling real time implementation

No QPs solved = 1224, Max. No QPs/step = 326

Total optimization time = 29.24 s., Max. time/step = 7.92 s.

The technique of MPC for SPMLD systems has been examined also with a simple automata, where automata of Figure 7 have been added to V_1 and V_2 valves of the three tanks benchmark. We assumed for a simplification purpose that $\delta_{03} = 0$ i.e. the level in the third tank is always behind the h_v level.

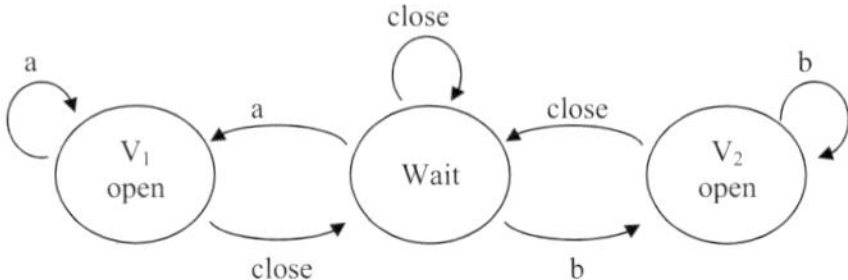

Figure 7: Added Automata to the three tanks benchmark.

The automata of Figure 7 can be presented as follows

$$V_{1open} = (wait \,\&\, a) \big| (V_{1open} \,\&\, a)$$

$$V_{2open} = (wait \,\&\, b) \big| (V_{2open} \,\&\, b) \tag{23}$$

$$Wait = close$$

The SPMLD technique succeeds to reduce the total optimization time to arrive to the specifications, from 5691.4 s for the classical MLD technique to a 173.5 s, solving 3990 QPs instead of 60468 QPs where for each sequence I, δ variables as well as the logical control variables that control the automata are known.

6 CONCLUSION

This paper presents the SPMLD formalism. It is developed by partitioning the feasible region according to the auxiliary binary elements δ_1 of the MLD model that depends on the state variables. A reformulation of the MPC strategy for this formalism has been presented. It is shown that the SPMLD model successfully improves the computational problem of the mixed Logical Dynamical (MLD) model and Piecewise Affine (PWA) model. Moreover, the partition into several sub-regions enables to define particular weighting factors according to the priority of each region. Future work may consider examining δ variables that depends on the control inputs, by partitioning the feasible region according to those variables also instead of leaving them free included in the optimization vector.

REFERENCES

Bemporad, A., 2003. "A Recursive Algorithm for Converting Mixed Logical Dynamical Systems into an Equivalent Piecewise Affine Form", *IEEE Trans. Autom. Contr.*

Bemporad, A., Ferrari-Trecate G., and Morari M., 2000. Observability and controllability of piecewise affine and hybrid systems. *IEEE Trans. Automatic Control, 45(10): 1864–1876.*

Bemporad, A., and Mignone, D., 2000.: Miqp.m: A Matlab function for solving mixed integer quadratic programs. *Technical Report.*

Bemporad, A., Mignone, D. and Morari, M., 1999. Moving horizon estimation for hybrid systems and fault detection. In *Proceedings of the American Control Conference, San Diego.*

Bemporad, A. and Morari, M., march 1999. Control of systems integrating logical, dynamics, and constraints. *Automatica, 35(3): 407-427.*

Branicky, M.S., Borkar, V.S. and Mitter, S.K., January 1998. A unified framework for hybrid control: model and optimal control theory. *IEEE Transaction. on Automatic. Control, 43(1): 31-45.*

Dolanc, G., Juricic D., Rakar A., Petrovcic J. and Vrancic, D., 1998. Three-tank Benchmark Test. *Technical Report Copernicus Project Report* CT94-02337.

Dumur, D. and Boucher, P.,1998. A Review Introduction to Linear GPC and Applications. *Journal A, 39(4), pp. 21-35.*

Fletcher, R. and Leyffer, S., 1995. Numerical experience with lower bounds for MIQP branch and bound. *Technical report, Dept. of Mathematics, University of Dundee, Scotland.*

Kerrigan, E., 2000. Robust Constraint Satisfaction: Invariant sets and predictive control. *PhD thesis University of Cambridge.*

Pena, M., Camacho, E. F. and Pinon, S., 2003. Hybrid systems for solving model predictive control of piecewise affine system. *IFAC conference, Hybrid system analysis and design Saint Malo, France.*

Sontag, E.D., April 1981: Nonlinear regulation: the piecewise linear approach. *IEEE Transaction. on Automatic. Control, 26(2): 346-358.*

Torrisi, F., Bemporad, A. and Mignone, D., 2000. Hysdel – a tool for generating hybrid models. *Technical report, AUT00-03, Automatic control laboratory, ETH Zuerich.*

EFFICIENT SYSTEM IDENTIFICATION FOR MODEL PREDICTIVE CONTROL WITH THE ISIAC SOFTWARE

Paolino Tona

Institut Français du Pétrole

1 & 4, avenue de Bois-Préau, 92852 Rueil-Malmaison Cedex - France

paolino.tona@ifp.fr

Jean-Marc Bader

Axens

BP 50802 - 89, boulevard Franklin Roosevelt, 92508 Rueil-Malmaison Cedex - France

jean-marc.bader@axens.net

Keywords: System identification, model predictive control.

Abstract: ISIAC (as Industrial System Identification for Advanced Control) is a new software package geared to meet the requirements of system identification for model predictive control and the needs of practicing advanced process control (APC) engineers. It has been designed to naturally lead the user through the different steps of system identification, from experiment planning to ready-to-use models. Each phase can be performed with minimal user intervention and maximum speed, yet the user has every freedom to experiment with the many options available. The underlying estimation approaches, based on high-order ARX estimation followed by model reduction, and on subspace methods, have been selected for their capacity to treat the large dimensional problems commonly found in system identification for process control, and to produce fast and robust results. Models describing parts of a larger system can be combined into a composite model describing the whole system. This gives the user the flexibility to handle complex model predictive control configurations, such as schemes involving intermediate process variables.

1   INTRODUCTION

It is generally acknowledged that finding a dynamic process model for control purposes is the most cumbersome and time-consuming step in model predictive control (MPC) commissioning. This is mainly due to special requirements of process industries that make for difficult experimental conditions, but also to the relatively high level of expertise needed to obtain empirical models through the techniques of system identification. The vast majority of MPC vendors (and a few independent companies) have recognized the need for efficient system identification and model building tools and started providing software and facilities to ease this task.

In these pages, we present ISIAC (as *Industrial System Identification for Advanced Control*), a product of the Institut Français du Pétrole (IFP). This new software package is geared to meet the requirements of system identification for model predictive control and the needs of practicing advanced process control (APC) engineers.

In section 2, we discuss the peculiarities of system identification for process control. Then we explain how these peculiarities have been taken into account in ISIAC design (section 3). Section 4 illustrates the user workflow in ISIAC. Finally, section 5 presents an example taken from an industrial MPC application, carried out by Axens, process licensor ans service provider for the refining and petrochemical sectors.

2   SYSTEM IDENTIFICATION AND MODEL PREDICTIVE CONTROL

With several thousands applications reported in the literature, model predictive control (Richalet et al., 1978; Cutler and Ramaker, 1980) technology has been widely and successfully adopted in the process industries, and more particularly, in the petrochemical sector (see (Qin and Badgwell, 2003) for an overview of modern MPC techniques).

The basic ingredients of any MPC algorithm are:

- a dynamic model, which is used to make an open-loop prediction of process behavior over a chosen future interval (the *control model*);

- an optimal control problem, which is solved at each

225

J. Braz et al. (eds.), Informatics in Control, Automation and Robotics I, 225–232.
© 2006 *Springer. Printed in the Netherlands.*

control step, via constrained optimization, to minimize the difference between the predicted process response and the desired trajectory;

- a *receding horizon* approach (only the first step of the optimal control sequence is applied).

Most commonly, control models employed by industrial MPC algorithms are linear time-invariant (LTI) models, or, in same cases, combinations of LTI models and static nonlinearities. The vast majority of reported industrial applications have been obtained utilizing finite impulse response (FIR) or finite step response (FSR) models. Modern MPC packages are more likely to use state-space, transfer function matrix, or autoregressive with exogenous input (ARX) models. Linear models of dynamic process behavior can be obtained from a linearized first-principles model, or more commonly, from experimental modeling, applying system identification (Ljung, 1999) techniques to obtain black-box models from input-output data gathered from the process. Those models can be subsequently converted to the specific form required by the MPC algorithm.

Several researchers have pointed out that process identification is the most challenging and demanding part of a MPC project ((Richalet, 1993; Ogunnaike, 1996)). Although system identification techniques are basically domain independent, process industries have special features and requirements that complicate their use: slow dominant plant dynamics, large scale units with many, strongly interacting, manipulated inputs and controlled outputs, unmeasured disturbances, stringent operating constraints. This makes for difficult experimental conditions, long test durations and barely informative data. But even the subsequent step of performing system identification using one of the available software packages may prove lengthy and laborious, especially when the relatively high level of expertise needed is not at hand.

Designing such software for efficiency and manageability requires taking into account the peculiarities of system identification for process control.

- System identification is a complex process involving several steps (see Fig. 1). The user should be guided through it with a correct balance between structure and suppleness.

- When identifying an industrial process, a significant number of data records must be dealt with. Data may contain thousands of samples of tens (or hundreds) measured variables. Several different multivariable models can be identified for the whole process, or for parts of it. It is important to allow the user to handle multiple models and data sets of any size, to seamlessly visualize, evaluate, compare and combine them, to arrange and keep track of the work done during an identification session.

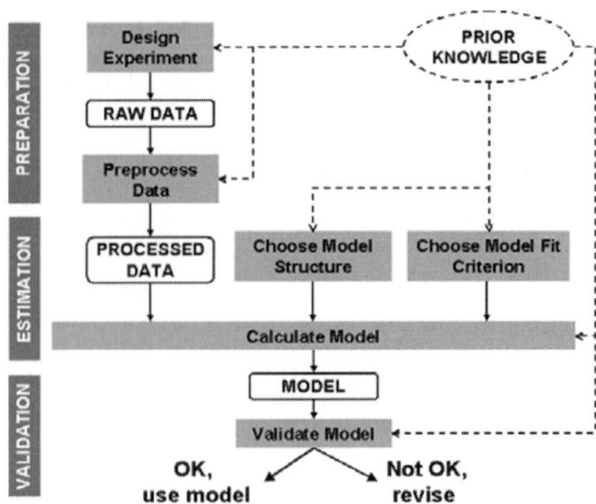

Figure 1: Steps of the system identification process (adapted from (Ljung, 1999)).

- Estimation and validation methods at the heart of the identification process must be chosen carefully. When dealing with multi-input multi-output (MIMO) models, model structure choice and parametrization may prove challenging even for experienced users. Moreover, large data and model sizes, utterly common in the context of system identification for process control, may easily lead to numerical difficulties and unacceptable computation times. Methodologies giving systematic answers to these problems exist (Juricek et al., 1998; Zhu, 1998) and have been incorporated into some commercial packages (Larimore, 2000; Zhu, 2000).

- MPC algorithms usually need more information to define their internal control structure, than a plain linear model. As a minimum, the user has to sort model inputs into *manipulated variables* (MV) and *disturbance variables* (DV), and to choose which model outputs are to be kept as *controlled variables* (CV). With modern MPC packages, control configuration may become really complex, including observers and unmeasured disturbance models, which can be used, among other things, to take into account the presence of *intermediate variables* i.e., measured output variables that influence controlled variables) for control calculation. Without a suitable control model building tool, supplying the additional pieces of information turns out to be a laborious task, even for mildly complex control configurations.

3 ISIAC

ISIAC is primarily meant to support the model-based predictive multivariable controller MVAC, a part of the APC suite developed by IFP and its affiliate RSI.

MVAC has first been validated on a challenging pilot process unit licensed by IFP (Couenne et al., 2001), and is currently under application in several refineries world-wide. Its main features are:

- state-space formulation;
- observer to take into account unmeasured disturbances, intermediate variables, integrating behavior;
- ranked soft and hard constraints;
- advanced specification of trajectories (funnels, set-ranges);
- static optimization of process variables.

Though ISIAC is intended to be the natural companion tool to MVAC, it is actually flexible enough to be used as a full-fledged system identification and model building tool or to support other APC packages.

3.1 Approaches to Model Estimation

The model estimation approaches selected for inclusion in ISIAC combine accuracy and feasibility, both in term of computational requirements and of user choices. We have decided to favor non-iterative methods over *prediction error methods* (Ljung, 1999), to avoid problems originating from a demanding minimization routine and a complicated underlying parametrization.

3.1.1 Two-Stage Method

The benefits of high-order ARX estimation in industrial situations have been advocated by several researchers ((Zhu, 2001; Rivera and Jun, 2000)). Indeed, using a model order high enough, and with sufficiently informative data, ARX estimation yields models that can approximate any linear system arbitrarily well (Ljung, 1999). Ljung's *asymptotic blackbox theory* also provides (asymptotic) expressions for the transfer function covariance which can be used for model validation purposes.

A model reduction step is necessary to use these models as process control models, since they usually are over-parameterized (i.e., an unbiased model with much lower order can be found) and have high variance (which is roughly proportional to model order). Different schemes have been proposed ((Hsia, 1977; Wahlberg, 1989; Zhu, 1998; Rivera and Jun, 2000; Tjärnström and Ljung, 2003)) to perform model reduction. For a class of reduction schemes (Tjärnström

and Ljung, 2003), it has been demonstrated that the reduction step actually implies variance reduction, and a resulting variance which is nearly optimal (that is, close to the Cramer-Rao bound).

In ISIAC, truly multi-input multi-output (MIMO) ARX estimation is possible, using the structure

$$A(q)y(t) \;=\; B(q)u(t) + e(t)$$

where $y(t)$ is the p-dimensional vector of outputs at time t, $u(t)$ is the m-dimensional vector of inputs at time t, $e(t)$ is a p-dimensional white noise vector, $A(q)$ and $B(q)$ polynomial matrices respectively of dimensions $p \times p$ and $p \times m$. For faster results, in the two-stage method, the least-square estimation problem is decomposed into p multi-input single-output (MISO) problems. The Akaike information criterion (AIC) (Ljung, 1999) is used to find a "high enough" order in the first step, and the resulting model is tested for unbiasedness (whiteness of the residuals and of the inputs-residuals cross-correlation). To find a reduced order model, we adopt a frequency-weighted balanced truncation (FWBT) technique (Varga, 1991). The calculated asymptotic variance is used to set the weights and to automatically choose the order of the reduced model, following an approach inspired by (Wahlberg, 1989). The obtained model is already in the state-space form needed by the MPC algorithm.

As a result, we get an estimation method which is totally automated, and provides accurate results in most practical situations, with both open-loop and closed-loop data. Yet, this method might not work correctly when dealing with short data sets and (very) ill-conditioned systems.

3.1.2 Automated Subspace Estimation

The term subspace identification methods (SIM) refers to a class of algorithms whose main characteristic is the approximation of subspaces generated by the rows or columns of some block matrices of the input/output data (see (Bauer, 2003) for a recent overview). The underlying model structure is a state-space representation with white noise (*innovations*) entering the state equation through a Kalman filter and the output equation directly

$$\begin{aligned} x(t+1) &= Ax(t) + Bu(t) + Ke(t) \\ y(t) &= Cx(t) + Du(t) + e(t) \end{aligned}$$

Simplifying (more than) a bit a fairly complex theory, input and output data are used to build an extended state-space system, where both data and model information are represented as matrices, and not just vector and matrices. Kalman filter state sequences are then identified and used to estimate system matrices A, C, and, if a disturbance model is needed, K (B and D can be subsequently estimated in several different ways). This can be assimilated (Ljung, 2003) to the estimation of a high-order ARX model, which is then reduced using weighted Hankel-norm model reduction. The three most well-known algorithms, CVA, N4SID and MOESP, can be studied under an unified framework (Van Overschee and DeMoor, 1996), where each algorithm stems from a different choice of weightings in the model reduction part. Subspace identification methods can handle the large dimensional problems commonly found in system identification for process control, producing (very) fast and robust results. However, it must be pointed out that their estimates are generally less accurate than those from prediction error methods and that standard SIM algorithms are biased under closed-loop conditions.

Even though a lower accuracy is to be expected, it is important to have a viable alternative to the two-step method. This is why, we have implemented in ISIAC two estimation procedures based on subspace algorithms taken from control and systems library SLICOT (Benner et al., 1999):

- a *combined method*, where MOESP is used to estimate A and C and N4SID is used to estimate B and D;

- a *simulation method*, where MOESP is used to estimate A and C and linear regression is used to estimate B and D.

The second method is usually more accurate (though a little slower) and is presented as the default choice. High-level design parameters for these two methodologies are the final model order n, and the prediction horizon r used in the model reduction step. ISIAC can select both parameters automatically: the latter using a modified AIC criterion based on a high-order ARX estimation, the former using a combined criterion taking into account the relative importance of singular values and errors on simulated outputs (output errors).

3.2   General Structure and Layout

Fig. 2 shows ISIAC graphical user interface (GUI). The tree on the left (the *session tree*) highlights the relationships between the different elements the user deals with during a typical system identification session.

- A *System* object representing the whole process the user is working on. It includes a list of process *Variables* and a list of *Subsystems*, which can be used to store partial measurements and dynamic models relating groups of input and output variables.

- *Data* objects in two flavors: *Raw Data* and *IO Data*. The former are defined from raw process

data records and do not carry any structure information. They are mainly used for preliminary appraisal and processing of available data sets. IO data are defined by selecting subsets of fields of raw data objects, and can be used for system identification. Notice that data objects in ISIAC may include measurements coming from different experiments (*multi-batch* data).

- *Models* include all the dynamic models estimated or defined during a session. ISIAC handles discrete-time state-space, transfer function (matrix), FIR, ARX and general polynomial models. State-space and transfer function models are also available in continuous time. Simple process models, such as first-order plus time delay (FOPTD) models in gain-time constant-delay form, are handled as specializations of transfer function models.

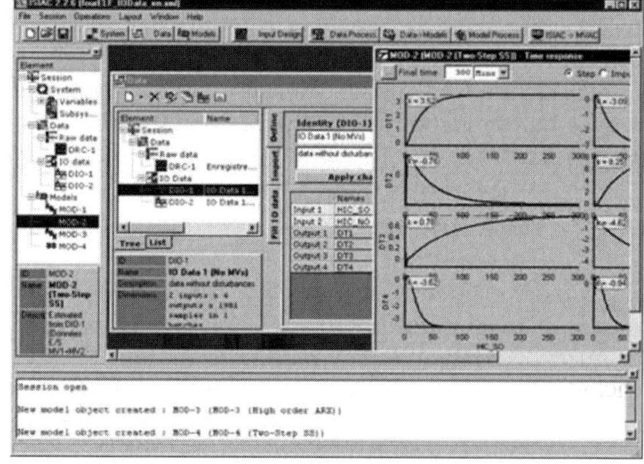

Figure 2: ISIAC GUI.

This structure provide a powerful and flexible support to the user:

- no restriction is put on the number of models and data objects, nor on their sizes;

- system identification of the whole process can be decomposed in smaller problems whose results can be later recombined;

- it is straightforward to keep track of all the work done during an identification session.

Three special windows are dedicated to definition and basic handling of system, data and model objects. The *Input Design Window* is intended to help the user to design appropriate test signals for system identification. More advanced operations on data and models are available in the *Data Processing Window* and in the *Model Processing Window*. The most important window is certainly the *Data To Models Window*, where model estimation and validation take place. Last, the *ISIAC To MVAC Window* hosts the graphical control model builder.

ISIAC GUI implements a *multi-document interface* (MDI) approach: it is possible to have several windows opened at once in the *child window* area. Furthermore, thanks to drag-and-drop operations and pop-up menus, the same action (say, plotting a model time response) is accessible from different locations. This means that, although ISIAC layout clearly underlines the different steps of the identification process, the user is never stuck into a fixed workflow.

4 WORKING WITH ISIAC

4.1 Experiment Design

Whenever the APC engineer has the freedom to choose other input moves than classical step-testing (not often, unfortunately), ISIAC offers support to generate test signals which are more likely to yield informative data. The *Input Design Window* lets the user design signals such as pseudo-random binary signals (PRBS), using few high level parameters.

4.2 Working with Data

As mentioned before, ISIAC data objects provide a structure to handle measurements of process variables. Input files containing raw measurements do not need to carry any special information (other than including delimited columns of values), and can be imported without any external spreadsheet macro, since data object formatting is done interactively into ISIAC *Data Window*. Data visualization tools, which include stacked plots, single-axis plots and various statistical plots, have been designed with particular care.

The more advanced functions of the *Data Processing Window* are those commonly found in industrial system identification packages: normalization, detrending, de-noising, re-sampling, filtering, nonlinear transformations, data slicing, data merging. Notice that a set of these data processing operations is incorporated in the automated model identification procedure (see section 4.4).

4.3 Working with Models

Most commonly, dynamic process models in ISIAC are estimated from data or obtained combining or processing existing (estimated) models. Models can be also directly defined by the user (in FOPTD form, for instance) or imported from other packages. ISIAC *Model Window* also provides transformations between different LTI representations and time domain conversions, as well as several analysis and visualization tools. Model response plots, both in time

domain and in frequency domain, are extremely flexible. An unlimited number of models (not necessarily sharing the same inputs or outputs) can be compared on the same chart, and an unlimited number of charting windows can be opened at once.

Advanced model processing (in the *Model Processing Window*) includes model reduction and model building tools. Time domain techniques (step response fitting of simple process models, see Fig. 3) or frequency weighted model reduction techniques are available. Model building can be performed through simple connections (cascade, parallel) or through a full-fledged graphical model builder which closely resembles the one described in section 4.5.

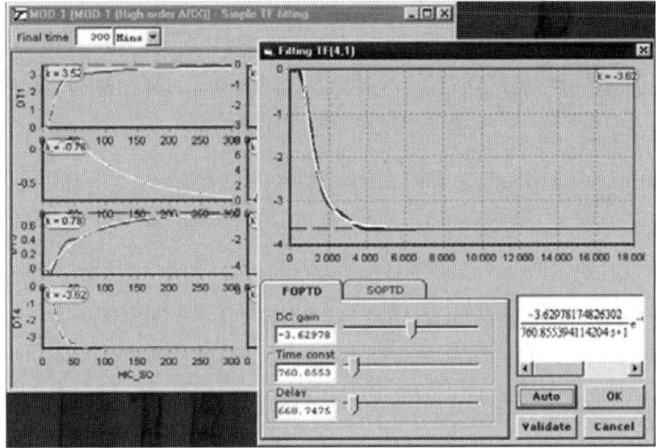

Figure 3: Step response fitting.

4.4 Estimating and Validating Models

Model estimation must be preceded by some amount of data processing, namely offset removal, normalization and detrending (that is, removal of drifts and low frequency disturbances). In ISIAC, these basic but essential transformation are automatically applied (unless the user does not want to), in a transparent manner, before model estimation. Actually the *Data Processing Window*, is only necessary when more advanced data processing is needed. Moreover, prior information about certain characteristics of the system (integrating behavior, input-output delay) can be also incorporated to help the estimation algorithms.

As explained in section 3.1, the default estimation method in ISIAC is the two-stage method. This method, combined with the transparent basic data processing, results in a "click&go" approach that is greatly appreciated by industrial practitioners. The industrial example of Fig. 4 shows that with this method the user can really make the most of the available data, even when the inputs are not very informative. Alternatively, subspace estimation can be selected. It is also possible to estimate FIR models or general ARX models.

Beside the comparison between measured outputs and simulated outputs (as in Fig. 4), model validation can be performed by checking the confidence bounds and visualizing and comparing time and frequency responses.

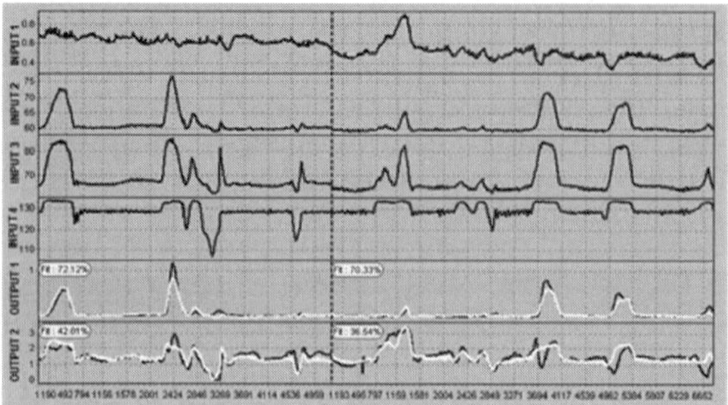

Figure 4: Model validation through simulation (simulated outputs in white).

4.5   Building the Control Model

One of the most interesting features of ISIAC is the graphical control model builder, in the *ISIAC To MVAC Window* (Fig. 5). With a few mouse clicks, it is possible to build a complex control model from a combination of sub-models, identified from different data sets or extracted from existing models. The user is only required to indicate the role of each input and output in the control scheme. The model graph can be then translated into a control model with the appropriate format, or into a plant model for off-line simulations. The resulting plant model can be also transferred back to ISIAC workspace and applied to the existing data sets to verify its correctness.

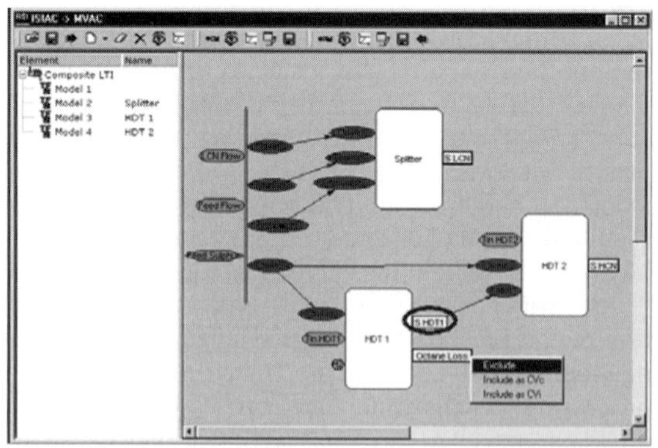

Figure 5: The graphical control model builder.

The model builder proves particularly helpful when

intermediate process variables are to be included. In figure 5, which depicts part of the control configuration for a unit involving cascaded reactors, the variable denoted S HDT 1 is one of those variables.

5   AN INDUSTRIAL APPLICATION: MODEL PREDICTIVE CONTROL OF A MTBE UNIT

To prove the usefulness of ISIAC in an industrial context, we examine some aspects of a model predictive control project, carried out by IFP affiliate Axens on a petrochemical process unit.

The process under consideration is an etherification unit producing methyl-tert-butyl ether (MTBE), from a reaction between isobutene (IB) and methanol (MeOH).

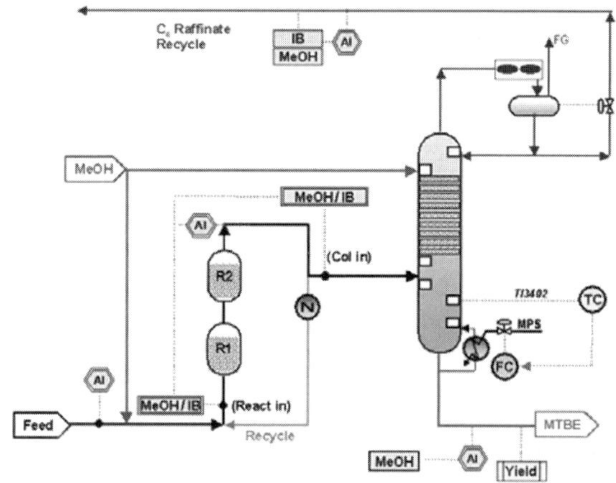

Figure 6: The MTBE unit.

The key control objectives are:

- maximize MTBE yield;
- increase IB recovery;
- reduce steam consumption;
- control MTBE purity.

The MVAC-based control system includes 7 MVs, 3 DVs, 6 CVs, with 5 intermediate variables. In the following, we only consider a subset corresponding to the control of MeOH percentage in MTBE (last item of the objective list). Fig. 7 shows, from a system viewpoint, how the controlled variable MEOH IN MTBE is influenced by others process variables of the control configuration:

- feed flow, MeOH flow and sensitive temperature of the catalytic column as CVs;

- IB and MeOH percentages in feed to first reactor as DVs;

- the ratios of MeOH over IB, respectively entering the first reactor and the catalytic column, as intermediate variables.

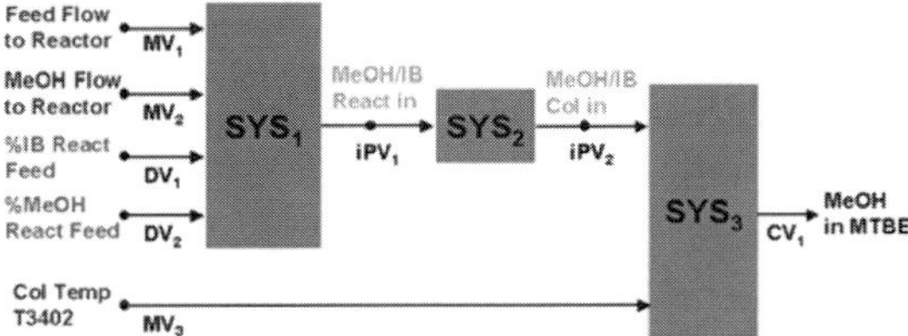

Figure 7: Part of the MTBE control scheme.

There are several avantages in introducing intermediate variables in the control configuration, instead of considering only direct transfer functions between input variables (MVs plus DVs) and the CV:

- intermediate variables can be bounded, for tighter control;

- unavoidable uncertainties in cascaded models (upstream from intermediate variables) can be compensated for;

- deviations from predicted behavior can be detected long before they affect the CV.

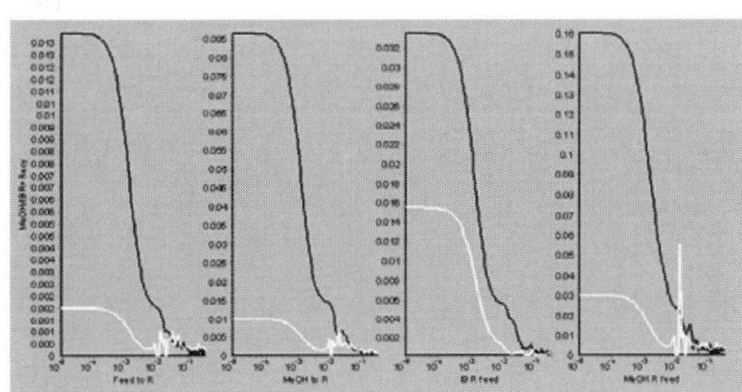

Figure 8: Frequency response of an identified model for subsystem SYS1.

ISIAC has been used for data visualization and analysis, as well as for identification of sub-models later included in the overall control configuration. As an example, we present some identification results relating to subsystem SYS1 of Fig. 7. Fig. 8 shows the frequency response of a high order ARX model together with its error bounds. The estimates of the first two transfer functions ($MV_1 \rightarrow iPV_1$, $MV_2 \rightarrow iPV_1$) appear to be fairly accurate, since their error bounds are comparatively quite small. The overall quality of estimation is confirmed by the comparison between measured and predicted output (figure 9).

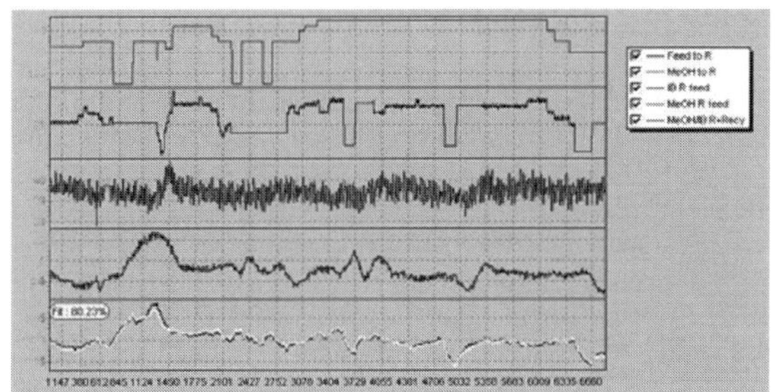

Figure 9: Measured vs. predicted (white) output for subsystem SYS1.

As for control model building, Fig. 10 shows how naturally the dedicated graphical tool translates block diagrams like the one in Fig. 7. From this graphical representation, it takes only one mouse-click to generate scripts for simulation purposes or for final MPC implementation.

6   CONCLUSION

ISIAC proposes a modern, flexible and efficient framework to perform system identification for advanced process control. Its main strengths are:

- a graphical user interface which emphasizes the multi-step nature of the identification process, without trapping the user into a fixed workflow;

- fast and robust estimation methods requiring minimal user intervention;

- no restriction on the number or on the size of data sets and models the user can work with;

- full support for the specification of complex model predictive control schemes, by means of block diagram combination of (linear) models.

Through an exemple taken from an industrial MPC application, we have illustrated the advantages of using our software in a concrete situation.

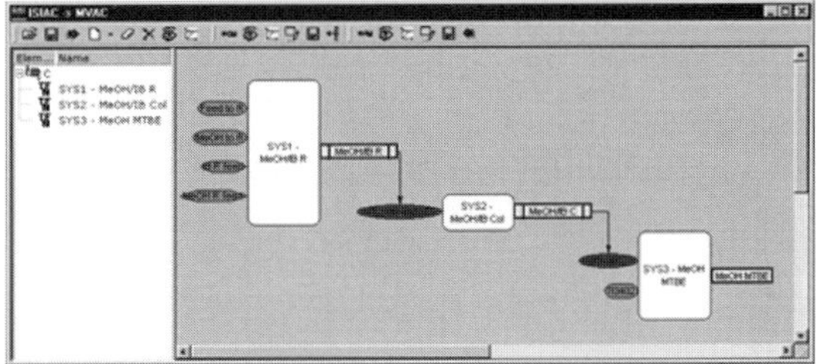

Figure 10: Building the partial MTBE control scheme in ISIAC.

REFERENCES

Bauer, D. (2003). Subspace algorithms. In *Proc. of the 13th IFAC Symposium on System Identification*, Rotterdam, NL.

Benner, P., Mehrmann, V., Sima, V., Van Huffel, S., and Varga, A. (1999). SLICOT, a subroutine library in systems and control theory. *Applied and Computational Control, Signals and Circuits*, 1: 499–539.

Couenne, N., Humeau, D., Bornard, G., and Chebassier, J. (2001). Contrôle multivariable d'une unité de séparation des xylènes par lit mobile simulé. *Revue de l'Electricité et de l'Electronique*, 7-8.

Cutler, C. R. and Ramaker, B. L. (1980). Dynamic matrix control: a computer control algorithm. In *Proc. of the American Control Conference*, San Francisco, CA, USA.

Hsia, T. C. (1977). *Identification: Least Square Methods*. Lexington Books, Lexington, Mass., USA.

Juricek, B. C., Larimore, W. E., and Seborg, D. E. (1998). Reduced-rank ARX and subspace system identification for process control. In *Proc. IFAC DYCOPS Sympos.*, Corfu, Greece.

Larimore, W. (2000). The ADAPTx software for automated multivariable system identification. In *Proc. of the 12th IFAC Symposium on System Identification*, Santa Barbara, CA, USA.

Ljung, L. (1999). *System Identification, Theory for the User*. Prentice-Hall, Englewood Cliffs, NJ, USA, second edition.

Ljung, L. (2003). Aspects and experiences of user choices in subspace identification methods. In *Proc. of the 13th IFAC Symposium on System Identification*, Rotterdam, NL.

Ogunnaike, B. A. (1996). A contemporary industrial perspective on process control theory and practice. *A. Rev. Control*, 20: 1–8.

Qin, S. J. and Badgwell, T. A. (2003). A survey of industrial model predictive control technology. *Control Engineering Practice*, 11: 733–764.

Richalet, J. (1993). Industrial applications of model based predictive control. *Automatica*, 29(5): 1251–1274.

Richalet, J., Rault, A., Testud, J. L., and Papon, J. (1978). Model predictive heuristic control: applications to industrial systems. *Automatica*, 14: 414–428.

Rivera, D. E. and Jun, K. S. (2000). An integrated identification and control design methodology for multivariable process system applications. *IEEE Control Systems Magazine*, 20: 2537.

Tjärnström, F. and Ljung, L. (2003). Variance properties of a two-step ARX estimation procedure. *European Journal of Control*, 9: 400 –408.

Van Overschee, P. and DeMoor, B. (1996). *Subspace Identification of Linear Systems: Theory, Implementation, Applications*. Kluwer Academic Publishers.

Varga, A. (1991). Balancing-free square-root algorithm for computing singular perturbation approximations. In *Proc. of 30th IEEE CDC*, Brighton, UK.

Wahlberg, B. (1989). Model reduction of high-order estimated models: The asymptotic ML approach. *International Journal of Control*, 49: 169–192.

Zhu, Y. (1998). Multivariable process identification for MPC: the asymptotic method and its applications. *Journal of Process Control*, 8(2): 101–115.

Zhu, Y. (2000). Tai-Ji ID: Automatic closed-loop identification package for model based process control. In *Proc. of the 12th IFAC Symposium on System Identification*, Santa Barbara, CA, USA.

Zhu, Y. (2001). *Multivariable System Identification for Process Control*. Elsevier Science Ltd.

IMPROVING PERFORMANCE OF THE DECODER FOR TWO-DIMENSIONAL BARCODE SYMBOLOGY PDF417

Hee Il Hahn and Jung Goo Jung

Department of Information and Communications Eng., Hankuk University of Foreign Studies, Korea
Email: hihahn@hufs.ac.kr, jgchong@hanmail.net

Keywords: Two-dimensional barcode, Segmentation, Error correction code, Warping

Abstract: In this paper we introduce a method to extract the bar-space patterns directly from the gray-level two-dimensional barcode images, which employs the location and the distance between extreme points of profiles scanned from the barcode image. This algorithm proves to be very robust from the high convolutional distortion environments such as defocussing and warping, even under badly illuminating condition. The proposed algorithm shows excellent performance and is implemented in real-time.

1 INTRODUCTION

Linear barcodes have been used globally in the various fields such as supermarkets and other stores for several decades. Usually, they do not have detailed information but just carry a key to database, because they can hold only several bytes of information. The need to increase the amount of data in a symbol brought the introduction of a new form of barcodes with much higher density called the two-dimensional (2-D) barcodes. They have been introduced since 1990's. While conventional or one-dimensional barcodes usually function as keys to databases, the new or two-dimensional barcodes meet a need to encode significantly more data than the conventional codes and would act as a portable data file because the information could be retrieved without access to a database. They have additional features such as error correction and the ability to encode in various languages like English, Korean and Chinese, etc., besides their increased capacity. Thus, the traditional concept of barcode as a key to a database is changing towards a "portable data file" in which all the relevant information accompanies the item without access to a database.

There are two types of 2-D symbologies - stacked and matrix-type symbologies. The stacked barcodes, to which Code49, PDF417, etc. belong, have the structure of rectangular block comprising numbers of rows, each of which is like 1-D symbology. The matrix-type barcodes are essentially a form of binary encoding in which a black or white cell can represent either binary 1 or 0. These cells are arranged in an array structured on a grid of rectangular block. DataMatrix, Maxicode, and QR code, etc. are representative of Matrix-type symbology (Pavlidis, 1992).

In this paper, we focuss only on PDF417 (AIM USA, 1994), known as the most widely used 2-D stacked symbology. PDF417 is a multi-row, variable-length symbology offering high data capacity and error-correction capability. A PDF417 symbol is capable of encoding more than 1,100 bytes, 1,800 ASCII characters, or 2,700 digits, depending on the selected data compaction mode. Every PDF417 symbol is composed of a stack of rows, from a minimum of 3 to a maximum of 90 rows. Each PDF417 row contains start and stop patterns, left and right row indicators. Fig. 1 shows the high-density scanned barcode image of PDF417 code.

We explain about our barcode decoder and propose a novel method of extracting codewords directly from gray-level barcode image by analyzing their profiles instead of binarizing the image.

2 LOCALIZING THE DATA REGION

The data region is localized to extract the bar-space patterns and to detect the codewords, in the following way. Firstly, the start pattern or stop pattern, or both are searched by scanning horizontally and vertically. The start pattern is 81111113 and the stop pattern is 711311121, both beginning with bar. Fig. 2 shows the segments

233

J. Braz et al. (eds.), Informatics in Control, Automation and Robotics I, 233–237.
© 2006 *Springer. Printed in the Netherlands.*

corresponding to the start patterns, which are extracted from the clear barcode image and the highly blurred, high-density one respectively, together with their profiles. Although the start patterns shown in Fig. 2-(a) can be easily identified, it might not be easy to detect them in case of Fig. 2-(b) because the width of one module is less than 1.5 pixels and the dynamic range of the profile at the region corresponding to narrow bar or space is severely reduced due to convolution with the point spread function of the camera. After the start pattern and/or stop pattern are detected, line equations corresponding to them are estimated through line fitting to localize left and/or right row indicators. Secondly, the header information such as the number of rows and columns, error correction level, row number, error correction level and the width of module, etc. are obtained by extracting and decoding the left and/or right row indicators (Hahn, 2002).

Figure 1: The scanned PDF-417 barcode image.

Finally, four vertices of the data region are detected by scanning in the directions perpendicular to the start pattern line and the stop pattern line. To ensure the vertices obtained are correct, row numbers are checked by decoding the left or right row indicators around the vertices. If the row numbers corresponding to the detected four vertices are not the first row or the last row, the true four vertices are predicted by considering the proportion of the row number to the number of row, as depicted in Fig. 3. Since barcode image is usually scanned using a digital camera, it should be warped due to the nonlinearity of lens and the viewing angle of the camera. In this paper, Affine transform is adopted to warp the data region of the scanned image (Gonzalez, 1993). Fig. 4 shows the warped result of the data region inside the barcode in Fig. 1.

3 DECODING CODEWORDS FROM BAR-SPACE PATTERNS

After the data region is localized and warped as mentioned above, bar-space patterns are extracted

from the data region and are decoded to obtain the corresponding codewords. The problem is how we get bar-space pattern from the barcode data region.

Usually, image thresholding is employed for this goal. Many researchers developed the image thresholding algorithms,most of which reported to date are based on edge information or histogram analysis using information-theoretic

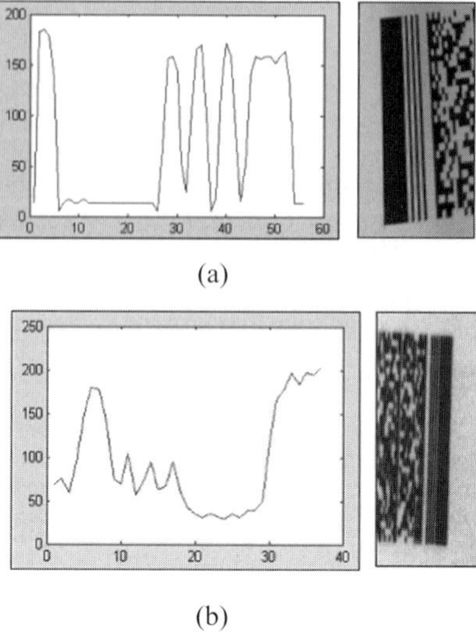

(a)

(b)

Figure 2: Profiles and segments corresponding to the start patterns of the scanned barcode, which are excerpted from (a) clear image and (b) highly blurred, high-density image.

approaches (Parker,1991). The selection of optimal thresholds has remained a challenge over decades. However, they can not be applied to decode the barcode images, because the widths of bars or spaces can be 2 pixels or even less in case of high-density barcode images and even slight variation of threshold can cause severe errors (Eugene, 1991).

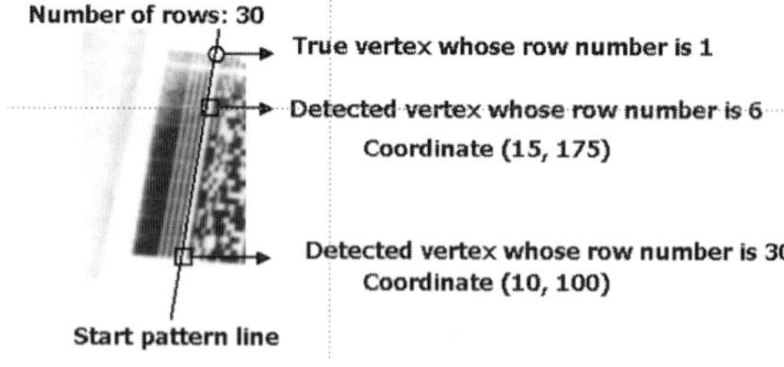

Figure 3: Locating true vertices by checking row numbers.

Fig. 5 shows the segmented bar-space patterns and their corresponding profiles obtained from the focussed clean image and the outfocussed image. Although the widths of bars and spaces of Fig. 5-(a) can be easily measured, it might not be a simple matter to measure them in case of Fig. 5-(b) and the obvious thing is single global threshold can not even discriminate them. Fig. 6 shows the intensity profile of the codeword (824 : 1, 5, 1, 2, 1, 1, 1, 5) comprised of four bars and four spaces, together with their widths and classification results.

As shown in Fig. 6-(a), it is impossible to select the single optimal threshold for detecting the widths of four bars and four spaces because the pixel values change dynamically according to the widths of bars and spaces.

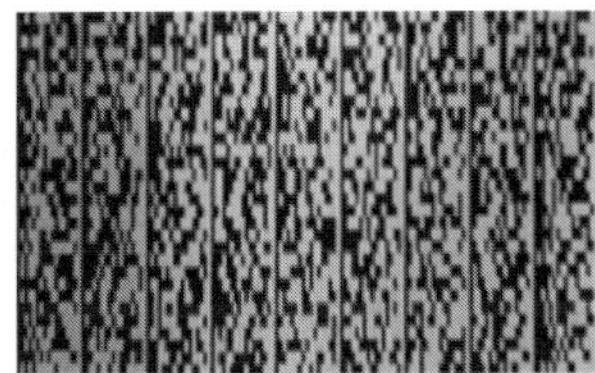

Figure 4: Warping the data region of Fig. 2 using Affine transform.

The widths and peaks of narrow bars or spaces corresponding to 1 or 2 module values get smaller compared to the wide ones even under the same illumination, due to convolution with the point spread function. The proposed algorithm employs the high curvature points and local extreme points to extract four bars and four spaces from the warped barcode image. The points of high curvature on the waveform represent the midpoints of the bar or space, whereas the local extreme points are estimates of their centers.

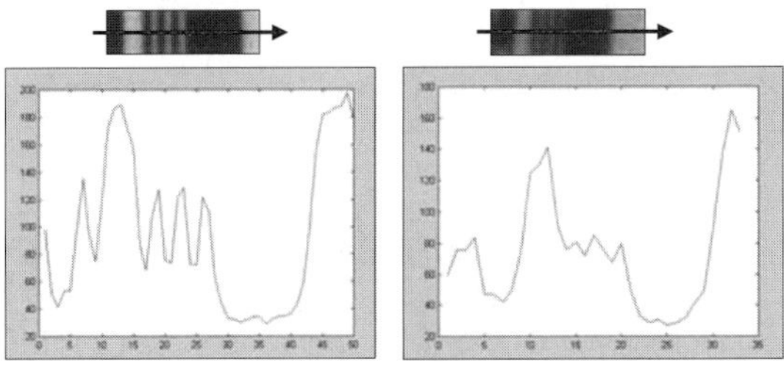

Figure 5: Segmented bar-space patterns and their corresponding profiles obtained from (a) the focussed image, (b) the outfocussed image.

At first, our algorithm localizes four local minimum points and four local maximum points by scanning the profile of each bar-space pattern, as shown in Fig. 6-(b). The local minimum point whose value is large compared to the other minimum points or the maximum point smaller than the adjacent ones are compensated to increase their dynamic range as depicted in Fig. 6-(c). Then, the break regions are detected between the compensated extreme points,

where the break region means the region whose profile is horizontal. The regions marked as rectangular box in Fig. 6-(d) represent the break regions. Finally, the break regions are partitioned according to the following rules. Usually, the edges between bars and spaces lie in the break regions. In Fig. 7, we define $\Delta a = x2 - x1$ and $\Delta b = x3 - x2$. If Δa is greater than Δb, the more part of the break region belongs to the space region, vice versa. The ratios are obtained by experimenting with several hundreds of bar-space patterns, extracted from the various barcode images scanned under varying conditions. Thus, the widths of bars and spaces are represented as real values rather than integer ones to measure the bar-space patterns as correctly as possible.

4 EXPERIMENTAL RESULTS

The "edge to similar edge" estimation method is employed to check whether the detected bar-space pattern is correct. The detected bar-space patterns can be converted to the encoded codewords by using the lookup table specified in (AIM USA, 1994). The codewords are checked whether there are any errors through Reed-Solomon error correction coding algorithm. Given the error-corrected codewords, they are decoded in the manner as specified in (AIM USA, 1994), to get the message.

In order to verify our algorithm , we benchmark-tested our decoder with the test barcode images. Our database is composed of 153 barcode images, which were taken under various conditions. In other words, they are rotated, outfocussed, warped or even severely damaged by cutting off some region, as shown in Fig. 8. Almost of them are taken under badly illuminated conditions. At first, 2,000 pieces of profiles corresponding to the bar-space patterns are extracted from our database. Each profile is tested to extract the bar-space pattern and decode the corresponding codeword. Among them, 1,466 profiles are detected correctly. As an example, the image of Fig. 8-(a) is obtained by hardcopying the barcode image several times, whose upper part is severely degraded. When it is applied to our decoder, 42 erasures and 32 errors are detected among total 329 codewords, which can be decoded correctly through Reed-Solomon error correction algorithm. Fig. 8-(b) is an image reassembled but aligned incorrectly after tearing it into two parts. Fig. 8-(c) is obtained from Fig. 8-(a) by cutting off the right part of it and Fig. 8-(d) are taken outfocussed under too bright illuminating condition. Our algorithm decoded 138 images correctly among total 153 images. This result is expected to be good for manufacturing purpose although there might be no public images for benchmarking test.

5 CONCLUSION

We have proposed algorithms to decode two-dimensional barcode symbology PDF417 and implemented a barcode reader in real-time using ARM core. Our decoder employs a method to extract the bar-space patterns directly from the profiles of the gray-level barcode image. Our algorithm shows performance improved further than the method of extracting the bar-space patterns after binarizing the barcode image, when we experiment with the test images of variable resolution and error correction levels. In order to improve the performance further, it is needed to extract bar-space patterns more accurately from the barcode image, which might be defocused or taken under badly illuminating conditions.

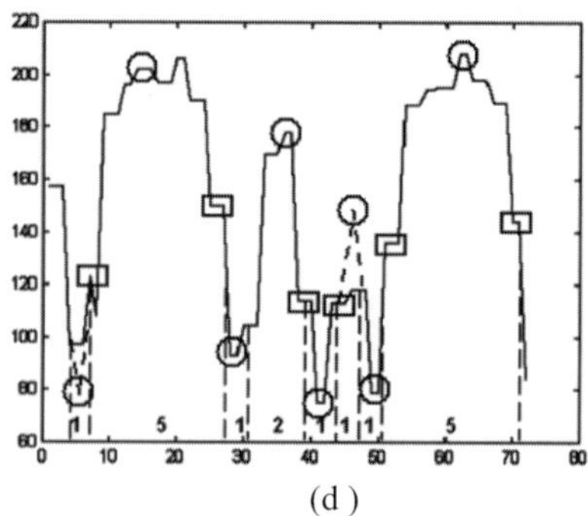

(d)

Figure 6: The intensity profiles of the codeword comprised of four bars and four spaces

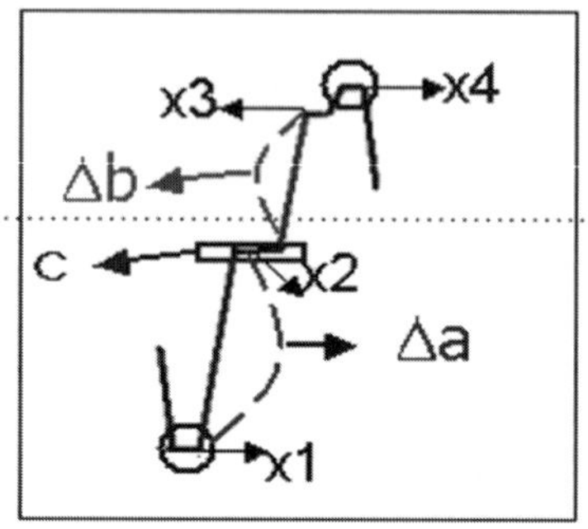

Figure 7: Partition of break region according to the ratio of Δa to Δb

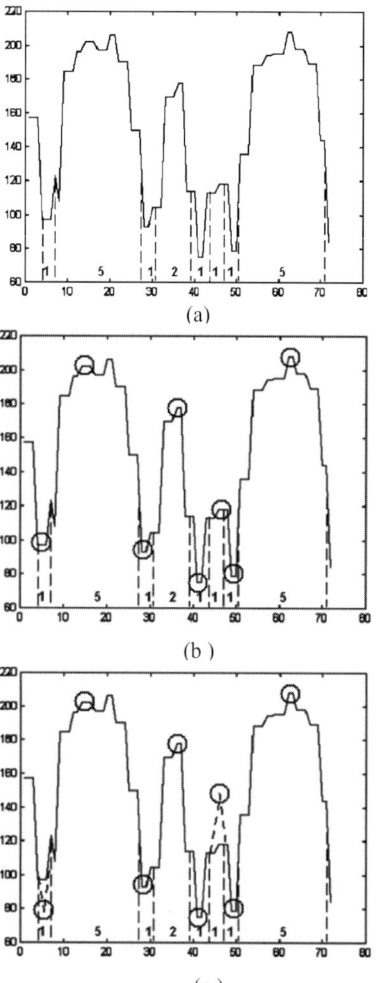

(a)

(b)

(c)

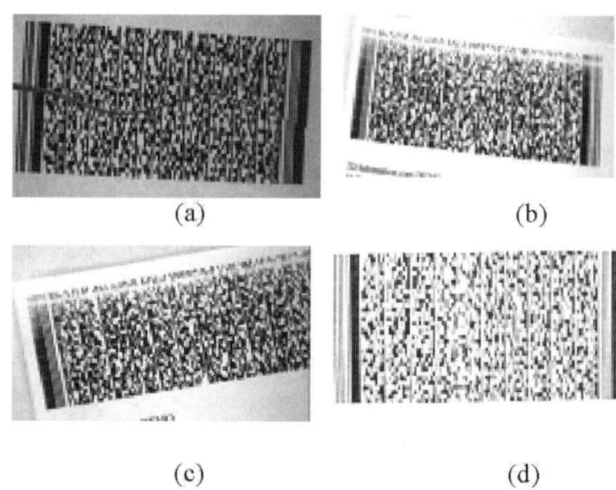

(a)   (b)

(c)   (d)

Figure 8: Sample images for measuring the performance of our decoder.

REFERENCES

AIM USA, 1994. Uniform Symbology Specification PDF417.

Eugene Joseph and Theo Pavlidis, 1991. Waveform Recognition with Application to Barcodes, *Symbol Technologies Inc.* 116 Wilbur Place, Bohemia, NY.

R. C. Gonzalez, R.E.Woods, 1993. *The book*, Digital Image Processing, Addison Wesley.

Hee Il Hahn, Joung Goo Joung, 2002. Implementation of Algorithm to Decode Two-Dimensional Barcode PDF-417, 6th International Conference on Signal Processing ICSP'02.

J. R. Parker, 1991. Gray level thresholding in badly illuminated images, IEEE Trans. on PAMI, vol. 13, No. 8.

T. Pavlidis, J. Swartz, Y. P. Wang, 1992. Information encoding with two-dimensional barcodes, *IEEE Computer*.

CONTEXT IN ROBOTIC VISION
Control for real-time adaptation

Paolo Lombardi

Istituto Trentino di Cultura ITC-irst, via Sommarive 18, Trento, Italy
(formerly with Dip. Informatica e Sistemistica, Università di Pavia)
Email:lombardi@itc.it

Virginio Cantoni

Dip. Informatica e Sistemistica, Università di Pavia, via Ferrata 1, Pavia, Italy
Email:virginio.cantoni@unipv.it

Bertrand Zavidovique

Institut d'Electronique Fondamentale, Université de Paris Sud-11, bât. 220, Campus d'Orsay, Orsay, France
Email: zavido@ief.u-psud.fr

Keywords: computer vision, contextual adaptation, context definition, Bayesian opportunistic switching.

Abstract: Nowadays, the computer vision community conducts an effort to produce canny systems able to tackle unconstrained environments. However, the information contained in images is so massive that fast and reliable knowledge extraction is impossible without restricting the range of expected meaningful signals. Inserting a priori knowledge on the operative "context" and adding expectations on object appearances are recognized today as a feasible solution to the problem. This paper attempts to define "context" in robotic vision by introducing a summarizing formalization of previous contributions by multiple authors. Starting from this formalization, we analyze one possible solution to introduce context-dependency in vision: an opportunistic switching strategy that selects the best fitted scenario among a set of pre-compiled configurations. We provide a theoretical framework for "context switching" named *Context Commutation*, grounded on Bayesian theory. Finally, we describe a sample application of the above ideas to improve video surveillance systems based on background subtraction methods.

1 INTRODUCTION

Computer vision was always considered a promising sensor for autonomous robots (e.g. domestic assistant robots, autonomous vehicles, video surveillance robotic systems, and outdoor robotics in general). Such applications require fast and reliable image processing to ensure real-time reaction to other agents around. Meanwhile, robots operating in varying and unpredictable environments need flexible perceptive systems able to cope with sudden context changes. To a certain extent, in robotics flexibility and robustness may be intended as synonyms.

Conciliating real-time operation and flexibility is a major interest for the vision community today. Traditionally, flexibility has been tackled by increasing the complexity and variety of processing stages. Voting schemes and other data fusion methods have been widely experimented. Still, such methods often achieve flexibility at the expense of real time.

Contextual information may open possibilities to improving system adaptability within real-time constraints. A priori information on the current world-state, scene geometry, object appearances, global dynamics, etc may support a concentration of system computational and analytical resources on meaningful components of images and video sequences. The recognition of the current operative "context" may allow a reconfiguration of internal parameters and active processing algorithms so as to maximize the potential of extractable information, meanwhile constraining the total computational load. Hence, "context" recognition and managing has attracted much interest from the robotic vision community in the last two decades.

239

J. Braz et al. (eds.), Informatics in Control, Automation and Robotics I, 239–246.
© 2006 *Springer. Printed in the Netherlands.*

A necessary step to implement context-dependency in practical vision system is defining the notion of "context" in robotic vision. Various authors have covered different aspects of this matter. A summarizing operative definition may serve as an interesting contribution and a reference for future work. Furthermore, it helps in identifying possible "context changes" that a system should cope with.

Overall, context managing represents a replacement of parallel image processing with less computationally expensive control. Controlling internal models and observational modalities by swapping among a finite set of pre-compiled configurations is probably the fastest and yet more realistically realizable solution.

In Section 2, we present a wide range of works related to "context" in computer vision. Section 3 details our proposal of formalization of such contributions by describing an operative definition. Then, Section 4 applies these concepts to a realistic implementation of real-time context-dependent adaptation within the scope of Bayesian theory. Finally, Section 5 concludes by suggesting some discussion and presenting future work.

2 CONTEXT IN COMPUTER VISION

In earlier works, contextual information referred to image morphology in pixel neighborhoods, both spatial and temporal. Methods integrating this information include Markov Random Fields (Dubes, 1989), and probabilistic relaxation (Rosenfeld, 1976). More recent works have moved the concept to embrace environmental and modeling aspects rather than raw signal morphology. General typologies of "context" definitions include:

1. *physical world models*: mathematical description of geometry, photometry or radiometry, reflectance, etc – e.g. (Strat, 1993), (Merlo, 1988).
2. *temporal information*: tracking, temporal filtering (e.g. Kalman), previous stable interpretations of images in a sequence, motion behavior of objects, etc – e.g. (Kittler, 1995), (Tissainayagam, 2003).
3. *site knowledge*: specific location knowledge, geography, terrain morphology, topological maps, expectations on occurrence of objects and events, etc – e.g. (Coutelle, 1995), (Torralba, 2003).
4. *scene knowledge*: scene-specific priors, illumination, accidental events (e.g. current weather, wind, shadows), obstacles in the viewfield, etc – e.g. (Strat, 1993).
5. *interpretative models and frames*: object representations (3d-geometry-based, appearance-based), object databases, event databases, color models, etc – e.g. (Kruppa, 2001).
6. *relations among agents and objects*: geometrical relationships, possible actions on objects, relative motion, split-and-merge combinations, intentional vs. random event distinctions, etc – e.g. (Crowley, 2002).
7. *acquisition-device parameters*: photo-grammetric parameters (intrinsic and extrinsic), camera model, resolution, acquisition conditions, daylight/infrared images, date and time of day, etc – e.g. (Strat, 1993), (Shekhar, 1996).
8. *observed variables*: observed cues, local vs. global features, original image vs. transformed image analysis, etc – e.g. (Kittler, 1995).
9. *image understanding algorithms*: observation processes, operator intrinsic characteristics, environmental specialization of individual algorithms, etc – e.g. (Horswill, 1995).
10. *intermediate processing results*: image processing quality, algorithm reliability measures, system self-assessment, etc – e.g. (Draper, 1999), (Rimey, 1993), (Toyama, 2000).
11. *task-related planning and control*: observation tasks, global scene interpretation vs. specialized target or event detection, target tracking, prediction of scene evolution, etc – e.g. (Draper, 1999), (Strat, 1993).
12. *operation-related issues*: computational time, response delay, hardware breakdown probabilities, etc – e.g. (Strat, 1993).
13. *classification and decision techniques*: situation-dependent decision strategies, features and objects classifiers, decision trees, etc – e.g. (Roli, 2001).

Despite definitions of "context" in machine vision have appeared under multiple forms, they all present "context" as an interpretation framework for perceptive inputs, grounding perception with expectation.

Probably a definition of *context* in computer vision, yet rather a non-operative one, could be given by dividing a perceptive system into an *invariant part* and a *variable part*. The *invariant part* includes structure, behaviors and evolutions that are inherent to the system itself, and that are not subject to a possible change, substitution or control. Examples may be the system very hardware, acquisition sensors, and fixed links between them, etc.; basic sub-goals like survival; age, endemic breakdowns, mobility constraints, etc. The *variable part* is all parameters, behaviors, and relations between components, which can be controlled. By means of these parts, the system may acquire dependence from the outer world and situation, with the purpose of better interacting with other agents and objects. In this view, *context* is what imposes changes to the *variable part* of a system. When mapped into the system through its variable parts, *context* becomes a particular configuration of internal parameters.

3 AN OPERATIVE DEFINITION OF CONTEXT

Inspired by the partial definitions from the previous references, we propose the following formalization (see (Lombardi, 2003) for details).

Definition D.1: *Context Q* in computer vision is a triplet $Q = (M, Z, D)$, where:
- *M* is the *model set* of object classes in the environment;
- *Z* is the *operator set*, i.e. the set of visual modules used in the observation process;
- *D* is the *decision policy* to distinguish between different classes of objects.

The rationale is that in perceptive systems, elements that can be parameterized and thus controlled are prior models of external objects, models of system components, and the relations among them. In short, *D* includes all prior assumptions on the strategy for inter-class separation and intra-class characterization. Essentially, it stands for point 13 in the above list. Hereafter, we further specify the definitions of *M* and *Z*.

3.1 Model Set M

The *model set M* contains all a priori knowledge of the system regarding the outer scene, object/agent appearances, and relations among objects, agents and events (essentially, points 1-6). We explicitly list three groups of knowledge inside *M*.

Definition D.2: A *model set* is a triplet $M = (\{m\}, P_{\{m\}}, V_{\{m\}})$, where:
- $\{m\}$ is the *entity knowledge* describing their appearance;
- $P_{\{m\}}$ is the prior expectation of occurrence in the scenario;
- $V_{\{m\}}$ is the *evolution functions* describing the dynamics.

Entity knowledge m indicates the set of features and/or attributes that characterize an object type.

Here, we call "entity" (Crowley, 2002) any object, agent, relation, or global scene configuration that is known, and thus recognizable, by the perceptive system. The set of all entity descriptions $\{m\}$ is the total scene-interpretation capability of the system, namely the set of all available a priori models of object classes that the system can give semantics to raw data with. Minsky frames and state vectors containing geometrical information are examples of descriptors. Moreover, the image itself can be thought of as an object, thus $\{m\}$ includes a description of global scene properties.

P_m is the prior expectations on the presence of *entity m* in the scene. We distinguish P_m from *m* because object descriptions are inherently attached to an *entity*, while its probability of occurrence depends on causes external to objects. *Evolution functions* $V_{\{m\}}$ indicate the set of evolution dynamics of an *entity* state parameters, e.g. object motion models.

3.2 Operator Set Z

The *operator set Z* gathers all prior self-knowledge on the perceptive system, available algorithms and hardware, feature extraction and measurement methods, observation matrixes, etc (points 7-12). We explicitly list three descriptors in *Z*.

Definition D.3: An *operator set* is a triplet $Z = (\{z\}, H_{\{z\}}, C_{\{z\}})$, where:
- $\{z\}$ is the *operator knowledge* describing their mechanisms;
- $H_{\{z\}}$ are the *operative assumptions* of operators;
- $C_{\{z\}}$ is the *operation cost* paid by system performance to run operators.

Operator knowledge z contains all parameters, extracted features, tractable elaboration noise, and other relevant features of a given visual operator. The set $\{z\}$ spans all visual modules in a system and their relative connections and dependencies. Operators constitute a grammar that allows matching data and semantics (*model set M*). Set $\{z\}$ includes logical operators, relation operators (e.g. detectors of couples), and events detectors.

Operative assumptions H_z is the set of hypotheses for the correct working of a visual module *z*. Implicit assumptions are present in almost every vision operator (Horswill, 1995). A misuse of *z* in situations where H_z do not hold true may cause abrupt performance degradation. Parameter C_z is a metrics depending on average performance ratings (e.g. computational time, delay, etc) useful to optimize system resources.

3.3 Contextual Changes

The explicit formulation of D.1 allows for a deeper understanding of *contextual adaptability* problems and of "context changes".

Definition D.4: A *context change* is a change in any component of a *context Q*, and we write it with $\Delta Q = (\Delta\{m\} \| \Delta P_{\{m\}} \| \Delta V_{\{m\}} \| \Delta\{z\} \| \Delta H_{\{z\}} \| \Delta C_{\{z\}} \| \Delta D)$, where $\|$ is a logical or.

Each component of ΔQ generates a class of *adaptability problems* analyzed in the literature under an application-specific definition of "context change". Here follow some examples:

a) $\Delta\{m\}$ may occur when i) the camera dramatically changes its point of view, ii) a perceptive system enters a completely different environment of which it lacks some object knowledge, iii) object description criteria become inappropriate.

b) $\Delta P_{\{m\}}$ means that the frequency of an *entity* class occurrence has changed, e.g. i) a camera enters a new geographical environment, ii) stochastic processes of object occurrence are non-stationary in time.

c) $\Delta V_{\{m\}}$ may occur when agents change trajectory so that hybrid tracking is needed – see (Tissainayagam, 2003), (Dessoude, 1993).

d) $\Delta\{z\}$ may consist in i) inappropriate modeling of operator mechanisms, ii) inappropriate self-assessment measures, etc.

e) $\Delta H_{\{z\}}$ indicates a failure of assumptions underlying $\{z\}$. For instance, a skin color detector whose color model is inappropriate to lighting conditions – see (Kruppa, 2001).

f) $\Delta C_{\{z\}}$ turns into a resource management problem. Dynamic programming, task planning, parametric control are examples of methods to find the best resource reallocation or sequencing.

g) ΔD may occur when i) assumptions for separation of object classes become inappropriate, ii) critical observed feature become unavailable, iii).

Definition D.5: The problem of insuring reliable system processing in presence of a *context change* is called an *adaptability problem*.

4   BAYESIAN CONTEXT SWITCHING

Two are the solutions to cope with *context changes*: i) a system has available alternative perceptive modalities; ii) a system can develop new perceptive modalities. The latter solution would involve on-line learning and trial-and-error strategies. Although some works have been presented – e.g. genetic programming of visual operators (Ebner, 1999) –, this approach is likely beyond the implementation level at present.

The first solution may be implemented either by using "parallelism" or by "opportunistic switching" to a valid configuration. "Parallelism" consists in introducing redundancy and data fusion by means of alternative algorithms, so that failures of one procedure be balanced by others working correctly. However, parallelism is today often simulated on standard processors, with the inevitable effect of dramatically increasing the computational load at the expense of real time. This feature conflicts with the requirements of machine vision for robotics. "Opportunistic switching" consists in evaluating the applicability of a visual module or in pointing out a change in the environmental context, to commuting the system configuration accordingly. Opposite to parallelism and data fusion, this swapping strategy conciliates robustness and real time. Here we further develop the latter option (4.1), we describe a Bayesian implementation of it (4.2), and finally we exemplify an application to contextual video surveillance (4.3).

4.1 Opportunistic Switching

Opportunistic switching among a set of optimized configurations may ensure acceptable performance over a *finite* range N of pre-identified situations (i.e. "contexts").

Definition D.6: Designing a system for context-dependent *opportunistic switching* consists in building and efficiently controlling a mapping ζ between a set of *contexts Q* and a set of *sub-systems S*, i.e. (1). The switching is triggered by *context changes* D.4.

$$\zeta : Q(t) \to S(t) \qquad (1)$$

Building the map is an application-dependent engineering task: for each typical situation, the perceptive system must be engineered to deliver acceptable results. Control is performed by detecting the current *context Q(t)*, or equivalently by detecting *context changes* ΔQ. A context-adaptable system must be endowed with context-receptive processing, i.e. routines capable of classifying N different. *context states* $\{q_1, q_2, ... q_N\}$. Essentially, such routines detect "context features", and *context* recognition can be thought of as an object

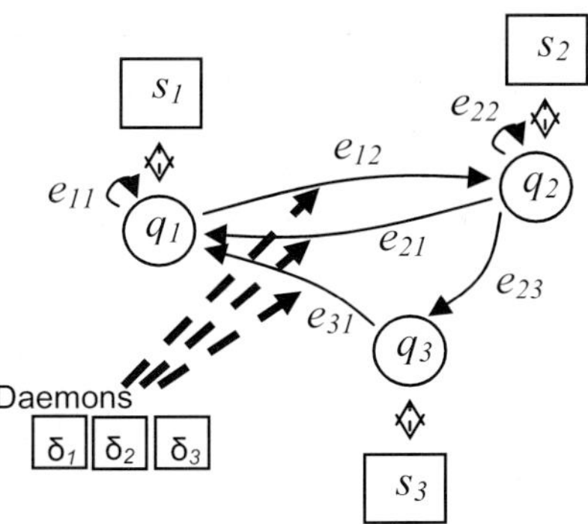

Figure 1: An oriented graph may easily accommodate all the elements of an *opportunistic switching* structure as defined in Section 3: *context/sub-system* pairs in nodes, and *events* in arcs. *Daemons* trigger global state change.

recognition task. The design of such routines appears to be an application-dependent design issue

Definition D.7: Let us name *daemon* an algorithm or sensor δ exclusively dedicated to estimating *context states q*.

Opportunistic switching has two advantageous features: i) *flexibility and real-time*, because multiple configurations run one at a time, and ii) *software reuse*, because an increased flexibility can be achieved by integrating current software with *ad-hoc* configurations for uncovered *contexts*. Assumptions for its use are: i) there exists a rigid (static) mapping from problems to solutions, ii) reliable *context* detection.

4.2 Context Commutation

The mapping ζ and its control may assume the form of parametric control, of knowledge-based algorithm selection, of neural network controlled systems, etc. Hereafter we present a Bayesian implementation of the opportunistic switching strategy, named *Context Commutation* (*CC*) (Lombardi, 2003). It is inspired by hybrid tracking –e.g. (Dessoude, 1993) –, where a swapping among multiple Kalman filters improves tracking of a target moving according to changing regimes.

Context Commutation represents context switching by means of a Hidden Markov Model – e.g. (Rabiner, 1989) –, where the hidden process is *context evolution* in time, and the stochastic observation function is provided by appropriate probabilistic sensor models of *daemons*. Time is ruled by a discrete clock *t*. Each clock step corresponds to a new processed frame.

Definition D.8: *Context Commutation* represents *context evolution* by means of a discrete, first-order HMM with the following components (Figure 1):

1. A *set of states* $Q = \{q_1, q_2, ...q_N\}$. Each state q_i corresponds to a *context* and gets an associated optimized system configuration s_i. For every i, s_i is such that the perceptive system works satisfactorily in $q_i = \{M_i, Z_i, D_i\}$ i.e. M_i, Z_i, D_i are the appropriate models, operators and decision policies in the *i*-th situation.

2. An *observation feature space* Φ composed of *daemon* outputs φ. If there are *K daemons*, φ is a *K*-dimensional vector.

3. A *transition matrix E*, where elements E_{ij} correspond to the a priori probability of transition from q_i to q_j, i.e. (2).

$$E_{ij} = P[e_{ij}] = P[Q(t) = q_j \mid Q(t\text{-}1) = q_i] \qquad (2)$$

4. An *observation probability distribution* $b_i(\varphi)$ for each context q_i, defined in (3). Thus, the *N* different $b_i(\varphi)$ define the global *daemon* sensor model of Bayesian signal analysis theory.

$$b_i(\varphi) = P(\Phi(t) = \varphi \mid Q(t) = q_i) \qquad (3)$$

5. An *initial state distribution* function $\pi = \{\pi_1, \pi_2, ...\pi_N\}$, where $\pi_i \in [0, 1]$ for $i = 1, 2, ...N$, and (4) holds true.

$$\sum_{i=1}^{N} \pi_i = 1 \qquad (4)$$

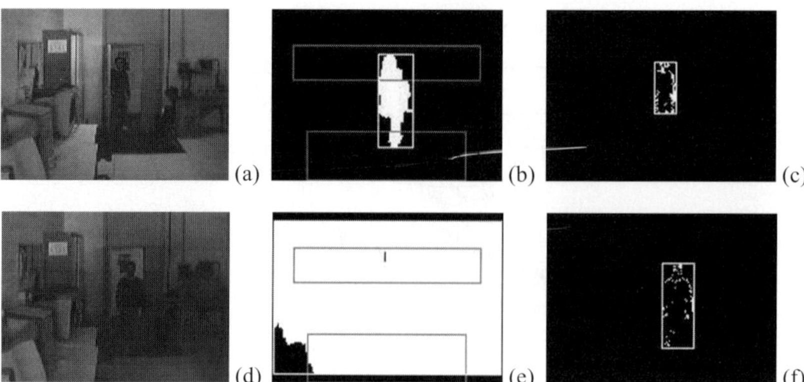

Figure 2: When a reliable background reference model is available (a), background subtraction methods deliver more meaningful motion information (b) than simple frame differencing (c). However, if the lighting conditions suddenly change, e.g. an artificial light is turned off (d), BS fails (e) while FD still works properly.

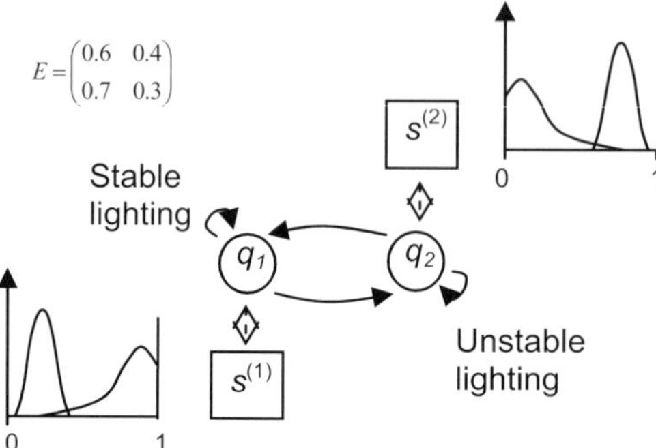

Figure 3: The simple CC system for "light switch" problems has two states and one daemon. The picture shows the transition matrix E used in the experiments (top left), and a representation of daemon models (next to the s_i boxes).

6. The *current context* q_v is estimated by the Maximum A Posteriori on $\Psi(t)$ (5), (6).

$$\Psi(t) = (P(q_1), P(q_2), \dots P(q_N)) \qquad (5)$$

$$v = \text{argmax}_i [\Psi_i(t)] \qquad (6)$$

4.3 A Practical Implementation

As a final illustration, we demonstrate an application of *Context Commutation* to tackle the "light switch" problem affecting *background subtraction* (BS) for motion detection in automatic video surveillance. In indoor environments, when artificial lights are turned on or off, the reference background model used in BS looses validity in one frame-time. Modern time-adaptive background systems (Stauffer, 1999) usually take around 10÷100 frames to recover. An alternative solution involves the use of a second algorithm that degrades less its performance in case of abruptly changing lighting conditions. For instance, *frame differencing* (FD) algorithms deliver motion information like BS does,

and they recover from "light switch" just after 1 frame (Figure 2).

A context-adaptable system based on opportunistic switching would feature two system states: i) using BS when appropriate, ii) using FD otherwise. In the general case, BS delivers a more informing motion map than FD. However, when lighting conditions are unstable, the system swaps to FD – which recovers more quickly.

Here, we design a CC system as shown in Table 1 and Figure 3. The two *contexts*, corresponding to "stable" and "unstable" global lighting, cope with a *context change* ΔH_{bs} which corresponds to a failure of a basic *operative assumption* founding BS's correct working – i.e. stable lighting –. The *daemon* δ_1 apt to detecting ΔH_{bs} is modeled with two truncated Gaussians of the kind shown in Figure 3, with parameters tuned by training. *Daemon* δ_1 counts the pixels n_a and n_b showing a luminance change that breaks thresholds $\theta_{\delta 1}$ and $-\theta_{\delta 1}$, respectively: $n_a + n_b$ represents all pixels showing substantial luminance change. The output (7) is then a measure of the luminance unbalance over the last two images. In stable lighting conditions φ_1 would be 0. The closer φ_1 to 1, the more likely switched the light.

$$\varphi_1 = 2\frac{\max(n_a, n_b)}{n_a + n_b} - 1 \tag{7}$$

Table 1

Q	Situation	S
q_1	stable lighting	BS active if in ready state
		FD active if BS in recovering state
q_2	unstable lighting	FD active

To assess context estimation performance, δ_1 was tested on over 1500 images containing about 50 light switches. The test was done on sequences indexed by a human operator. Figure 4 shows the results on one test sequence: when the confidence rating breaks 0.5, q_2 is estimated. Bold dots on the top line show the ground truth for q_2 occurrence. Model parameters $G_i \sim (\mu_i, \sigma_i)$ in q_i are in Table 2.

Table 2

	μ_1	μ_2	σ_1	σ_2
δ_1	0.09	0.71	0.17	0.36

We measured an average correct estimation rate of 0.95. The percentage goes up to 0.98 if a 3-frame-range error is allowed in locating the contextual switch. In effect, this error allowance accounts for human mistakes in indexing the test videos.

The motion detection system with and without *CC* was tested on several sequences. No tracking was performed, only motion detection. The graph of Figure 5 shows the improvement provided by *CC* in terms of such distance (when BS failed because of inappropriate background model – e.g. Figure 2 –, the corresponding estimation error was set to 100). Figure 6 shows some results for one sequence where light switches twice: on-off on frame 327, and off-on on frame 713. The distance of the barycentre of motion between automatic detection and human labeling was computed for BS only, and for BS/FD combined by means of *CC*.

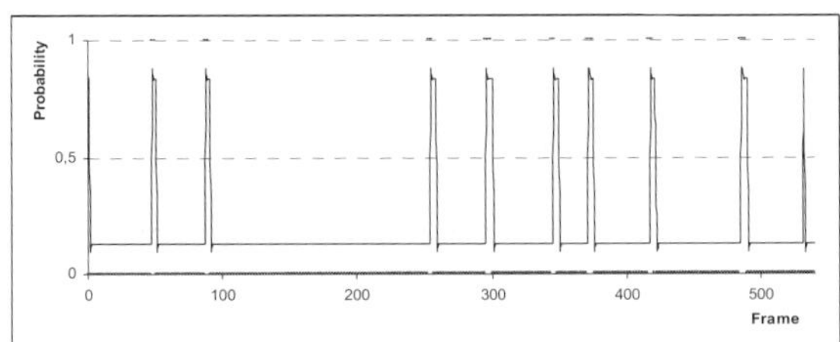
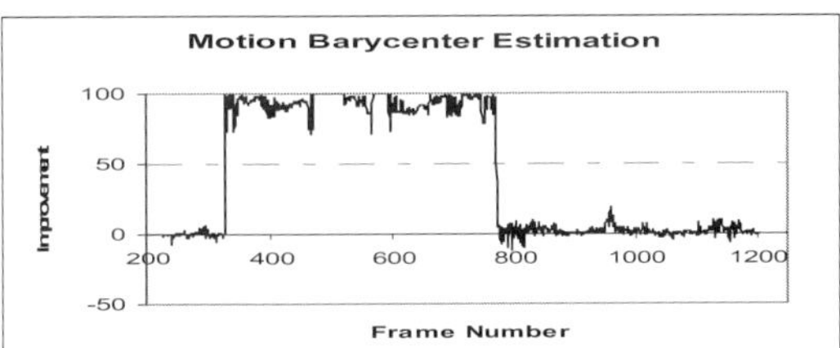

Figures 4, 5: Probability that the current *context state* be q2 as estimated by δ_1 in a test sequence (left). Improvement in the estimation error provided by context switching (*CC*) with respect to BS alone (right).

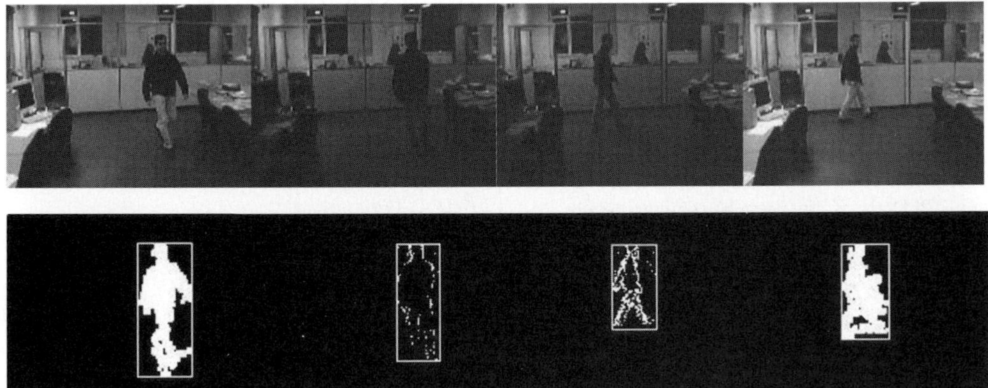

Figure 5: Frames no. 322, 332, 702, and 932 from a test sequence: original images (first row), motion detection by BS and FD managed opportunistically by *CC* (second row).

5 CONCLUSIONS

In this paper we foster deeper studies in the management of contextual information in robotic vision. In the first part, we proposed an operative definition of "context" to identify the variable parts of a perceptive system susceptible of becoming inappropriate in case of contextual changes: models, operators, and decision policies.

In the second part, we described a novel Bayesian framework (i.e. *Context Commutation*) to implement contextual opportunistic switching. Dedicated algorithms, called *daemons*, observe some environmental features showing a correlation with system performance ratings rather than with the target signal (e.g. people tracking). When such features change, the system commutes its state to a more reliable configuration.

Critical points in *Context Commutation* are mainly related to its Bayesian framework. Parameters like sensor models of *daemons* and coefficients of the *transition matrix* need thorough tuning and massive training data. An error in such parameters would corrupt correct contextual switching.

Possible points for future work are: i) exploring switching reliability with incorrect parameters, ii) studying *Context Commutation* with more than eight states, iii) extending the framework to perceptive systems including sensors other than solely vision.

REFERENCES

Coutelle, C., 1995. *Conception d'un système à base d'opérateurs de vision rapides*, PhD thesis (in French), Université de Paris Sud (Paris 11), Paris, France.

Crowley, J.L., J. Coutaz, G. Rey, P. Reignier, 2002. Perceptual Components for Context Aware Computing. In *Proc. UBICOMP2002*, Sweden, available at http://citeseer.nj.nec.com/541415.html.

Dessoude, O., 1993. *Contrôle Perceptif en milieu hostile: allocation de ressources automatique pour un système multicapteur*, PhD thesis (in French), Université de Paris Sud (Paris 11), Paris, France.

Draper, B.A., J.Bins, K.Baek, 1999. ADORE: Adaptive Object Recognition. In *Proc. ICVS99*, pp. 522-537.

Dubes, R. C., Jain, A. K., 1989. Random Field Models in Image Analysis. In *J. Applied Statistics*, V. 16, pp. 131-164.

Ebner, M., A. Zell, 1999. Evolving a task specific image operator. In *Proc. 1st European Wshops on Evolutionary Image Analysis, Signal Processing and Telecommunications*, Göteborg, Sweden, Springer-Verlag, pp. 74-89.

Horswill, I., 1995. Analysis of Adaptation and Environment. In *Artificial Intelligence*, V. 73(1-2), pp. 1-30, 1995.

Kittler, J., J. Matas, M. Bober, L. Nguyen, 1995. Image interpretation: Exploiting multiple cues. In *Proc. Int. Conf. Image Processing and Applications*, Edinburgh, UK, pp. 1-5.

Kruppa, H., M. Spengler, B. Schiele, 2001. Context-driven Model Switching for Visual Tracking. In *Proc. 9th Int. Symp. Intell. Robotics Sys.*, Toulouse, France.

Lombardi, P., 2003. *A Model of Adaptive Vision System: Application to Pedestrian Detection by Autonomous Vehicles*. PhD thesis (in English), Università di Pavia (Italy) and Université de Paris XI (France).

Merlo, X., 1988. *Techniques probabilistes d'intégration et de contrôle de la perception en vue de son exploitation par le système de décision d'un robot*, PhD thesis (in French), Université de Paris Sud (Paris 11), Paris, France.

Rabiner, L.R., 1989. A tutorial on hidden Markov models. In *Proceedings of the IEEE*, V. 77, pp. 257-286.

Rimey, R.D., 1993. Control of Selective Perception using Bayes Nets and Decision Theory. Available at http://citeseer.nj.nec.com/rimey93control.html.

Roli, F., G. Giacinto, S.B. Serpico, 2001. Classifier Fusion for Multisensor Image Recognition. In *Image and Signal Processing for Remote Sensing VI*, Sebastiano B. Serpico, Editor, Proceedings of SPIE, V. 4170, pp.103-110.

Rosenfeld, A., R.A. Hummel, S.W. Zucker, 1976. Scene labeling by relaxation operations. In *IEEE Trans. Syst. Man Cybern.*, V. 6, pp. 420-433.

Shekhar, C., S. Kuttikkad, R. Chellappa, 1996. KnowledgeBased Integration of IU Algorithms. In *Proc. Image Understanding Workshop, ARPA*, V. 2, pp. 1525-1532, 1996.

Stauffer, C., W.E.L. Grimson, 1999. Adaptive Background Mixture Models for Real-Time Tracking. In *Proc. IEEE Conf. Comp. Vis. Patt. Rec. CVPR99*, pp. 246-252.

Strat, T.M., 1993. Employing Contextual Information in Computer Vision. In *Proc. DARPA93*, pp. 217-229.

Tissainayagam, P., D. Suter, 2003. Contour tracking with automatic motion model switching. In *Pattern Recognition*.

Torralba, A., K.P. Murphy, W.T. Freeman, M.A. Rubin, 2003. Context-based vision system for place and object recognition. In *Proc. ICCV'03*, available at http://citeseer.nj.nec.com/torralba03contextbased.html.

Toyama, K., E.Horvitz, 2000. Bayesian Modality Fusion: Probabilistic Integration of Multiple Vision Algorithms for Head Tracking. In *Proc. ACCV'00, 4th Asian Conf. Comp. Vision*, Tapei, Taiwan, 2000.

DYNAMIC STRUCTURE CELLULAR AUTOMATA
IN A FIRE SPREADING APPLICATION

Alexandre Muzy, Eric Innocenti, Antoine Aïello, Jean-François Santucci, Paul-Antoine Santoni
University of Corsica
SPE – UMR CNRS 6134
B.P. 52, Campus Grossetti, 20250 Corti. FRANCE.
Email: a.muzy@univ-corse.fr

David R.C. Hill
ISIMA/LIMOS UMR CNRS 6158
Blaise Pascal University
Campus des Cézeaux BP 10125, 63177 Aubière Cedex. FRANCE.
Email: drch@isima.fr

Keywords: Systems modeling, Discrete event systems, Multi-formalism, Fire spread, cellular automata.

Abstract: Studying complex propagation phenomena is usually performed through cellular simulation models. Usually cellular models are specific cellular automata developed by non-computer specialists. We attempt to present here a mathematical specification of a new kind of CA. The latter allows to soundly specify cellular models using a discrete time base, avoiding basic CA limitations (infinite lattice, neighborhood and rules uniformity of the cells, closure of the system to external events, static structure, etc.). Object-oriented techniques and discrete event simulation are used to achieve this goal. The approach is validated through a fire spreading application.

1   INTRODUCTION

When modeling real systems, scientists cut off pieces of a biggest system: the world surrounding us. Global understanding of that world necessitates connecting all these pieces (or subsystems), referencing some of them in space. When the whole system is complex, the only way to study its dynamics is simulation.

Propagation phenomena as fire, swelling, gas propagation, (…) are complex systems. Studying these phenomena generally leads to divide the propagation space in cells, thus defining a cellular system.

Developed from the General Systems Theory (Mesarovic and Takahara, 1975), the Discrete Event Structure Specification (DEVS) formalism (Zeigler et al., 2000) offers a theoretical framework to map systems specifications into most classes of simulation models (differential equations, asynchronous cellular automata, etc.). For each model class, one DEVS sub-formalism will allow to faithfully specify one simulation model. As specification of complex systems often needs to grasp different kinds of simulation models,

connections between the models can be performed using DEVS multi-formalism concepts.

Another DEVS advantage relates to its ability in providing discrete event simulation techniques, thus enabling to concentrate the simulation on active components and resulting in performance improvements.

Precise and sound definition of propagation needs to use models from physics as Partial Differential Equations (PDEs). These equations are then discretized leading to discrete time simulation models. These models are generally simulated from scientists by using specific Cellular Automata (CA). As defined in (Wolfram, 1994), standard CA consist of an infinite lattice of discrete identical sites, each site taking on a finite site of, for instance, integer values. The values of the sites evolve in discrete time steps according to deterministic rules that specify the value of each site in terms of the values of neighboring sites. CA may thus be considered as discrete idealizations of PDEs. CA are models where space, time and states are discrete (Jen, 1990).

However, definition of basic CA is too limited to specify complicated simulations (infinite lattice, neighborhood and rules uniformity of the cells,

247

J. Braz et al. (eds.), Informatics in Control, Automation and Robotics I, 247–254.
© 2006 *Springer. Printed in the Netherlands.*

closure of the system to external events, discrete state of the cells, etc.). Scientists often need to modify CA's structure for simulation purposes (Worsch, 1999).

We extend here basic CA capabilities by using object-oriented techniques and discrete event simulation (Hill, 1996). A mathematical specification of the approach is defined using the Dynamic Structure Discrete Time System Specification (DSDTSS) formalism (Barros, 1997). This formalism allows dynamically changing the structure of discrete time systems during the simulation. These new CA are called the Dynamic Structure Cellular Automata (DSCA).

DSCA have been introduced in (Barros and Mendes, 1997) as a formal approach allowing to dynamically change network structures of asynchronous CA. Using an asynchronous time base, cells were dynamically instantiated or destroyed during a fire spread simulation.

The scope here is to extend basic CA capabilities using a discrete time base. If previous DSCA were dedicated to discrete event cellular system specification, we extend here the DSCA definition to discrete time cellular system specification.

Table 1 sums up the advantages of the DSCA in relation to basic CA. In DSCA, each cell can contain different behaviors, neighborhoods and state variables. Rules and neighborhoods of the cells can dynamically change during the simulation. Each cell can receive external events. During the simulation, the computing of state changes is limited to active. Finally, a global transition function allows the specification of the DSCA global behaviors.

Table 1: DSCA extensions

	Basic CA	DSCA
Time	discrete	Discrete
Space	discrete	Discrete
State	discrete	Continuous
Closure to external events	-	+
Different variables per cell	-	+
Rules uniformity*	-	+
Neighborhood uniformity*	-	+
Activity tracking*	-	+
Global function	-	+

*at simulation time

The DSCA definition is validated against a fire spreading application. Recent forest fires in Europe (Portugal, France and Corsica) and in the United States (California) unfortunately pinpoint the necessity of increasing research efforts in this domain. Fires are economical, ecological and human catastrophes. Especially as we know that present rising of wild land surfaces and climate warming will increase forest fires.

Modeling such a huge and complex phenomenon obviously leads to simulation performance overloadings and design problems. Simulation model reusability has to face to the complicated aspects of both model implementations and model modifications. Despite a large number of cells, simulation has to respect real time deadlines to predict actual fire propagations. Hence, this kind of simulation application provides a powerful validation to our work.

This study is organized as follows. First some formalisms background is provided. Then two sections present the DSCA modeling and simulation principles. After, simulation results of a fire spreading application are provided. Finally, we conclude and make some prospects.

2 BACKGROUND

A formalism is a mathematical description of a system allowing to guide a modeler in the specification task. The more a formalism fits to a system class, the more simple and accurate it will be.

Efficiently modeling complex systems often implies the need to define subsystems using different formalisms. Connections between the different formalisms can then be achieved through a multi-formalism to perform the whole system specification.

In this study, subsystems are specified using DEVS, DTSS (Discrete Time System Specification) and DSDTSS formalisms. Connections between the different models are achieved using a Multi-formalism Network (*MFN*). A structure description of each model is provided hereafter.

A DEVS atomic model is a structure:

$$DEVS = (X, Y, Q, q_0, \delta_{int}, \delta_{ext}, \lambda, t_a)$$

where X is the input events set, Q is the set of state, q_0 is the initial state, Y is the output events set, $\delta_{int}: Q \rightarrow Q$ is the internal transition function, $\delta_{ext}: Q \times X \rightarrow Q$ is the external transition function, $\lambda: Q \rightarrow Y$ the output function, t_a is the time advance function.

DSDTSS basic models are DTSS atomic model:

$$DTSS = (X, Y, Q, q_0, \delta, \lambda, h)$$

where X, Y are the input and output sets, Q is the set of state, q_0 is the initial state, $\delta: Q \times X \to Q$ is the state transition function, $\lambda: Q \to Y$ is the output function (considering a Moore machine) and h is a constant time advance.

At a periodic rate, this model checks its inputs and, based on its state information, produces an output and changes its internal state.

The network of simple DTSS models is referred to as a Dynamic Structure Discrete Time Network (DSDTN) (Barros, 1997). We introduce here input and output sets to allow connections with the network. Formally, a DSDTN is a 4-tuple:

$$DSDTN = (X_{DSDTN}, Y_{DSDTN}, \chi, M_\chi)$$

where X_{DSDTN} is the network input values set, Y_{DSDTN} is the network input values set, χ is the name of the DSDTN executive, M_χ is the model of the executive χ.

The model of the executive is a modified *DTSS* defined by the 8-tuple:

$$M_\chi = (X_\chi, Q_\chi, q_{0,\chi}, Y_\chi, \gamma, \Sigma^*, \delta_\chi, \lambda_\chi)$$

where $\gamma: Q_\chi \to \Sigma^*$ is the structure function, and Σ^* is the set of network structures. The transition function δ_χ computes the executive state q_χ. The network executive structure Σ, at the state $q_\chi \in Q_\chi$ is given by $\Sigma = \gamma(q_\chi) = (D, \{M_{i}\}, \{I_{i}\}, \{Z_{i,j}\})$, for all $i \in D$, $M_i = (X_i, Q_i, q_{0,i}, Y_i, \delta_i, \lambda_i)$, where D is the set of model references, I_i is the set of influencers of model i, and $Z_{i,j}$ is the i to j translation function.

Because the network coupling information is located in the state of the executive, transition functions can change this state and, in consequence, change the structure of the network. Changes in structure include changes in model interconnections, changes in system definition, and the addition or deletion of system models.

Formally, a multiformalism network (Zeigler et al., 2000) is defined by the 7-tuple:

$$MFN = (X_{MFN}, Y_{MFN}, D, \{M_{i}\}, \{I_{i}\}, \{Z_{i,j}\}, select)$$

where $X_{MFN} = X^{discr} \times X^{cont}$ is the network input values set, X^{discr} and X^{cont} are discrete and continuous input sets, $Y_{MFN} = Y^{discr} \times Y^{cont}$ is the network input values set, Y^{discr} and Y^{cont} are discrete and continuous output sets, D is the set of model references,

For each $i \in D$,

  M_i is are DEVS, DEVN, DTSN, DTSS, DESS, DEV&DESS or other MFN models. As DSDTSS proved to be closed under coupling, M_i can also be dynamic structure models or networks,

I_i is the set of influencers of model i,

$Z_{i,j}$ is the i to j translation function,

select is the tie-breaking function.

3 DSCA MODELLING

Models composing a DSCA are specified here using the previous model definitions. As described in the modeling part of Figure 1, external events are simulated using a DEVS atomic model: the *Generator*. The latter can asynchronously generate data information to the DSCA during the simulation. The cell space is embedded in a DSDTN. Each cell is defined as a DTSS model. Using its transition function, the DSCA executive model (containing every cells) achieves changes in structure directly accessing to the attributes of cells. A mathematical description of each model is provided here after.

We define the *MFN* by the structure:

$$MFN = (X_{MFN}, Y_{MFN}, D, \{M_{i}\}, \{I_{i}\}, \{Z_{i,j}\}, select)$$

where $D = \{G, DSDTN\}$, $M_G = (X_G, Q_G, q_{0,G}, Y_G, \delta_G, \lambda_G, \tau_G)$, $M_{DSDTN} = DSDTN$, $I_G = \{\}$, $I_{DSDTN} = \{G\}$, and $Z_{DSDTN,MFN}: Y_{DSDTN} \to Y_{MFN}$, $Z_{G,DSDTN}: Y_G \to X_{DSDTN}$.

We define the *DSDTN* by the structure:

$$DSDTN = (X_{DSDTN}, Y_{DSDTN}, DSCA, M_{DSCA})$$

where $\Sigma = \gamma(q_{0,\chi}) = (D, \{M_{i}\}, \{I_{i}\}, \{Z_{i,j}\})$, where $D = \{(i,j) \ / \ (i,j) \in \mathfrak{I}^2 \}$, $M_{DSCA} = (X_{DSCA}, Q_{DSCA}, q_{0,DSCA}, Y_{DSCA}, \delta_{DSCA}, \lambda_{DSCA})$, $I_{DSCA} = \{DSDTN\}$, $I_{cell} = \{cell^n, DSDTN\}$. Where $I_{cell} = \{I_{kl} \ / \ k \in [0,m], l \in [0,n]\}$ is the neighbourhood set (or the set of influencers) of the cell as defined in (Wainer and Giambiasi, 2001). It is a list of pairs defining the relative position between the neighbours and the origin cell. $I_{kl} = \{(i_p, j_p) \ / \ \forall \ p \in I_{cell}, p \in [1, \eta_{kl}], i_p, j_p \in Z \ ; \ |k - i_p| \geq 0 \ \wedge \ |l - j_p| \geq 0 \ \wedge \ \eta_{kl} \in I_{cell} \}$, and $\eta \in I_{cell}$ is the neighborhood size.

$Z_{cell,DSCA}:$ $Y_{cell} \to Y_{DSCA}$, $Z_{DSCA,DSDTN}:$ $Y_{DSCA} \to Y_{DSDTN}$, $Z_{DSDTN,DSCA}:$ $X_{DSDTN} \to X_{DSCA}$, $Z_{DSCA,cell}: X_{DSCA} \to X_{cell}$.

For the implementation, ports are defined:

$$Py_G = Px_{DSDTN} = Px_{DSCA} = \{data\}$$

$$Py_{DSCA}=Py_{DSDTN}=Py_{MFN}=\{state\}$$
$$Px_{cell}=Py_{cell}=\{(i,j)\}$$

We specify each cell as a special case of DTSS model:

$$cell=(X_{cell},\ Q_{cell},\ q_{0,cell},\ Y_{cell},\ \delta_{cell},\ \lambda_{cell}\ )$$

where X_{cell} is an arbitrary set of input values, Y_{cell} is an arbitrary set of output values, $q_{0,cell}$ is the initial state of the cell and

$q \in Q_{cell}$ is given by:

$q=((i,j),\ state,\ N,\ phase)$,

$(i,j) \in \mathfrak{I}^2$, is the position of the cell,

state is the state of the cell,

$N = \{N_{kl}\ /\ k \in [0,m],\ l \in [0,n]\}$. N is a list of states N_{kl} of the neighboring cells of coordinates (k,l),

phase = *{passive, active}* corresponds to the name of the corresponding dynamic behavior. For numerous adjacent active cells, the *active* phase can be decomposed in *'testing'* and *'nonTesting'* phases. The use of these phase is detailed in section 5.

$$\delta_{cell}:\ Q_{cell}{\times}X_{cell} \rightarrow Q_{cell}$$
$$\lambda_{cell}:\ Q_{cell} \rightarrow Y_{cell}$$

4 DSCA SIMULATION

As depicted in Figure 2, implementation of discrete event models consists in dividing a transition function δ_d of a model according to event types ev_n issued from a set of possible event types S_d. The transition function then depends on the event types the model receives.

```
Model d
   S_d = {ev_1, ev_2, ... , ev_n}
   δ_d (S_d)
        case S_d
        ev_1 : call event-routine_1
        ev_2 : call event-routine_2...
        ev_n : call event-routine_n
```
Figure 2: Discrete event model implementation
(Zeigler et al., 2000)

The DSCA receives data from the *Generator*. These data represent external influences of the DSCA. During the simulation, information is embedded in messages and transits through the *data* and *state* ports. Messages have fields *[Message type, Time, Source processor, Destination port, Content]*, where *Content* is a vector of triplets *[Event type,*

Value, Coordinate port]. When a DSCA receives a message on its *data* port, the corresponding Aggregated Network Simulator (*AN Simulator*) scans the *Content* vector and according to the *Coordinate port*, sends the *[Event type, Value]* pairs to the concerned cells. A vector of pairs *[Event type, Value]* can be sent to a cell port. Then, according to the *Event type* a cell receives, it will update the concerned attributes, executing the concerned transition function.

The simulation tree hierarchy is described in the simulation part on the right side of Figure 1. Except for the *Root* and *DTSS interface,* all nodes of the tree are processors attached to models. The processors manage with message exchanges and execution of model functions. Each processor is automatically generated when the simulation starts.

The *Coordinator* pilots the *MFN* model, the simulator *SimG* pilots the *Generator*, the *DSDTN* model and the *DSDTN* models are piloted by the Aggregated Network Simulator (*AN Simulator*). Algorithms of the DSCA simulators can be found in (Muzy et al., 2003). Algorithms of basic DEVS simulators and DTSS interfaces can be found in (Zeigler et al., 2000).

The Root processor supervises the whole simulation loop. It updates the simulation time and activates messages at each time step. For the Coordinator, the DTSS interface makes the Aggregated Network simulator seen as a DEVS atomic simulator. This is done by storing all messages arriving at the same time step and then by calculating the new state and output of the DSCA when receiving an internal transition message. The simulation tree thus respects the DEVS bus principle. That means that whatever DEVS model can be appended to the simulation tree.

5 ACTIVITY TRACKING

Using discrete event cellular models, activity tracking can be easily achieved (Nutaro et al., 2003). Active cells send significant events to be reactivated or to activate neighbors at next time step. However, pure discrete event models proved to be inefficient for discrete time system simulation (Muzy et al., 2002). Interface configurations and message management produce simulation overheads, especially for numerous active components.

For discrete time systems, we know that each component will be activated at each time step. Moreover, in CA, states of cells directly depend on

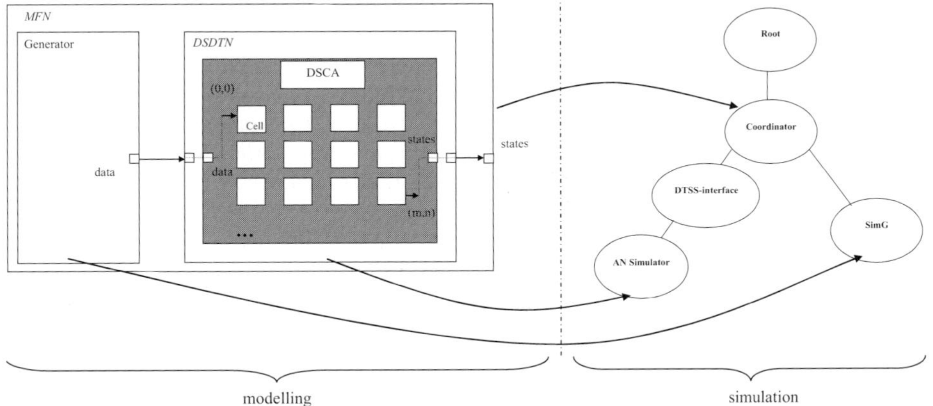

Figure 1: DSCA modeling and simulation.

the states of their neighbors. To optimize the simulation, messages between the cells have to be canceled and simulation time advance has to be discrete. However, a new algorithm has to be defined to track active cells.

To focus the simulation on active cells, we use the basic principles exposed in (Zeigler et al., 2000) to predict whether a cell will possibly change state or will definitely be left unchanged in a next global state transition: "*a cell will not change state if none of its neighboring cells changed state at the current state transition time*".

Nevertheless, to obtain optimum performance the entire set of cells cannot be tested. Thereby, an algorithm, which consists in testing only the neighborhood of the active bordering cells of a propagation domain, has been defined for this type of phenomena.

To be well designed, a simulation model should be structured so that all information relevant to a particular design can be found in the same place. This principle enhances models modularity and reusability making easier further modifications.

Pure discrete event cells are all containing a micro algorithm, which allows to focus the whole simulation loop on active cells. We pinpointed above the inefficiency of such an implementation for discrete time simulation. An intuitive and efficient way to achieve activity tracking in discrete time simulation is to specify this particular design at one place. As depicted in Figure 3, the activity tracking algorithm is located in the global transition function of the DSCA, in charge of the structure evolution of the cell space.

Cells are in '*testing*' phases when located at the edge of the propagation domain, '*nonTesting*' when not tested at each state transition and '*quiescent*' when inactive.

A propagation example is sketched in Figure 4 for cardinal and adjacent neighborhoods. In our

algorithm, only the bordering cells test their neighborhood, this allows to reduce the number of testing cells.

The result of the *spreadTest(i,j,nextState)* function of Figure 3, depends on the state of the tested cells. If this state fulfils a certain condition defined by the user, the cell becomes '*nonTesting*' and new tested neighboring are added to the set of active cells. The transition function receives x_χ messages from the Generator corresponding to external events. The x_χ messages contain the coordinates of the cells influenced by the external event. If the coordinates are located in the domain calculation, the state of the cell is changed by activating its transition function with the new value. Otherwise, new cells are added to the propagation domain.

```
//'Q' is for the quiescent phase,
//'T' for the testing one and 'N' for //the nonTesting
one

Transition Function(x_χ)
 For each cell(i,j) Do
  If(cellPhase(i,j)=='Q') Then
    removeCell(i,j)
  Endif

  If(cellPhase(i,j)=='T') Then
   If (cellNearToBorder(i,j)) Then
    setSpreadStateCell(i,j,'N')
   Else
    If(spreadTest(i,j,nextState))Then
     setCellPhase(i,j,'N')
     addQuiescentNeighboringCells(i,j)
    EndIf
   EndIf
  EndIf
 EndFor

 If(x_χ message is not empty)
  If(cells in the propagation domain) Then
     cell(i,j).transitionF(newState)
     //change the cell states
  Else
   addNewCells()
  EndIf
 EndIf
EndTransitionFunction()
```

Figure 3: Transition function of the DSCA.

For efficiency reasons, the simulation engine we developed has been implemented in C++ and dynamic allocation has been suppressed for some classes. Indeed, for significant numbers of object instantiation/deletion dynamic allocation is inefficient and we have designed a specialized static allocation (Stroustrup, 2000). A pre-dimensioning via large static arrays can be easily achieved thanks to current modern computer memory capabilities.

The state of the executive model is a matrix of cellular objects. References on active cells are stored in a vector container. A start-pointer and an end-pointer are delimiting the current calculation domain on the vector. Thus initial active cells that are completely burned during a simulation run can be dynamically ignored in the main loop. At each time step, by modifying the position of pointers, new tested cells can be added to the calculation domain and cells that return in a quiescent state are removed from the former.

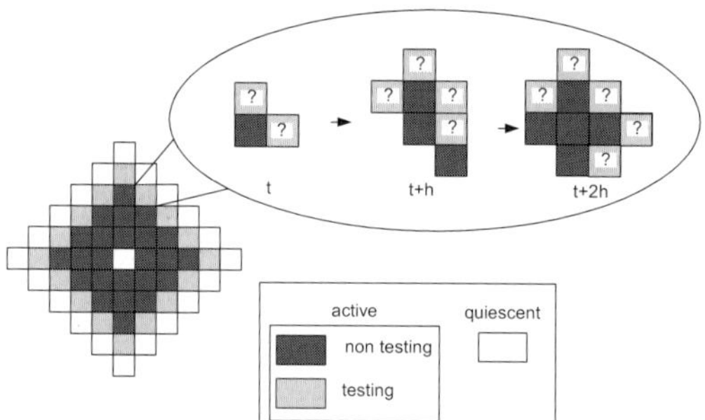

Figure 4: Calculation domain evolution.

6 FIRE SPREADING APPLICATION

The simulation engine we use has been proved to achieve real-time simulation (Muzy et al., 2003). Moreover, we use a mathematical fire spread model already validated and presented in (Balbi et al., 1998). In this model, a Partial Differential Equation (PDE) represents the temperature of each cell. A CA is obtained after discretizing the PDE. Using the finite difference method leads to the following algebraic equation:

$$T_{i,j}^{k+1} = a(T_{i-1,j}^{k} + T_{i+1,j}^{k}) + b(T_{i,j-1}^{k} + T_{i,j+1}^{k}) + \sigma_{v,0}e^{-a(t-t_{ig})} + dT_{i,j}^{k} \quad (1)$$

where T_{ij} is the grid node temperature. The coefficients a, b, c and d depend on the time step and mesh size considered, t is the real time, t_{ig} the real ignition time (the time the cell is ignited) and $\sigma_{v,0}$ is the initial combustible mass.

Figure 5 depicts a simplified temperature curve of a cell in the domain. We consider that above a threshold temperature T_{ig}, the combustion occurs and below a temperature T_f, the combustion is finished.

The end of the real curve is purposely neglected to save simulation time. Four states corresponding to the behavior of each cell behavior are defined from these assumptions. The four states are: *'unburned'*, *'heating'*, *'onFire'* and *'burned'*.

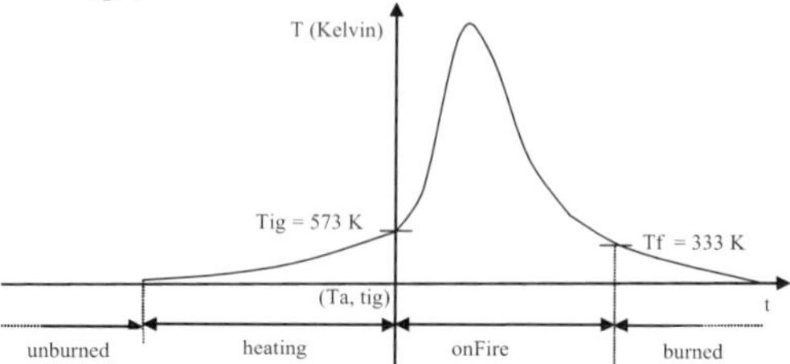

Figure 5: Simplified temperature curve of a cell behavior.

Figure 6 depicts a fire spreading in a Corsican valley, generated using the OpenGL graphics library. By looking at this picture we easily understand that simulation has to focus only on a small part of the whole land. Actually, areas of activity just correspond to the fire front, and to the cells in front of the latter (corresponding to cells in one of the following states: *'heating'* or *'onFire'*)

Figure 6: 3D Visualization of fire spreading.

During a fire spreading, flying brands ignite new part of lands away from the fire. This is an important cause of fire spreading. However, tracking activity of flying brands is difficult. Firstly, because flying brands occur whenever during the simulation time.

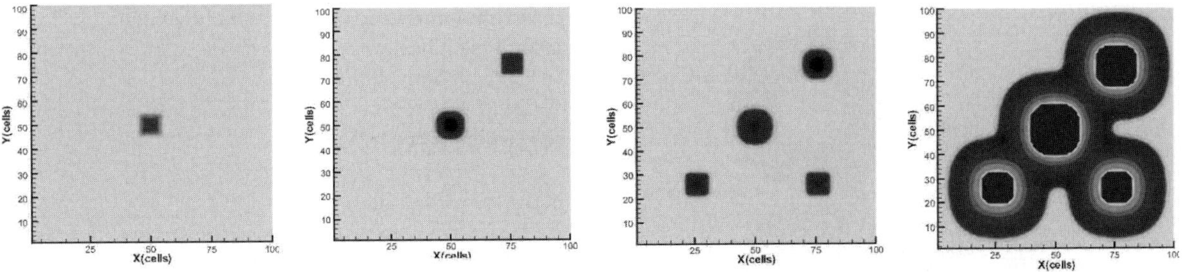

Figure 7: Fire ignitions and propagation.

Secondly, because they occur far away from the calculation domain, thus new calculation allocations need to be created dynamically.

Figure 7 represents a case of multi-ignitions during the simulation. Simulation starts with one ignition on the center of the propagation domain. Then, at time *t=7s*, a second ignition occurs on the top right corner of the propagation domain. Finally, at *t=12s*, two new ignitions occur on the right and left bottom of the propagation domain. The last picture shows the multiple fire fronts positions at *t=70s*.

Figure 8 describes the DSCA state transitions in a fire spread simulation. First the simulation starts with the first ignition, which is simulated by an output external event of the Generator of Figure 1. Then the main simulation loop calculating the fire front position is activated. The latter consists in calculating the temperature of cells. After, according to the calculated temperatures, the calculation domain is updated. For each cell of the calculation domain, the temperature is calculated using equation *(1)*, according to the state of the cells. The calculation domain is updated using the algorithm described in Figures 3 and 4. Phase transitions depend on the temperature of cells.

At the initialization, only one calculation domain corresponding to the one described in Figure 4 is generated. Bordering cells of the calculation domain are in a '*testing*' phase and non-bordering cells in a '*nonTesting*' one. Remaining cells are '*quiescent*'. If the temperature of a '*testing*' cell fulfills a certain threshold temperature T_t, the testing cell will pass in the '*nonTesting*' phase and neighboring '*testing*' cells will be added to the calculation domain. In the fire spread case, this threshold can be fixed slightly over the ambient temperature.

During the simulation, the Generator simulates the flying brands by sending external events. When the DSCA receives the events and updates the calculation domain.

The resulting activity tracking is showed in Figure 9. We can notice that active cells correspond to the fire front lines, not to the burned and non-heated areas.

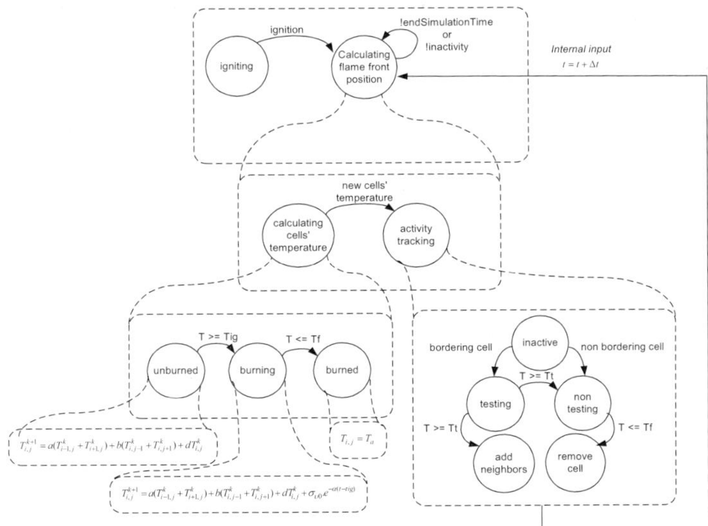

Figure 8: Transition state diagram.

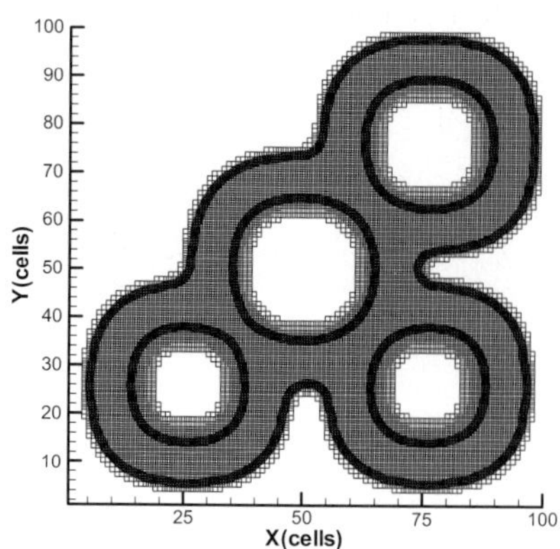

Figure 9: Activity tracking.

7 CONCLUSION

Considering the previous discrete event DSCA (Barros and Mendes, 1997), new well-designed and complementary discrete time DSCA have been defined here. These two methodologies allow to faithfully guide modelers for modeling and simulating discrete event and discrete time cellular simulation models.

DSCA allow to simulate a large range of complicated cellular models. Complex phenomena can be simulated thanks to basic CA simplicity. We hope that even more complex phenomena will be able to be simulated thanks to DSCA. To be well understood and widely applied, DSCA definition has to be as clear and simple as possible. Clearness and simplification of DSCA specification will remain our objective.

Another objective will be to improve DSCA specification using new experiments. To achieve this goal, complexity of fire spread remains an infinite challenge for DSCA simulation. We plan now to extend the DSCA specification to the simulation of implicit-time models.

Another important validation of our approach concerns network structure changes. Here again, a fire spread model taking into account wind effects (Simeoni et al., 2003) will allow us to validate DSCA network structure changes. This model actually needs to dynamically change the neighborhood of burning cells according to the fire front shape.

ACKNOWLEDGEMENTS

We would like to thank our research assistant Mathieu Joubert, for his technical support and for his C++ implementation of the 3D visualization tool using OpenGL.

REFERENCES

Balbi, J. H., P. A. Santoni, and J. L. Dupuy, 1998. Dynamic modelling of fire spread across a fuel bed. *Int. J. Wildland Fire*, p. 275-284.

Barros, F. J., 1997. Modelling Formalisms for Dynamic Structure Systems. *ACM Transactions on Modelling and Computer Simulation*, v. 7, p. 501-515.

Barros, F. J., and M. T. Mendes, 1997. Forest fire modelling and simulation in the DELTA environment. *Simulation Practice and Theory*, v. 5, p. 185-197.

Hill, D. R. C., 1996. *Object-oriented analysis and simulation*, Addison-Wisley Longman, UK, 291 p.

Jen, E., 1990. A periodicity in one-dimensional cellular automata. *Physica*, v. 45, p. 3-18.

Mesarovic, M. D., and Y. Takahara, 1975. *General Systems Theory: A mathematical foundation*, Academic Press. New York.

Muzy, A., E. Innocenti, F. Barros, A. Aïello, and J. F. Santucci, 2003. Efficient simulation of large-scale dynamic structure cell spaces. *Summer Computer Simulation Conference*, p. 378-383.

Muzy, A., G. Wainer, E. Innocenti, A. Aiello, and J. F. Santucci, 2002. Comparing simulation methods for fire spreading across a fuel bed. *AIS 2002 - Simulation and planning in high autonomy systems conference*, p. 219-224.

Nutaro, J., B. P. Zeigler, R. Jammalamadaka, and S. Akrekar, 2003. Discrete event solution of gaz dynamics winthin the DEVS framework: exploiting spatiotemporal heterogeneity. *Intrnational conference for computational science*.

Simeoni, A., P. A. Santoni, M. Larini, and J. H. Balbi, 2003. Reduction of a multiphase formulation to include a simplified flow in a semi-physical model of fire spread across a fuel bed. *International Journal of Thermal Sciences*, v. 42, p. 95–105.

Stroustrup, B., 2000. *The C++ Programming Language*, 1029 p.

Wainer, G., and N. Giambiasi, 2001. Application of the Cell-DEVS paradigm for cell spaces modeling and simulation. *Simulation*, v. 76, p. 22-39.

Wolfram, S., 1994. *Cellular automata and complexity: Collected papers*, Addison-Wesly.

Worsch, T., 1999. Simulation of cellular automata. *Future Generation Computer Systems*, v. 16, p. 157-170.

Zeigler, B. P., H. Praehofer, and T. G. Kim, 2000. *Theory of modelling and simulation*, Academic Press.

SPEAKER VERIFICATION SYSTEM
Based on the stochastic modeling

Valiantsin Rakush, Rauf Kh. Sadykhov

Byelorusian State University of Informatics and Radioelectronics, 6, P. Brovka str., Minsk, Belarus
Email: rsadykhov@gw.bsuir.unibel.by

Keywords: Speaker verification, vector quantization, Gaussian mixture models.

Abstract: In this paper we propose a new speaker verification system where the new training and classification algorithms for vector quantization and Gaussian mixture models are introduced. The vector quantizer is used to model sub-word speech components. The code books are created for both training and test utterances. We propose new approaches to normalize distortion of the training and test code books. The test code book quantized over the training code book. The normalization technique includes assigning the equal distortion for training and test code books, distortion normalization and cluster weights. Also the LBG and K-means algorithms usually employed for vector quantization are implemented to train Gaussian mixture models. And finally, we use the information provided by two different models to increase verification performance. The performance of the proposed system has been tested on the Speaker Recognition database, which consists of telephone speech from 8 participants. The additional experiments has been performed on the subset of the NIST 1996 Speaker Recognition database which include.

1 INTRODUCTION

The speaker verification systems so far has been based on the different methods. There is a category of the algorithms that are using back-end models to facilitate the speaker traits extraction (Roberts, W.J.J. et al., 1999) (Burton, D., 1987) (Pelecanos, J. et al., 2000) (Homayounpour, M.M., Challet, G., 1995). The neural networks, vector quantization (VQ), and Gaussian mixture models (GMM) are constructed directly or indirectly for subword or subspeech units modeling. Those units can be compared to make a verification decision. Also there is a class of the speaker verification systems that employ long term statistics computation over the speech phrase (Zilca, R.D., 2001) (Moonsar, V., Venayagamorthy, G.K., 2001). In some systems authors use a combination of the methods to improve system performance. The methods can be combined in two ways. First way is to use one model to improve performance of another one (Chun-Nan Hsu et al., 2003) (Singh, G. et al., 2003) (Sadykhov et al., 2003). Second way is to use recognition decision from both models to perform a data fusion to calculate a final score (Farrell, K. R. et al., 1998) (Farrell, K. R. et al., 1997). The data fusion methods can be interpreted using normalization and/or Bayessian approach.

Units comparison requires normalization to be applied. In case of VQ models the test and the reference codebooks have different structure, different distortion as well as units of measure for distortion. To compare two codebooks, which were created on different phrases, we need to normalize distortions and their units of measure. In the (Rakush V.V., Sadykhov R.H., 1999) authors proposed to create reference and test codebooks with equal distortion. Here we investigate two additional approaches that transform distortions so they can be compared.

The GMM model has the problem with parameters initialization. We propose to solve that problem using VQ codebook or applying LBG algorithm to split Gaussian mixture model starting from the single component. Also we use VQ codebook for GMM parameters initialization.

This paper is organized as follows. The following section describes modelling approach using VQ and GMM models. We will propose new algorithms combining VQ and GMM. Then we will discuss several techniques for data normalization and fusion, and will describe the structure of the experimental system, speech corpus and performance measures. Finally, we will show our experimental results, that will be followed by summary and conclusions.

255

2 BASIC IDEA OF THE VQ-VERIFICATION

The sub-word units, created during signal transformation from scalar to vector representation can be used as structural elements of the speaker voice model. Let $\bar{x} = [x_1, x_2, ..., x_N]^T$ - N-dimensional vector, coordinates of which $x_k, \{1 \le k \le N\}$ are real random values and represent temporal speech spectrum. It can be displayed into N-dimensional vector $\bar{y}$. The set $\bar{Y} = \bar{y}_i, \{1 \le i \le M\}$ is the code book, where M - the code book size and $\{\bar{y}_i\}$ - the set of code vectors. The N-dimensional space of vectors $\bar{x}$ is divided on M areas $c_i, 1 \le i \le M$ to create the code book. The vector $\bar{y}_i$ corresponds to each area c_i. If $\bar{x}_i$ lays in c_i, then $\bar{x}_i$ is quantized to a value of code vector $\bar{y}_i$. It is evident, that we get the error of quantization. The deviation $\bar{x}$ from $\bar{y}$ can be determined by a measure of closeness $d(\bar{x}, \bar{y})$

$$d(\bar{x}, \bar{y}) = \frac{1}{N}(\bar{x} - \bar{y})^T (\bar{x} - \bar{y})$$
$$= \frac{1}{N}\sum_{i=1}^{K}(x_i - y_i)^2,$$

(1)

where N – dimension of the parameters vector.

The basic idea of the VQ based verification system is to build two codebooks using the reference and test phrases. Definitely, reference and test phrases will be similar in the linguistic sense and will be modeling the features of the speaker voice. We assume that codebook clusters are modeling the sub-word units of speech so the test and reference codebooks should have approximately similar clusters for the two phrases pronounced by same speaker. The verification decision can be made comparing two codebooks using following expression

$$D_{compare} = \frac{1}{MK}\sum_{i=1}^{M}\sum_{j=1}^{K}(y_i - z_j)^2,$$

(2)

where $\bar{Y} = \bar{y}_i, \{1 \le i \le M\}$ - set of code vectors for reference codebook; $\bar{Z} = \bar{z}_i, \{1 \le j \le K\}$ - set of code vectors for test codebook; $D_{compare}$ - quantization distortion of test on the reference code book. In case of the speaker verification, if the codebooks distortion does not exceed predefined

threshold, then test and training utterances belong to the same person. When the recognition is applied to arbitrary speech then duration of the reference and test phrases has a huge difference. The reference phrase should contain as much as possible linguistic material from the speaker. The test phrase should be as small as possible but enough to provide acceptable verification performance. The reference code book should have more code vectors, and the test code book should have variable number of vectors K, depending on duration and linguistic content of the test phrase. Based on the idea that every cluster models sub-word component we assume that reference codebook presents model of all possible pronunciations for given speaker. We will quantize test codebook over reference codebook using expression (2) and will expect that distortion for right speaker will be minimal. Unfortunately, the distortion for the shortest test phrase can be smaller. Also the linguistic content of the phrase can influence on the distortion value. The distortion will be smaller for phrase with less sub-word components. To avoid phrase duration and content impact we propose the

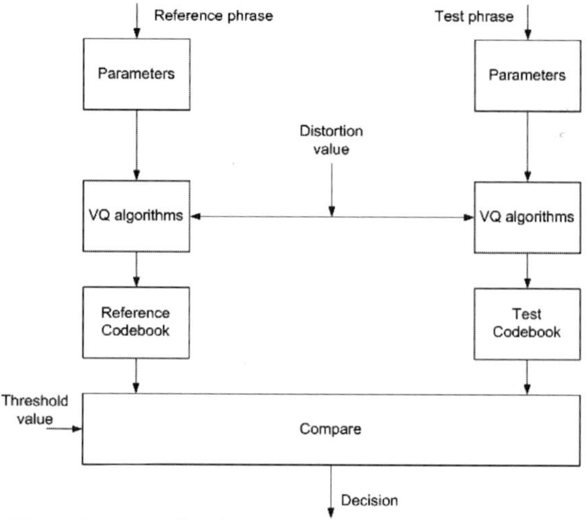

Figure 1: Normalization using predefined distortion value.

normalization techniques. First approach Fig. 1 described in (Rakush V.V., Sadykhov R.H., 1999) is based on the equal distortion for the reference and test code books. It has main assumption that two different code books with equal distortion do model same sub-word components.

Second approach is to use test codebook distortion for normalization. In that approach when test codebook created on the test phrase the final distortion is stored together with code vectors and used for decision normalization

$$D_{final} = D_{compare} \Big/ D_{norm} \tag{3}$$

The Fig.2 shows algorithm for normalization using distortion of the test phrase code book.

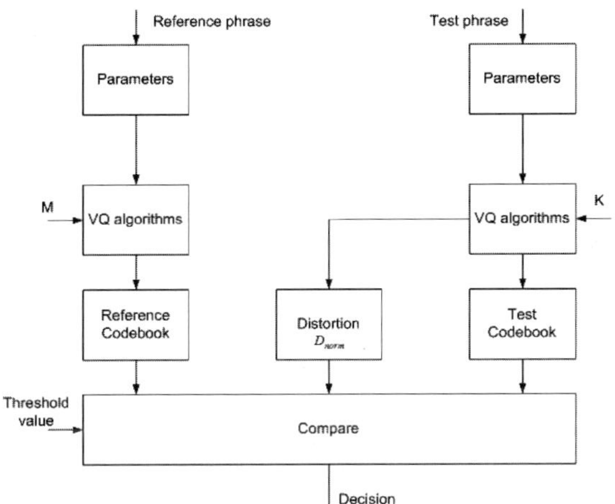

Figure 2: Normalization using the test codebook distortion value.

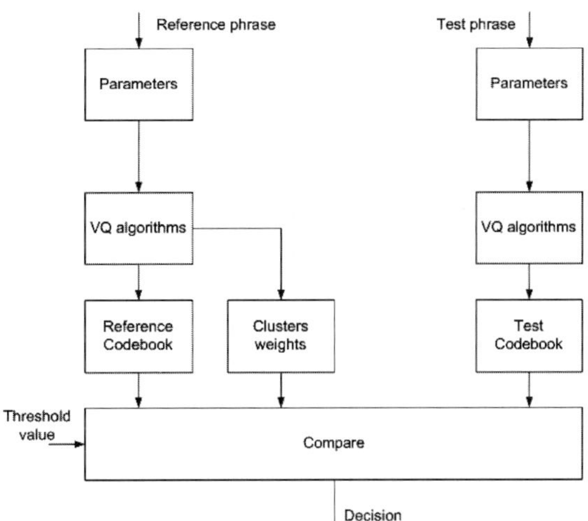

Figure 3: Normalization using cluster weigths

The third and last approach is to use number of vectors distributed in codebook clusters as a weight coefficients for normalization. The empirical theoretical assumption for that type of normalization can be defined as follows. If one cluster has more vectors then another one then it should have greater weight. Therefore test vectors that fall into it should be more meaningful and more significant for verification. This approach is not a pure normalization but can increase performance of the system because it uses more information from the code book then previous ones. The Figure 3 shows this normalization method. The VQ algorithm is used to calculate code book vectors. It is modified to produce cluster weights which will be stored along with cluster's center vector and will be used to weighted distance during testing phase.

3 THE GMM BASED SPEAKER VERIFICATION

The Gaussian mixture model is given by equation

$$p(\bar{x}|\lambda) = \sum_{i=1}^{M} p_i b_i(\bar{x}), \tag{4}$$

where λ - defines a Gaussian mixture density, $\bar{x}$ - N-dimensional feature vector, $b_i(\bar{x}), i = 1,..., M$ - probability distribution functions for model components, and $p_i, i = 1,..., M$ - components weights. Every component is a D-dimensional Gaussian probability distribution function

$$b_i(\bar{x}) = \frac{1}{(2\pi)^{D/2}|\delta_i|^{1/2}} \exp\left\{ -\frac{1}{2}(\bar{x} - \bar{\mu}_i)'\delta_i^{-1}(\bar{x} - \bar{\mu}_i) \right\} \tag{5}$$

where $\bar{\mu}_i$ - mean vectors, δ_i - covariance matrixes. The mixture weights values are constrained by the equality

$$\sum_{i=1}^{M} p_i = 1 \tag{6}$$

The GMM is a good tool, which can virtually approximate almost any statistical distribution. Due to that property mixture models are widely used to create speaker recognition systems. Unfortunately, the expectation-maximization (EM) algorithm has huge computational time so training procedure takes long time. The EM algorithm needs parameters to be initialized also. The number of components of the GMM is the same for all speaker voices stored in the system. Those are serious disadvantages of the EM algorithm that can fixed by applying vector quantization technique to GMM models training.

The initialization step is based on the vector quantization algorithm and uses codebook to initialize parameters of the GMM. There is another algorithm useful for initializing GMM. Initially that algorithm was developed for vector quantization and had name LBG algorithm. We will introduce new implementation of that algorithm for Gaussian mixture models.

The initial GMM model has only one component. The component, which gives maximum probability is split into two parts and new model parameters are estimated.

Step 1. Initialization

Component weight $p_1 = 1$. The mean vector is the mean of all feature vectors $\overline{\mu}_1 = \dfrac{1}{N}\sum_{i=1}^{N}\overline{x}_i$.

Covariance matrix is a diagonal matrix of variances calculated from the training set of feature vectors.

Step 2. Splitting component

Select the mixture component which has maximum probability. Increment the mean vector parameters on small value $\Delta\overline{\mu}$ will give two mean vectors.

$$\overline{\mu}_2 = \overline{\mu}_1 + \Delta\overline{\mu} \qquad (7)$$

Step 3. Optimization

Using EM algorithm estimate new GMM model. The EM algorithm can use fixed number of iterations or threshold condition.

Step 3. Iteration

Steps 2 and 3 can be performed until some threshold will be reached.

Both LBG and K-means initialization algorithms showed good performance acceptable for ASV systems. The system built on combination of the LBG and EM algorithms are shown on Figure 1.

4 EXPERIMENTAL RESULTS

The experiments have been performed on two speaker recognition databases. First one is the speech database proposed by the Centre of Spoken Language Understanding of the Oregon Institute of Science and Technology. The data set had 4 female and 4 male speakers with 50 utterances for each speaker. The speech was recorded on telephone channel with sampling rate 8 kHz. The duration of the test and train utterances was approximately equal 10 sec. The second database is the SWITCHBOARD speaker recognition corpus created by the National Institute of Technology in 1996. This database represents data in the Microsoft WAV files compressed using $\mu - law$. The subset of the development including 20 males and 20 females

The preliminary step used linear prediction coding and cepstral analysis to build vectors of spectral features. Analysis used 30 ms Hamming window with 10 ms shift to weight original speech signal. There were used vectors with 24 cepstral coefficients. Also as recommended in (Homayounpour, M.M., Challet, G., 1995) first derivative and second derivative of the cepstral coefficients have been used along with cepstr. The resulting feature vector had size $N=72$ parameters.

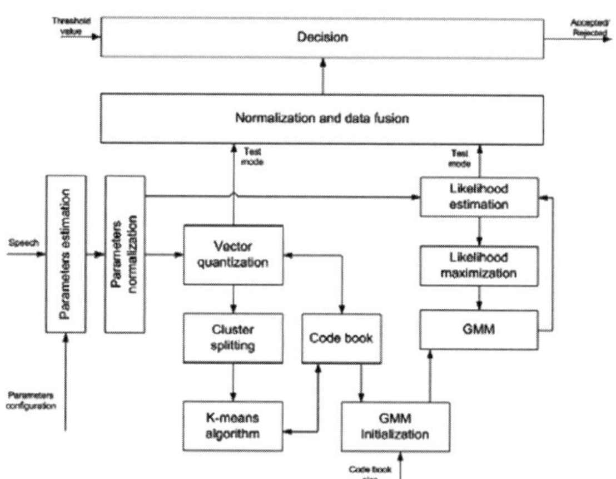

Figure 4: The structure of the speaker verification system.

The GMM models had maximum 32 components. The code book for GMM initialization and for verification had 32 and 256 clusters correspondingly. The system is working in two modes: training and testing mode. In training mode parameter vectors from both models are used to build the code book and GMM model for every speaker. In the test mode those models are used to verify speaker identity. Normalization and data fusion module uses following expression to combine results from both models.

$$P(x) = \sum_{i=1}^{n}\alpha_i p_i(x), \qquad (8)$$

where $P(x)$ is a probability of combined system, α_i are weights, $p_i(x)$ is a probability output of the

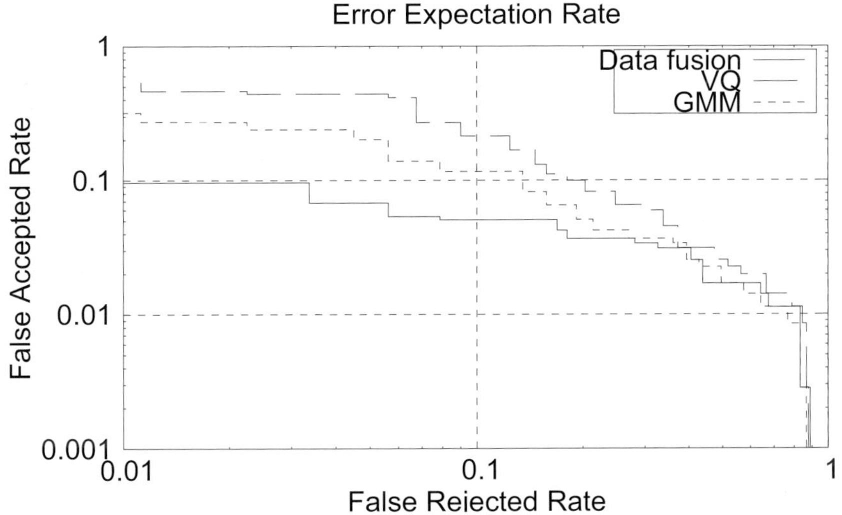

Figure 5: The DET curve of the ASV system performance.

i^{th} model, and n is a number of models (two models in our case).

The GMM and code book models weights have values $\alpha_1 = 0,545$ and $\alpha_2 = 0,455$. Experimental results shown almost identical performance for VQ and GMM algorithms. The data fusion of both algorithms improved overall performance of the system. The DET curve for LBG initialization and EM algorithm is shown on Figure 5.

In the second section of this paper we were discussing the normalization approaches to the vector quantization based speaker verification. The experimental results for the first approach with equal distortion for reference and test codebooks has been described in (Rakush V.V., Sadykhov R.H., 1999). In this paper we provide experimental result comparing second and third normalization approaches on figure 6.

Additional experiment using NIST 1996 Speaker Verification database SWITCHBOARD shown results printed on the figure 7.

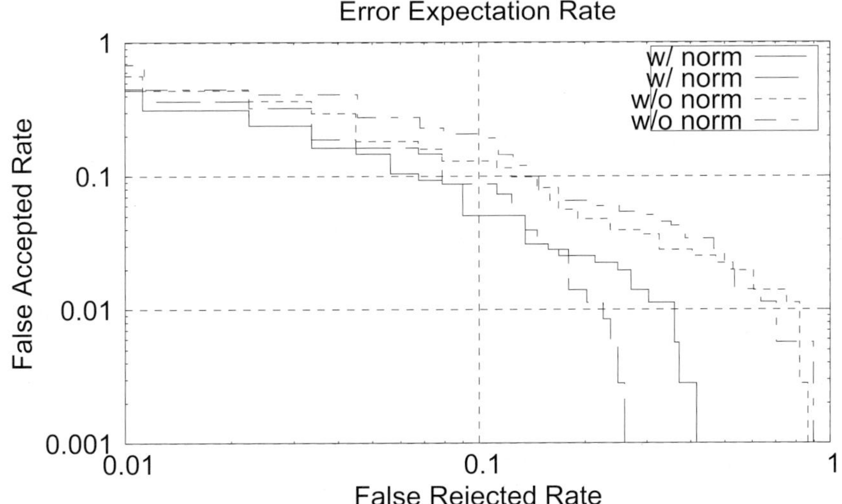

Figure 6: The DET curve for weight normalization.

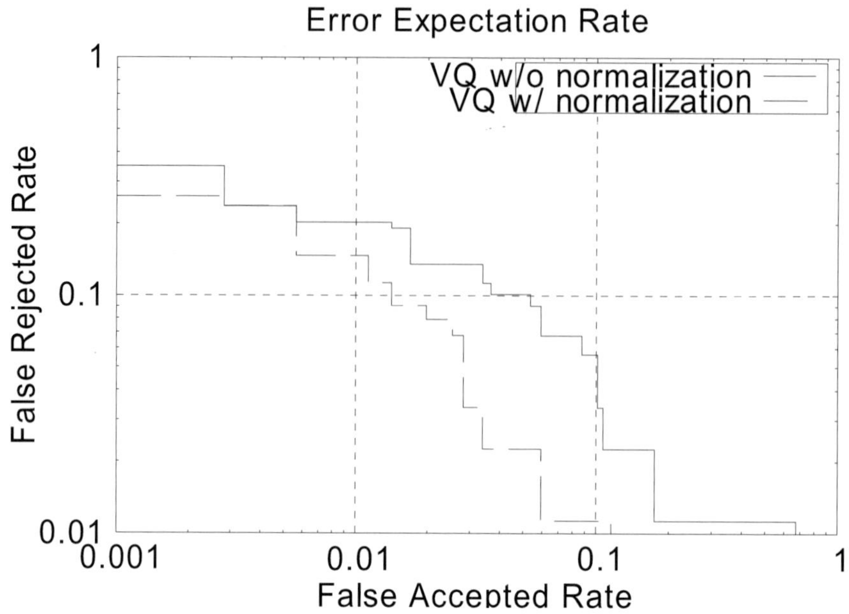

Figure 7: The weight normalization results tested on the SWITCHBOARD'96 corpus.

5 CONCLUSION

The first conclusion is that the speaker verification system based on voice modeling is showing acceptable performance even for speech degraded with telephone channel. Both VQ and GMM models are suitable for different statistical noise reduction techniques such as mean cepstral subtraction. That makes both algorithms are good choice for building automatic speaker verification systems for noisy signal.

The performance measure for the NIST speaker detection tasks is the Detection Cost Function (DCT) defined as a weighted sum of probability of the False Accepted Rate (FAR) and the probability of the False Rejected Rate (FRR) (NIST, 2002)

$$C_{Norm} = 0,1 FRR + 0,9 FAR \tag{9}$$

The minimal value for the DCF has been obtained for the best possible detection threshold and has value 0,1 for verification system created with data fusion methods and value 0,269 for verification system created with VQ algorithms only.

It is obvious from experiments that VQ speaker modeling performance is comparable to the GMM performance but time required for training is much less. In case of the VQ based modeling the number of clusters can be determined automatically from quantization distortion.

REFERENCES

Roberts, W.J.J., Wilmore J.P., 1999. Automatic speaker recognition using Gaussian mixture models. In *Proceedings of Information, Decision and Control*, IDC 99.

Farrell, K., Kosonocky, S., Mammone, R., 1994. Neural tree network/vector quantization probability estimators for speaker recognition. In *Proceedings of the Neural Networks for Signal Processing*, IEEE Workshop.

Burton, D., 1987. Text-dependent speaker verification using vector quantization source coding. In *Acoustics, Speech, and Signal Processing*, IEEE Transactions.

Zilca, R.D., 2001. Text-independent speaker verification using covariance modeling. In *Signal Proceesing Letters*, IEEE.

Moonsar, V., Venayagamorthy, G.K., 2001. A committee of neural networks for automatic speaker recognition (ASR) systems. *In Proceedings of International Joint Conference on Neural Networks*, IJCNN'01.

Pelecanos, J., Myers, S., Shridharan, S., Chandran, V., 2000. Vector quantization based Gaussian modeling for speaker verification, In *Proceedings of 15th International Conference on Pattern Recognition.*

Chun-Nan Hsu, Hau-Chang Yu, Bo-Han Yang, 2003. Speaker verification without background speaker models, In *Acoustics, Speech, and Signal Processing*, IEEE International Conference, ICASSP'03.

Homayounpour, M.M., Challet, G., 1995. Neural net approach to speaker verification: comparison with second order statistics measures, In *Acoustics, Speech, and Signal Processing*, IEEE International conference, ICASSP-95.

Singh, G., Panda, A., Bhattacharyga, S., Srikanthan, T., 2003. Vector quantization techniques for GMM based speaker verification, In *Acoustics, Speech, and Signal Processing*, IEEE International Conference, ICASSP'03.

Farrell, K. R., Ramachandran, R.P., Mammone, R.J., 1998. An analysis of data fusion methods for speaker verification, In *Acoustics, Speech, and Signal Processing*, IEEE International Conference, ICASSP'98.

Farrell, K.R., Ramachandran, R.P., Sharman, M., Mammone, R.J., 1997. Sub-word speaker verification using data fusion methods. In Neural Networks for Signal Processing, Proceedings of the IEEE Workshop.

Sadykhov, R. Kh., Rakush, V.V., 2003, Training Gaussian models with vector quantization for speaker verification, In *Proceedings of the 3rd International Conference on Neural Networks and Artificial Intelligence.*

Rakush V.V., Sadykhov R.H., 1999, Speaker Identification System on Arbitrary Speech In *Pattern Recognition and Information Processing. Proc. Of 5th International Conference.*

The NIST year 2002 speaker recognition evaluation plan, 2002, *http://www.nist.gov/speech/tests/spk/2002/doc.*

MOMENT-LINEAR STOCHASTIC SYSTEMS

Sandip Roy
Washington State University, Pullman, WA
sroy@eecs.wsu.edu

George C. Verghese
Massachusetts Institute of Technology, Cambridge, MA
verghese@mit.edu

Bernard C. Lesieutre
Lawrence Berkeley National Laboratory, Berkeley, CA
BCLesieutre@lbl.gov

Keywords: jump-linear systems, linear state estimation and control, stochastic network models.

Abstract: We introduce a class of quasi-linear models for stochastic dynamics, called *moment-linear stochastic systems* (MLSS). We formulate MLSS and analyze their dynamics, as well as discussing common stochastic models that can be represented as MLSS. Further studies, including development of optimal estimators and controllers, are summarized. We discuss the reformulation of a common stochastic hybrid system—the Markovian jump-linear system (MJLS)—as an MLSS, and show that the MLSS formulation can be used to develop some new analyses for MJLS. Finally, we briefly discuss the use of MLSS in modeling certain stochastic network dynamics. Our studies suggest that MLSS hold promise in providing a framework for modeling interesting stochastic dynamics in a tractable manner.

1   INTRODUCTION

As critical networked infrastructures such as air traffic systems, electric power systems, and communication networks have become more interdependent, the need for models for large-scale and hybrid network dynamics has grown. While the dramatic improvement in computer processing speeds in recent years has sometimes facilitated predictive simulation of these networks' dynamics, the development of models that allow not only prediction of dynamics but also network control and design remains a challenge in several application areas (see, e.g., (Bose, 2003)). In this article, we introduce and give the basic analysis for a class of models called *moment-linear stochastic systems* (MLSS) that can represent some interesting stochastic and hybrid system/network dynamics, and yet are sufficiently structured to allow computationally-attractive analyses of dynamics, state estimation, and control. Our studies suggest that MLSS hold promise as a modeling tool for a variety of stochastic and hybrid dynamics, especially because they provide a framework for several common stochastic and/or hybrid models in the literature, and because they can capture some network dynamics in a tractable manner.

Our research is partially concerned with hybrid (mixed continuous- and discrete-state) dynamics.

Stochastic hybrid models whose dynamics are constrained to Markovian switching between a finite number of linear time-invariant models have been studied in detail, under several names (e.g., Markovian jump-linear systems (MJLS) and linear jump-Markov systems) (Loparo et al., 1991; Mazor et al., 1998). Techniques for analyzing the dynamics of MJLS, and for developing estimators and controllers, are well-known (e.g., (Fang and Loparo, 2002; Costa, 1994; Mazor et al., 1998)). Of particular interest to us, a linear minimum-mean-square error (LMMSE) estimator for the continuous state of an MJLS was developed by (Costa, 1994), and quadratic controllers have been developed by, e.g., (Chizeck and Ji, 1988). We will show that similar estimation and control analyses can be developed for MLSS, and hence can be applied to a wider range of stochastic dynamics.

We are also interested in modeling stochastic network interactions. There is wide literature on stochastic network models that is largely beyond the scope of this article. Of particular interest to us, several models from the literature on queueing and stochastic automata can be viewed as stochastic hybrid networks (see (Kelly, 1979; Rothman and Zaleski, 1997; Durrett, 1981) for a few examples). By and large, the aims of the analyses for these models differ from our aims, in that transient dynamics are not characterized, and estimators and controllers are not developed. One

263

J. Braz et al. (eds.), Informatics in Control, Automation and Robotics I, 263–271.
© 2006 *Springer. Printed in the Netherlands.*

class of stochastic automata of particular interest to us are Hybrid Bayesian Networks (e.g., (Heskes and Zoeter, 2003)). These are graphical models (i.e., models in which stochastic interactions are specified using edges on a graph) with hybrid (i.e., mixed continuous-state and discrete-state) nodal variables. Dynamic analysis, inference (estimation), and parameter learning have been considered for such networks, but computationally feasible methods are approximate.

We observe that stochastic systems models with certain linear or quasi-linear structures (e.g., linear systems driven by additive white Gaussian noise, MJLS) are often widely tractable: statistics of transient dynamics can be analyzed, and LMMSE estimators and optimal quadratic controllers can be constructed. Several stochastic network models with quasi-linear interaction structures have also been proposed—examples include the *voter model* or *invasion process* (Durrett, 1981), and our *influence model* (Asavathiratham, 2000). For these network models, the quasi-linear structure permits partial characterization of state occupancy probabilities using low-order recursions. In this article, we view these various linear and quasi-linear models as examples of *moment-linear models*—i.e., models in which successive moments of the state at particular times can be inferred from equal and lower moments at previous times using linear recursions. This common representation for quasi-linear system and network dynamics motivates our formulation of MLSS, which are similarly tractable to the examples listed above. The MLSS formulation is further useful, in that it suggests some new analyses for particular examples and elucidates some other types of stochastic interactions that can be tractably modeled.

The remainder of the article is organized as follows: in Section 2, we describe the formulation and basic analysis—namely, the recursive computation of state statistics—of MLSS. We also list common models, and types of stochastic interactions, that can be captured using MLSS. Section 3 contains a summary of further results on MLSS. In Section 4, we reformulate the MJLS as an MLSS, and apply the MLSS analyses to this model. We also list other common hybrid models that can be modeled as MLSS. Section 5 summarizes our studies on using MLSS to model network dynamics. In this context, a discrete-time flow network model is developed in some detail.

2   MLSS: FORMULATION AND BASIC ANALYSIS

An MLSS is a discrete-time Markov model in which the conditional distribution for the next state given the current state is specially constrained at each time-step. These conditional distributions are constrained so that successive moments and cross-moments of state variables at each time-step can be found as linear functions of equal and lower moments and cross-moments of state variables at the previous time-step, and hence can be found recursively.

Formally, consider a discrete-time Markov process with an m-component real state vector. The state (i.e., state vector) of the process at time k is denoted by $\mathbf{s}[k]$. We write $\{\mathbf{s}[k]\}$ to represent the state sequence $\mathbf{s}[0], \mathbf{s}[1], \ldots$ and $s_i[k]$ to denote the ith element of the state vector at time k. We stress that we do not in general enforce any structure on the state variables, other than that they be real; the state vector may be continuous-valued, discrete-valued, or hybrid.

For this Markov process, we consider the conditional expectation $E(\mathbf{s}[k+1]^{\otimes r} \,|\, \mathbf{s}[k])$, for $r = 1, 2, \ldots$, where the notation $\mathbf{s}[k+1]^{\otimes r}$ refers to the Kronecker product of the vector $\mathbf{s}[k+1]$ with itself r times and is termed the rth-*order state vector* at time k. This expectation vector contains all rth moments and cross-moments of the state variables $s_1[k+1], \ldots, s_n[k+1]$ given $\mathbf{s}[k]$, and so we call the vector the *conditional rth (vector) moment* for $\mathbf{s}[k+1]$ given $\mathbf{s}[k]$. We say that the process $\{\mathbf{s}[k]\}$ is *rth-moment linear* at time k if the conditional rth moment for $\mathbf{s}[k+1]$ given $\mathbf{s}[k]$ can be written as follows:

$$E(\mathbf{s}[k+1]^{\otimes r} \,|\, \mathbf{s}[k]) = H_{r,0}[k] + \sum_{i=1}^{r} H_{r,i}[k]\mathbf{s}[k]^{\otimes i},$$

$$(1)$$

for some set of matrices $H_{r,0}[k], \ldots, H_{r,r}[k]$.

The Markov process $\{\mathbf{s}[k]\}$ is called a *moment-linear stochastic system* (MLSS) of degree $\hat{r}$ if it is rth-moment linear for all $r \leq \hat{r}$, and for all times k. If a Markov model is moment linear for all r and k, we simply call the model an MLSS. We call the constraint (1) the *rth-moment linearity condition* at time k, and call the matrices $H_{r,0}[k], \ldots, H_{r,r}[k]$ the *rth-moment recursion matrices* at time k. These recursion matrices feature prominently in our analysis of the temporal evolution of MLSS.

MLSS are amenable to analysis, in that we can find statistics of the state $\mathbf{s}[k]$ (i.e., moments and cross-moments of state variables) using linear recursions. In particular, for an MLSS of degree $\hat{r}$, $E(\mathbf{s}[k+1]^{\otimes r})$ (called the rth *moment* of $\mathbf{s}[k+1]$) can be found in terms of the first r moments of $\mathbf{s}[k]$ for any $r \leq \hat{r}$. To find these rth moments, we use the law of iterated expectations and then invoke the rth-moment linearity condition:

$$E(\mathbf{s}[k+1]^{\otimes r}) = E(E(\mathbf{s}[k+1]^{\otimes r} \mid \mathbf{s}[k])) \qquad (2)$$

$$= E\left(H_{r,0}[k] + \sum_{i=1}^{r} H_{r,i}[k]\mathbf{s}[k]^{\otimes i}\right)$$

$$= H_{r,0}[k] + \sum_{i=1}^{r} H_{r,i}[k]E(\mathbf{s}[k]^{\otimes i}).$$

We call (2) the rth-*moment recursion* at time k. Considering equations of the form (2), we see that the first r moments of $\mathbf{s}[k+1]$ can be found as a linear function of the first r moments of $\mathbf{s}[k]$. Thus, by applying the moment recursions iteratively, the rth moment of $\mathbf{s}[k]$ can be written in terms of the first r moments of the initial state $\mathbf{s}[0]$.

The recursions developed in equations of the form (2) can be rewritten in a more concise form, by stacking rth and lower moment vectors into a single vector. In particular, defining $\mathbf{s}'_{(r)}[k] = \begin{bmatrix} \mathbf{s}'[k]^{\otimes r} & \ldots & \mathbf{s}'[k]^{\otimes 1} & 1 \end{bmatrix}$, we find that

$$E(\mathbf{s}_{(r)}[k+1]) = \tilde{H}_{(r)}[k]E(\mathbf{s}_{(r)}[k]), \qquad (3)$$

where

$$\tilde{H}_{(r)}[k] = \qquad\qquad\qquad (4)$$

$$\begin{bmatrix} H_{r,r}[k] & H_{r,r-1}[k] & \cdots & H_{r,1}[k] & H_{r,0}[k] \\ 0 & H_{r-1,r-1}[k] & \cdots & H_{r-1,1}[k] & H_{r-1,0}[k] \\ \vdots & \ddots & \ddots & \vdots & \vdots \\ 0 & \cdots & 0 & H_{1,1}[k] & H_{1,0}[k] \\ 0 & \cdots & 0 & 0 & 1 \end{bmatrix}.$$

Interactions and Examples Captured by MLSS
MLSS provide a convenient framework for representing several models that are prevalent in the literature. These MLSS reformulations are valuable in that they expose similarity in the moment structure of several types of stochastic interactions, and in that they permit new analyses (e.g., development of linear state estimators, or characterization of higher moments) in the context of these examples. Common models that can be represented as MLSS are listed:

- A linear system driven by independent, identically distributed noise samples with known statistics is an MLSS.

- A finite-state Markov chain can be formulated as an MLSS, by defining the MLSS state to be a 0–1 *indicator vector* of the state of the Markov chain.

- An MJLS is a hybrid model that can be reformulated as an MLSS. This reformulation of an MJLS is described in Section 4.

- A Markov-modulated Poisson process (MMPP) is a stochastic model that has commonly been used

to represent sources in communications and manufacturing systems (e.g., (Baiocchi et al., 1991), (Nagarajan et al., 1991), (Ching, 1997)). It consists of a finite-state *underlying Markov chain*, as well as a Poisson arrival process with (stochastic) rate modulated by the underlying chain. An MMPP can be formulated as an MLSS using a state vector the indicates the underlying Markov chain's state and also tracks the number of arrivals over time (see (Roy, 2003) for details). The MLSS reformulation of MMPPs highlights that certain state-parametrized stochastic updates, such as a Poisson generator with mean specified as a linear function of the current state, can be represented using MLSS. More generally, various stochastic updates with state-dependent noise can be represented.

- Certain infinite server queues can be represented as MLSS (see (Roy, 2003) for details).

Observations and Inputs Observations and external inputs are naturally incorporated in MLSS. These are structured so as to preserve the moment-linear structure of the dynamics, and hence to allow development of recursive linear estimators and controllers for MLSS.

At each time k, we observe a real vector $\mathbf{z}[k]$ that is independent of the past history of the system, (i.e., $\mathbf{s}[0], \ldots, \mathbf{s}[k-1]$ and $\mathbf{z}[0], \ldots, \mathbf{z}[k-1]$), given the current state $\mathbf{s}[k]$. The observation $\mathbf{z}[k]$ is assumed to be first- and second-moment linear given $\mathbf{s}[k]$. That is, $\mathbf{z}[k]$ is generated from $\mathbf{s}[k]$ in such a way that the first moment (mean) for $\mathbf{z}[k]$ given $\mathbf{s}[k]$ can be written as an affine function of $\mathbf{s}[k]$:

$$E(\mathbf{z}[k] \mid \mathbf{s}[k]) = C_{1,1}\mathbf{s}[k] + C_{1,0} \qquad (5)$$

for some $C_{1,1}$ and $C_{1,0}$, and the second moment for $\mathbf{z}[k]$ given $\mathbf{s}[k]$ can be written as an affine function of $\mathbf{s}[k]$ and $\mathbf{s}[k]^{\otimes 2}$:

$$E(\mathbf{z}[k]^{\otimes 2} \mid \mathbf{s}[k]) = C_{2,2}\mathbf{s}[k]^{\otimes 2} + C_{2,1}\mathbf{s}[k] + C_{2,0} \qquad (6)$$

for some $C_{2,2}$, $C_{2,1}$, and $C_{2,0}$.

We have restricted observations to a form for which analytical expressions for LMMSE estimators can be found, yet relevant stochastic interactions that are not purely linear can be captured. Observation generation from the state in MLSS admits the same variety of stochastic interactions—e.g., linear interactions, finite-state Markovian transitions, certain state-parameterized stochastic updates—as captured by the MLSS state update. This generality allows us to model, e.g., Hidden Markov Model observations (Rabiner, 1986), random selection among multiple linear observations, and observations of a Poisson arrival process with mean modulated by the state.

The inputs in our model are structured in such a way that the next state is first- and second-moment linear with respect to both the current state and the current input. Specifically, a system is a 2nd-degree MLSS with state sequence $\{s[k]\}$ and input $\{u[k]\}$ if the conditional distribution for the next state is independent of the past history of the system, given the current state and input, and if there exist matrices $H_{1,1}$, $D_{1,1}$, $H_{1,0}$, $H_{2,2}$, $G_{2,2}$, $D_{2,2}$, $H_{2,1}$, $D_{2,1}$, and $H_{2,0}$ such that

$$E(s[k+1] \mid s[k], u[k]) = H_{1,1}s[k] + D_{1,1}u[k] + H_{1,0}$$

and

$$\begin{aligned}
E(&s[k+1]^{\otimes 2} \mid s[k], u[k]) \\
&= H_{2,2}s[k]^{\otimes 2} + G_{2,2}(s[k] \otimes u[k]) + D_{2,2}u[k]^{\otimes 2} \\
&\quad + H_{2,1}s[k] + D_{2,1}u[k] + H_{2,0}.
\end{aligned} \tag{7}$$

That is, a Markov process is a 2nd-degree MLSS with input $u[k]$ if the first and second moments and cross-moments of the next state, given the current state and input, are first- and second-degree polynomials, respectively, of current state and input variables.

We have restricted MLSS inputs to a form for which analytical expressions for optimal quadratic controllers can be found, yet several relevant types of input interactions can nevertherless be represented. The dependence of the state on the input has the same generality as the state update.

3 FURTHER RESULTS

In the following paragraphs, we summarize four results on MLSS that are elaborated further in (Roy, 2003).

Cross Moments   Cross-moments of state variables across multiple time-steps can be calculated, by recursively rewriting cross-moments across time-steps as linear functions of moments and cross-moments at single times. Our expressions for cross-moments are similar in flavor to the Kronecker product-based expressions for higher-order statistics of linear systems given by, e.g., (Mendel, 1975; Swami and Mendel, 1990), but apply to MLSS.

Asymptotics   We develop necessary and sufficient conditions for convergence (approach to a steady-state value) of MLSS moments[1] in (Roy, 2003). Conditions for moment convergence are useful because they can help to characterize asymptotic dynamics: they can in some instances be used to prove convergence of the state itself, or to prove distributional convergence, for example.

Because the moments satisfy linear recursions, conditions for convergence can straightforwardly be expressed in terms of the modes of the moment recursion matrices. However, we note that redundancy in the moment vectors, which is inherent to the MLSS formulation, complicates development of good convergence conditions because it allows introduction of spurious unstable modes that do not actually alter the moments. We therefore develop reduced forms of the moment recursions to construct the necessary and sufficient conditions for moment convergence. Details are in (Roy, 2003).

Estimation   We have developed a recursive algorithm for linear minimum mean square error (LMMSE) filtering of MLSS. Because tractability of estimation and control is a primary goal in our formulation of MLSS, it is worthwhile for us to present our estimator, and to connect it to related literature.

Our LMMSE filter for an MLSS is a generalization of the discrete-time Kalman filter (see, e.g., (Catlin, 1989)), in which the state update and observation processes are constrained to be moment-linear rather than purely linear. Equivalently, we can view our filter as applying to a linear system in which certain quasi-linear state-dependence of state and observation noise is permitted. From this viewpoint, our filter is related to the LMMSE filter introduced in (Zehnwirth, 1988), which allows for state-dependent observation variance. Also of interest to us are linear estimation techniques for arrival processes whose underlying rate processes are themselves random, and/or arrival-dependent (e.g., (Snyder, 1972),(Segall et al., 1975)). Segall and Kailath describe a broad class of arrival processes of this type, for which a martingale formulation facilitates nonlinear recursive estimation (Segall et al., 1975). We also can capture some arrival processes with stochastic rates (e.g., MMPPs), and hence develop recursive state estimators for these processes. The arrival processes that are MLSS are a subset of those in (Segall et al., 1975), but for which linear filtering is possible, and hence exact finite-dimensional estimators can be constructed.

The derivation of our LMMSE filter for MLSS, which closely follows the derivation of the discrete-time Kalman filter, can be found in (Roy, 2003). Here, we only present the results of our analysis. We use the following notation: we define $\widehat{s}_{k|k}$ as the LMMSE estimate for $s[k]$ given $z[0], \ldots, z[k]$, and define $\Sigma_{k|k} \stackrel{\triangle}{=} E\left((s[k] - \widehat{s}_{k|k})(s[k] - \widehat{s}_{k|k})'\right)$ as the error covariance of this estimate. Also, we define $\widehat{s}_{k+1|k}$ as the LMMSE estimate for $s[k+1]$ given $z[0], \ldots, z[k]$, and let define $\Sigma_{k+1|k} \stackrel{\triangle}{=} E\left((s[k+1] - \widehat{s}_{k+1|k})(s[k+1] - \widehat{s}_{k|k})'\right)$ as the error covariance of this estimate.

---

[1] Our study of moment convergence is limited to the state update of an MLSS; input-output stability has not yet been considered.

As with the Kalman filter, the estimates are found recursively, in two steps. First, a *next-state update* is used to determine $\widehat{\mathbf{s}}_{k+1|k}$ and $\Sigma_{k|k}$ in terms of $\widehat{\mathbf{s}}_{k|k}$, $\Sigma_{k|k}$, and the *a priori* statistics of $\mathbf{s}[k]$. The next-state update for our filter is

$$\widehat{\mathbf{s}}_{k+1|k} = H_{1,1}\widehat{\mathbf{s}}_{k|k} + H_{1,0} \tag{8}$$

$$\Sigma_{k+1|k}$$
$$= H_{1,1}\Sigma_{k|k}H'_{1,1} + M_k(E(\mathbf{s}[k], E(\mathbf{s}[k]^{\otimes 2}))) \tag{9}$$

In (9), $M_k(E(\mathbf{s}[k], E(\mathbf{s}[k]^{\otimes 2})))$ is the (*a priori*) expectation for the conditional variance of $z[k]$ given $s[k]$; an explicit expression is given in (Roy, 2003).

Second, a *measurement update* is used to determine $\widehat{\mathbf{s}}_{k|k}$ and $\Sigma_{k|k}$ in terms of $\mathbf{z}[k]$, $\widehat{\mathbf{s}}_{k|k-1}$, $\Sigma_{k|k-1}$, and the *a priori* statistics of $\mathbf{s}[k]$. The measurement update for our filter is

$$\widehat{\mathbf{s}}_{k|k} = \widehat{\mathbf{s}}_{k|k-1} + \Sigma_{k|k-1}C'_{1,1}(C_{1,1}\Sigma_{k|k-1}C'_{1,1}$$
$$+ N_k(E(\mathbf{s}[k], E(\mathbf{s}[k]^{\otimes 2})))^{-1}(\mathbf{z}[k] - C_{1,1}\widehat{\mathbf{s}}_{k|k-1}C_{1,0})$$
$$- \Sigma_{k|k} = \Sigma_{k|k-1}$$
$$- \Sigma_{k|k-1}C'_{1,1}(C_{1,1}\Sigma_{k|k-1}C'_{1,1}$$
$$+ N_k\left(E(\mathbf{s}[k]), E(\mathbf{s}[k]^{\otimes 2}))\right)^{-1}C_{1,1}\Sigma_{k|k-1}. \tag{10}$$

In (10), $N_k\left(E(\mathbf{s}[k]), E(\mathbf{s}[k]^{\otimes 2})\right)$ is the *a priori* expectation for the conditional variation of $\mathbf{s}[k+1]$ given $\mathbf{s}[k]$; an explicit expression is given in (Roy, 2003).

Control   Considering the duality between linear estimation and quadratic control, it is not surprising that optimal dynamic quadratic control can be achieved for MLSS, given full state information. In (Roy, 2003), we use a dynamic programming formulation for the quadratic control problem to find a closed-form recursion for the optimal control. The optimal control at each time is linear with respect to the current state, in analogy to the optimal quadratic control for a linear system. The reader is referred to (Roy, 2003) for presentation and discussion of the optimal controller, as well as a description of relevant literature.

4   EXAMPLE: MJLS

We describe the reformulation of MJLS as MLSS, and then present some analyses of MJLS using the MLSS formulation. For simplicity, we only detail the reformulation of a Markovian jump-linear state-update equation here; reformulation of the entire input-output dynamics is a straightforward extension.

Let's consider the update equation

$$\mathbf{x}[k+1] = A(\mathbf{q}[k])\mathbf{x}[k] + \mathbf{b}_k(\mathbf{q}[k]), \tag{11}$$

where $\{\mathbf{q}[k]\}$ is a 0–1 indicator vector sequence representation for an underlying Markov chain with finite state-space. We denote the transition matrix for the underlying Markov chain by $\Theta = [\theta_{ij}]$. We denote the number of components of $\mathbf{x}[k]$ as n, and the number of statuses in the underlying Markov chain as m.

For convenience, we rewrite (11) in an extended form as

$$\tilde{\mathbf{x}}[k+1] = \tilde{A}(\mathbf{q}[k])\tilde{\mathbf{x}}[k], \tag{12}$$

where $\tilde{\mathbf{x}}[k] = \begin{bmatrix} \mathbf{x}[k] \\ 1 \end{bmatrix}$ and $\tilde{A}(\mathbf{q}[k]) = \begin{bmatrix} A(\mathbf{q}[k]) & \mathbf{b}(\mathbf{q}[k]) \\ \mathbf{0} & 1 \end{bmatrix}$.

To reformulate the jump-linear system as an MLSS, we define a state vector that captures both the continuous state and underlying Markov dynamics of the jump-linear system. In particular, we define the state vector as $\mathbf{s}[k] = \mathbf{q}[k] \otimes \tilde{\mathbf{x}}[k]$, and consider the first conditional vector moment $E(\mathbf{s}[k+1]\,|\,\mathbf{s}[k])$. Since $\mathbf{s}[k]$ uniquely specifies $\mathbf{x}[k]$ and $\mathbf{q}[k]$, we can determine this first conditional vector moment as follows:

$$E(\mathbf{s}[k+1]\,|\,\mathbf{s}[k]) = E(\mathbf{s}[k+1]\,|\,\mathbf{q}[k], \mathbf{x}[k]) \tag{13}$$
$$= E(\mathbf{q}[k+1] \otimes \tilde{\mathbf{x}}[k+1]\,|\,\mathbf{x}[k], \mathbf{q}[k])$$
$$= E(\mathbf{q}[k+1]\,|\,\mathbf{q}[k]) \otimes \tilde{A}(\mathbf{q}[k])\tilde{\mathbf{x}}[k]$$
$$= \Theta'\mathbf{q}[k] \otimes \tilde{A}(\mathbf{q}[k])\tilde{\mathbf{x}}[k]$$

With a little bit of algebra, we can rewrite Equation 13 as

$$E(\mathbf{s}[k+1]\,|\,\mathbf{s}[k]) = \tag{14}$$
$$\begin{bmatrix} \theta_{11}\tilde{A}(\mathbf{q}[k]=\mathbf{e}(1)) & \cdots & \theta_{m1}\tilde{A}(\mathbf{q}[k]=\mathbf{e}(m)) \\ \vdots & \ddots & \vdots \\ \theta_{1m}\tilde{A}(\mathbf{q}[k]=\mathbf{e}(1)) & \cdots & \theta_{mm}\tilde{A}(\mathbf{q}[k]=\mathbf{e}(n)) \end{bmatrix} \mathbf{s}[k],$$

where $\mathbf{e}(i)$ is an indicator vector with the ith entry equal to 1.

Equation (14) shows that the first-moment linearity condition holds for $\{\mathbf{s}[k]\}$. To justify that higher-moment linearity conditions hold, let's consider entries of the rth-conditional moment vector $E(\mathbf{s}[k+1]^{\otimes r}\,|\,\mathbf{s}[k]) = E((\mathbf{q}[k+1] \otimes \tilde{\mathbf{x}}[k+1])^{\otimes r}\,|\,\mathbf{x}[k], \mathbf{q}[k])$. Because $\mathbf{q}[k]$ is an indicator, the non-zero entries of the rth-conditional moment vector can all be written in the form $E(q_i[k+1]\prod_{i=1}^{n}x_i[k+1]^{\alpha_i}\,|\,\mathbf{x}[k], \mathbf{q}[k])$, where each $\alpha_i \geq 0$, and $\sum_{i=1}^{n}\alpha_i = \widehat{r} \leq r$. Given that $\mathbf{q}[k] = \mathbf{e}(i)$, $1 \leq i \leq m$, $E(\prod_{i=1}^{n}x_i[k+1]^{\alpha_i}\,|\,\mathbf{x}[k], \mathbf{q}[k])$ is an $\widehat{r}$th degree polynomial of $x_1[k], \ldots, x_n[k]$, say $p_i[k]$. Using that $\mathbf{q}[k]$ is an indicator vector, we can rewrite $E(q_i[k+1]\prod_{i=1}^{n}x_i[k+1]^{\alpha_i}\,|\,\mathbf{x}[k], \mathbf{q}[k])$ as $\sum_{i=1}^{m}q_i[k]p_i[k]$. Hence, we see that each entry in the rth-conditional moment vector is linear with respect to the entries in $\mathbf{s}[k]^{\otimes r} = (\mathbf{q}[k+1] \otimes \tilde{\mathbf{x}}[k+1])^{\otimes r}$, and so the state vector is rth-moment linear. Some bookkeeping is required to express the higher-moment linearity conditions in vector form; these expressions are omitted.

We believe that the MLSS reformulation of MJLS is valuable, because it places MJLS in the context of a broader class of models, and because it facilitates several analyses of MJLS dynamics. Some analyses of MJLS dynamics that can be achieved using the MLSS formulation are listed below. Our analyses are illustrated using an example MJLS with a two-status underlying Markov chain and scalar continuous-valued state. The underlying Markov chain for this example has transition matrix $\Theta = \begin{bmatrix} 0.9 & 0.1 \\ 0.3 & 0.7 \end{bmatrix}$. The scalar continuous state is updated as follows: if the Markov chain is in the first status at time k, then the time-$(k+1)$ continuous state is $x[k+1] = -0.9x[k] + 0.5$; if the Markov chain is in the second status, the time-$(k+1)$ continuous state is $x[k+1] = x[k] + 1$.

The MLSS formulation allows us to compute statistics (moments and cross-moments) for both the continuous-valued state and underlying Markov status, as well as conditional statistics for the continuous-valued state given the underlying Markov status (at one or several times). Recursions for the first- and second-moments of the continuous-valued state are well-known (see, e.g., (Costa, 1994)), though our second-moment recursion differs in form from the recursion on the covariance matrix that is common in the literature. We have not seen general computations of higher-order statistics, or of statistics across time-steps: the MLSS formulation provides a concise notation in which to develop recursions for these statistics. The higher-moment recursions are especially valuable because they can provide characterizations for MJLS asymptotics. We can specify conditions for δ-*moment stability* (see, e.g., (Fang et al., 1991)) for all even integer δ in terms of the eigenvalues of the higher moment recursions. We can also characterize the asymptotics of MJLS in which the state itself does not stabilize, by checking for convergence of moments (to non-zero values). We are currently exploring whether the methods of (Meyn and Tweedie, 1994) can be used to prove distributional convergence from moment convergence.

For illustration, first- and second-order statistics of the example MLSS are shown along with a fifty time-step simulation in Fig. 1. Additionally, the steady-state values for the first three moments of the continuous-valued state have been obtained from the moment recursions, and are shown along with the corresponding steady-state distribution. We note that the first three moments provide significant information about the steady-state distribution and, in this example, require much less effort to compute than the full distribution.

The MLSS formulation allows us to develop conditions for moment convergence in MJLS. In (Roy,

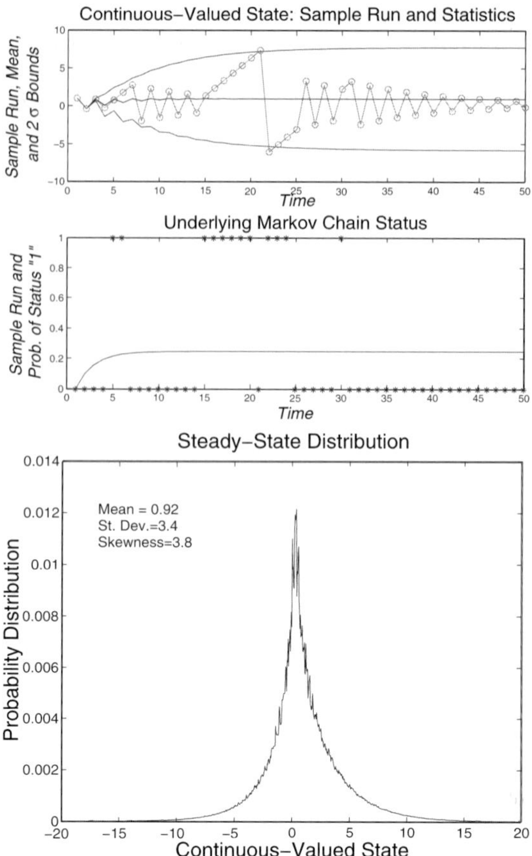

Figure 1: This figure illustrates the MLSS-based analysis of moments for an example MJLS. The top plot in this figure specifies the continuous-valued state during a 50 time-step simulation, along with the computed mean value and two standard deviation intervals for this continuous-valued state. The middle plot indicates the status of the underlying Markov chain during the simulation and also shows the probability that the Markov chain is in status "1". The bottom plot shows the steady-state distribution for the continuous-valued state (found through iteration of the distribution) and lists the moments of this continuous-valued state (found with much less computation using the moment recursions).

2003), we have presented necessary and sufficient conditions for moment convergence of an MJLS with scalar continuous-valued state and two-status underlying Markov chain, in terms of its parameters. Our example MJLS, with statistics shown in Fig. 1, can be shown to be moment-convergent. Our study of moment convergence complements stability studies for MJLS (e.g., (Fang and Loparo, 2002)), by allowing identification of MJLS whose moments reach a steady-state but that are not stable (in the sense that state itself does not reach a steady-state).

The MLSS reformulation can be used to develop LMMSE estimators for MJLS. LMMSE estimation of the continuous-valued state of an MJLS from noisy observations has been considered by (Costa, 1994). The observation structure (i.e., the generator of the observation from the concurrent state) assumed by (Costa, 1994) can be captured using our MLSS formulation, and in that case our estimator is nearly identical to that of (Costa, 1994); only the squared error that is minimized by the two estimators is slightly different.

Our MLSS formulation allows for estimation for a variety of observation structures. For instance, we can capture observations that are Poisson random variables, with parameter equal to a linear function of the state variables. (Such an observation model may be realistic, for example, if the state process modulates an observed Poisson arrival process.) We can also capture other types of state-dependent noise in the observation process, as well as various discrete and continuous-valued state-independent noise. Further, hybrids of multiple observation structures can be captured in the MLSS formulation.

Another potential advantage of the MLSS formulation is that, because the underlying Markov status is part of the state vector, this underlying status can be estimated. The accuracy of our estimator for the underlying state remains to be assessed. A direction for future study is to compare our estimator for the underlying status with the nonlinear estimators of, e.g., (Sworder et al., 2000; Mazor et al., 1998).

For illustration, we have developed an LMMSE filter for our example MJLS. Here, we observe the (scalar) continuous-valued state upon corruption by additive white Gaussian noise. Fig. 2 illustrates the performance of the LMMSE estimator, for a particular sample run. Our analysis shows that the error covariance of our estimate is about half the measurement error covariance, verifying that our estimate is usually a more accurate estimate for the state than the unfiltered observation sequence.

5 MLSS MODELS FOR NETWORK DYNAMICS: SUMMARY

We believe that MLSS representations for networks are of special interest because analysis of stochastic network dynamics is quite often computationally intensive (e.g., exponential in the number of vertices), while MLSS representations can permit analysis at much lower computational cost. Below, we summarize our studies of MLSS representations for network dynamics. More detailed description of MLSS models for network dynamics can be found in (Roy, 2003).

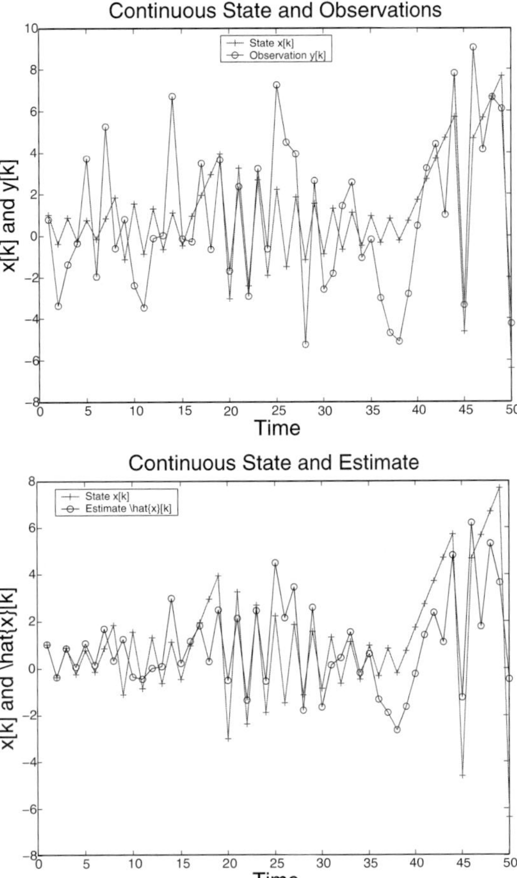

Figure 2: In the left plot, the continuous-valued state and observations during a 50 time-step simulation of the example MJLS are shown. In the right plot, the LMMSE estimate for the continuous-valued state is compared with the actual continuous-valued state. The LMMSE estimate is a better approximation for the continuous-valued state than the observation sequence.

The following are examples of MLSS models for network dynamics that we have considered.

- The *influence model* was introduced in (Asavathiratham, 2000) as a network of interacting finite-state Markov chains, and is developed and motivated in some detail in (Asavathiratham et al., 2001). We refer the reader to (Basu et al., 2001) for one practical application, as a model for conversation in a group. The influence model consists of **sites** with discrete-valued statuses that evolve in time through stochastic interaction. The influence model's structured stochastic update permits formulation of the model as an MLSS, using 0–1 indicator vector representations for each site's status. The rth moment recursion permits computation of the joint probabilities of the statuses of groups of r sites, and hence the configurations of small groups of sites can be characterized using low-order recursions. The MLSS formulation for the influence model can also be used to develop good state estimators for the model, and to prove certain results on the convergence of the model.

- MLSS can be used to represent single-server queueing networks with linear queue length-dependent arrival and service rates, operating under heavy-traffic conditions. The MLSS formulation allows us to find statistics and cross-statistics of the queue lengths. Of particular interest is the possibility for using the MLSS formulation to develop queue-length estimators for these state-dependent models and, indirectly, for Jackson networks (see, e.g., (Kelly, 1979)).

- In (Roy et al., 2004), we have developed an MLSS model for aircraft counts in regions of the U.S. airspace, and have used the MLSS formulation to compute statistics of aircraft counts in regions. In the context of this MLSS model, we have also developed techniques for parameter estimation from data.

REFERENCES

Asavathiratham, C. (2000). *The Influence Model: A Tractable Representation for the Dynamics of Networked Markov Chains, Ph.D. Thesis*. EECS Department, Massachusetts Institute of Technology.

Asavathiratham, C., Roy, S., Lesieutre, B. C., and Verghese, G. C. (2001). The influence model. *IEEE Control Systems Magazine*.

Baiocchi, A., Melazzi, N. B., Listani, M., Roveri, A., and Winkler, R. (1991). Loss performance analysis of an ATM multiplexer loaded with high-speed on off sources. *IEEE Journal on Selected Areas of Communications*, 9:388–393.

Basu, S., Choudhury, T., Clarkson, B., and Pentland, A. (2001). Learning human interactions with the influence model. *MIT Media Lab Vision and Modeling TR#539*.

Bose, A. (2003). Power system stability: new opportunities for control. *Stability and Control of Dynamical Systems with Applications*.

Catlin, D. E. (1989). *Estimation, Control, and the Discrete Kalman Filter*. Springer-Verlag, New York.

Ching, W. K. (1997). Markov-modulated Poisson processes for multi-location inventory problems. *International Journal of Production Economics*, 52:217–233.

Chizeck, N. J. and Ji, Y. (1988). Optimal quadratic control of jump linear systems with Gaussian noise in discrete-time. *Proceedings of the 27th IEEE Conference on Decision and Control*, pages 1989–1999.

Costa, O. L. V. (1994). Linear minimum mean square error estimation for discrete-time markovian jump linear systems. *IEEE Transactions on Automatic Control*, 39:1685–1689.

Durrett, R. (1981). An introduction to infinite particle systems. *Stochastic Processes and their Applications*, 11:109–150.

Fang, Y. and Loparo, K. A. (2002). Stabilization of continuous-time jump linear systems. *IEEE Transactions on Automatic Control*, 47:1590–1603.

Fang, Y., Loparo, K. A., and Feng, X. (1991). Modeling issues for the control systems with communication delays. *Ford Motor Co., SCP Research Report*.

Heskes, T. and Zoeter, O. (2003). Generalized belief propagation for approximate inference in hybrid bayesian networks. *Proceedings of the Ninth International Workshop on Artificial Intelligence and Statistics*.

Kelly, F. P. (1979). *Reversibility and Stochastic Networks*. John Wiley and Sons, New York.

Loparo, K. A., Buchner, M. R., and Vasuveda, K. (1991). Leak detection in an experimental heat exchanger process: a multiple model approach. *IEEE Transactions on Automatic Control*, 36.

Mazor, E., Averbuch, A., Bar-Shalom, Y., and Dayan, J. (1998). Interacting multiple model methods in target tracking: a survey. *IEEE Transactions on aerospace and electronic systems*, 34(1):103–123.

Mendel, J. M. (1975). Tutorial on higher-order statistics (spectra) in signal processing and system theory: theoretical results and some applications. *Proceedings of the IEEE*, 3:278–305.

Meyn, S. and Tweedie, R. (1994). *Markov Chains and Stochastic Stability*. http://black.csl.uiuc.edu/ meyn/pages/TOC.html.

Nagarajan, R., Kurose, J. F., and Towsley, D. (1991). Approximation techniques for computing packet loss in finite-buffered voice multiplexers. *IEEE Journal on Selected Areas of Communications*, 9:368–377.

Rabiner, L. R. (1986). A tutorial on hidden markov models and selected applications in speech recognition. *Proceedings of the IEEE*, 77(2):257–285.

Rothman, D. and Zaleski, S. (1997). *Lattice-Gas Cellular Automata: Simple Models of Complex Hydrodynamics*. Cambridge University Press, New York.

Roy, S. (2003). *Moment-Linear Stochastic Systems and their Applications*. EECS Department, Massachusetts Institute of Technology.

Roy, S., Verghese, G. C., and Lesieutre, B. C. (2004). Moment-linear stochastic systems and networks. *Submitted to the Hybrid Systems Computation and Control conference*.

Segall, A., Davis, M. H., and Kailath, T. (1975). Nonlinear filtering with counting observations. *IEEE Transactions on Information Theory*, IT-21(2).

Snyder, D. L. (1972). Filtering and detection of doubly stochastic poisson processes. *IEEE Transactions on Information Theory*, IT-18:91–102.

Swami, A. and Mendel, J. M. (1990). Time and lag recursive computation of cumulants from a state space model. *IEEE Transactions on Automatic Control*, 35:4–17.

Sworder, D. D., Boyd, J. E., and Elliot, R. J. (2000). Modal estimation in hybrid systems. *Journal of Mathematical Analysis and Applications*, 245:225–247.

Zehnwirth, B. (1988). A generalization of the Kalman filter for models with state-dependent observation variance. *Journal of the American Statistical Association*, 83(401):164–167.

Filipe Morais and J.M. Sá da Costa
Instituto Superior Técnico, Technical University of Lisbon,
Dept. of Mechanical Engineering / GCAR - IDMEC
Av. Rovisco Pais, 1049-001Lisboa, Portugal
Email: morais@dem.ist.utl.pt , sadacosta@dem.ist.utl.pt

Keywords: Active noise control, feedforward control, filtered-reference LMS, modified filtered-reference LMS, filtered-u, frequency domain filtered-reference LMS.

Abstract: In this paper existing classical and advanced techniques of active acoustic noise cancellation (ANC) in ducts are collected and compared. The laboratory plant used in experience showed a linear behaviour and so the advanced techniques were not used. Due to delay on the plant, the feedback classical techniques could not be applied. The best results were obtained with the modified filtered-reference LMS (MFX-LMS) and filtered-u techniques. A very important conclusion is that the quadratic normalisation is needed to maintain the algorithms always stable. In this paper 18dB of attenuation of white noise and 35 dB of attenuation of tonal noise were achieved. Thus, ANC can be applied in a real situation resulting in important noise attenuations.

1   INTRODUCTION

Acoustic noise is since a long time considered as pollution due to the diverse problems that it causes in the human being, both physical, as for instance deafness, and psychological. As a consequence, competent authorities tend to enforce restrictive laws on the allowed sound levels, and it is thus necessary to look for solutions leading to its fulfilment. On the other hand, acoustic noise is a cause of lack of productivity. By these reasons, there is a pressing need to solve the problem of acoustic noise.

In practice passive solutions for the cancellation of acoustic noise can be found by simple use of absorption and reflection phenomena. However, they are of little use for frequencies below 1000Hz. In these other cases acoustic noise cancellation (ANC) based on the principle of interference, should be used.

The idea of the ANC is 70 years old. One of the first references remounts to 1934 when P. Lueg patented some ideas on the subject (Elliot, 2001 and Tokhi et al., 1992). Lueg addressed ANC in ducts and in the three-dimensional space. For ducts, Lueg considered a microphone that captured the acoustic noise. The signal from the microphone would pass through the controller and feed the loudspeaker as shown in fig. 1. The controller would result in acoustic waves emitted by the loudspeaker with the same amplitude of the acoustic noise but in phase opposition, so that the two waves would cancel each other (interference principle). This configuration is nowadays the most used in ANC applications in ducts.

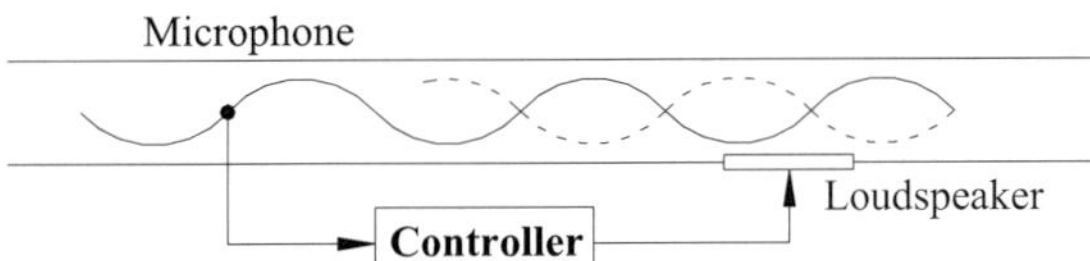

Figure 1: Single-channel feedforward control in a duct.

The purpose of using ANC in ducts is to cancel the plane waves that are propagated in the duct due to fans, like in an air conditioner installation. The ANC mostly used techniques were developed to control stochastic disturbances, because acoustic noise can be considered as a disturbance with significant spectral richness. Furthermore, techniques for stochastic disturbances can be applied in deterministic disturbances but the inverse is not feasible. ANC techniques for stochastic disturbances can be divided into two main groups: classical or advanced. Those of the first group are based on plant linearity, and consequent validity of the superposition principle (Ogata, 1997).

Linear techniques can also be applied to nonlinear systems, but they usually have bad performance. Advanced techniques were developed

273

J. Braz et al. (eds.), Informatics in Control, Automation and Robotics I, 273–280.
© 2006 *Springer. Printed in the Netherlands.*

to nonlinear plants, although they can be applied to linear systems with good performance. However, they are also more complex and demand more computational power than the classic ones. For that reason advanced techniques are not preferred instead of classic ones when linear plants are concerned.

Both classic and advanced techniques can be divided according to the type of control: feedforward or feedback. In the feedforward control information is collected in advance about the disturbance and so the controller can act in anticipation; while the feedback control has no information in advance about the disturbance and thus the controller reacts to the disturbance. The feedback control is useful when the acoustic noise is created by several different sources, or by distributed sources, or when it is not practical or possible to get information in advance concerning all the noise sources. However, this is not the case of ducts because the noise source is well defined and acoustic waves are plane and travel in a single direction.

In this paper existing feedforward techniques for ANC in ducts are compared to assess the performance of these techniques in a real situation.

In ducts it is possible to have only plane acoustic waves, rending ANC much simpler since some acoustic effects are not to be found, as for instance the diffraction of acoustic waves. In this work the range of frequencies to be deal with ANC is limited to the interval [200 Hz; 1000 Hz] since ANC is not effective for frequencies above 1000 Hz and the actual set-up used does not allow to go below 200 Hz.

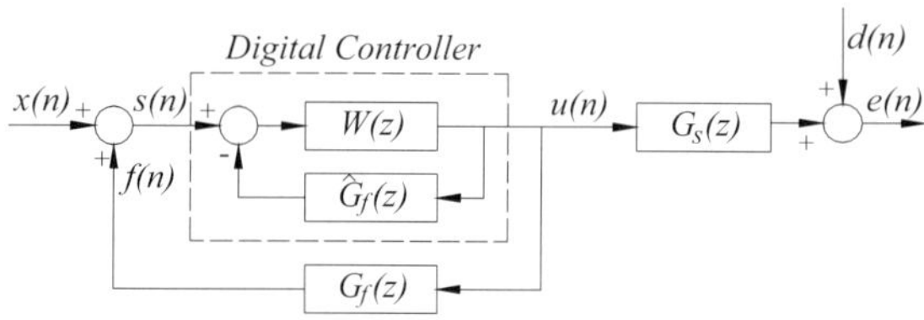

Figure 2: Block diagram of feedforward control.

2 FEEDFORWARD CONTROL

The general block diagram of the feedforward control of acoustic plane waves in a duct is found in fig. 2. The signal $x(n)$ is the reference signal measured by the reference microphone, $d(n)$ is the primary noise signal passed through the primary path, $e(n)$ is the error signal given by the error microphone, and $G_s(z)$ is the secondary path between the secondary source and the error sensor. It is assumed that the controller is digitally implemented and made up by a direct filter $W(z)$ and a feedback filter $\hat{G}_f(z)$. The feedback filter consists of an

estimation of the natural feedback path of the system $G_f(z)$, i.e., reproduces the influence of the secondary source to the reference sensor. When $\hat{G}_f(z) = G_f(z)$, the two feedback loops cancel each other and the signal that feeds the controller is equal to $x(n)$. In this situation the control is purely feedforward. In the situation in which the estimate of $G_f(z)$ is not perfect, a residue appears from the cancellation of two loops. If $\hat{G}_f(z)$ is a good estimate of the path $G_f(z)$, the residue has a small value and will not affect the performance of the control. If the estimate of $G_f(z)$ is poor, this can influence the performance of the control, that may become unstable. In this situation it might be necessary to use feedback control techniques to improve the performance or to stabilize the control (Elliot, 2001).

Assuming that the two feedback loops cancel each other completely and that the plants are linear and time invariant (LTI), so that the filter $W(z)$ and the discrete transfer function $G_s(z)$ can be interchanged, the error signal $e(n)$ comes

$$e(n) = d(n) + \mathbf{w}^T \mathbf{r}(n) = d(n) + \mathbf{r}(n)^T \mathbf{w}, \qquad (1)$$

where $\mathbf{w}$ is a vector with the coefficients of the filter and $\mathbf{r}(n)$ the vector with the last samples of the filtered reference signal $r(n)$ given by:

$$r(n) = \sum_{i=0}^{I-1} g_i x(n-i), \qquad (2)$$

where the g_i are the I coefficients of the discrete transfer function $G_s(z)$, assuming that has a FIR structure.

2.1 Filtered-reference LMS (FX-LMS) Algorithm

This algorithm is based on the steepest descent algorithm, which is mostly used for adapting FIR controllers (Elliot, 2001). The expression for adapting the coefficients of controller $W(z)$ of fig. 2 is given by:

$$\mathbf{w}(n+1) = \mathbf{w}(n) - \mu \frac{\partial J}{\partial \mathbf{w}} \qquad (3)$$

where J is a quadratic index of performance, equal to the error signal squared $e^2(n)$, and $\partial \cdot / \partial \mathbf{w}$ is the gradient:

$$\frac{\partial J}{\partial \mathbf{w}} = 2E\left[\mathbf{r}(n)e(n)\right] \qquad (4)$$

For this algorithm a simpler version than the one given by eq. (4) is used, since the expected value of the product is not reckoned, but only the current value of the gradient. Thus,

$$\frac{\partial e^2(n)}{\partial \mathbf{w}} = 2e(n)\frac{\partial e(n)}{\partial \mathbf{w}} = 2e(n)\mathbf{r}(n) \qquad (5)$$

The expression for adapting the coefficients of the controller is given by:

$$\mathbf{w}(n+1) = \mathbf{w}(n) - \alpha\hat{\mathbf{r}}(n)e(n) \qquad (6)$$

where $\alpha = 2\mu$ is the convergence coefficient and $\hat{r}(n)$ is the estimate of the filtered reference signal, obtained with the estimate of the $G_s(z)$ model. The algorithm is called filtered-reference LMS because the filtered reference signal is used to adapt the coefficients. The block diagram of the algorithm is in given in fig. 3.

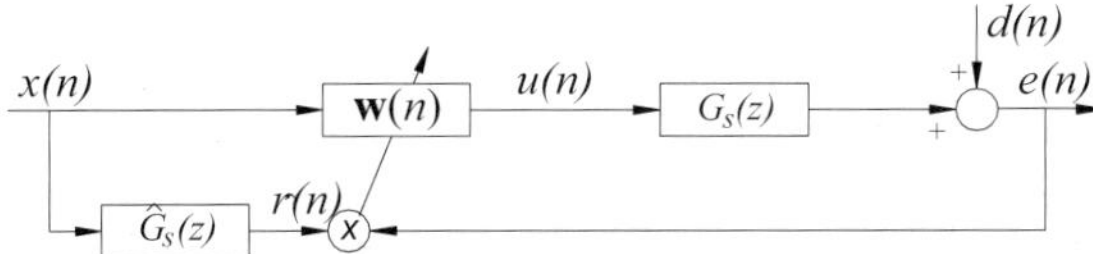

Figure 3: Block diagram for FX-LMS algorithm.

If the reference signal $x(n)$ were used instead of the filtered reference $r(n)$ to adapt the coefficients, the adaptation would be wrong because there is a time shift between the signal $x(n)$ and the error signal $e(n)$. This is a consequence of the existence of a time delay in $G_s(z)$. This algorithm is rather simple to implement and is numerically stable, being therefore frequently used (Elliot, 2001).

2.2 Normalized Filtered-reference LMS Algorithm (NFX-LMS)

In the previous approach the adaptation of the coefficients of the controller $W(z)$ is directly proportional to the coefficient of convergence α and the vector $r(n)$. Sometimes, when $r(n)$ has large values, the FX-LMS algorithm has a problem of amplification of the gradient noise (Haykin, 2002). The coefficients of the vector $r(n)$ are normalized in order to solve this problem. Haykin (2002) suggests dividing the coefficients by the Euclidean norm of vector $r(n)$. The expression for adapting the coefficients becomes:

$$\mathbf{w}(n+1) = \mathbf{w}(n) - \frac{\tilde{\alpha}}{\delta + \|\mathbf{r}(n)\|^2}\mathbf{r}(n)e(n) \qquad (7)$$

where δ it is a very small and positive number. This term allows preventing numerical difficulties when $r(n)$ is small because the Euclidean norm takes small values. Elliot (2001) suggests another solution where the coefficients of vector $r(n)$ are divided by the inner product of vector $r(n)$, $\mathbf{r}^{\mathrm{T}}\mathbf{r}$. Whatever the option is, algorithm NFX-LMS presents the following advantages over algorithm FX-LMS: faster

convergence rate and sometimes better performance of the obtained controller; the algorithm is more stable when there is a change of the spectral richness of the reference signal $x(n)$. This normalization of the filtered reference signal can be applied to other algorithms.

2.3 Leaky LMS Algorithm

For this algorithm another index of quadratic performance is used:

$$J_2 = E\left[e^2(n)\right] + \beta\mathbf{w}^{\mathrm{T}}\mathbf{w} \qquad (8)$$

where β is a positive constant. This performance index weighs both the average of the error signal $e(n)$ squared as well as the sum of the squares of the coefficients of the controller. This performance index prevents the coefficients of the controller from taking large values that can render the algorithm unstable when both the amplitude of the reference signal, $x(n)$, and its spectral components undergo variations (Elliot, 2001). The adaptation becomes:

$$\mathbf{w}(n+1) = (1-\alpha\beta)\mathbf{w}(n) - \alpha\hat{\mathbf{r}}(n)e(n). \qquad (9)$$

Eq. (9) is different for the FX-LMS algorithm because of term $(1-\alpha\beta)$, which is called leakage factor. This term must take values between 0 and 1 and is normally 1. When it takes another value the error signal $e(n)$ is not zero and the value of coefficients decreases with each iteration. Adding the term $(1-\alpha\beta)$ to the coefficients adapting equation allows the increasing of the robustness of the algorithm. On the other hand the term $(1-\alpha\beta)$ reduces the noise attenuation that can be reached. Thus, the choice of the value for beta must take into account the robustness of the algorithm and the reduction of the attenuation. In most applications, the use of a small value of beta allows a sufficient increase of robustness and the attenuation of the acoustic noise suffers little (Elliot, 2001). The modification introduced in the FX-LMS algorithm can also be implemented in the other algorithms.

2.4 Modified Filtered-reference LMS Algorithm (MFX-LMS)

The FX-LMS algorithm requires a rather slow adaptation compared with the plant dynamics so that the error may be given by eq. (1). This is because adapting the coefficients is somehow a nonlinearity which influence depends on the speed of adaptation (Elliot, 2001). Thus, to make this influence negligible the adaptation of the coefficients must be very slow when compared with the dynamics of the plant. This should be regarded as a disadvantage. The arrangement shown in fig. 4 allows overcoming this limitation. In this scheme, the estimated filtered reference signal, $\hat{r}(n)$, in the adaptation path of the controller is common to the adaptive filter and to the adaptation, and has no time shift in relation to the modified error, $e'_m(n)$.

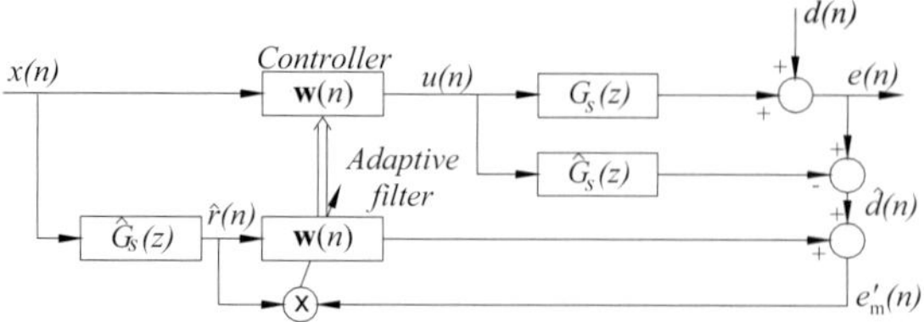

Figure 4: Block diagram for MFX-LMS algorithm.

For this algorithm the adaptation is given by:

$$\mathbf{w}(n+1) = \mathbf{w}(n) - \alpha\hat{\mathbf{r}}(n)e_m(n) \qquad (10)$$

where $e_m(n)$ is the modified error, given by:

$$e_m(n) = \hat{d}(n) + \sum_{i=0}^{I-1}\sum_{j=0}^{J-1} w_i(n)\hat{g}_j x(n-i-j) \qquad (11)$$

The modified error can be seen as a prediction of the error for the case where the coefficients of the controller do not change at each instant. The MFX-LMS algorithm usually presents convergence rates larger than those of the FX-LMS algorithm (Elliot, 2001). This is because the adaptive filter and the plant estimate were interchanged and thus the delay between the exit of the controller and the error signal was eliminated. For this reason it is no longer necessary to consider the delay in the restriction of the convergence coefficient, and larger steps may be used with the MFX-LMS algorithm. However the MFX-LMS algorithm has the disadvantage of requiring more computational means.

2.5 Frequency Domain Filtered-reference LMS Algorithm (FX-LMS Freq)

For the FX-LMS algorithm the estimate of the gradient of eq. (5) was used to adapt the coefficients of the controller. The estimate of the gradient will be assumed to be given by the average of the product $r(n)e(n)$ during N instants. Thus, the adaptation is given by:

$$\mathbf{w}(n+N) = \mathbf{w}(n) + \frac{\alpha}{N}\sum_{l=n}^{n+N-1} \mathbf{r}(l)e(l) \qquad (12)$$

In this case, the adaptation is carried only after N time samples. The use of the average of the product $r(n)e(n)$ during N instants can be considered as a more precise estimate of the gradient than the use of the product $r(n)e(n)$ for each time sample. In practice adaptation with eq. (12) has a convergence rate very similar to the FX-LMS algorithm, since though the adaptation for eq. (12) has a lower frequency, the value of the update of the coefficients is larger (Elliot, 2001). The summation in eq. (12) can be thought of as an estimate of the crossed correlation between the filtered reference $r(n)$ and the error signal $e(n)$. The estimate must be reckoned from $i = 0$ up to $I-1$, where I is the number of coefficients of the adaptive filter. For long filters the reckoning of the estimate can be inefficient in the time domain, requiring a large computational effort. For large values of I it is more efficient to calculate the cross correlation in the frequency domain. If discrete Fourier transform (DFT) with 2N points for the signals $e(n)$ and $r(n)$ are considered, an estimate of the cross spectral density can be calculated through:

$$\hat{S}_{re}(k) = R^*(k)E(k) \qquad (13)$$

where k is the index of discrete frequency and * means the complex conjugate. Some care must be taken to prevent the effect of circular convolution. Thus, before reckoning the DFT of the error signal, $e(n)$, with 2N points, in the block with 2N points of the error signal the first N points must be zero. This will eliminate the non-causal part of the cross correlation (Elliot, 2001). The expression that gives the adaptation of the coefficients is:

$$\mathbf{w}(m+1) = \mathbf{w}(m) - \alpha IFFT\left\{R_m^*(k)E_m(k)\right\}_+ \qquad (14)$$

where $\{\ \}_+$ means the causal part of the cross correlation, IFFT is the inverse fast Fourier transform and α is the convergence coefficient. $R_m(k)$ is directly obtained multiplying the DFT of the reference signal $X(k)$ by the frequency response estimate of the system. This algorithm is called fast LMS. Fig. 5 shows the block diagram of this algorithm.

The advantage of the fast LMS algorithm over the FX-LMS is that it requires few computations. Assuming that the implementation of the DFT requires $2N\log_2 2N$ multiplications, the FX-LMS algorithm requires $2N^2$ calculations per iteration while fast LMS needs $(16 + 6\log_2 2N)N$.

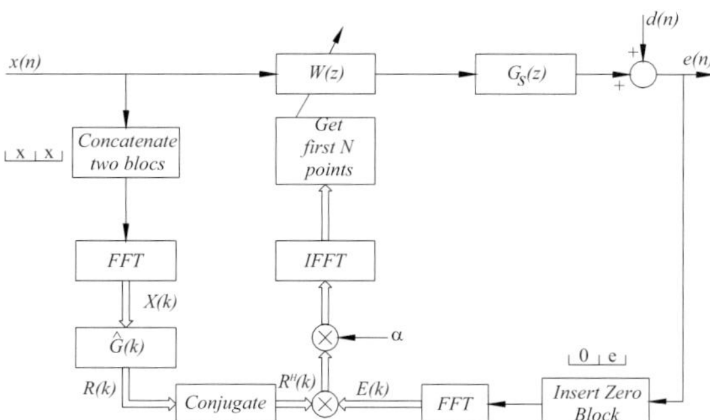

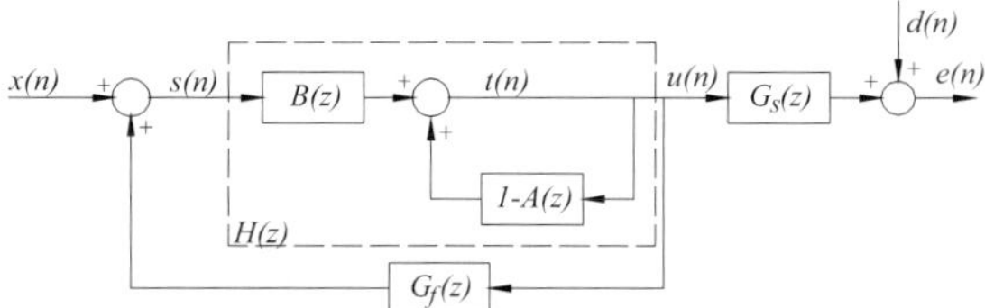

Figure 5: Block diagram for FX-LMS Freq algorithm.

2.6 Filtered-u Algorithm

Up to now Finite Impulse Response (FIR) filters have been considered to build the controllers. However, Infinite Impulse Response (IIR) filters can be used as well. In this case, the equivalent to fig.2 for IIR controllers is shown in fig. 6.

Figure 6: Block diagram for IIR controller.

Compared with the block diagram of fig. 2 for the FIR controllers, we can notice that this does not possess a specific feedback to cancel the natural feedback path of the system. In this case the recursive characteristic of IIR controllers is assumed to deal with the feedback path problem. However, practice shows that if this estimation is included numerical stability is guaranteed and the performance is improved.

The filtered-u algorithm uses IIR filter as controller. It is based on the recursive LMS (RLMS) algorithm (see Elliot (2001) or Haykin (2002)). Fig. 7 shows the block diagram of the filtered-u algorithm. The adaptation of the coefficients a_j and b_i is given by:

$$\mathbf{a}(n+1) = \gamma_1\mathbf{a}(n) - \alpha_1 e(n)\mathbf{t}(n) \quad (15)$$

$$\mathbf{b}(n+1) = \gamma_2\mathbf{b}(n) - \alpha_2 e(n)\mathbf{r}(n) \quad (16)$$

where α_1, α_2 are the convergence coefficients, $t(n)$ and $r(n)$ are respectively the filtered output and the filtered reference, and γ_1 and γ_2 are the forgetting factors.

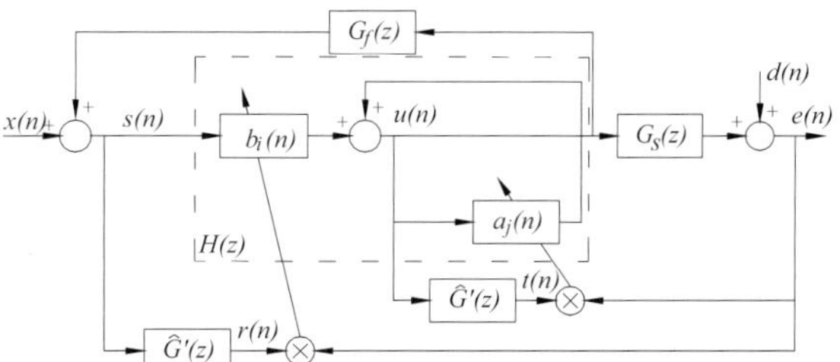

Figure 7: Block diagram for filtered-u algorithm.

The use of different convergence coefficients may be shown in practice to allow for higher convergence rates and the use of leakage factors slightly under 1 allows for a greater robustness of the algorithm (Elliot, 2001). The plants modified response, $G'(z)$, is equal to $G_s(z)$. For that purpose, the coefficients of the controller $H(z)$ are assumed to be very slowly adapted in comparison to the dynamics of the system of the system $G_s(z)$. The same had already been assumed for the adaptation of the FIR controller, but for the adaptation of the IIR controller this is even more necessary since the controller is recursive. One of the interesting characteristics of the filtered-u algorithm is that it presents a self-stabilising behaviour that is also to be found in RLMS algorithms (Elliot, 2001). During the adaptation of the controller, if a pole leaves the unit-radius circle, the natural evolution of filtered-u algorithm brings it back inside. Although some researchers have addressed this behaviour, still it was not possible to discover the mechanism that results in this self-stabilising property (Elliot, 2001). The self-stabilising behaviour is found in many practical applications, and that is why the filtered-u algorithm is the most used in active cancellation of noise applications (Elliot, 2001).

3 EXPERIMENTAL SET-UP

The experimental set-up used is shown in Fig. 8. A PVC pipe with 0.125 m of diameter and 3 m of length was used for simulating the cylindrical duct. Given the diameter of the duct, the cut-on frequency, which is the frequency above which waves may no longer be considered plane, is 1360 Hz. To simulate the acoustic noise to cancel a conventional loudspeaker was placed in one of the ends of the duct. At 1.25 m away from this end two loudspeakers are placed to act as source of acoustics waves for noise cancellation. For the detection of acoustic noise a microphone, placed 0.08 m away from the primary noise source, is used. The error microphone is placed at the opposite end of the primary noise source.

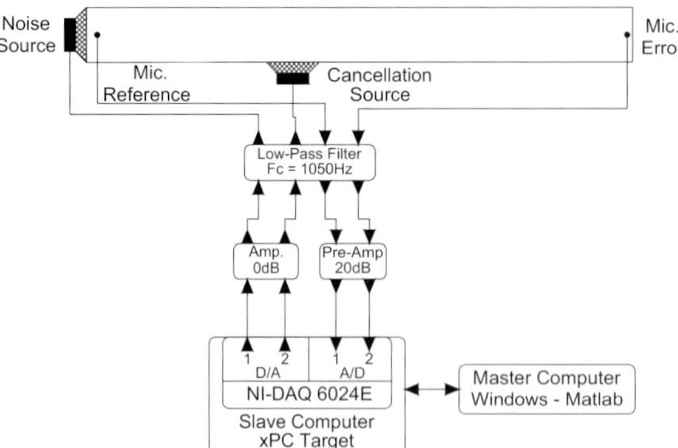

Figure 8: Block diagram of experimental setup.

Besides the duct, loudspeakers and microphones, the experimental set-up consists of: four low-pass filters that allow filtering the signals to remove the effects of aliasing and zero-order-hold; an amplifier that allows amplifying the signals that feed the loudspeakers; pre-amplifiers for the microphones; and two computers, one the slave act has a digital controller and the other the master is used for data analysis. The slave computer is a Pentium III 733MHz with 512MB of RAM memory, running on xPC Target, having a data acquisition board NI-DAQ 6024E. Algorithms have been implemented as S-Functions in the Matlab/Simulink environment.

Due to hardware restrictions on the cancellation source this set-up cannot generate relevant signals for frequencies below 200 Hz. Therefore, the frequency range where acoustic noise cancellation is intended is restricted to the frequency bandwidth of [200 Hz - 1000 Hz].

4   IDENTIFICATION

The models used are discrete in time since the implementation of the controller is made using a digital computer. Therefore, the simulations will be based on discrete models. This requires the models to include the devices associated with the discretisation and restoration of the signals, A/D and D/A conversions, anti-aliasing and reconstruction filters, and the dynamic of the microphones, loudspeakers and amplifiers associated to the experimental set-up. Assuming that the behaviour of these devices is linear, then each one can be represented by a discrete transference function. The necessary models are:

$G_s(z)$ - secondary acoustic path: includes computer - secondary source - error microphone - computer;
$G_f(z)$ - acoustic feedback path: includes computer- secondary source - reference microphone-computer.

Models have been obtained for the sampling frequency of 2500 Hz (sampling time 0.4ms) because that allows the Nyquist frequency of 1250Hz, to be slightly larger than the superior limit of the frequency range to cancel, 1000 Hz. FIR and ARX models have been obtained. Variance account for (VAF) criterion and root mean square (RMS) have been used for models validation. Table 1 shows the results obtained in these identifications.

Table 1: Order, VAF and RMS of the obtained models.

Model	Order			VAF (%)		RMS (V)	
	FIR	ARX					
	I	n_a	n_b	FIR	ARX	FIR	ARX
$G_s(z)$	500	150	150	99.96	99.94	0.0193	0.0195
$G_F(z)$	450	150	150	99.60	99.57	0.0363	0.0373

As shown above the obtained models have excellent performances. This shows the plant to have a linear behaviour being unnecessary to appeal to ANC advanced techniques.

5   EXPERIMENTAL RESULTS

The previously mentioned algorithms have been implemented and test for different noise conditions in the duct. However, before presenting the results it must be point out that the use of the normalisation of the filtered reference signal was very important. Experiences have shown that the normalised LMS technique has a significant influence in the behaviour of the algorithms. In fig. 9 the evolution of the attenuation is shown for the FX-LMS algorithm when the variance of white noise changed, for the following cases: the filtered reference signal was not normalized, was normalised using the Euclidean norm, and was normalised using quadratic normalization. The behaviour of the other algorithms is similar. In the figures that follow, attenuation is given by the expression

$$Attenuation\ (dB)\ =\ 10\log_{10}\left(\frac{E\left[e^2\right]}{E\left[d^2\right]}\right) \qquad (18)$$

where e is the error signal, d the disturbance and $E[\ ]$ is the expected value operator. In this case the expected value is given by the average of last 50 samples.

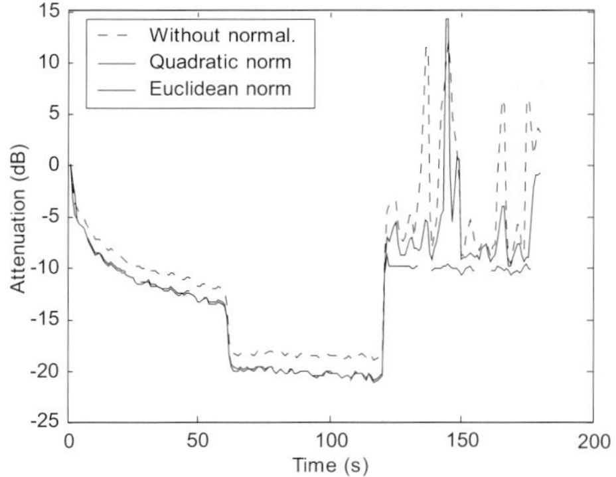

Figure 9: Evolution of attenuation for the FX-LMS algorithm.

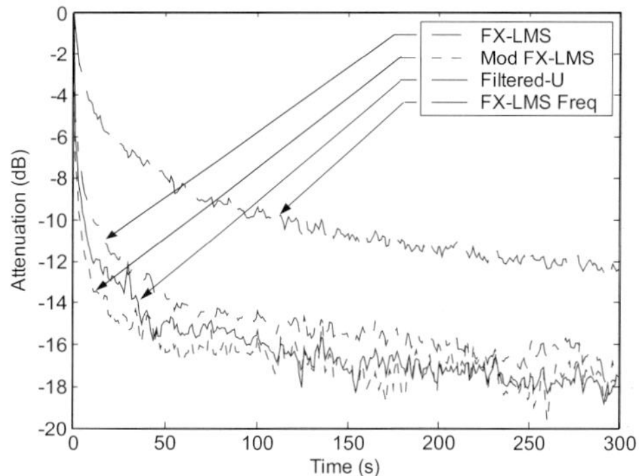

Figure 10: Evolution of attenuation for white noise.

Table 2: Execution time of each iteration for the white noise case.

Algorithm	FX-LMS	MFX-LMS	Filtered-u	FX-LMS Freq
Average time *(ms)*	0.044	0.067	0.081	0.027
Maximum time *(ms)*	0.047	0.081	0.089	0.065

As can be observed the normalization of the filtered reference signal allows obtaining higher attenuations. The quadratic norm is the only one that ensures the stability of the algorithms when the spectral power changes. If this were not the case different adaptation steps would have to be used to keep the algorithms stable.

For the comparison of the algorithms two types of disturbances had been considered: white noise and pure tones. The frequency range of the white noise is [200 Hz; 1000 Hz], for the reason explained before. Tones under 200 Hz have also not been used. Parameters in the algorithms were chosen based upon other experiences that had shown the influence of parameters in algorithms performance. These values are:

- FX-LMS: $w = 200$, $\mu = 0.10$;
- MFX-LMS: $w = 400$, $\mu = 0.1$;
- Filtered-u: $n_a = 150$, $n_b = 100$, $\mu_a = 0.01$, $\mu_b = 0.025$;
- FX-LMS Freq: $w = 256$, $\mu = 0.16$.

Common to all the algorithms are the leakage factor, equal to one, and the normalization method, which was the quadratic norm.

Results are shown in fig. 10-13 for different types of noise to be cancelled, and Table 2 that indicates the computational burden for the white noise case.

- White noise
- Pure tones: 320 Hz + 640 Hz + 960Hz.

All pure tones have the same spectral power. The adaptation steps of FX-LMS and FX-LMS Freq algorithms had to be reduced so that they would remain stable. Steps used were $\mu = 0,03$ for the FX-LMS and $\mu = 0,06$ for the FX-LMS Freq.

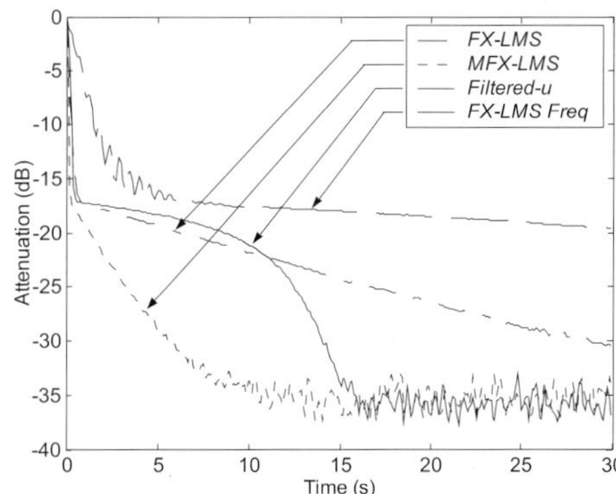

Figure 11: Evolution of attenuation for pure tones.

The two previous figures show that the MFX-LMS algorithm obtains a larger attenuation sooner but the filtered-u algorithm obtains slightly larger attenuations. These two algorithms get the best performances of the four. Worst of them all is the FX-LMS Freq, even though it presents the most reduced average time for executing each iteration. This shows how efficient algorithms are in the frequency domain. However, the execution time of each iteration is not important in this case since all times are clearly under the sampling time of 0.4ms.

This is because of the high computational power of the slave computer.

Robustness to the variations of the model of the feedback path

An important question is the robustness to the degradation of the model of the acoustic feedback path $G_f(z)$, since when this model becomes poor the simplification assumed on point 2.1 (that the model cancels the feedback path exactly) is no longer verified. If the residual of the cancellation is large, the performance of the algorithms based on scheme of Fig. 2 will degrade and may even be unstable.

The filtered-u algorithm can deal with the feedback path problem. However, using the model of Fig. 6, this algorithm has revealed to be unstable on start. To solve this problem the adaptation steps had to be reduced, and thus, have a slower evolution of attenuation. Using the scheme of fig. 2 with filtered-u algorithm has proved to be more robust and have a faster and more regular evolution of attenuation.

That is why two experiences have been carried out in which the performance of estimated model of $G_f(z)$ was reduced. In the two following figures the results for the MFX-LMS algorithms and filtered-u algorithms are shown. Only those are shown because they are the ones with better performances, as was seen above. Parameters used in the algorithms are those given above.

Figures 12 and 13 show that the filtered-u algorithm is more robust to variations of the estimated model of $G_f(z)$ model even though it leads to more irregular evolutions. This shows that the filtered-u algorithm is the one that should be applied in practice since it has a performance identical to the MFX-LMS but is more robust to modelling errors.

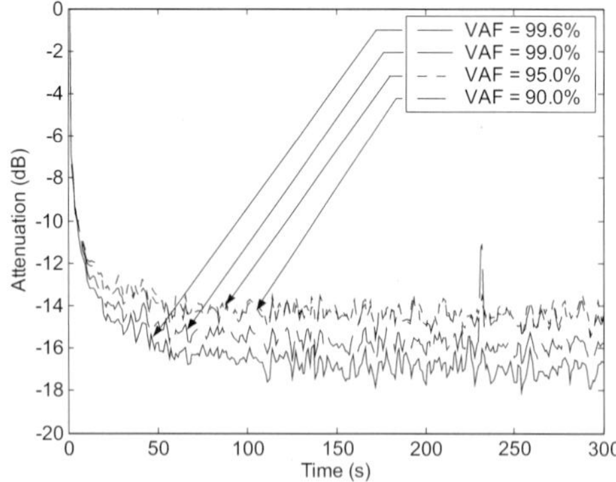

Figure 12: Evolution of attenuation for MFX-LMS algorithm for different estimated models of $G_f(z)$.

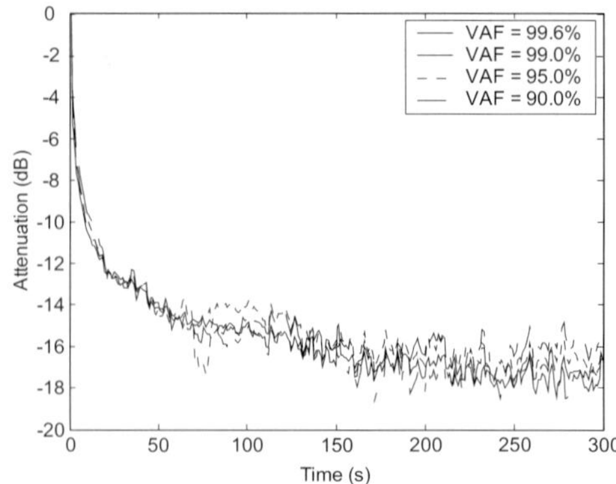

Figure 13: Evolution of attenuation for filtered-u algorithm for different estimated models of $G_f(z)$.

6 CONCLUSIONS

This paper evaluates the use of feedforward ANC to cancel noise in ducts. The *FX-LMS, NFX-LMS, Leaky LMS, MFX-LMS, FX-LMS Freq* and the *Filtered-u* algorithms have been considered. The best performance was achieved with the filtered-u algorithm. Active cancellation of acoustic noise was seen to be possible in practice since attenuations obtained were about 18 dB for white noise and 35 dB for pure tones. Moreover, algorithms were seen to be robust when models degrade.

In what concerns the algorithms it was shown that the normalization of the filtered reference signal is of extreme importance allowing to ensure the stability of the algorithms as well as better attenuations. However this happens only for the quadratic norm.

REFERENCES

Elliot, S. J., 2001. *Signal Processing for Active Control.* Academic Press, London.

Haykin, Simon, 2002. *Adaptive Filter Theory.* Prentice Hall, New Jersey, 4th edition.

Ogata, Katsuhiko, 1997. *Modern Control Engineering.* Prentice Hall, New Jersey, 3rd edition.

Oppenheim, Alan V., Schafer, Ronald W. and Buck, John R., 1999. *Discrete-time Signal Processing.* Prentice Hall, New Jersey, 2nd edition.

Tokhi, M. and Leitch, R. R., 1992. *Active Noise Control.* Oxford University Press, New York.

HYBRID UML COMPONENTS FOR THE DESIGN OF COMPLEX SELF-OPTIMIZING MECHATRONIC SYSTEMS[*]

Sven Burmester[†] and   Holger Giese
Software Engineering Group, University of Paderborn
Warburger Str. 100, D-33098 Paderborn, Germany
burmi@upb.de, hg@upb.de

Oliver Oberschelp
Mechatronic Laboratory Paderborn, University of Paderborn
Pohlweg 98, D-33098 Paderborn, Germany
Oliver.Oberschelp@mlap.de

Keywords: Mechatronic, Self-optimization, Control, Hybrid Systems, Components, Reconfiguration, Unified Modelling Language (UML), Real-Time

Abstract: Complex technical systems, such as mechatronic systems, can exploit the computational power available today to achieve an automatic improvement of the technical system performance at run-time by means of self-optimization. To realize this vision appropriate means for the design of such systems are required. To support self-optimization it is not enough just to permit to alter some free parameters of the controllers. Furthermore, support for the modular reconfiguration of the internal structures of the controllers is required. Thereby it makes sense to find a representation for reconfigurable systems which includes classical, non-reconfigurable block diagrams. We therefore propose hybrid components and a related hybrid Statechart extension for the Unified Modeling Language (UML); it is to support the design of self-optimizing mechatronic systems by allowing specification of the necessary flexible reconfiguration of the system as well as of its hybrid subsystems in a modular manner.

1   INTRODUCTION

Mechatronic systems are technical systems whose behavior is actively controlled with the help of computer technology. The design of these systems is marked by a combination of technologies used in mechanical and electrical engineering as well as in computer science. The focus of the development is on the technical system whose motion behavior is controlled by software.

The increasing efficiency of microelectronics, particularly in embedded systems, allows the development of mechatronic systems that besides the required control use computational resources to improve their long term performance.   These forms of self-optimization allow an automatic improvement of a technical system during operation which increases the operating efficiency of the system and reduces the operating costs.

A generally accepted definition of the term self-optimization is difficult to find. In our view, the core function of self-optimization in technical systems is generally an automatic improvement of the behavior of the technical system at run-time with respect to defined target criteria. In a self-optimizing design, development decisions are being shifted from the design phase to the system run-time.

There are two opportunities of optimization during runtime. The first is to optimize parameters (Li and Horowitz, 1997) the second is to optimize the structure. However, alteration of the free parameters of the system will not lead very far because many characteristics, in particular those of the controller, can be altered only in the internal structures and not just by a modification of parameters (Föllinger et al., 1994; Isermann et al., 1992).

While most approaches to hybrid modeling (Henzinger et al., 1995; Bender et al., 2002; Alur et al., 2001) describe how the continuous behavior can be modified when the discrete control state of the system is altered, we need an approach that allows the continuous behavior as well as its topology to be altered in a modular manner to support the design of self-optimizing systems.

---

[*]This work was developed in the course of the Special Research Initiative 614 - Self-optimizing Concepts and Structures in Mechanical Engineering - University of Paderborn, and was published on its behalf and funded by the Deutsche Forschungsgemeinschaft.

[†]Supported by the International Graduate School of Dynamic Intelligent Systems. University of Paderborn

281

J. Braz et al. (eds.), Informatics in Control, Automation and Robotics I, 281–288.
© 2006 *Springer. Printed in the Netherlands.*

Our suggestion is to integrate methods used in mechanical engineering and software engineering to support the design of mechatronic systems with self-optimization. We therefore combine component diagrams and state machines as proposed in the forthcoming UML 2.0 (UML, 2003) with block diagrams (Föllinger et al., 1994) usually employed by control engineers. The proposed component-based approach thus allows a decoupling of the domains: A control engineer can develop the continuous controllers as well as their reconfiguration and switching in form of hybrid components. A software engineer on the other hand can integrate these components in his design of the real-time coordination. As this paper focusses the modeling aspect we set simulation aside. Simulation results can be found in (Liu-Henke et al., 2000).

In Section 2 we will examine related work. Section 3 discusses problems resulting from reconfiguration by means of an application example. In Section 4, our approach to hybrid modeling with an extension of UML components and Statecharts will be described. Thereafter we describe our model's runtime platform in Section 5 and sum up in Section 6 with a final conclusion and an outlook on future work.

2   RELATED WORK

A couple of modeling languages have been proposed to support the design of hybrid systems (Alur et al., 1995; Lamport, 1993; Wieting, 1996). Most of these approaches provide models, like linear hybrid automata (Alur et al., 1995), that enable the use of efficient formal analysis methods, but lack of methods for structured, modular design, that is indispensable in a practical application (Müller and Rake, 1999).

To overcome this limitation, hybrid automata have been extended to hybrid Statecharts in (Kesten and Pnueli, 1992). Hybrid Statecharts reduce the visual complexity of a hybrid automaton through the use of high-level constructs like hierarchy and parallelism, but for more complex systems further concepts for modularization are required.

The hybrid extensions HyROOM (Stauner et al., 2001), HyCharts (Grosu et al., 1998; Stauner, 2001) and Hybrid Sequence Charts (Grosu et al., 1999) of ROOM/UML-RT integrate domain specific modeling techniques. The software's architecture is specified similar to ROOM/UML-RT and the behavior is specified by statecharts whose states are associated with systems of ordinary differential equations and differential constraints (HyCharts) or Matlab/Simulink block diagrams (HyROOM). HyROOM models can be mapped to HyCharts (Stauner et al., 2001). Through adding tolerances to the continuous behavior this interesting specification technique enables automatic implementation, but support for the modular reconfiguration is not given.

In (Conrad et al., 1998) guiding principles for the design of hybrid systems are sketched. It describes how to apply techniques that are used in automotive engineering, like the combination of statecharts, blockdiagrams and commercial tools. Following this approach hybrid systems need to be decoupled into discrete and continuous systems in the early design phases. Therefore a seamless support and a common model are not provided.

Within the Fresco project the description language Masaccio (Henzinger, 2000) which permits hierarchical, parallelized, and serially composed discrete and continuous components has been developed. A Masaccio model can be mapped to a Giotto (Henzinger et al., 2003) model, that contains sufficient information about tasks, frequencies, etc. to provide an implementaion. The project provides a seamless support for modeling, verification and implementation, but our needs for advanced modeling techniques that support dynamic reconfiguration are not addressed.

3   MODELING RECONFIGURATION

We will use the switching between three controller structures as a running example to outline the resulting modeling problems. The concrete example is an active vehicle suspension system with its controller which stems from the *Neue Bahntechnik Paderborn*[3] research project. The project has been initiated and worked upon by several departments of the University of Paderborn and the Heinz Nixdorf Institute. In the project, a modular rail system will be developed; it is to combine modern chassis technology with the advantages of the Transrapid[4] and the use of existing rail tracks. The interaction between information technology and sensor/actuator technology paves the way for an entirely new type of mechatronic rail system. The vehicles designed apply the linear drive technology used in the Transrapid, but travel on existing rail tracks. The use of existing rail tracks will eliminate an essential barrier to the proliferation of new railbound transport systems (Lückel et al., 1999).

Figure 1 shows a schema of the physical model of the active vehicle suspension system. The suspension system of railway vehicles is based on air springs which can be damped actively by a displacement of their bases. The active spring-based displacement is effected by hydraulic cylinders. Three vertical hydraulic cylinders, arranged on a plane, move the bases

---

[3]http://www-nbp.upb.de/en
[4]http://www.transrapid.de/en

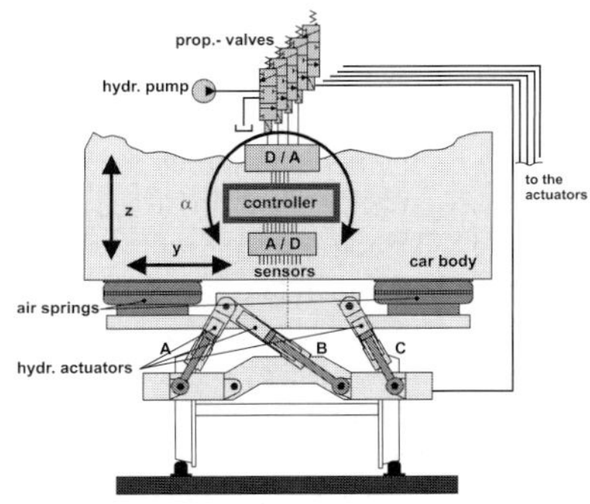

Figure 1: Scheme of the suspension/tilt module.

of the air springs via an intermediate frame, the suspension frame. This arrangement allows a damping of forces in lateral and vertical directions. In addition, it is also possible to regulate the level of the coach and add active tilting of the coach body. Three additional hydraulic cylinders allow a simulation of vertical and lateral rail excitation (Hestermeyer et al., 2002). The vital task for the control system is to control the dynamical behavior of the coach body. In our example, we will focus only on the vertical dynamic behavior of the coach body. The overall controller structure comprises different feedback controllers, e.g., for controlling the hydraulic cylinder position and the dynamics of the car body.

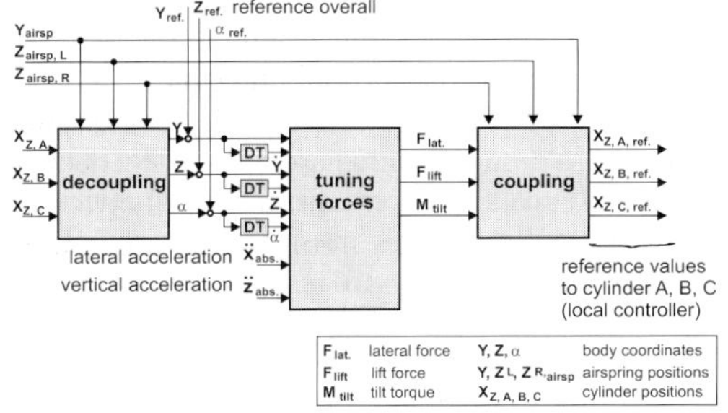

Figure 2: Reference controller.

The schema of the reference controller for the overall body controller is depicted in Figure 2. The subordinated controller is not displayed here for reasons of clarity. The controller essentially consists of the blocks decoupling, tuning forces, and coupling. In the block decoupling the kinematics of the cylinders is converted into Cartesian coordinates for the description of the body motion. The actual control algorithm is in the block tuning forces. Here the forces are computed which are to affect the mechanical structure. In the block coupling the forces and torques are converted again into reference values for the subordinated cylinder controller (Liu-Henke et al., 2000).

A common representation for the modeling of mechatronic systems which have been employed here are hierarchical block diagrams. This kind of representation has its seeds in control engineering, where it is used to represent mathematic transfer functions graphically. It is widely-used in different CAE-tools. Block diagrams consist generally of function blocks, specifying function resp. behavior and hierarchy blocks grouping function and hierarchy blocks. This allows a structured draft and reduces the overall complexity of a block diagram. Between single blocks exists connections or couplings, which can have the shape of directed or non-directed links. With directed links data is exchanged whereas non-directed ones often describe functional relations or physical links, such as a link between mass and spring in multi-body system representation. While parameter optimization can be described using simply an additional connection for the parameter, the static structure of block diagrams does not permit to model structural modifications (Isermann et al., 1992).

We can, however, use special switches or fading blocks to describe the required behavior. The controller in our example has two normal modes: Reference and absolute. The controller reference uses a given trajectory that describes the motion of the coach body $z_{ref} = f(x)$ and the absolute velocity of the coach body $\dot{z}_{abs}$ (derived from $\ddot{z}_{abs}$). The z_{ref} trajectory is given for each single track section. A track section's registry communicates this reference trajectory to the vehicle. In case a reference trajectory is not available, another controller which requires only the absolute velocity of the coach body $\dot{z}_{abs}$ for the damping of the coach-body motion has to be used.

Besides the regular modes another controller named robust is required for an emergency; it must be able to operate even if neither the reference trajectory nor the measurement of the coach-body acceleration are available (see Figure 3).

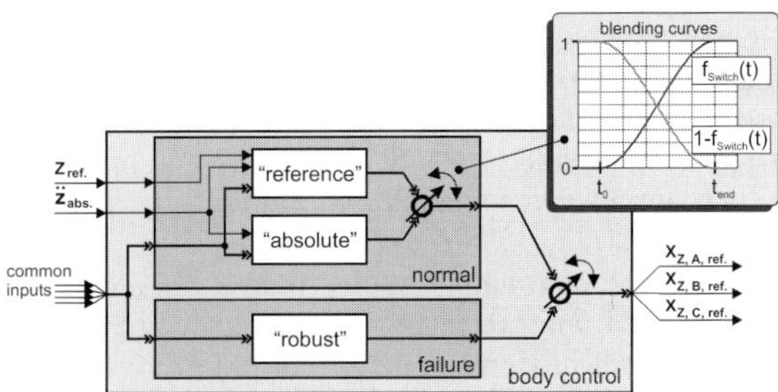

Figure 3: Fading between different control modes.

For a switching between two controllers one must distinguish between two different cases: *atomic switching* and *cross fading*.[5] In the case of atomic

---

[5]The structure and type of cross fading depends on the

switching the change can take place between two computation steps. To start the new controller up, it is often necessary to initialize its states on the basis of the states of the old controller. In our example, the switching from the normal block to the failure block (see Figure 3) can be processed atomically because the robust controller actually has no state. Different theoretical works deal with the verification of stability in systems with any desired switching processes (Lygeros et al., 2003). In the simple case of a switch to a robust controller, stability can be maintained with the help of condition monitoring (Deppe and Oberschelp, 2000).

If, however, the operating points of the controllers are not identical it will be necessary to cross-fade between the two controllers. This handling is required in the normal block depicted in Figure 3, where a transition between the reference and the absolute controller is realized. The cross fading itself is specified by a fading function $f_{switch}(t)$ and an additional parameter which determines the duration of the cross fading.

While the outlined modeling style allows to describe the required behavior, the result is inadequate because of the following two problems: (1) the different operation modes and the possible transitions between them are more appropriately modeled using a techniques for finite state systems rather than a set of blocks scattered all around in the block diagrams and (2) this style of modeling would require to execute all alternative controllers in parallel even though a straight forward analysis of the state space of the resulting finite state system would permit to run only the blocks which realize the currently active controllers.

To overcome these problems, hybrid modeling approaches such as hybrid automata (Henzinger et al., 1995) can be employed. When modeling the fading and switching between the different controllers in our example according to Figure 3, a hybrid automaton with at least three discrete states – one for each of the controllers– might be used. These are the locations Robust, Absolute and Reference in Figure 4. The locations' continuous behavior is represented by the associated controllers. Contrary to the white-filled arrows, black-filled ones denote the common inputs which are always available and required by all controllers.

When the automaton resides in the start location Robust and for instance the $\ddot{z}_{abs}$ signal becomes available (as indicated by the discrete zAbsOK signal), the location and the controller change to the absolute mode. As this change requires cross fading an additional location (FadeRobAbs) is introduced in which the cross fading is comprised. To specify a fading duration $d_1 = [d^1_{low}, d^1_{up}]$ an additional state variable

---

controller types and could lead to complex structures. In our example we use only output fading.

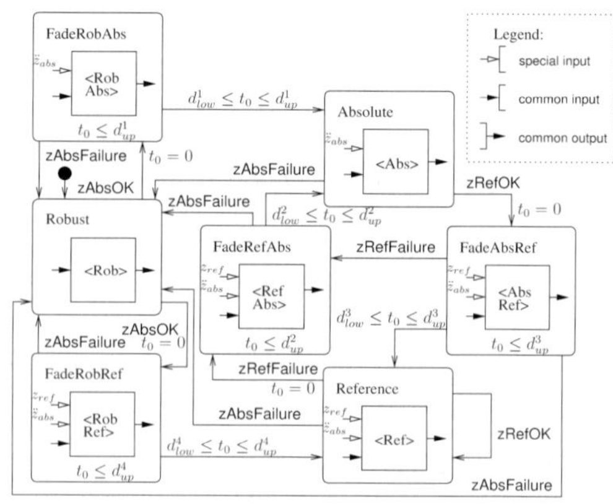

Figure 4: Hybrid body control with fading locations.

t_c with $\dot{t}_c = 1$ is introduced. The reset when entering the fading location FadeRobAbs, its invariant $t_c \leq d^1_{up}$ and the transition's guard $d^1_{low} \leq t_c \leq d^1_{up}$ guarantee that the fading will take d^1_{low} at a minimum and d^1_{up} at a maximum. After completing the fading, the target location Absolute is entered. The duration of the other fading transitions is specified similarly. If the $\ddot{z}_{abs}$ signal is lost during fading or during the use of the absolute-controller, the default location with its robust control is entered immediately (cf. Figure 4).

In an appropriate self-optimizing design the aim must be to use the most comfortable controller while still maintaining the shuttle's safety. If, for instance, the absolute controller is in use and the related sensor fails, the controller may become instable. Thus there is need for a discrete control monitor which ensures that only safe configurations are allowed. The monitor which controls the correct transition between the discrete controller modes must also be coordinated with the overall real-time processing. In our example, the reference controller can only be employed when the data about the track characteristics has been received in time by the real-time shuttle control software. Trying to also integrate these additional discrete state elements into our hybrid description of the body control would obviously result in a overly complex description by means of a single hybrid Statechart which lacks modularity.

4   THE APPROACH

To support the design of complex mechatronic systems and to overcome the problems of the available modeling techniques as outlined in the preceding section, we introduce in the following informally our approach for modeling with the UML in Section 4.1, our notion for hybrid Statecharts in Section 4.2, our notion for hybrid components in Section 4.3, and the modular reconfiguration in Section 4.4. The exact formalization is presented in (Giese et al., 2004).

4.1 Hybrid UML Model

In Figure 5c the overall structural view of our example is presented by means of a UML component diagram. The Monitor component, that embeds three components, communicates with the Registry component through ports and a communication channel. The Shuttle-Registration pattern specifies the communication protocol between Monitor and Registry. The behavior of the track section registry which is frequently contacted by the monitor to obtain the required z_{ref} is depicted in Figure 5b. In Figure 5a the sensor's behavior is described by a Statechart.

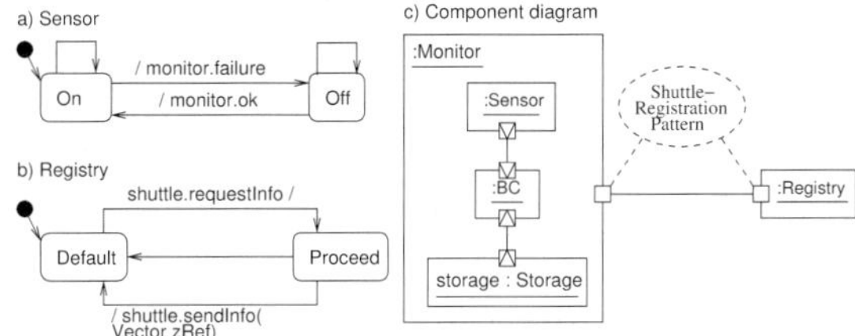

Figure 5: Monitor and its environment (incl. behavior).

The embedded components communicate continuous signals through so called *continuous ports*, depicted by framed triangles whose orientation indicates the direction of the signal flow, and discrete events through discrete, non-filled ports.

4.2 Hybrid Statecharts

A look at the hybrid automaton from Figure 4 reveals that the explicit fading locations considerably increase the number of visible locations of the automaton and make it unnecessarily difficult to understand.

Therefore we propose an extension of UML Statecharts towards Hybrid Statecharts that provide a short-form to describe the fading. The short-form contains the following parameters: A source- and a target-location, a guard and an event trigger, information on whether or not it is an atomic switch, and, in the latter case, a fading strategy (here cross fading is used for all *fading transitions*), a fading function (f_{fade}) and the required fading duration interval $d = [d_{low}, d_{up}]$ specifying the minimum and maximum duration of the fading. This short-form is displayed in Figure 6. The fading-transitions are visualized by thick arrows while atomic switches have the shape of regular arrows.

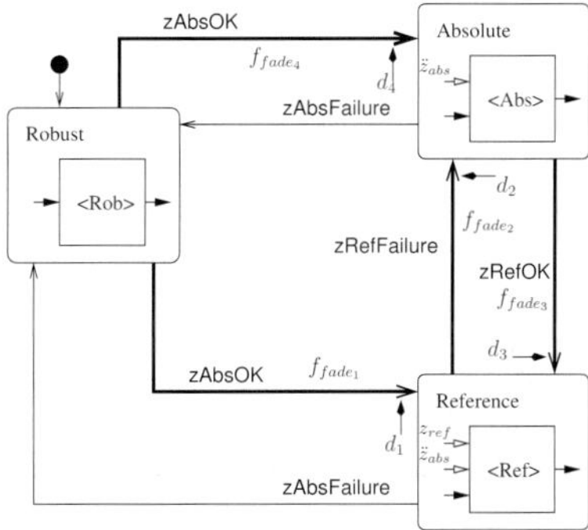

Figure 6: Behavior of the body control component.

An atomic transition, leaving the source or the target state of a currently active fading transition, interrupts the execution of the fading transition. In case of conflicting atomic transitions of the source and target state, the source state's transitions have priority by default.

4.3 Hybrid Components

One serious limitation of today's approaches for hybrid systems is due to the fact that the continuous part of each location has to have the same set of required input and provided output variables (continuous interface). To foster the reconfiguration we propose to describe the different externally relevant continuous interfaces as well as the transition between them using hybrid interface Statecharts.

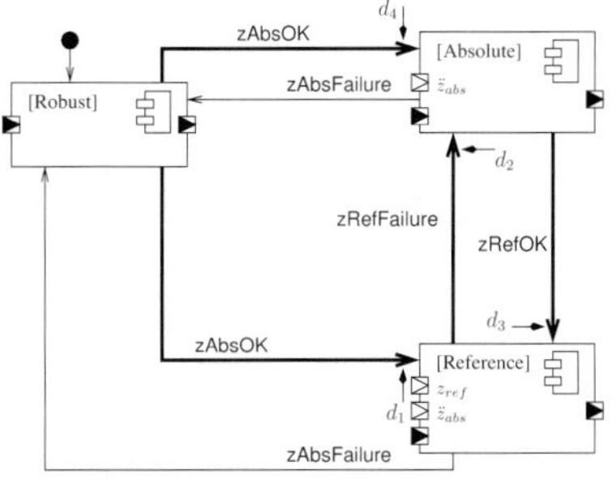

Figure 7: Interface Statechart of the BC component.

The related interface Statechart of the body control component of Figure 6 is displayed in Figure 7. It shows that the body control component has three possible different interfaces. The (continous) ports that are required in each of the three interfaces are filled black, the ones that are only used in a subset of the states are filled white. For all possible state changes, only the externally relevant information, such as durations and the signals to initiate and to break the transitions, are present.

Interface Statecharts can be employed to abstract from realization details of a component. A component in our approach can thus be described as a UML component – with ports with distinct quasi-continuous and discrete signals and events– by

- a hybrid interface Statechart which is a correct abstraction of the component behavior (cf. (Giese et al., 2004)) which determines what signals are used in what state,
- the dependency relation between the output and input signals of a component per state of the interface Statechart in order to ensure only acyclic dependencies between the components, and
- the behavior of the component usually described by a single Hybrid Statechart and its embedded subcomponents (see Section 4.4).

In our example, the BC component is described by its hybrid interface Statechart presented in Figure 7, the additionally required information on which dependencies between output and input variables exist which is not displayed in Figure 7, and its behavior described by the Hybrid Statechart of Figure 6 where the required quasi-continuous behavior is specified by controllers.

4.4 Modular Reconfiguration

The hybrid statechart from Figure 6, which supports state-dependent continuous interfaces, does still not include the case that the employed controllers show hybrid behavior themselves. Instead, complete separation of discrete and continuous behavior like in (Alur et al., 2001; Bender et al., 2002; Henzinger et al., 1995; Kühl et al., 2002) is still present.

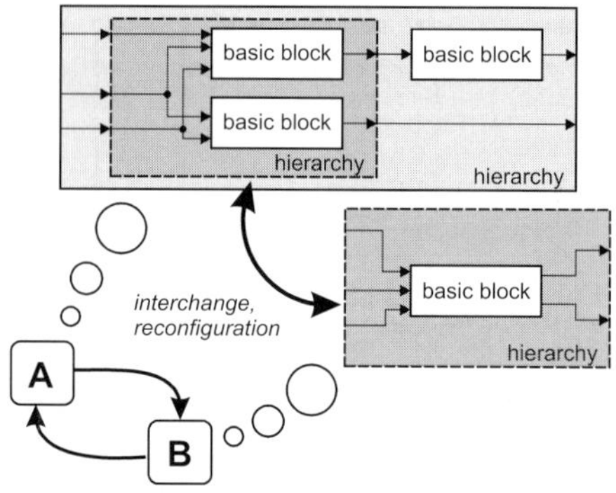

Figure 8: Reconfigurable block diagram.

To overcome this limitation, we propose to assign the required configuration of embedded subcomponents (not only quasi-continuous blocks) to each state of a Hybrid Statechart by means of UML instance diagrams. This idea is motivated by the observation

that the topology of hierarchical block diagrams could be seen as a tree. With the leafs of this tree representing the behavior whereas the inner nodes describe the structure of the system. This distinction between structure (hierarchy) and function (block) can be used for the required modular reconfigurable systems. In our context, a reconfiguration can be understood as a change in the structure resp. substructure of a block diagram. It alters the topology of the system; functions are added and/or interlinked anew. Thus to realize modular reconfiguration we only have to provide a solution to alter the hierarchical elements (cf. Fig. 8). In this manner the required coordination of aggregated components can be described using a modular Hybrid statechart which alter the configurations of its hybrid subcomponents (see Figure 9).

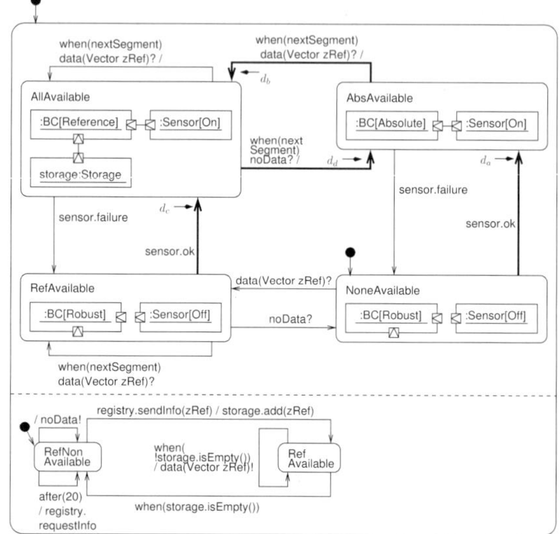

Figure 9: Monitor behavior with modular reconfiguration of the subcomponent BC.

Figure 9 specifies the behavior of the control monitor software. The upper AND-state consists of four locations indicating which of the two signals $\ddot{z}_{abs}$ and z_{ref} are available. Some transitions can fire immediately, others are associated with a deadline interval $d = [d_{low}, d_{up}]$ specifying how much time a transition may take minimal and maximal. These transitions are thicker and marked with an arrow and the intervals $(d_a, \ldots, d_d)$. The lower branch of the modular Hybrid statechart communicates with the track section registry (Figure 5c), frequently requests the z_{ref} function, and puts it in the storage.

In the Hybrid Statechart, every discrete state has been associated with a configuration of the subcomponents (BC, Sensor, Storage). In the design of these associations, only the interface description of the embedded component BC (see Figure 7) is relevant and the inner structure can be neglected. Therefore, as shown in Figure 9, we can assign to each location of the upper AND-state of the statechart the BC component in the appropriate state. E.g., the BC component instance in state Reference has been (via a visual embedding) assigned to the location AllAvailable of the monitor where z_{ref} as well as $\ddot{z}_{abs}$ are available. The required structure is specified by instances of the Sensor and the Storage component and the communication links.

The Hybrid Statechart of Figure 9 defines a mapping of states to required configurations of the subcomponents. The required synchronization between the components is accomplished through explicit raising of the discrete signals defined in Figure 7.

5 RUN-TIME ARCHITECTURE

In order to reach interoperability for mechatronic systems, which are only alterable within certain limits, one can use appropriate middleware solutions like IPANEMA (Honekamp, 1998), which allows abstraction from hardware details. IPANEMA is a platform concept for distributed real-time execution and simulation to support rapid prototyping. It allows a modular-hierarchical organization of tasks or processes on distributed hardware.

In order to make interoperability possible also for hybrid components, which contain the kinds of alteration capability described above, the support of alteration capability by the middleware must be considerably extended. First it is necessary to generate the models in accordance to their modular-hierarchical structure. This is the basis for a reconfiguration.

In each discrete location of the system the equations, that implement the currently active controllers, have to be employed to compute the correct continuous behavior. Thus in every location of this kind only the relevant equations have to be evaluated. To reach this aim, the architecture provides means for every component to adjust the set of active equation blocks in such a way that the required reconfiguration of the component system is efficiently managed.

In the modular execution framework outlined, the required execution order results from the local evaluation dependencies within each component as well as from their interconnection. It must thus be determined at deployment-time or run-time.

The continuous nonlinear differential equations are solved by applying suitable numeric solvers. Computation is time-discrete. Incrementation depends on the solver and the dynamics of the continuous system. A time-accurate detection of a continuous condition is not possible if the controller is coupled with a real technical system. Thus, we restrict the urgent reaction

to continuous conditions in the hybrid statecharts to a detection within the desired time slot (cf. (Henzinger et al., 2003; Stauner, 2002)).

6 CONCLUSION AND FUTURE WORK

Complex mechatronic systems with self-optimization are hybrid systems which reconfigure themselves at run-time. As outlined in the paper, their modeling can hardly be done by the approaches currently available. Therefore, we propose an extension of UML components and Statecharts towards reconfigurable hybrid systems which supports the modular hierarchical modeling of reconfigurable systems with hybrid components and hybrid Statecharts.

The presented approach permits that the needed discrete coordination can be designed by a software engineer with extended Statecharts. In parallel, a control engineer can construct the advanced controller component which offers the technical feasible reconfiguration steps. These two views can then be integrated using only the component interface of the controller component.

It is planned to support the presented concepts by both the CAE tool CAMeL (Richert, 1996) and the CASE tool Fujaba (Giese et al., 2003). With both tools, the result of every design activity will be a hybrid component. Each of these hybrid components itself can integrate hybrid components. The integration only requires the information given by the component interface. Additional automatic and modular code generation for the hybrid models and run-time support for the reconfiguration will thus result in support for the modular and component-based development of self-optimizing mechatronic systems from the model level down to the final code.

REFERENCES

(2003). *U L 2.0 Superstructure Specification*. Object Management Group. Document ptc/03-08-02.

Alur, R., Courcoubetis, C., Halbwachs, N., Henzinger, T., Ho, P.-H., Nicollin, X., Olivero, A., Sifakis, J., and Yovine, S. (1995). The algorithmic analysis of hybrid systems. *Theoretical Computer Science*, 138(3-34).

Alur, R., Dang, T., Esposito, J., Fierro, R., Hur, Y., Ivancic, F., Kumar, V., Lee, I., Mishra, P., Pappas, G., and Sokolsky, O. (2001). Hierarchical Hybrid Modeling of Embedded Systems. In *irst Workshop on mbedded Software*.

Bender, K., Broy, M., Peter, I., Pretschner, A., and Stauner, T. (2002). Model based development of hybrid systems. In *odelling, nalysis, and Design of Hybrid*

Systems, volume 279 of *Lecture Notes on Control and Information Sciences*, pages 37–52. Springer Verlag.

Conrad, M., Weber, M., and Mueller, O. (1998). Towards a methodology for the design of hybrid systems in automotive electronics. In *Proc. of the International Symposium on utomotive Technology and utomation (IS T '98)*.

Deppe, M. and Oberschelp, O. (2000). Real-Time Support For Online Controller Supervision And Optimisation. In *Proc. of DIP S 2000. Workshop on Distributed and Parallel mbedded Systems, echatronics Laboratory Paderborn, University of Paderborn*.

Föllinger, O., Dörscheid, F., and Klittich, M. (1994). *egelungstechnik - inführung in die ethoden und ihre nwendung*. Hüthig.

Giese, H., Burmester, S., Schäfer, W., and Oberschelp, O. (2004). Modular Design and Verification of Component-Based Mechatronic Systems with Online-Reconfiguration. In *Proc. of 12th C SIGSO T oundations of Software ngineering 2004 (S 2004), Newport Beach, US*. ACM. (accepted).

Giese, H., Tichy, M., Burmester, S., Schäfer, W., and Flake, S. (2003). Towards the Compositional Verification of Real-Time UML Designs. In *Proc. of the uropean Software ngineering Conference (S C), Helsinki, inland*. Copyright 2003 by ACM, Inc.

Grosu, R., Krueger, I., and Stauner, T. (1999). Hybrid sequence charts. Technical Report TUM-I9914, Technical University Munich, Munich.

Grosu, R., Stauner, T., and Broy, M. (1998). A modular visual model for hybrid systems. In *Proc. of ormal Techniques in eal-Time and ault-Tolerant Systems (T T T'98)*, LNCS 1486. Springer-Verlag.

Henzinger, T. A. (2000). Masaccio: A Formal Model for Embedded Components. In *Proceedings of the irst I IP International Conference on Theoretical Computer Science (TCS), Lecture Notes in Computer Science 1872, Springer-Verlag, 2000, pp. 549-563*.

Henzinger, T. A., Ho, P.-H., and Wong-Toi, H. (1995). HyTech: The Next Generation. In *Proc. of the 16th I eal-Time Symposium*. IEEE Computer Press.

Henzinger, T. A., Kirsch, C. M., Sanvido, M. A., and Pree, W. (2003). From Control Models to Real-Time Code Using Giotto. In *I Control Systems agazine 23(1):50-64, 2003*.

Hestermeyer, T., Schlautmann, P., and Ettingshausen, C. (2002). Active suspension system for railway vehicles-system design and kinematics. In *Proc. of the 2nd I C - Confecence on mechatronic systems*, Berkeley, California, USA.

Honekamp, U. (1998). *IP N - Verteilte chtzeit-Informationsverarbeitung in mechatronischen Systemen*. PhD thesis, Universität Paderborn, Düsseldorf.

Isermann, R., Lachmann, K.-H., and Matko, D. (1992). *daptive Control Systems*. Prentice Hall, Herfordshire.

Kesten, Y. and Pnueli, A. (1992). Timed and hybrid statecharts and their textual representation. In *Proc. ormal Techniques in eal-Time and ault-Tolerant Systems, 2nd International Symposium, LNCS 571*. Springer-Verlag.

Kühl, M., Reichmann, C., Prötel, I., and Müller-Glaser, K. D. (2002). From object-oriented modeling to code generation for rapid prototyping of embedded electronic systems. In *Proc. of the 13th I International Workshop on apid System Prototyping (SP'02)*, Darmstadt, Germany.

Lamport, L. (1993). Hybrid systems in tla+. Springer.

Lückel, J., Grotstollen, H., Jäker, K.-P., Henke, M., and Liu, X. (1999). Mechatronic design of a modular railway carriage. In *Proc. of the 1999 I / S International Conference on dvanced Intelligent echatronics (I 99)*, Atlanta, GA, USA.

Li, P. Y. and Horowitz, R. (1997). Self-optimizing control. In *Proc. of the 36th I Conference on Decision and Control (CDC)*, pages 1228–1233, San Diego, USA.

Liu-Henke, X., Lückel, J., and Jäker, K.-P. (2000). Development of an Active Suspension/Tilt System for a Mechatronic Railway Carriage. In *Proc. of the 1st I C-Conference on echatronics Systems (echatronics 2000)*, Darmstadt, Germany.

Lygeros, J., Johansson, K. H., Simic′, S. N., Zhang, J., and Sastry, S. S. (2003). Dynamical Properties of Hybrid Automata. In *I T NS CTIONS ON UTO-TIC CONT OL, VOL. 48, NO. 1, J NU Y 2003*, volume 48.

Müller, C. and Rake, H. (1999). Automatische Synthese von Steuerungskorrekturen. In *KONDISK-Kolloquium Berlin*.

Richert, J. (1996). Integration of mechatronic design tools with camel, exemplified by vehicle convoy control design. In *Proc. of the I International Symposium on Computer ided Control System Design*, Dearborn, Michigan, USA.

Stauner, T. (2001). *Systematic Development of Hybrid Systems*. PhD thesis, Technical University Munich.

Stauner, T. (2002). Discrete-time refinement of hybrid automata. In Tomlin, C. and Greenstreet, M., editors, *Proceedings of the 5th International Workshop on Hybrid Systems: Computation and Control (HSCC 2002)*, volume 2289 of *Lecture Notes in Computer Science*, page 407ff, Stanford, CA, USA.

Stauner, T., Pretschner, A., and Péter, I. (2001). Approaching a Discrete-Continuous UML: Tool Support and Formalization. In *Proc. U L'2001 workshop on Practical U L-Based igorous Development ethods – Countering or Integrating the eXtremists*, pages 242–257, Toronto, Canada.

Wieting, R. (1996). Hybrid high-level nets. In *Proceedings of the 1996 Winter Simulation Conference*, pages 848–855, Coronado, CA, USA.